Drag Reduction

Papers presented at the Second International Conference on Drag Reduction

Conference held at University of Cambridge
August 31 - September 2, 1977

Published by BHRA Fluid Engineering

EDITORIAL NOTE

This volume contains the papers presented at the Second International Conference on Drag Reduction, held at the University of Cambridge August 31 - September 2, 1977.

The Conference was sponsored and organised by BHRA Fluid Engineering.

Papers were reproduced from the author's original typescript in order to minimise delay.

The views expressed in these papers are those of the Authors and do not necessarily represent the views of BHRA Fluid Engineering.

It is recommended that in citing work reported in this collection of papers the following convention be adopted, e.g. Drag Reduction, 2nd International Conference on Drag Reduction, (August 1977).

Editors: H.S. Stephens
Miss J.A. Clarke

Published by BHRA Fluid Engineering

Cranfield, Bedford MK43 0AJ, England.

ISBN 0 900983 71 X

Printed by Inprint of Luton (Designers & Printers) Ltd.

Acknowledgements

BHRA Fluid Engineering is indebted to the Organising Committee and the Panel of Referees for their invaluable assistance in the organisation of this Conference.

ORGANISING COMMITTEE

Dr. R.H.J. Sellin	University of Bristol, U.K. (Chairman)
Dr. E. Dunlop	I.C.I. (Agricultural Division) U.K.
Dr. E.J. Hinch	University of Cambridge, U.K.
Dr. J.W. Hoyt	U.S. Naval Academy, U.S.A.
Dr. O. Scrivener	Universite Louis de Strasbourg, France
Professor L.I. Sedov	University of Moscow, U.S.S.R.
Mr. H.S. Stephens	BHRA Fluid Engineering, U.K.
Dr. A. White	Middlesex Polytechnic, U.K.
Professor J.L. Zakin	University of Missouri Rolla, U.S.A.
Dr. G. Zimmerman	Max Planck Institut fur Stromungs-forschung, German Federal Republic
Mr. N.G. Coles	BHRA Fluid Engineering, U.K. (Organising Secretary)

Cover: Photograph by courtesy of J.W. Hoyt and J.J. Taylor of the Naval Undersea Center, U.S.A. *(A photographic study of polymer solution: jets in air, 1st International Conference on Drag Reduction, Paper E3, 1974)*

DRAG REDUCTION

Papers presented at the Second International Conference on Drag Reduction

Contents

Second International Conference on

Drag Reduction

August 31st - September 2nd, 1977

PAPER A1

POLYMER DRAG REDUCTION - A LITERATURE REVIEW, 1975-6

J.W. Hoyt,

U.S. Naval Academy, U.S.A.

Summary

Polymer drag reduction, the phenomenon whereby a few parts per million of dissolved macromolecules reduce fluid friction to one-half or less that of the pure solvent, is currently a very active research area, with new contributions appearing in the literature almost daily.

The goal of this review is to summarize and highlight some of the major findings of the two-year period 1975-6, and to provide a fairly extensive compilation of the literature of the period. This then supplements the masterful summary and bibliography by White and Hemmings (Ref.1), who, under the sponsorhip of BHRA Fluid Engineering, brought together the world literature on polymer drag reduction up to the beginning of 1975.

Although no claim is made as to completeness, the number of citations involved (over 170) indicates the scope of involvement by research workers in many countries in studying this fascinating area of fluid mechanics.

During this period, several summary articles and bibliographies have appeared (Refs. 2 - 7) which present the status of polymer drag reduction, and its various facets, from somewhat different viewpoints. An especially thorough and incisive summary by Virk (Ref.8) is recommended as the most comprehensive of the recent reviews.

In addition, two major international conferences have been held recently covering the polymer additive area. The papers from a CNRS Colloquium on "Polymers et Lubrification" held at Brest in May 1974 have been released in a bound volume. Papers from the IUTAM Symposium on "Structure of Turbulence and Drag Reduction", held in Washington in June 1976 are not yet available in printed form, it being intended that they appear as a special edition of Physics of Fluids. The titles of the papers as given at the Symposium are listed here, however, for completeness. The literature has been divided into six rather loosely defined categories:

- Drag reduction in various experimental geometries
- New experimental techniques
- Structure and conformation of polymers in solution
- Theories and experiments to study turbulence structure and the mechanism of drag reduction by additives
- Drag reducing suspensions
- Heat transfer in drag-reducing fluids

A significant effort is apparent, world-wide, in each of these categories.

Held at St. John's College, Cambridge, England.
Organised and sponsored by BHRA Fluid Engineering, Cranfield, Bedford, MK43 0AJ.

Drag Reduction in Various Experimental Geometries

Pipe Flow

Although polymer drag reduction in pipe flow is one of the oldest and most established techniques of observing the friction-reducing effects of dilute polymer solutions, a few new and interesting topics in this area are found in the current literature. Tullis, Ramu, and their associates (Refs. 9, 10) have further documented their studies of the details of flow associated with the injection of poly(ethylene oxide) and polyacrylamide solutions into the boundary layer of a 12-inch (0.305m) diameter pipe. Spectacular friction reductions, 80% or so, are reported at poly (ethylene oxide) concentrations of 24 parts per million (ppm) and a Reynolds number of 10^6. The effectiveness of polymers in other large-scale pipe-flow experimentation (Refs. 11,12) has also been reported.

The drag-reducing effect of polymer solutions reflects on the wall-pressure fluctuations and thus the flow noise is reduced (Ref. 13). By using a tapered pipe, Tomita and Mochimaru (Ref. 14) show that polymer solutions, in addition to the drag-reducing effect, improve the diffuser action with flow in the expanding tapered direction, and when the flow is reversed, an enhanced laminarizing effect of the contraction is noted.

The drag reduction effect can be used at flow velocities to 120 m/sec, according to Russian tests (Ref. 15), without serious degradation due to wall shear stress affecting the polymer molecules. However, Culter, et al (Ref. 16) indicate that some degradation does take place in high velocity turbulent flow.

Clark and Rodriguez (Ref. 17) compared pipe-flow measurements with flows between parallel plates; a similarity in behavior was noted. Other recent pipe-flow data are found in (Refs. 18-20).

Airfoils, Streamline Bodies, and Flat Plates

The controversy over whether lift on airfoils in polymer solutions decreases or increases compared with pure water is not yet completely resolved, but what is certain is that the lift forces are fundamentally changed if polymer solutions are ejected from one side of the foil. As shown in Fig. 1 (from Ref. 21), the lift change is a function of velocity, angle of attack, and polymer characteristics. Also, the location on the foil is a factor (Ref. 22). Substantial increases in lift-drag ratio can be realized under favorable conditions (Refs. 23, 24). Ejection of polymer at any point on the airfoil seems to change the pressure distribution all around the foil (Refs. 21, 22), not just on the side where ejection takes place.

Measurements of pressure distribution on an axisymmetric body of revolution (Refs. 25, 26) show that polymer solutions greatly change the pressure distribution all around the body: all local pressure (except at the stagnation point) tending to be higher with increasing polymer concentration. The changed turbulence levels around a small body of revolution was noted by Sirmalis (Ref. 27). Dye studies showed that 2 ppm polymer in the boundary layer eliminated fine-scale turbulent structure; higher concentrations suppressed the larger turbulent fluctuations, and skin friction reductions of 70% were achieved. Sarpkaya (Ref. 28) has reported turbulence measurements in the wake of a circular cylinder with polymer-solution flow.

Wetzel and Ripken (Ref. 29) in a very large-scale experiment show drag reductions of up to 70% on the flat bottom of a flow channel at Reynold's numbers up to 4×10^7. At the other Reynold's number extreme, Cantwell and Coles (Ref. 30) have studied the growth of a turbulent spot in a laminar, polymer-filled boundary layer.

Turning to a somewhat more exotic aspect of polymer drag reduction on external flows, Breder (Ref. 31) has suggested that one reason that many types of fish swim in large schools is to take advantage of the increased local concentration of fish mucus, believed to be long chain polymers capable of reducing the drag (Ref. 32). Breder further showed that fish swim faster and with less effort in poly (ethylene oxide) than in pure water.

Other Geometries

Mashelkar, et al (Ref. 33) has shown that drag-reducing additives are effective in reducing torque on rotating disks, agitators, and paddle-type stirrers. Surprisingly, the presence of entrained air in the stirred fluids greatly increased the drag-reduction effect of dissolved polymers.

The torque-reduction effect was also noted in a rotating-cylinder apparatus, arranged to permit a variable eccentricity (Ref. 34). Torque decreases were greatest at the highest eccentricity, and the rotative speed for the appearance of Taylor vortices was increased with the polymer solutions. Similar drag decreases were noted by Keller et al (Ref. 35) but an increase in critical Taylor number was not seen. The Taylor vortices were visualized by tracer particles: a slight increase in rotational speed was noted with drag reduction. That the situation is much more complicated is indicated by the work of Friebe (Ref. 36), who identified the polymer effect on four different types of rotating-cylinder instabilities, and accompanied the text with beautiful photographs. Some of the instabilities appear stabilized with increasing polymer concentration; others are destabilized. Drag reduction in rotating-cylinder flow is also accompanied by reduced mass transfer from the rotating cylinder in polymer solutions (Ref. 37).

Turning now to laminar flow, we see that in many cases of oscillatory or elongational-type flows, drag reduction has been noted when polymer solutions are tested. Thus Driels and Ayyash (Ref. 38) found drag reduction in an oscillating liquid column; Popadić (Ref. 39) measured reduced drag in film flows; vortex flow was inhibited (Ref. 40); pulsed flow of shear-thinning non-Newtonian fluids shows a higher mean flow rate (Ref. 41); and flow in helical channels is increased (Refs. 42,43,44).

In other laminar flow situations, the higher elongational viscosity of polymer solutions acts to lower the Reynold's number at which Karman vortices are shed (Ref. 45), this effect changing with type of polymer and the age of solution; and flow through orifices and around pitot tubes is changed (Ref. 46-48). A much greater pressure loss than the pure solvent is noted in flow through porous beds. The remarkable increase in pressure drop of dilute polymer solutions through porous beds, elegantly shown by James and McLaren (Ref. 49), leads Laufer, et al (Ref. 50) to propose the technique as a means of measuring solution elasticity. Daoudi (Ref. 51) suggests the increased pressure drop in porous beds may be due to the polymer molecules going from a coil to a stretched configuration in this flow geometry. Some practical problems involved in filtering polymer solutions through a diatomaceous-earth filter are outlined in (Ref. 52), and polymer degradation in flow through porous rock formations are discussed in(Ref. 53).

New Experimental Techniques

The search for more information on the properties of drag-reducing flows has lead to the introduction of new techniques for visualizing and quantifying the flow, and brought forth marked improvements on older techniques. Methods for measuring viscoelastic properties of fluids, flow visualization, and use of electrochemical and laser-Doppler techniques have been reported.

Before covering those, however, an important development in measuring low concentration of aqueous solutions of high polymers has been reported. A difficult, practical problem is often involved in determining the actual concentration of a sample of drag-reducing fluid taken somewhere in the flow, and Attia and Rubio (Ref. 54) show how with proper reagents, simple turbidity measurements (or nephelometry) can be used to detect concentrations of polyacrylamide or poly (ethylene oxide) to as little as 0.1 ppm.

A new device for measuring normal stresses in dilute polymer solutions has been described by James (Ref. 55). His technique uses a rotating disk, closely spaced to a stationary disk, with fluid introduced at the center. The key measurement is efflux time of the liquid through the narrow gap; viscoelastic fluids are retarded due to normal stresses and a fluid relaxation time can be found. Other clever extensional-flow devices for viscoelastic fluids have been reported: a "triple-jet" arrangement (Ref. 56); recoil in an orifice flow (Ref. 57); a four-roll mill (Ref. 58); and spinning the solution in the same manner synthetic threads are spun (Ref. 59).

Tiederman (Refs. 60, 61) has shown how, using fluorescent dyes, the flow visualization of the low-speed, streaky fluid in near-wall flows can be improved. More importantly, he has shown that the non-dimensionalized streak spacing, as shown by various visualization techniques, depends considerably on the non-dimensional distance out from the wall where the visualization takes place and amount of drag reduction in polymer flows as shown in Fig. 2. He has thus been able to correlate results from various investigators which had previously shown unaccountable differences, a major advance in understanding the significance of the streak-spacing measurements. Both Tiederman and Donohue (Ref. 62) note that hydrogen-bubble techniques are unusable when visualizing drag-reducing flows; the bubbles seem to coalesce rather than flowing smoothly off the generating wire.

A series of publications (Refs. 63-65) give the details of ingenious electrochemical methods to deduce the flow very close to the wall (about 1 micron) in drag-reducing polymer solutions. Hanratty (Ref. 66) gives results of electrochemical techniques in pipe flow.

Turning now to laser-Doppler anemometry, several new papers have appeared. Reischman and Tiederman (Ref. 67) give velocity profile and turbulence measurements in a wall geometry unaffected by secondary flows; other measurements (Ref. 68) may have some interpretation difficulties due to measurements in a square pipe and hence some secondary-flow effect. In (Ref. 69) the early appearance of turbulence in laminar flow is examined.

Structure and Conformation of Polymers in Solution

Any student of the conformational properties of polymers in solution can profit by reading the Nobel Laureate lecture of P. J. Flory (Ref. 70). A masterpiece of conciseness, it presents an outline of the salient facts, history and state of knowledge of the area in an outstanding way.

The most interesting new idea in the polymer-conformation approach to study of polymer drag reduction is that of DeGennes (Ref. 71), who shows that under high shear, the molecules can abruptly unwind from the random coil configuration to the extended or stretched configuration. This can happen because the hydrodynamic interactions are reduced by stretching. This concept, Fig. 3, would provide the order-of-magnitude or two extensions required to allow the drag-reducing molecules to physically interfere with the smallest hydrodynamic eddies. Clearly, the approach of DeGennes is of great importance in the analysis of polymer drag reduction.

If the polymer molecule extends in a shear flow, the computations of Hinch (Ref. 72,73) may be of special interest. Hinch represents the molecule as a flexible thread and studies its distortion in a shear flow. Experimental studies confirm that the extended configuration of polyelectrolytes gives the best drag reduction (Ref. 74-77), while an external electric charge has no effect (Ref. 78). Other conformational changes due to solvent (Ref. 79), and due to concentration (Ref. 80) have been reported. Virk (Ref. 81) interprets these conformational changes as an excluded-volume effect. Layec-Raphalen and Wolff (Ref. 82) found shear-thickening behavior in solutions of polystyrene in decalin, which they regard as being due to chain-extension at higher shear rates. A lattice model for polymer chains in dilute solution is discussed in (Ref. 83). Finally, Tanner (Ref. 84) has given a theoretical treatment to some extensional flows, indicating the effect of viscoelasticity.

Several papers discuss molecular weight distributions of polymer samples; Hunston (Ref. 85) relates molecular weight distributions to the onset of drag reduction, maximum drag reduction, and shear stability. Ting (Ref. 86) shows how the minimum molecular weight for the occurrence of drag reduction is a function of pipe diameter. Little and Ting (Ref. 87) give a useful catalog of molecular weight distributions for commercially available drag-reducing polymers.

Turning to studies specifically directed to polymers of special interest for drag reduction, Maxfield and Shepherd (Ref. 88) have shown that poly (ethylene oxide) hydrogen bonds in water solution, three water molecules to each repeating polymer unit. Powell and Schwartz (Ref. 89,90) have studied poly (ethylene oxide) on the Wissenberg rheogoniometer to measure normal stresses; these seem to be a complicated function of concentration and strain rate. The dipole moments of poly (ethylene oxide) have been measured (Ref. 91); the effects of the polymer on the cloud point when various inorganic salts are added has been given (Ref. 92); and evidence given for molecular aggregation in solutions of poly (ethylene oxide) in dimethylformamine (Ref. 93). Little, et al (Ref. 94) have characterized the drag-reducing characteristics of poly (ethylene oxide) and also polyacrylamide in a rotating-disk apparatus, while Huang and Santelli (Ref. 95) have measured the drag reduction of mixtures of the two polymers.

Studies specifically related to polyacrylamide include Patterson, et al (Ref. 96) who show that the polymer is rather insensitive to added salts; and Nicodemo, et al (Ref. 97) who measured elongational flows of polyacrylamide.

In drag reduction with soap colloids, several new papers have appeared (Refs. 98, 99). In addition, mixtures of poly (ethylene oxide) and soaps can have drag reductions higher than that of the polymer in water only (Ref. 100).

Theories and Experiments Relating to Turbulence Structure and the Mechanism of Drag-Reduction by Additives

Much effort in the last few years has been devoted to attempts to elucidate the mechanism of drag reduction, either by theory or ingenious experiment. In turn, this has led to substantial improvement in computing ability and general experimental technique in the drag-reduction field.

Turning first to theoretical studies in drag-reducing flows, Tiederman and Reischman (Ref. 101) developed a computer code which develops velocity profiles in polymer flows from very simple input data; Poreh and Hassid (Ref. 102-104) have improved their flow models; and Landahl (Ref. 105, 106), Lumley (Ref. 107), Saffman (Ref. 108), and Hinch (Ref. 109) have each contributed to our theoretical understanding of the complex flow situation when polymer additives are present.

In experimental studies of velocity profiles, Frederick and White (Ref. 110) have confirmed the maximum drag reduction asymptote of Virk, Fig. 4; Nikitin (Ref. 111) measured velocity profiles using a photographic technique, and Scrivener (Ref. 112) used both photographic and laser-Doppler techniques for velocity profiles. Other experiments on velocity profiles are given in (Refs. 113-114), while the kinetic energy profile is discussed in (Ref. 115).

A thorough discussion of turbulence structure was held at the IUTAM Symposium on Structure of Turbulence and Drag Reduction held in Washington in June 1976. A review of these important papers (Ref.. 116-124) must await their appearance in published form. In other work, Greshilov (Ref. 125) shows that polymers suppress transverse fluctuations, but the longitudinal fluctuations are somewhat enhanced; the bursting events are also reduced (Ref. 126).

In other studies of the effect of polymers on the flow, Bazilevich (Ref. 127) suggests using the Kolmogorov eddy size as a factor in correlating polymer test data; Hlávacek, et al (Refs. 128,129) look for a structuring of the fluid due to the presence of polymers; the effect of the molecular weight distribution on polymer flows is covered in (Refs. 130, 131); and other aspects of turbulence structure on polymer drag reduction are discussed in (Ref. 132).

A good review of the extensive program of the Naval Research Laboratory in the polymer drag reduction area has been given (Ref. 133,134); work in organic solvents in(Ref. 135); and calculations on drag reduction as a special case in turbulent fluid flow are reported in (Ref. 136). Papers relating to viscoelasticity and elongational flows of polymer solutions are given in (Refs. 137-140). The large effects of polymers on turbulent diffusion are given in a number of papers (Refs. 141-145). Finally, Ayyash and McComb (Ref. 146) discuss their attempts to form an absorbed layer of polymer on the wall of a pipe.

Drag-Reducing Suspensions

Two major reviews highlight the work on drag-reducing suspensions. Jeffrey and Acrivos (Ref. 147) review the viscosity and other rheological properties of suspensions, while Radin, Zakin and Patterson (Ref. 148) cover drag-reduction in suspensions. Both reviews form a thorough basis for further work on drag-reducing suspensions. A key finding of the latter review is that drag reduction can always be obtained if the particle l/d ratio is 25 or greater.

In other work, Fedors (Ref. 149) has discussed the viscosity relationships for a particular particle configuration. From the drag-reduction aspect, Drew (Ref. 150) presents results on energy production dissipation relationships in suspension flow; while Bark (Ref. 151)introduces stability calculations for fiber suspensions. Several papers give the results of extensive tests on wood-pulp suspensions (Refs. 152-155); wood pulp appears to be a very effective drag-reducing suspension, and (Ref. 152) is an especially lucid introduction to the subject. Other drag-reducing suspensions flows are treated in(Refs. 156, 157).

The flows of suspensions can be augmented and made less resistive by adding polymers. This effect has been noted in asbestos-polymer suspensions (Refs. 158,159) where it is found that polymer and suspended fibers are additive in their drag-reducing effect; flow of iron ore (Ref. 160) and coal (Ref. 161) suspensions is increased when polymers are added.

Finally, drag reduction is observed in two-phase gas-liquid suspensions when polymers are dissolved in the liquid (Refs. 162,163).

Heat Transfer

Reductions in heat transfer accompany friction reductions due to polymers and fiber suspensions; Dimant and Poreh (Ref. 164) have recently given an excellent review of the state-of-the-art. Ghajar and Tiederman (Ref. 165) show how heat transfer can be predicted using their computing program; Yoo and Hartnett (Ref. 166) give an estimate of the increased thermal entrance lengths for drag-reducing polymers; Mizushina, et al (Ref. 167,168) give additional heat transfer data for polymer flows; and Taylor and Sabersky (Ref. 169) extrapolate their experimental polymer heat-transfer data to various pipe sizes.

The final topic in heat transfer is that of suspension flows, and Moyls and Sabersky (Ref. 170,171) have studied heat transfer in asbestos-water suspensions. As shown in Fig. 5, spectacular reductions in heat transfer are obtained from this highly drag-reducing suspension.

Conclusion

The large amount of work in the polymer drag-reduction shows that the field is lively and active. The survey shows, however, that much is still to be done, notably in the explanation of the occurrence of the drag-reducing effect. Clever experiments, and even more clever analyses and interpretations leading to definite understanding of the mechanics of polymer and suspension flows are still goals for our fluid-mechanics community.

References

1. White, A. and J.A.G. Hemmings, "Drag Reduction by Additives - Review and and Bibliography," BHRA Fluid Engineering, Cranfield (1976).

2. Bark, F.H., E.J. Hinch and M.T. Landahl, "Drag Reduction in Turbulent Flow Due to Additives: A Report on Euromech 52," J.Fluid Mechanics, 68, pp. 129-138(1975).

3. Block, H. "Reduction of Turbulent Drag by Polymers," in Ledwith, A., and A.M. North, eds., Molecular Behavior and the Development of Polymeric Materials, pp. 492-527, Wiley & Sons, New York (1975).

4. Granville, P.S. "Progress in Frictional Drag Reduction - Summer 1974 to Summer 1975," D.W. Taylor NSRDC Report SPD-569-03 (1975).

5. Granville, P.S. "Progress in Frictional Drag Reduction - Summer 1975 to Summer 1976," D.W. Taylor NSRDC Report SPD 569-04 (1976).

6. Hoyt, J.W. "Recent Progress in Polymer Drag Reduction," in Colloques Internationaux du C.N.R.S. No. 233 - Polymères et Lubrification, pp. 193-215, Paris (1975).

7. Lehmann, E.J. and G.H. Adams, "Drag Reducing Fluids (A Bibliography with Abstracts 1969-June 1975)," National Technical Information Service, NTIS/PS-75/497/86A (1975).

8. Virk, P.S. "Drag Reduction Fundamentals," AIChE J. 21, p. 625 (1975).

9. Tullis, J.P., M. Poreh and J.A. Hooper, "Polymer Injection Into a Developing Boundary Layer," Engineering Research Center,Colorado State Univ., Hydro Machinery Laboratory Report No. 71 (1976).

10. Ramu, K.L.V. and J.P. Tullis, "Drag Reduction and Velocity Distribution in Developing Pipe Flow," J. Hydronautics, 10, pp. 55-61 (1976).

11. Treiber, K.L., J.M. Nash and V.F. Neradka, "Military Pipeline Experience with Drag Reduction Additives," in Fluid Mechanics in the Petroleum Industry, C. Dalton and E. Denison, Eds., American Society of Mechanical Engineers (1975).

12. Savins, J.G. and F.A. Seyer, "Drag Reduction Scale-Up Criteria," IUTAM Symposium on Structure of Turbulence and Drag Reduction, Washington, DC (Jun 1976).

13. Greshilov, E.M., A.V. Evtushenko and L.M. Lyamshev, "Hydrodynamic Noise and the Toms Effect," Soviet Physics-Acoustics, 21, pp. 247-251 (1975).

14. Tomita, Y. and Y. Mochimaru, "A Study on a Tapering Pipe Flow of a Dilute Polymer Solution," in Colloques Internationaux du C.N.R.S. No. 233 - Polymères et Lubrification, pp. 193-215, Paris (1975).

15. Nikitin, I.K., N.G. Poznyaya and A.V. Dzevaltovskiy, "High-Velocity Flow of Dilute Polymer Solutions," Fluid Mechanics - Soviet Research, 4, 2, pp. 47-51 (Mar-Apr 1975).

16. Culter, J.D., J.L. Zakin and G.K. Patterson, "Mechanical Degradation of Dilute Solutions of High Polymers in Capillary Tube Flow," J. Applied Polymer Science, 19, pp. 3235-3240 (1975).

17. Clark, D.B. and F. Rodriguez, "A Comparison of Drag Reduction for Flows in Circular Tubes with Flow Between Parallel Planes," J. Applied Polymer Science, 20, pp. 315-325 (1976).

18. Chodorkowskij, J.S. "Turbulent Flow of a Visco-elastic Fluid Through Smooth and Rough Pipes," (Turbulente Strömung einer Visko-elastischen Flüssigkeit Durch Glatte und Rauhe Röhre) ZAMM, 56, p. T388-T391 (1976). (In German).

19. Jones, W.M. "The Flow of Dilute Aqueous Solutions of Macromolecules in Various Geometries: I. Introduction; II. Straight Pipes of Circular Cross-Section; II. Bent Pipes and Porous Materials," J. Phys. D.: Appl. Phys., 9, pp. 721-770 (1976).

20. Oosterveld, M.W.C. and P. van Oossanen, "Ship Research Activities in the Netherlands (1969-1972)," International Shipbuilding Progress, 22, pp. 35-55 (1975).

21. Lehman, A.F. "Additional Investigations of the Hydrodynamic Characteristics of a Hydrofoil Employing Polymer Ejection," Oceanics, Inc. Report 75-122 (1975).

22. Fruman, D.H., M.P. Tulin and H.L. Liu, "Lift, Drag, and Pressure Distribution Effects Accompanying Drag-Reducing Polymer Injection on Two-Dimensional Hydrofoil," Hydronautics, Inc. Report 7101-5 (1975).

23. Sinnarwalla, A.M. and T.R. Sundaram, "Lift and Drag Effects Due to Polymer Injections on the Surface of Symmetric Hydrofoils," Hydronautics, Inc. Technical Report 7603-1 (1976).

24. Fruman, D.H. "Lift Effects Associated with Drag-Reducing Polymer Injection on Two-Dimensional Hydrofoils," J. Ship Research, 20, pp. 145-151 (1976).

25. Lang, B. "Investigation of Resistance Decrease on an Ideal Ship Body by Polymer Solutions," (Untersuchungen zur Widerstandsverminderung durch Polymerlösungen an idealisierten Schiffskörpern) Mitteilungen der Versuchsanstalt für Wasserbau und Schiffbau (Berlin) No. 53 (1975). (In German).

26. Lang, B. "The Influence of Drag Reducing Polymers on the Turbulent Boundary Layer of a Body of Revolution," IUTAM Symposium on Structure of Turbulence and Drag Reduction, Washington, DC (Jun 1976).

27. Sirmalis, J.E. "A Study of the Drag Characteristics and Polymer Diffusion in the Boundary Layer of an Axisymmetric Body," Naval Underwater Systems Center, Newport Lab., Tech. Report 4860 (1976).

28. Sarpkaya, T. "Turbulence Measurements in the Near-Wake Flow Field of a Circular Cylinder in a Dilute-Polymer Solution Flow," IUTAM Symposium on Structure of Turbulence and Drag Reduction, Washington, DC (Jun 1976).

29. Wetzel, J.M. and J.F. Ripken, "The Influence of Polymer Injection on a Developing Boundary Layer," Proc. 17th American Towing Tank Conference, Pasadena, pp. 52-64 (1974).

30. Cantwell, B. and D. Coles, "Growth of a Turbulent Spot in a Polymer-Filled Laminar Boundary Layer," IUTAM Symposium on Structure of Turbulence and Drag Reduction, Washington, DC (Jun 1976).

31. Breder, C.M. "Fish Schools as Operational Structures," Fishery Bulletin, 74, pp. 471-502 (1976).

32. Hoyt, J.W. "Hydrodynamic Drag Reduction Due to Fish Slimes," in Wu, T. Y.-T., C.J. Brokaw and C. Brennan, eds. Swimming and Flying in Nature. Plenum Press, New York, pp.653-672 (1975).

33. Mashelkar, R.A., D.D. Kale, and J. Ulbrecht, "Rotational Flows of Non-Newtonian Fluids," Transactions of Institution of Chemical Engineers, 53, p. 143 (1975).

34. Aoki, H., Y. Tomita, T. Kusaka, and Y. Mochimaru: "A Study on Drag Reduction of Dilute Polymer Solutions in Annular Flow Between Rotating Cylinders," in Colloques Internationaux du C.N.R.S. No. 233 - Polymères et Lubrification, pp. 345-348, Paris (1975).

35. Keller, A., G. Kiss, and M.R. Mackley, "Polymer Drag Reduction in Taylor Vortices," Nature, 257, pp. 304-305 (1975).

36. Friebe, H.W. "Destabilization by Dilute, Long Chain High Polymer Solutions in Couette Flow," (Das Stabilitätsverhalten verdünnter Lösungen sehr lankettiger Hochpolymer in der Couette-Strömung), Rheol.Acta, 15, pp. 329-355 (1976). (In German).

37. Sedahmed, G.H., A.M. Al-Taweel, and A. Abdel-Khalik, "Turbulent Mass-Transfer From a Rotating Cylinder to Drag-Reducing Fluids," Trans. Instn. Chem. Engrs., 53, pp. 191-193 (1975).

38. Driels, M.R. and S. Ayyash, "Drag Reduction in Laminar Flow," Nature, 259, pp. 389-390 (1976).

39. Popadić, V.O. "Drag Reduction in Film Flow," AIChE. J. 21, pp. 610-612 (1975).

40. Chiou, C.S. and R.J. Gordon, "Vortex Inhibition: Velocity Profile Measurements," AIChE J. 22, pp. 947-950 (1976).

41. Sicardi, S., G. Baldi, and A. Gianetto, "Laminar Pulsed Flow of Non-Newtonian Fluids," Ing. Chim. Ital., 11, pp. 105-110 (1975).

42. Oliver, D.R. and S.M. Asghar, "The Laminar Flow of Newtonian and Viscoelastic Liquids in Helical Coils," Trans. Instn. Chem. Engrs., 53, pp. 181-186 (1975).

43. Kuo, J.T. and L.S.G. Kovasznay, "Drag Reducing Polymer in Helicoidal Flow," Johns Hopkins University, Depts. of Mechanics and Materials Science, Technical Report 75-1 (1975).

44. Kovasznay, L.S.G. and J.T. Kuo, "Drag Reducing Polymer in Helicoidal Flow," IUTAM Symposium on Structure of Turbulence and Drag Reduction, Washington, DC (Jun 1976).

45. Kalashnikov, V.N., A.M. Kudin, and S.A. Ordinartsev, "Influence of Polymeric Additives on the Vortex Generation in the Wakes of Thin Cylinders," DISA Information Bulletin, 19, pp. 30-33 (1976).

46. Piau, J.M. "Mechanics of Dilute Polymer Solutions-Application to Friction Reduction," (Mécanique des solutions diluées de polymères. Application à la réduction de frottement) in Colloques Internationaux du C.N.R.S. No. 233 - Polymères et Lubrification, pp.225-240, Paris (1975). (In French).

47. Gatski, T.B. "The Numerical Solution of the Steady Flow of Newtonian and Non-Newtonian Fluids Through a Contraction," Ph.D. Thesis, Pennsylvania State University (1976).

48. Halliwell, N.A. and A.K. Lewkowicz, "Investigation into the Anomalous Behavior of Pitot Tubes in Dilute Polymer Solutions," Physics of Fluids, 18, p. 1617 (1975).

49. James, D.F. and D.R. McLaren, "The Laminar Flow of Dilute Polymer Solutions Through Porous Media," J. Fluid Mechanics, 70, pp. 733-752 (1975).

50. Laufer, G., C. Gutfinger, and N. Abuaf, "Flow of Dilute Polymer Solutions Through a Packed Bed," I&EC Fundamentals, 15, p. 74 (1976).

51. Daoudi, S. "Interpretation of Tests of Dilute Polymer Solutions in Porous Media," (Interprétation d'expériences sur des Solutions Diluées de Polymère en Milieu Poreux), J. de Physique-Lettres, 37, pp. L-41 - L-429 (1976). (In French).

52. Yost, M.E. and O.M. Stokke, " Filtration of Polymer Solutions," J. Pet. Technology, p. 1271 (1975).

53. Maerker, J.M. "Shear Degradation of Partially Hydrolyzed Polyacrylamide Solutions," Soc. Pet. Engrs. J., p. 311 (1975).

54. Attia, Y.A. and J. Rubio, "Determination of Very Low Concentrations of Polyacrylamide and Polyethleneoxide Flocculants by Nephelometry," Br. Polymer J., 7, p. 135 (1975).

55. James, D.F. "A Method for Measuring Normal Stresses in Dilute Polymer Solutions," Trans. Society of Rheology, 19, p. 67 (1975).

56. Oliver, D.R. "Development of the 'Triple Jet' System for Extensional Flow Measurements on Polymer Solutions," in Colloques Internationaux du C.N.R.S. No. 233 - Polymères et Lubrification, Paris, pp 325-330 (1975).

57. Balakrishnan, C. and R.J. Gordon, "Extensional Viscosity and Recoil in Highly Dilute Polymer Solutions," AIChE J. 21, p. 1226 (1975).

58. Crowley, D.G., F.C. Frank, M.R. Mackley and R.G. Stephenson, "Localized Flow Birefringence of Polyethylene Oxide Solutions in a Four Roll Mill," J. Polymer Science, 14, p. 1111 (1976).

59. Weinberger, C.B. and J.D. Goddard, "Extensional Flow Behavior of Polymer Solutions and Particle Suspensions in a Spinning Motion," Int. J. Multiphase Flow, 1, p. 465 (1974).

60. Tiederman, W.G. "Visualization of Drag-Reducing Turbulent Channel Flows," Oklahoma State Univ., School of Mechanical and Aerospace Engineering, Report ER-75-ME-12 (1975).

61. Oldalker, D.K. and W.G. Tiederman, "Spatial Structure of the Viscous Sublayer in Drag Reducing Channel Flows," IUTAM Symposium on Structure of Turbulence and Drag Reduction, Washington,DC (Jun 1976).

62. Donohue, G.L. "Hydrogen-Bubble Flow Visualization: Limitations in Drag Reducing Polymer Solutions," in "Turbulence in Liquids," G.K. Patterson and J.L. Zakin, eds., University of Missouri-Rolla, Dept. of Chemical Engineering, p.86-90 (1975).

63. Deslouis, C., I. Epelboin, B. Tribollet and L. Viet, "Study of the Reduction of Fluid Friction in the Turbulent Regime by Electrochemical Methods," (Etude par des méthodes electrochimiques de la réduction de frottement hydrodynamique en régime turbulent) in Colloques Internationaux du C.N.R.S. No. 233 - Polymères et Lubrification, Paris, pp. 283-294 (1975). (In French).

64. Deslouis, C. "A Study of Mass Transport in Turbulent Liquid Flow by an Electrochemical Method-Application to Hydrodynamic Drag Reduction," (Etude par une méthode électrochimique du transport de matière dans un liquide en éncoulement turbulent. Application à la réduction de la traînee hydrodynamique) Docteur ès Sciences thèse, Universitie Pierre et Marie Curie, Paris (1975). (In French).

65. Deslouis, C., I. Epelboin, B. Tribollet and L. Viet, "Electrochemical Methods in the Study of the Hydrodynamic Drag Reduction by High Polymer Additives," Electrochim. Acta, 20, p. 909 (1975).

66. Hanratty, T.J. "Details of Turbulent Fluctuations in the Viscous Wall Region," IUTAM Symposium on Structure of Turbulence and Drag Reduction, Washington, DC (Jun 1976).

67. Reischman, M.M. and W.G. Tiederman, "Laser-Doppler Anemometer Measurements in Drag-Reducing Channel Flows," J. Fluid Mechanics, 70, p. 369 (1975).

68. Logan, S.E. "Measurement of Reynolds Stress and Turbulence in Dilute Polymer Solution by Laser Velocimeter," in "Turbulence in Liquids," G.K. Patterson and and J.L. Zakin, eds., University of Missouri-Rolla, Dept. of Chemical Engineering, pp. 91-105 (1975).

69. Zakin, J.L., C.C. Ni, R.J. Hansen and M.M. Reischman, "Laser Doppler Velocimetry Studies of Early Turbulence," IUTAM Symposium on Structure of Turbulence and Drag Reduction, Washington, DC (Jun 1976).

70. Flory, P.J., "Spatial Configuration of Macromolecular Chains," Science, 188, p. 1268 (1975).

71. DeGennes, P.G. "Coil-Stretch Transition of Dilute Flexible Polymers Under Ultrahigh Velocity Gradients," J. Chem. Physics, 60, p. 5030 (1974).

72. Hinch, E.J. "The Distortion of a Flexible Inextensible Thread in a Shearing Flow," J. Fluid Mech., 74, p. 317 (1976).

73. Hinch, E.J. "The Deformation of a Nearly Straight Thread in a Shearing Flow With Weak Brownian Motions," J. Fluid Mech. 75, p. 765 (1976).

74. Merrill, E.W., K.A. Smith, and R.C. Armstrong, "Drag Reduction in Turbulent Flow of Polymer Solutions," Dept. of Chemical Engineering, Massachusetts Institute of Technology Report CED-DR-75-2 (1975).

75. White, D. Jr. and R.J. Gordon. "The Influence of Polymer Conformation on Turbulent Drag Reduction," AIChE J. 21, p. 1027 (1975).

76. Balakrishnan, C. and R.J. Gordon, "Influence of Molecular Configuration and Intermolecular Interactions on Turbulent Drag Reduction," J. Applied Polymer Science, 19, p. 909 (1975).

77. Wade, R.H. "A Study of Molecular Parameters Influencing Polymer Drag Reduction," Naval Undersea Center Report NUC TP-473 (1975).

78. Carter, L. "Drag Reduction Characteristics of Ionic Polymers in the Presence and Absence of an Electrical Charge," D.W. Taylor NSRDC Report 4729 (1975).

79. Gramain, P. and P. Philippides, "Principal Physico-Chemical Factors Characterizing the Drag-Reduction Efficiency of Polymer Solutions," (Principaux facteurs physico-chimiques caractérisant l'efficacité des solutions de polymères en réduction de frottement hydrodynamique) in Colloques Internationaux du C.N.R.S. No. 233-Polymères et Lubrification, Paris, pp. 349-356 (1975). (In French).

80. Moan, M. and C. Wolff, "Study of the Conformational Rigidity of Polyelectrolytes by Elastic Neutron Scattering: 1. Carboxymethyl-celluloses in the Intermediate Momentum Range," Polymer, 16, p. 776 (1975).

81. Virk, P.S. "Conformational Effects in Drag Reduction by Polymers," Nature, 262, p. 46 (1976).

82. Layec-Raphalen, M.-N., and C. Wolff, "On the Shear-Thickening Behavior of Dilute Solutions of Chain Molecules," J. Non-Newtonian Fluid Mechanics, 1, pp. 159-173 (1976).

83. Morita, T. "A Lattice Model for a Polymer Chain in Dilute Solution," J. Physics, A. 9, pp. 169-178 (1976).

84. Tanner, R.I. "A Test Particle Approach to Flow Classification for Viscoelastic Fluids," AIChE J. 22, pp. 910-918 (1976).

85. Hunston, D.L. "Effect of Molecular Weight Distribution in Drag Reduction and Shear Degradation," J. Polymer Science, 14, pp. 713-727 (1976).

86. Ting, R.Y. "Diameter Dependence of the Cutoff Molecular Weights of Drag-Reducing Polymers," J. Applied Polymer Science, 20, pp. 3017-3023 (1976).

87. Little, R.C. and R.Y. Ting, "Number-Average and Viscosity-Average Molecular Weights of Popular Drag-Reducing Polymers," J. Chem. & Engineering Data, 21, pp. 422-423 (1976).

88. Maxfield, J. and I. W. Shepherd, "Conformation of Poly (ethylene oxide) in the Solid State, Melt and Solution Measured by Raman Scattering," Polymer, 16, pp. 505-509 (1975).

89. Powell, R.L. and W.H. Schwarz, "Rheological Properties of Polyethylene Oxide Solutions," Rheol. Acta, 14, pp. 729-740 (1975).

90. Powell, R.L. and W.H. Schwarz, "Rheological Properties of Aqueous Poly (ethylene oxide) Solutions in Parallel Superposed Flows," Trans. Soc. Rheology, 19, pp. 617-643 (1975).

91. Riande, E. "Dipole Moments of Poly (ethylene oxide) and Poly (hexamethylene oxide)" J. Polymer Science, 14, pp. 2231-2240 (1976).

92. Boucher, E.A. and P.M. Hines, "Effects of Inorganic Salts on the Properties of Aqueous Poly (ethylene Oxide) Solutions," J. Polymer Science 14, p.2241 (1976).

93. Cuniberti, C. "Evidence for Agregation in Solutions of Poly (ethylene oxide)," Polymer, 16, p. 306 (1975).

94. Little, R.C., R.L. Patterson, and R.Y. Ting, "Characterization of the Drag Reducing Properties of Poly (ethylene oxide) and Poly (acrylamide) Solutions in External Flows," J. Chem. & Engineering Data, 21, pp. 281-283 (1976).

95. Huang, T.T. and N. Santelli, "Drag Reduction of Degraded and Blended Polymer Solutions," D.W. Taylor NSRDC Report 4311 (1975).

96. Patterson, R.L., D.L. Hunston, R.Y. Ting and R.C. Little, "Drag Reduction Characteristics of Poly (acrylamides) in Aqueous Magnesium Sulphate and Acetone Solutions," J. Chem. & Engrg. Data, 20, pp. 381-384 (1975).

97. Nicodemo, L., B. deCindio and D. Acierno, "Rheological Behavior of Concentrated Polyacrylamide Solutions," (Comportamento reologico di soluzioni concentrate di poliacrilammide), Ing. Chim. Ital, 11 , pp. 59-62 (1975). (In Italian).

98. Hershey, H.C., J.T. Kuo and M.L. McMillan, "Drag Reduction of Straight and Branched Chain Aluminum Disoaps," I&EC, Product R&D, 14, pp. 192-199 (1975).

99. Zakin, J.L. "Drag Reduction with Soap and Surfactant Additives," in Colloques Internationaux du C.N.R.S. No. 233-Polymères et Lubrification, Paris, pp. 295-304 (1975).

100. Patterson, R.L. and R.C. Little, "The Drag Reduction of Poly(ethylene oxide)-Carboxylate Soap Mixtures," J. Colloid and Interface Sci., 53, pp. 110-114(1975).

101. Tiederman, W.G. and M.M. Reischman, "Calculation of Velocity Profiles in Drag-Reducing Flows," Trans. ASME, J. Fluids Engineering, 98I, p. 563 (1976).

102. Poreh, M,, S. Hassid and Y. Dimant, "Turbulent Flows With Drag Reduction," Faculty of Civil Engineering Publication No. 218, Technion (1975).

103. Hassid, S. and M. Poreh, "A Turbulent Energy Model for Flows with Drag Reduction," Trans. ASME, J. Fluids Engineering, 97, p. 234 (1975).

104. Poreh, M. "Mean Velocity and Turbulent Energy Closures for Flows with Drag Reduction," IUTAM Symposium on Structure of Turbulence and Drag Reduction, Washington, DC (Jun 1976).

105. Landahl, M.T. and F.H. Bark, "Application of a Two-Scale Boundary Layer Turbulence Model to Drag Reduction," in Colloques Internationaux du C.N.R.S. No. 233-Polymères et Lubrification, Paris, pp. 249-258 (1975).

106. Landahl, M.T. "On the Dynamics of Boundary Layer Turbulence and the Mechanism of Drag Reduction," IUTAM Symposium on Structure of Turbulence and Drag Reduction, Washington, DC (Jun 1976).

107. Lumley, J.L. "Drag Reduction in Two Phase and Polymer Flows," IUTAM Symposium on Structure of Turbulence and Drag Reduction, Washington, DC (Jun 1976).

108. Saffman, P.G. "A Phenomenological Theory for the Calculation of the Dependence of the Effective Slip Velocity of a Dilute Polymer Solution on Wall Roughness," IUTAM Symposium on Structure of Turbulence and Drag Reduction, Washington, DC (Jun 1976).

109. Hinch, E.J. "Mechanical Models of Dilute Polymer Solutions for Strong Flows With Large Polymer Deformations," in Colloques Internationaux du C.N.R.S. No. 233-Polymères et Lubrification, Paris, pp. 241-247 (1975).

110. Frederick, P.S. and A. White, "Structure of Turbulent Boundary Layers at Maximum Drag Reduction," Nature, 256, p. 30-31 (1975).

111. Nikitin, I.K., N.G. Poznyaya, and A.V. Dzevaltovskiy, "Turbulent Flow of Dilute Polymer Solutions in Smooth Pipes," Fluid Mechanics-Soviet Research, 4, p.113 (1975).

112. Scrivener, O. "Velocity Distributions in Pipe Flow With Drag-Reducing Additives," (Champ des vitesses dans un écoulement turbulent interne en présence d'additifs réducteurs de frottement), in Colloques Internationaux du C.N.R.S. No. 233-Polymères et Lubrification, Paris, pp. 315-324 (1975). (In French).

113. Pilipenko, V.N. "On the Possibility of Modeling the Integral Characteristics of the Turbulent Boundary Layer," Soviet Physics-Doklady, 20, p. 395 (1975).

114. Gustavsson, H. "Drag Reduction Experiments With Polystyrene With Some Implications for the Mean Velocity Profile," IUTAM Symposium on Structure of Turbulence and Drag Reduction, Washington, DC (Jun 1976).

115. Virk, P.S. "Turbulent Kinetic Energy Profile During Drag Reduction," Physics of Fluids, 18, p. 415 (1975).

116. Berman, N.S. "Flow Time Scales and Drag Reduction," IUTAM Symposium on Structure of Turbulence and Drag Reduction, Washington, DC (Jun 1976).

117. Dunlop, E.H. and L.R. Cox, "The Influence of Molecular Aggregates on Drag Reduction," IUTAM Symposium on Structure of Turbulence and Drag Reduction, Washington, DC (Jun 1976).

118. Gyr, A. "Burst Cycle and Drag Reduction," IUTAM Symposium on Structure of Turbulence and Drag Reduction, Washington, DC (Jun 1976).

119. Lindgren, E.R. "Mechanical Structure of Drag Reducing Additives," IUTAM Symposium on Structure of Turbulence and Drag Reduction, Washington, DC(Jun 1976).

120. Patterson, G.K., J. Chosnek and J.L. Zakin, "Turbulence Structure in Drag Reducing Polymer Solutions," IUTAM Symposium on Structure of Turbulence and Drag Reduction, Washington, D.C (Jun 1976).

121. Scrivener, O., C. Kopp and M. Mondon, "Influence of Polymer Solutions on the Structure of Turbulence in a Pipe," IUTAM Symposium on Structure of Turbulence and Drag Reduction, Washington, DC (Jun 1976).

122. Tomita, Y. and T. Jotaki, "Effects of Elongational Viscosity of Polymer Solution on Taylor-Görtler Vortices," IUTAM Symposium on Structure of Turbulence and Drag Reduction, Washington, DC (Jun 1976).

123. Tulin, M.P. and J. Wu, "Additive Effects in Free Turbulent Flows," IUTAM Symposium on Structure of Turbulence and Drag Reduction, Washington, DC(Jun 1976).

124. Virk,P.S. and M. Ohara,"Triggered Transition in the Pipe Flow of Dilute Solutions of Random-Coiling Macromolecules," IUTAM Symposium on Structure of Turbulence and Drag Reduction, Washington, DC (Jun 1976).

125. Greshilov, E.M. and N.L. Shirokova, "Effect of Polymer Additives on Wall Turbulence," Journal of Engineering Physics, 26, p. 178 (1975).

126. Smith, A.J. "An Investigation of the Bursting Events in Drag Reduction Turbulent Channel Flows," M.S. Thesis, Oklahoma State University (1975).

127. Bazilevich, V.A. "Utilization of the Kolmogorov Scale as a Parameter Defining Drag Reduction in the Flow of Polymeric Solutions," Fluid Mechanics-Soviet Research, 3, p. 51 (1974).

128. Hlăvacek, B., L.A. Rollin and H.P. Schreiber, "Drag Reduction Effectiveness of Macromolecules," Polymer, 17, p. 81 (1976).

129. Hlávacek, B. and J. Sangster, "Drag Reduction by Long-Chain Polymers: An Interpretation of Liquid Structure," Canadian J. Chemical Engineering, 54, p. 115 (1976).

130. Berman, N.S. and W.K. George, "Time Scale and Molecular Weight Distribution Contributions to Dilute Polymer Solution Fluid Mechanics," Proc. 1974 Heat Transfer and Fluid Mechanics Institute, L.R. Davis and R.E. Wilson, eds., Stanford University Press, p. 348 (1974).

131. Hunston, D.L. and M.M. Reischman, "The Role of Polydispersity in the Mechanism of Drag Reduction," Physics of Fluids, 18, p. 1626 (1975).

132. Khabakhpasheva, E.M. "Polymers and Turbulent Drag Reduction," in Colloques Internationaux du C.N.R.S. No. 233-Polymères et Lubrification, Paris, pp. 217-224 (1975).

133. Little, R.C., R.J. Hansen, D.L. Hunston, O.K. Kim, R.L. Patterson and R.Y. Ting, "Drag Reduction: Polymer Homology, Solvent Effects and Proposed Mechanisms," in Colloques Internationaux du C.N.R.S. No. 233-Polymères et Lubrification, Paris, pp. 259-269 (1975).

134. Little, R.C., R.J. Hansen, D.L. Hunston, O.K. Kim, R.L. Patterson and R.Y. Ting, "The Drag Reduction Phenomenon. Observed Characteristics, Improved Agents, and Proposed Mechanisms," I&EC Fundamentals, 14, p. 283 (1975).

135. Ram, A. "Polymers and Their Effectiveness in Drag Reduction," in Colloques Internationaux du C.N.R.S. No. 233-Polymères et Lubrification, Paris, pp. 271-281 (1975).

136. Kolář, V. and J. Štastna, "Drag Reduction-A Special Case of Turbulent Flow of Liquid Systems," ZAMM, 56, pp.T436-T438 (1976).

137. Smith, K.A., E.W. Merrill, L.H. Peebles and H. Banijamali, "Elongation of Drag-Reducting Macromolecules by a Pure Straining Motion," in Colloques Internationaux du C.N.R.S. No. 233-Polymères et Lubrification, Paris, pp. 341-344(1975).

138. Hunston, D.L. and R.Y. Ting, "The Viscoelastic Response of Drag Reducing Polymer Solutions in Simple Flows," Trans. Society of Rheology, 19, p. 115 (1975).

139. Tomita, Y. and T. Jotaki, "Elongational Viscosity and Secondary Instability of Laminar Boundary Layer," Bulletin of JSME, 19, pp. 938-942 (Aug 1976).

140. Ting, R.Y. "An Analysis of Unsteady One-Dimensional Stretching of a Viscoelastic Fluid," J. Applied Polymer Science, 20, pp. 1231-1244 (1976).

141. Bhowmick, S.K. "Contribution to the Study of Turbulent Diffusion in Non-Newtonian Fluid," (Contribution a l'Étude de la Diffusion Turbulente de Fluides Non-Newtoniens en Ecoulement Interne-Étude de la Corrélation entre la Diffusion Turbulente et la Réduction du Frottement) Doctoral Thesis, Universite Louis Pasteur, Strasbourg, France (1975). (In French).

142. Bhowmick, S.K., C. Gebel and H. Reitzer, "Turbulent Diffusion and Drag Reduction of Drag-Reducing Fluids," Rheol. Acta., 14, p. 1026 (1975).

143. Latto, B. and O.K.F. El Riedy, "Diffusion of Polymer Additives in a Developing Turbulent Boundary Layer," J. Hydronautics, 10, p. 135 (1976).

144. Fruman, D.H. and M.P. Tulin, "Diffusion of a Tangential Drag-Reducing Polymer Injection on a Flat Plate at High Reynold's Numbers," J. Ship Research, 20, p. 171 (1976).

145. Sellin, R.H.J. "The Suppression of Turbulent Diffusion by Drag Reducing Polymer Additives," in Colloques Internationaux du C.N.R.S. No. 233-Polymères et Lubrification, Paris, pp. 331-340 (1975).

146. Ayyash, S. and W. D. McComb, "Some Anomalous Results in Drag Reduction by Absorbed Layers," Chem. Engineering Science, 31, p. 169 (1976).

147. Jeffrey, D.J. and A. Acrivos, "The Rheological Properties of Suspensions of Rigid Particles," AIChE J. 22, pp. 417-432 (1976).

148. Radin, I., J.L. Zakin and G.K. Patterson, "Drag Reduction in Solid-Fluid Systems," AIChE J. 21, p. 358 (1975).

149. Fedors, R.F. "Viscosity of Newtonian Suspensions," Polymer, 16,pp.305-306(1975).

150. Drew, D.A. "Production and Dissipation of Energy in the Turbulent Flow of a Particle-Fluid Mixture, With Some Results on Drag Reduction," ASME Preprint 76-WA/APM-24 (1976).

151. Bark, F.H. "Some Properties of the Viscous Sublayer and Three-Dimensional Effects for the Stability Properties of Fiber Suspensions," IUTAM Symposium on Structure of Turbulence and Drag Reduction, Washington, DC (Jun 1976).

152. Duffy, G.G., A.L. Titchener, P.F.W. Lee and K. Moller, "The Mechanisms of Flow of Pulp Suspensions in Pipes," Appita, 29, pp. 363-370 (1976).

153. Lee, P.F.W. and G.G. Duffy, "Velocity Profiles in the Drag Reducing Regime of Pulp Suspension Flow," Appita, 30, pp. 219-226 (1976).

154. Lee, P.F.W. amd G.G. Duffy, "Relationships Between Velocity Profiles and Drag Reduction in Turbulent Fiber Suspension Flow," AIChE J. 22, pp. 750-753(1976).

155. Moller, K."A correlation of Pipe Friction Data for Paper Pulp Suspensions," I&EC Process Design and Development, 15, pp. 16-19(1976).

156. Gust, G. "Observations on Turbulent Drag Reduction in a Dilute Suspension of Clay in Sea-Water," J. Fluid Mechanics, 75, pp. 29-47(1976).

157. Alonson, C.V., W. H. Klaus and K.F. Wylie, "Turbulent Characteristics of Drag-Reduction Flows," J. Hydraulic Research, 14, pp. 103-114(1976).

158. Kale, D.D. and A.B. Metzner, "Turbulent Drag Reducation in Dilute Fiber Suspensions: Mechanistic Considerations," AIChE J. 22, pp. 669-674 (1976).

159. Metzner, A.B. "Drag Reduction in Dilute Fiber Suspensions and in Mixtures of Suspended Fibers and Dissolved Polymers," IUTAM Symposium on Structure of Turbulence and Drag Reduction, Washington, DC (Jun 1976).

160. Fujimoto, H. and T. Tagori, "Friction Reduction of Pipe Flow of Iron Ore Slurry by Polymer Solution Injection," Proc. First Japanese Towing Tank Conf. (April 1974).

161. Chashchin, I.P., N.T. Shelavin and V.A. Saenko, "The Effect of Polymeric Additives on Drag Reduction," Int'l. Chem. Engineering, 15, pp.88-90 (1975).

162. Otten, L. and A.S. Fayed, "Pressure Drop and Drag Reduction in Two-Phase Non-Newtonian Slug Flow," Canadian J. Chemical Engineering, 54, p. 111 (1976).

163. Sylvester, N.D. and J.P. Brill, "Drag Reduction in Two-Phase Annular-Mist Flow of Air and Water," AIChE J. 22, No. 3, pp. 615-617 (1976).

164. Dimant, Y. and M. Poreh, "Heat Transfer in Flows With Drag Reduction," in Irvine, T.F. Jr. and J.P. Hartnett, eds. Advances in Heat Transfer, 12, Academic Press, New York, pp. 77-113 (1976).

165. Ghajar, A.J. and W.G. Tiederman, "Prediction of Heat Transfer Coefficients in Drag-Reducing Turbulent Pipe Flows," AIChE J. 23, pp.128-131 (1977).

166. Yoo, S.S. and J.P. Hartnett, "Thermal Entrance Lengths for Non-Newtonian Fluids in Turbulent Pipe Flow. Letters in Heat and Mass Transfer, 2, pp.189-197(1975).

167. Mizushina, T., H. Usui, and T. Yamamoto, "Turbulent Heat Transfer of Viscoelastice Fluids Flow in Pipe," Letters in Heat and Mass Transfer, 2, p.19-26(1975).

168. Mizushina, T. and H. Usui, "Reduction of Eddy Diffusion for Momentum and Heat in Viscoelastic Fluid Flow in a Circular Tube," IUTAM Symposium on Structure of Turbulence and Drag Reduction, Washington, DC (Jun 1976).

169. Taylor, D.D. and R.H. Sabersky, "Extrapolation to Various Tube Diameters of Experimental Data Taken With Dilute Polymer Solutions in a Smooth Tube," Letters in Heat and Mass Transfer, 1, p. 103(1974).

170. Moyls, A.L. "Friction and Heat Transfer Reduction in Turbulent Flow of Dilute Asbestos Fiber Suspensions in Smooth and Rough Tubes," Ph.D. Thesis, California Institute of Technology (1976).

171. Moyls, A.L. and R. Sabersky, "Heat Transfer to Dilute Asbestos Dispersions in Smooth and Rough Tubes," Letters in Heat and Mass Transfer, 2, pp.293-302(1975).

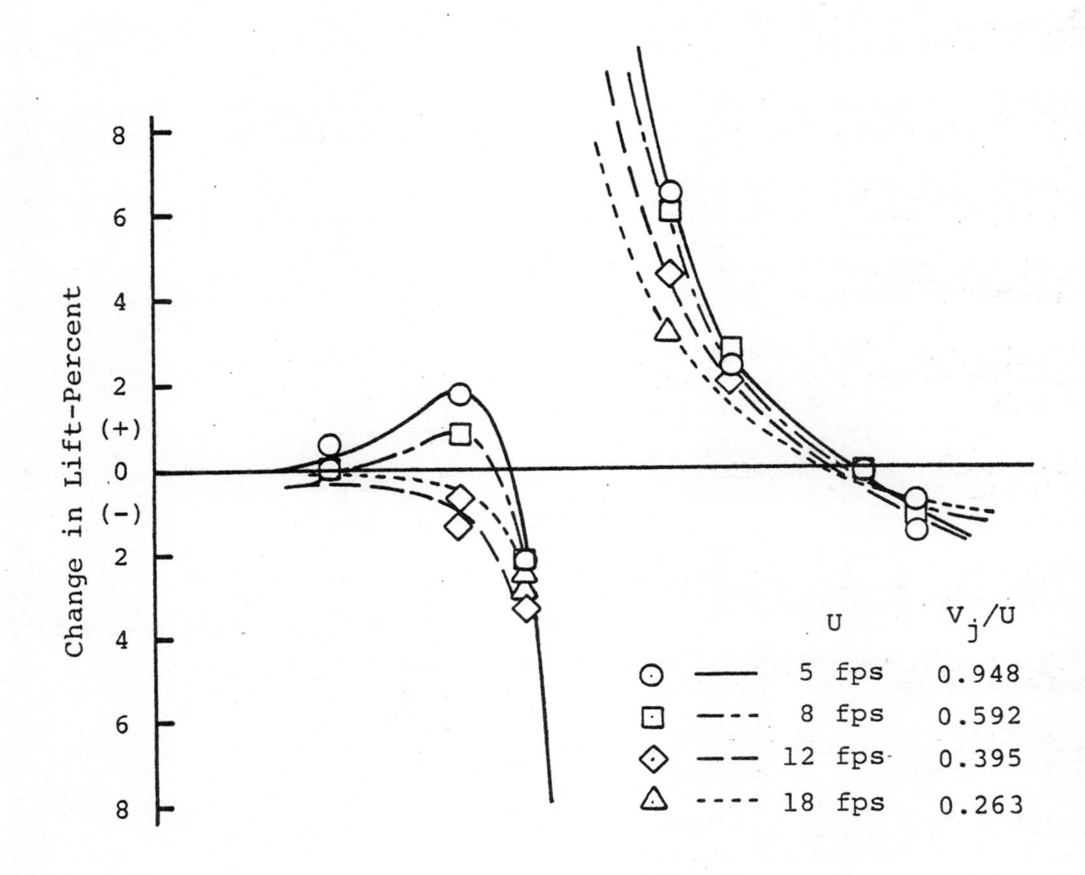

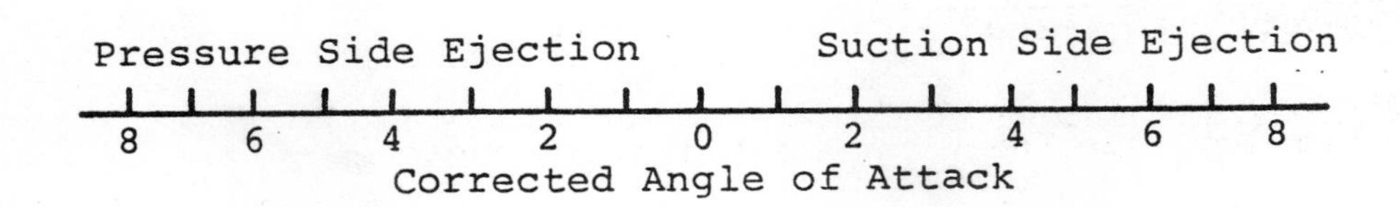

Figure 1. The effect of polymer ejection on lift of a NACA 65006 airfoil. 100 ppm poly (ethylene oxide) ejected at 4.74 ft/sec from the leading edge of of the airfoil. (From Lehman, Ref. 21).

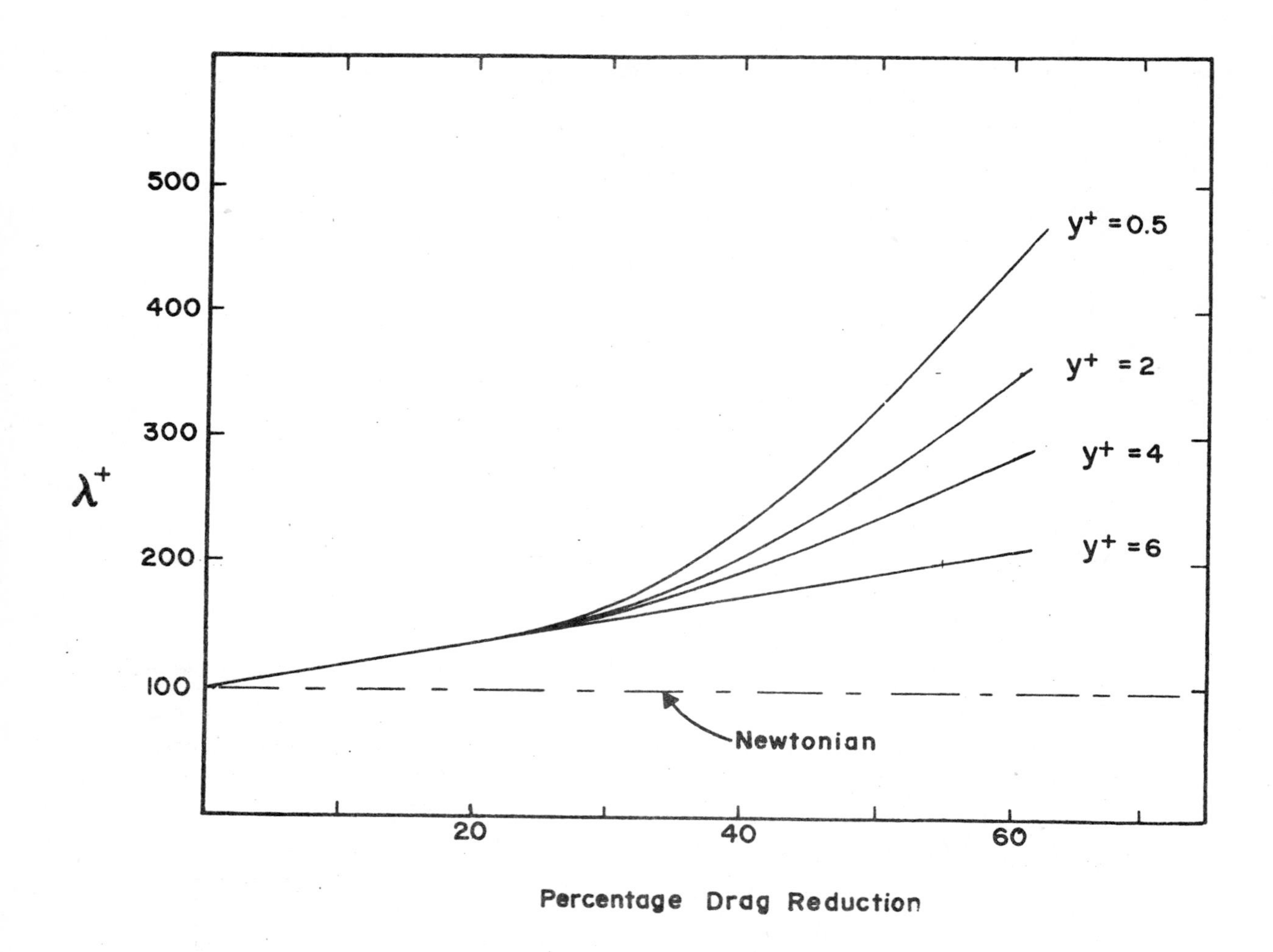

Figure 2. Variation in non-dimensional streak spacing, λ^+, as a function of drag reduction and distance from wall, y^+. (From Tiederman, Ref. 60).

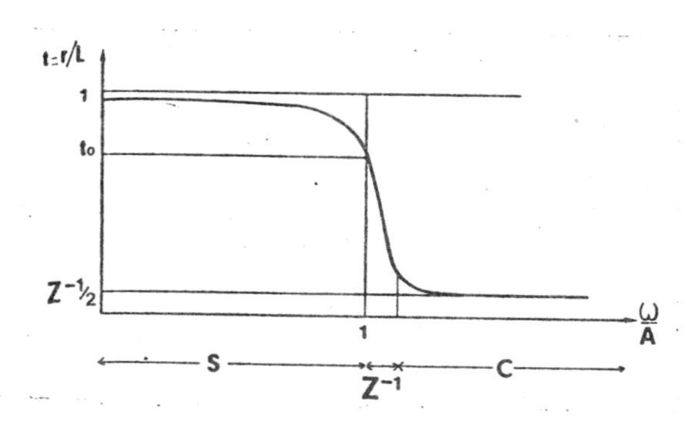

Figure 3. Coil (C) - Stretched (S) states of a polymer molecule as a function of vorticity (ω) to pure deformation rate (A) ratio. The deformation rate is assumed large. As ω/A becomes smaller, the ratio, t, of the molecule average end-to-end length (r) to its extended length (L) increases rapidly from its value of $Z^{-\frac{1}{2}}$ (where Z is the number of monomer links) to almost 1. This occurs in a $\Delta\omega/\omega$ interval proportional to Z^{-1}. (From deGennes, Ref. 71).

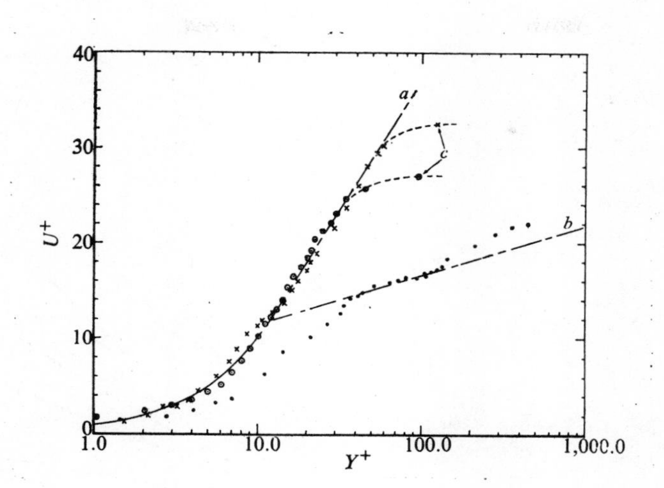

Figure 4. Experimental confirmation of Virks drag-reduction asymptote. Line a: $U^+ = 11.7 \ln Y^+ - 17.0$; line b: $U^+ = 2.44 \ln Y^+ + 4.9$; lines c are free-stream values. •: Water; x, o, CTAB-1-napthol solutions. (From Frederick and White, Ref. 110).

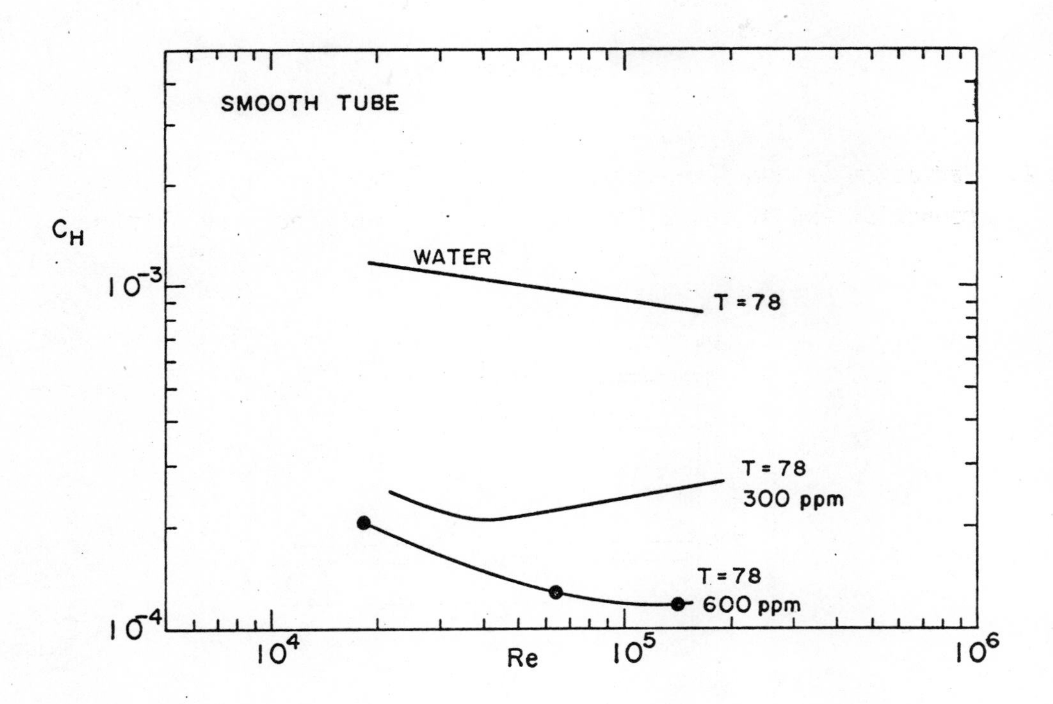

Figure 5. Reduction in heat-transfer coefficient, C_H, as a function of Reynold's number, for 300 and 600 ppm asbestos fiber suspension in a smooth tube. All tests at 78^0F. (From Moyls, Ref. 170).

Second International Conference on

Drag Reduction

August 31st - September 2nd, 1977

PAPER B1

SCALING-UP OF PIPE-FLOW FRICTIONAL DATA FOR DRAG-REDUCING POLYMER SOLUTIONS

P.S. Granville,

David W. Taylor Naval Ship Research and
Development Center, U.S.A.

Summary

The drag-reducing effects of dilute polymer solutions measured in one pipe may be scaled up or down to other pipe diameters by proper use of the velocity similarity laws of turbulent shear flows. The extension of the classical similarity laws to the flow of drag-reducing solutions by the addition of characteristic lengths, times and/or masses is restated. A newly developed wake-modification function for the outer similarity law is included in the derivation of a similarity-law logarithmic friction formula which then becomes the basis of a scaling-up method. Allowance is made for the presence of the interactive layer which leads to the condition of maximum drag reduction when the interactive layer extends to the pipe center. Finally, a simple but accurate method based on the similarity-law friction formula is presented for preparing a friction chart of drag-reducing polymer solutions when scaling up friction data from one pipe diameter to other pipe diameters.

Held at St. John's College, Cambridge, England.
Organised and sponsored by BHRA Fluid Engineering, Cranfield, Bedford, MK43 0AJ.

NOMENCLATURE

A	slope of logarithmic similarity laws, Eqs. (14), (16)
$\tilde{A}$	slope of interactive logarithmic law, Eq. (21)
B_1	intercept of logarithmic inner similarity law, Eq. (14)
$B_{1,0}$	B_1 for ordinary fluids
B_2	velocity-defect factor, Eq. (16)
$\tilde{B}$	intercept of interactive logarithmic law, Eq. (21)
C	concentration of polymer in solution
D	diameter of pipe
f	Fanning friction factor, Eq. (3)
$\ell,\ell_1,\ell_2,\ldots$	characteristic polymer lengths
$m,m_1,m_2,\ldots$	characteristic polymer masses
o	subscript denoting ordinary fluids
P	type of polymer mixture
q	wake modification function, Eq. (19)
R_D	diameter Reynolds number, Eq. (4)
$t,t_1,t_2,\ldots$	characteristic polymer times
U	velocity at pipe center
u	velocity at point in pipe
u_τ	shear velocity, $u_\tau = \sqrt{\tau_w/\rho}$
$u_{\tau,cr}$	onset value of u_τ
V	average velocity of pipe flow
w	wake function, Eq. (18)
y	radial distance measured inward from pipe wall
y^*	inner-law Reynolds number, $y^* = \frac{u_\tau y}{\nu}$
y^*_L	value of y^* for thickness of laminar sublayer
y^*_m	value of y^* for thickness of interactive layer
δ	pipe radius
ΔB	drag reduction characterization
η	value of y^* at pipe center, $\eta = \frac{u_\tau \delta}{\nu}$
λ	Darcy-Weisbach friction factor, Eq. (5)
ν	kinematic viscosity of solution
ρ	density of solution
ρ_p	density of polymer in solution
τ_w	wall shear stress

INTRODUCTION

It is elementary that any effect of diameter on the pressure loss due to friction of ordinary fluids flowing in pipes is accounted for by a nondimensional friction factor dependent on a nondimensional Reynolds number based on diameter. It is also well known that the situation is otherwise for the turbulent flow of drag-reducing polymer solutions in pipes. Different diameter pipes produce different relations between friction factor and Reynolds number except for the limiting case of maximum drag reduction. The effect of diameter is no longer completely absorbed by Reynolds number (Refs. 1, 2, 3).

For the same polymer species and concentration, the friction factor is now a function of two independent dimensionless ratios, a diameter Reynolds number and another dimensionless ratio based on diameter. Measurements of drag reduction in a pipe of one diameter are no longer simply applicable to pipes of other diameters. The prediction or scale-up of friction factors in such pipes may be accomplished, however, by use of logarithmic friction formulas derived from the velocity similarity laws for turbulent shear flows.

The velocity similarity laws are well established for ordinary fluids and have successfully correlated velocity data for pipe flow. The velocity similarity laws have been extended to the flow of drag-reducing polymer solutions by a consideration of added characteristic lengths, masses and/or times (Ref. 4). Logarithmic friction formulas for drag-reducing solutions have been derived for pipe flow from the velocity similarity laws (Refs. 4, 5).

The application of similarity-law logarithmic friction formulas to the scale-up problem seems to be not a straight-forward procedure. Taylor and Sabersky (Ref. 6) employ a trail-and-error method while Savins and Seyer (Ref. 7) substitute an approximate method based on an over-simplified empirical correlation. It will be shown that a similarity-law logarithmic friction formula for drag-reducing polymer solutions may be used in a simple but accurate manner to predict the friction factor for different pipe diameters. The method is based on the application of the similarity-law drag-reduction characterization. Consideration is also given to conditions near maximum drag reduction which are omitted in the procedures given by Taylor and Sabersky and by Savins and Seyer.

First the general problem of the diameter effect on pipe friction for drag-reducing solutions is made more understandable when characteristic polymer lengths, masses and/or times are considered. Then a brief exposition of the present state of development of the velocity similarity laws is presented which includes a new wake-modification function. Such velocity similarity laws are used to derive a logarithmic friction formula for the pipe flow of drag-reducing polymer solutions. The scale-up method is then developed from the similarity-law logarithmic friction formula. Finally the method is checked by a comparison of predictions with measurements.

PIPE FLOW

The general problem of the pipe flow of drag-reducing polymer solutions is reviewed. Resistance to the flow of ordinary fluids through pipes is well correlated by dimensionless ratios. For straight circular pipes with fully developed viscous flow, friction loss represented by wall shear stress τ_w is related to average velocity V, pipe diameter D and fluid properties of density ρ and kinematic viscosity ν. Symbolically

$$\tau_w = f\left[V, D, \rho, \nu\right] \tag{1}$$

A dimensional analysis results in the well-known Fanning representation of friction factor f as a function of Reynolds number R_D

or

$$f = f\left[R_D\right] \tag{2}$$

where

$$f \equiv \frac{\tau_w}{\frac{1}{2}\rho V^2} \tag{3}$$

and

$$R_D \equiv \frac{VD}{\nu} \tag{4}$$

Any effect due to diameter is absorbed by the Reynolds number. Another commonly used friction factor is the Darcy-Weisbach friction factor λ where

$$\lambda = 4f \tag{5}$$

The situation for the turbulent flow of drag-reducing polymer solutions is much more complicated (Ref. 8). The presence of polymer molecules may be represented by the addition of an unknown number of characteristic polymer lengths $\ell, \ell_1, \ell_2 \ldots$ and/or the addition of an unknown number of characteristic polymer masses $m, m_1, m_2 \ldots$ and/or the addition of an unknown number of characteristic polymer times $t, t_1, t_2, \ldots$ to the dimensional analysis. The density of the polymer molecules in the fluid ρ_p has to be considered also. Symbolically

$$\tau_w = f\,[V, D, \rho, \nu, \rho_p, \ell, \ell_1, \ell_2, \cdots, m, m_1, m_2, \cdots, t, t_1, t_2, \cdots] \tag{6}$$

ρ and ν now refer to the polymer solution. An additional condition is that the wall shear stress τ_w has to exceed a critical value $\tau_{w,cr}$ to achieve drag reduction or $\tau_w > \tau_{w,cr}$.

By dimensional analysis the variables may be grouped into any of the following nondimensional ratios:

$$f = f\,[R_D, D/\ell, C, P] \tag{7}$$

or

$$f = f\,[R_D, D/(m/\rho)^{\frac{1}{3}}, C, P] \tag{8}$$

or

$$f = f\,[R_D, D/\sqrt{t\nu}, C, P] \tag{9}$$

where $C = \frac{\rho_p}{\rho}$, concentration of polymer in the solution

$P =$ type of polymer including mixtures of molecular weights

P is formally given by ratios of the characteristic lengths, masses and times (Ref. 4).

In terms of added molecular characteristics the flow of a polymer solution of specific characteristics P and concentration C has then an added dimensionless ratio, D/ℓ or $D/(m/\rho)^{1/3}$ or $D/\sqrt{t\nu}$ involving diameter D. This accounts for the diameter effect observed experimentally whereby each pipe diameter has a separate f-R_D relation for the same type of polymer solution and the same concentration.

It is well known that a limiting condition of maximum drag reduction also exists where $f = f[R_D]$, independent of pipe diameter, polymer species and concentration (Ref. 2).

VELOCITY SIMILARITY LAWS

The present state of development of the velocity similarity laws for drag-reducing polymer solutions is now reviewed. The classical velocity similarity laws developed by Prandtl and by von Kármán for the flow of ordinary fluids in pipes have proved most successful in placing the hydraulics of pipe flow on a rational basis. The similarity laws provide the well-known Prandtl-von Kármán logarithmic formula for friction factor as a function of Reynolds number.

The velocity similarity laws have been extended to drag-reducing polymer solutions in the following manner (Ref. 4). In general, for turbulent shear flows such as fully-developed viscous flow in pipes or boundary-layer flow, the two laws which provide similarity to the mean velocity profile by relating it to the wall shear stress are:

1. The inner law or the law of the wall which applies to the flow immediately adjacent to the solid boundary.

2. The outer law or the velocity defect law which applies to the remaining outer region of the shear flow.

Consideration of an overlapping of the two similarity laws as analyzed by Millikan across the shear flow results in a logarithmic functional form for both similarity laws within the common region of overlap. The functional form for the outer law outside of the overlapping region is provided by the law of the wake and a wake modification factor.

Inner Law or Law of the Wall

A similarity law may be developed for the mean velocity u of the turbulent flow of dilute polymer solutions in the region close to the wall or solid boundary by considering additional parameters to account for the physical presence of polymer molecules in the solution such as characteristic lengths, masses, and/or times of the molecules (Ref. 4).

Symbolically

$$u = f\left[\tau_w, \rho, \nu, y, \rho_p, l, l_1, l_2, \ldots, m, m_1, m_2, \ldots, t, t_1, t_2, \ldots\right] \quad (10)$$

The no-slip boundary condition at the wall as usual is u = 0 at y = 0. By dimensional analysis, the variables may be grouped into the following nondimensional ratio:

$$\frac{u}{u_\tau} = f\left[y^*, C, \frac{u_\tau l}{\nu}, P\right] \quad (11)$$

where $u_\tau = \sqrt{\frac{\tau_w}{\rho}}$ is the friction or shear velocity

and

$y^* = \frac{u_\tau y}{\nu}$, an inner-law Reynolds number. $(m/\rho)^{\frac{1}{3}}$ or $\sqrt{t\nu}$ may be substituted for l.

Outer Law or Velocity-Defect Law

In general for turbulent shear flows, the velocity defect U-u has been experimentally found to be directly independent of viscosity except close to the wall and a function only of wall shear stress τ_w, density ρ and distance inward δ -y or

$$U - u = f\left[\tau_w, \rho, y, \delta\right] \quad (12)$$

Here U is the velocity at the center of a pipe or the velocity at the outer edge of a boundary layer and δ is the thickness of the shear layer. For fully-developed pipe flow δ = D/2, the pipe radius.

By dimensional analysis

$$\frac{U-u}{u_\tau} = f\left[\frac{y}{\delta}\right] \quad (13)$$

the statement of the outer law or velocity-defect law which has been found experimentally to be unaffected by drag-reducing polymer solutions.

Logarithmic Velocity Law

Both the inner and outer laws are considered to hold in a common overlapping region of a shear flow. Equating the y-derivatives of the inner and outer laws results in a logaithmic relation for both laws in overlapping regions.

For the inner law the logarithmic velocity law is

$$\frac{u}{u_\tau} = A \ln y^* + B_1 \tag{14}$$

where for drag-reducing polymer solutions

$$B_1 = B_{1,0} + \Delta B$$

$B_{1,0}$ = Value of B_1 for ordinary fluids, a constant

and

$$\Delta B = f\left[\frac{u_\tau l}{\nu}, C, P\right] \tag{15}$$

$(m/\rho)^{\frac{1}{3}}$ or $\sqrt{t\nu}$ may be substituted for l.

ΔB may be termed the similarity-law drag-reduction characterization since it represents a fundamental hydrodynamic property of polymer solutions in reducing frictional resistance.

For the outer law the logarithmic velocity law is

$$\frac{U-u}{u_\tau} = -A \ln \frac{y}{\delta} + B_2 \tag{16}$$

where B_2 is a constant for fully-developed pipe flow. Both Equations (14) and (16) hold then in a common overlapping region.

Law of the Wake

For the non-overlapping region of the outer law, the following relation has been found to apply (Ref. 10)

$$\frac{U-u}{u_\tau} = -A \ln \frac{y}{\delta} + B_2\left(1 - \frac{w}{2}\right) - Aq \tag{17}$$

where $\frac{w}{2}$ is the Coles law of the wake which may be represented by a polynomial expression

$$\frac{w}{2} = 3\left(\frac{y}{\delta}\right)^2 - 2\left(\frac{y}{\delta}\right)^3 \tag{18}$$

and q is a wake-modification function designed to provide a zero derivative of velocity at $y = \delta$. A polynomial expression may be derived (Ref. 10) as

$$q = \left(\frac{y}{\delta}\right)^2 - \left(\frac{y}{\delta}\right)^3 \tag{19}$$

Laminar Sublayer

The flow right next to the wall is considered laminar which is unaffected by the drag-reducing solutions. The usual relation holds

$$\frac{u}{u_\tau} = y^* \tag{20}$$

from $y^* = 0$ to $y^* = y^*_L$ the laminar sublayer thickness.

Interactive Sublayer

Next to the laminar sublayer for drag-reducing solutions is a layer termed the interactive sublayer where turbulence activity seems to be very intense. The velocity profile is given by (Refs. 2, 9)

$$\frac{u}{u_\tau} = \tilde{A} \ln y^* - \tilde{B} \tag{21}$$

between y_L^*, the laminar sublayer thickness and y_m^*, the interactive sublayer thickness. y_L^* is given by the intersection of Eqs. (14) and (20) or

$$y_L^* = A \ln y_L^* + B_{1,0} \tag{22}$$

and y_m^* is given by the intersection of Eqs. (14) and (22) or

$$y_m^* = \exp\left(\frac{B_{1,0} + \Delta B + \tilde{B}}{\tilde{A} - A}\right) \tag{23}$$

If the interactive sublayer extends to the pipe center, a condition of maximum drag reduction exists.

Similarity-Law Velocity Profile

In capitulation, the overall pipe flow may be considered to have a similarity-law velocity profile which for the various sublayers consists of (see Fig 1):

a) from the wall outward

a laminar sublayer $\quad 0 \le y^* \le y_L^*$

$$\frac{u}{u_\tau} = y^* \tag{20}$$

b) further towards the pipe center

an interactive layer $\quad y_L^* \le y^* \le y_m^*$

$$\frac{u}{u_\tau} = \tilde{A} \ln y^* - \tilde{B} \tag{21}$$

c) and for the remainder of the shear layer $\quad y_m^* \le y^* \le \eta = \frac{u_\tau \delta}{\nu}$

$$\frac{u}{u_\tau} = A \ln y^* + B_{1,0} + \Delta B + B_2 \frac{W}{2} + Aq \tag{24}$$

which incorporates the logarithmic overlapping layer and the non-overlapping outer layer.

SIMILARITY-LAW LOGARITHMIC FRICTION FORMULAS FOR PIPE FLOW

General

The average velocity V through a pipe may be determined from an integration of the similarity-law velocity profile in terms of inner-law variables as

$$\frac{V}{u_\tau} = \frac{2}{\eta}\left(\int_0^\eta \frac{u}{u_\tau}\, dy^* - \frac{1}{\eta}\int_0^\eta \frac{u}{u_\tau}\, y^*\, dy^*\right) \tag{25}$$

where $\eta \equiv \frac{u_\tau \delta}{\nu}$.

Use of Equations (20), (21) and (24) and assuming that the law of the wake holds up to the wall results in

$$\frac{V}{u_\tau} = A \ln \eta + B_{1,0} + \Delta B - \frac{43}{30} A + \frac{3}{10} B_2 + \frac{2\tilde{A} y_L^* - y_L^{*2} - 2(\tilde{A} - A) y_m^*}{\eta} + \frac{y_L^{*3}/3 - (\tilde{A}/2) y_L^{*2} + (\tilde{A} - A)(y_m^{*2}/2)}{\eta^2} \tag{26}$$

There a negligible error for using the law of the wake up to the wall (Ref. 11). From definitions of u_τ and of f in Equation (3)

$$\frac{V}{u_\tau} = \sqrt{\frac{2}{f}} \tag{27}$$

and

$$\eta = \frac{\sqrt{2}}{4}\sqrt{f}\, R_D \tag{28}$$

The similarity-law logarithmic friction formula for polymer solutions in pipe flow is then

$$\frac{1}{\sqrt{f}} = \frac{A}{\sqrt{2}} \ln \sqrt{f}\, R_D + \frac{1}{\sqrt{2}}\left(B_{1,0} + \Delta B - \frac{43}{30}A + \frac{3}{10}B_2 - \frac{A}{2}\ln 8\right) + \frac{2[2\tilde{A}y_L^* - y_L^{*2} - 2(\tilde{A}-A)\, y_m^*]}{\sqrt{f}\, R_D} + \frac{4\sqrt{2}[y_L^{*3}/3 - (\tilde{A}y_L^{*2})/2 + (\tilde{A}-A)(y_m^{*2}/2)]}{(\sqrt{f}\, R_D)^2} \tag{29}$$

For ordinary fluids B=0 and $y_m^* = y_L^*$, then the friction factor f_o is given by

$$\frac{1}{\sqrt{f_o}} = \frac{A}{\sqrt{2}} \ln \sqrt{f}\, R_D + \frac{1}{\sqrt{2}}\left(B_{1,0} - \frac{43}{30}A + \frac{3}{10}B_2 - \frac{A}{2}\ln 8\right) + \frac{2(2Ay_L^* - y_L^{*2})}{\sqrt{f}\, R_D} + \frac{4\sqrt{2}[y_L^{*3}/3 - (Ay_L^{*2})/2]}{(\sqrt{f}\, R_D)^2} \tag{30}$$

<u>Condition of Maximum Drag Reduction</u>

For small diameter pipes, a condition of maximum drag reduction (Ref. 9) results when the interactive layer becomes the outer layer or $y_m^* = \eta$. In this case the intersection of Equations (14) and (22) yields

$$\Delta B = (\tilde{A} - A)\ln \eta - \tilde{B} - B_{1,0} - B_2 \tag{31}$$

Then the logarithmic friction formula, Equation (29) reduces to

$$\frac{1}{\sqrt{f}} = \frac{\tilde{A}}{\sqrt{2}} \ln \sqrt{f}\, R_D + \frac{1}{\sqrt{2}}\left(-\tilde{B} + \frac{A}{15} - \frac{3}{2}\tilde{A} - \frac{7}{10}B_2 - \frac{\tilde{A}}{2}\ln 8\right) + \frac{2(2\tilde{A}y_L^* - y_L^{*2})}{\sqrt{f}\, R_D} + \frac{4\sqrt{2}[y_L^{*3}/3 - (\tilde{A}/2)y_L^{*2}]}{(\sqrt{f}\, R_D)^2} \tag{32}$$

Maximum drag reduction is thus independent of the type and concentration of polymer.

SCALE-UP METHOD

For drag-reducing polymer solutions of type P and concentration C it is desired to scale-up the friction lines, f versus R_D, from measured data for a test pipe of a fixed diameter to pipes of different fixed diameters. A simple method is now described which is based on the similarity-law characterization, $\Delta B = f\left[\frac{u_\tau \ell}{\nu}\right]$.

For convenience the coordinates $1/\sqrt{f}$ versus $\sqrt{f}\,R_D$ are to be used to plot the friction lines. Now it may be shown that each point of the measured friction line $(1/\sqrt{f}, \sqrt{f}\,R_D)$ corresponds to a value of ΔB and a value of $(u_\tau \ell)/\nu$. The problem which is posed is to determine the point, $[(1/\sqrt{f})_2, (\sqrt{f}\,R_D)_2]$, from a measured point, $[(1/\sqrt{f})_1, (\sqrt{f}\,R_D)_1]$ for the same values of ΔB and $(u_\tau \ell)/\nu$ in going from diameter D_1 to D_2. Since by definitions in Equations (3) and (27)

$$\frac{u_\tau \ell}{\nu} = \sqrt{\frac{1}{2}}\left(\sqrt{f}\,R_D\right)\left(\frac{\ell}{D}\right) \tag{33}$$

for the same value of $\frac{u_\tau \ell}{\nu}$ there results

$$\left(\sqrt{f}\,R_D\right)_2 = \left(\sqrt{f}\,R_D\right)_1 \left(\frac{D_2}{D_1}\right) \tag{34}$$

Substitution of Equation (33) into the logarithmic friction formula for polymer solutions, Equation (29), gives the following expression for $1/\sqrt{f}$:

$$\frac{1}{\sqrt{f}} = \frac{A}{\sqrt{2}} \ln \frac{D}{\ell} + \frac{A}{\sqrt{2}} \ln \frac{u_\tau \ell}{\nu} + \frac{1}{\sqrt{2}}\left(B_{1,0} + \Delta B - \frac{43}{20} A + \frac{3}{10} B_2 - A \ln 4\right) + \frac{2[2\tilde{A} y_L^* - y_L^{*2} - 2(\tilde{A}-A) y_m^*]}{\sqrt{f}\,R_D} + \frac{4\sqrt{2}[y_L^{*3}/3 - (\tilde{A} y_L^{*2})/2 + (\tilde{A}-A)(y_m^{*2}/2)]}{(\sqrt{f}\,R_D)^2} \tag{35}$$

For the same values of ΔB and $\frac{u_\tau \ell}{\nu}$, in going from $\left(\frac{1}{\sqrt{f}}\right)_1$ to $\left(\frac{1}{\sqrt{f}}\right)_2$, there results

$$\left(\frac{1}{\sqrt{f}}\right)_2 = \left(\frac{1}{\sqrt{f}}\right)_1 + \frac{A}{\sqrt{2}} \ln \frac{D_2}{D_1} + 2[2\tilde{A} y_L^* - y_L^{*2} - 2(\tilde{A}-A) y_m^*]\left[\frac{1}{(\sqrt{f}\,R_D)_2} - \frac{1}{(\sqrt{f}\,R_D)_1}\right] + 4\sqrt{2}[y_L^{*3}/3 - (\tilde{A} y_L^{*2})/2 + (\tilde{A}-A)(y_m^{*2}/2)]\left[\frac{1}{(\sqrt{f}\,R_D)_2^2} - \frac{1}{(\sqrt{f}\,R_D)_1^2}\right] \tag{36}$$

where y_L^* is a constant but y_m^* varies with ΔB as given in Equation (23)

$$y_m^* = \exp\left(\frac{B_{1,0} + \Delta B + \tilde{B}}{\tilde{A} - A}\right)$$

ΔB is determined by subtracting $1/\sqrt{f_0}$ for ordinary fluids, Equation (30), from $1/\sqrt{f_1}$ for drag-reducing solutions, Equation (29), at the same value of $(\sqrt{f}\,R_D)_1$ which yields

$$\frac{\Delta B}{\sqrt{2}} = \frac{1}{\sqrt{f_1}} - \frac{1}{\sqrt{f_0}} + \frac{4(\tilde{A}-A)(y_m^* - y_L^*)}{(\sqrt{f}\,R_D)_1} - \frac{2\sqrt{2}(\tilde{A}-A)(y_m^{*2} - y_L^{*2})}{(\sqrt{f}\,R_D)_1^2} \tag{37}$$

Since y_m^* also appears in this equation, ΔB is determined by reiteration starting with

$$\frac{\Delta B}{\sqrt{2}} \simeq \frac{1}{\sqrt{f_1}} - \frac{1}{\sqrt{f_0}} \tag{38}$$

and substituting into Equation (23). Values of $B_{1,0}$ = 5.5, $\tilde{B}$ = 17, $\tilde{A}$ = 11.7, A = 2.5 and y_L^* = 11.6 may be used (Ref. 2)

Method of Two Loci

For small thicknesses of the interactive layer, $y_L^* \ll \eta$, a simple graphical procedure may be employed which is based on the intersection of two loci as illustrated in Figure 2. Coordinates $1/\sqrt{f}$ and $\sqrt{f}\,R_D$ are to be considered. A locus of constant $(u_\tau \ell)/\nu$ is given by Equation (34) which represents a constant value of $\sqrt{f}\,R_D$ offset a constant distance $\log(D_2/D_1)$. A locus of constant ΔB is given by Equation (38) which represents a line parallel to the friction line for ordinary fluids.

COMPARISON OF PREDICTIONS WITH MEASUREMENTS

As an example, test data by Wang (Ref. 12) is considered for a type of Guar Guam, Jaguar A-20-D, in a 0.02% (200 ppm) concentration aqueous solution. Data is given for 1/2 inch (12.7 mm) and 1 inch (25.4 mm) diameter pipes. The exercise is to scale-up the 1/2-inch diameter data to a 1 inch diameter and to compare the results with test data.

The friction line for ordinary fluids, the Prandtl-von Kármán formula,

$$\frac{1}{\sqrt{f}} = 4.0 \log_{10} \sqrt{f}\,R_D - 0.4 \tag{39}$$

is drawn in Figure 3. Also the friction line for the condition of maximum drag reduction, given by the Virk formula (Ref. 2),

$$\frac{1}{\sqrt{f}} = 19.0 \log_{10} \sqrt{f}\,R_D - 32.4 \tag{40}$$

is drawn.

The test data of Wang for water flowing in the 1/2-inch and 1-inch pipes are also plotted in Figure 3. A mean line through the test points agrees with the Prandtl-von Kármán formula at high Reynolds numbers but deviates at low Reynolds numbers. The mean line is used as the reference friction line for the ΔB-locus.

Figure 3 shows excellent agreement of the scaled-up results by the method of two loci with measured values.

CONCLUSION

The measured velocity profiles of shear flows have been well correlated by the similarity laws for drag-reducing polymer solutions. Since the scale-up method presented here is based firmly on similarity laws, predictions should be on a sound fundamental basis. This is borne out by the example shown.

REFERENCES

1. Hoyt, J.W.: "The effect of additives on fluid friction". Journal of Basic Engineering, 94, 2 pp.258-285. (June 1972).

2. Virk, P.S.: "Drag reduction fundamentals". AIChE Journal, 21, 4 pp.625-656. (July 1975).

3. White, A. and Hemmings, J.A.G: "Drag reduction by additives - review and bibliography". BHRA Fluid Engineering, Cranfield. (1976).

4. Granville, P.S.: "Frictional resistance and velocity similarity laws of drag-reducing polymer solutions". Journal of Ship Research, 12, 3 pp.201-222. (September 1968).

5. Huang, T.T.: "Similarity laws for turbulent flow of dilute solutions of drag-reducing polymers". Physics of Fluids, 17, 2 pp.298-309. (February 1974).

6. Taylor, D.D. and Sabersky, R.H.: "Extrapolation to various tube diameters of experimental data taken with dilute polymer solutions in a smooth tube". Letters in Heat and Mass Transfer, 1, pp. 103-108. (1974).

7. Savins, J.G. and Seyer, F.A.: "Drag reduction scale-up criteria". IUTAM Symposium on Structure of Turbulence and Drag Reduction, Washington, D.C. (June 1976). (to appear in Physics of Fluids, 1977).

8. Granville, P.S.: "Hydrodynamic aspects of drag reduction with additives". Marine Technology, 10, 3 pp.284-292. (July 1973).

9. Granville, P.S.: "Limiting conditions to similarity-law correlations for drag-reducing polymer solutions". Naval Ship R&D Center (Bethesda, Maryland) Report 3635. (August 1971).

10. Granville, P.S.: "A modified law of the wake for turbulent shear flows". Journal of Fluids Engineering, 98, 3 pp.578-580 (September 1976).

11. Granville, P.S.: "The drag and turbulent boundary layer of flat plates at low Reynolds numbers". Journal of Ship Research, 21,1 pp.30-39 (March 1977).

12. Wang, C.-B.: "Pipe flow of dilute polymer solutions". PhD. Thesis, University of Wisconsin. (1969). University Microfilms, Ann Arbor, Michigan, No. 69-16,998.

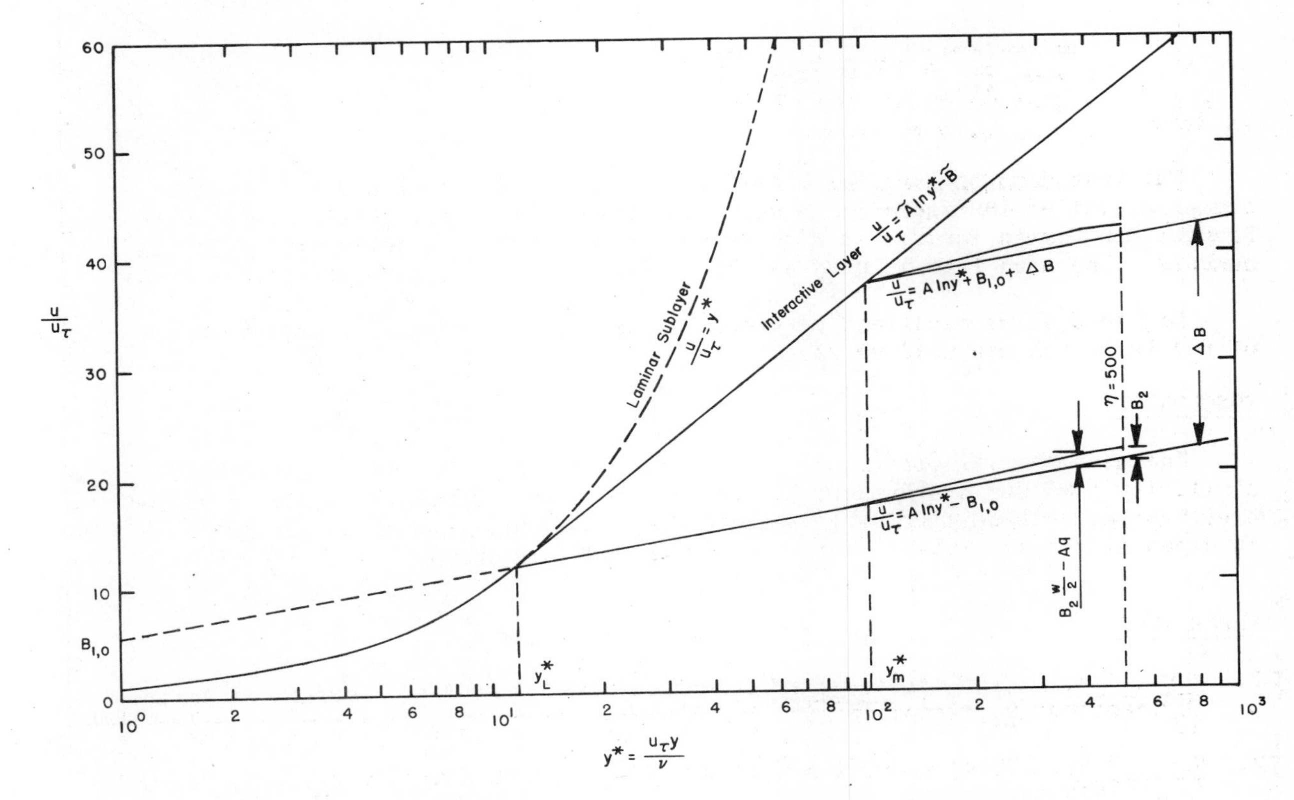

Fig.1 Inner similarity law diagram for drag-reducing solutions.

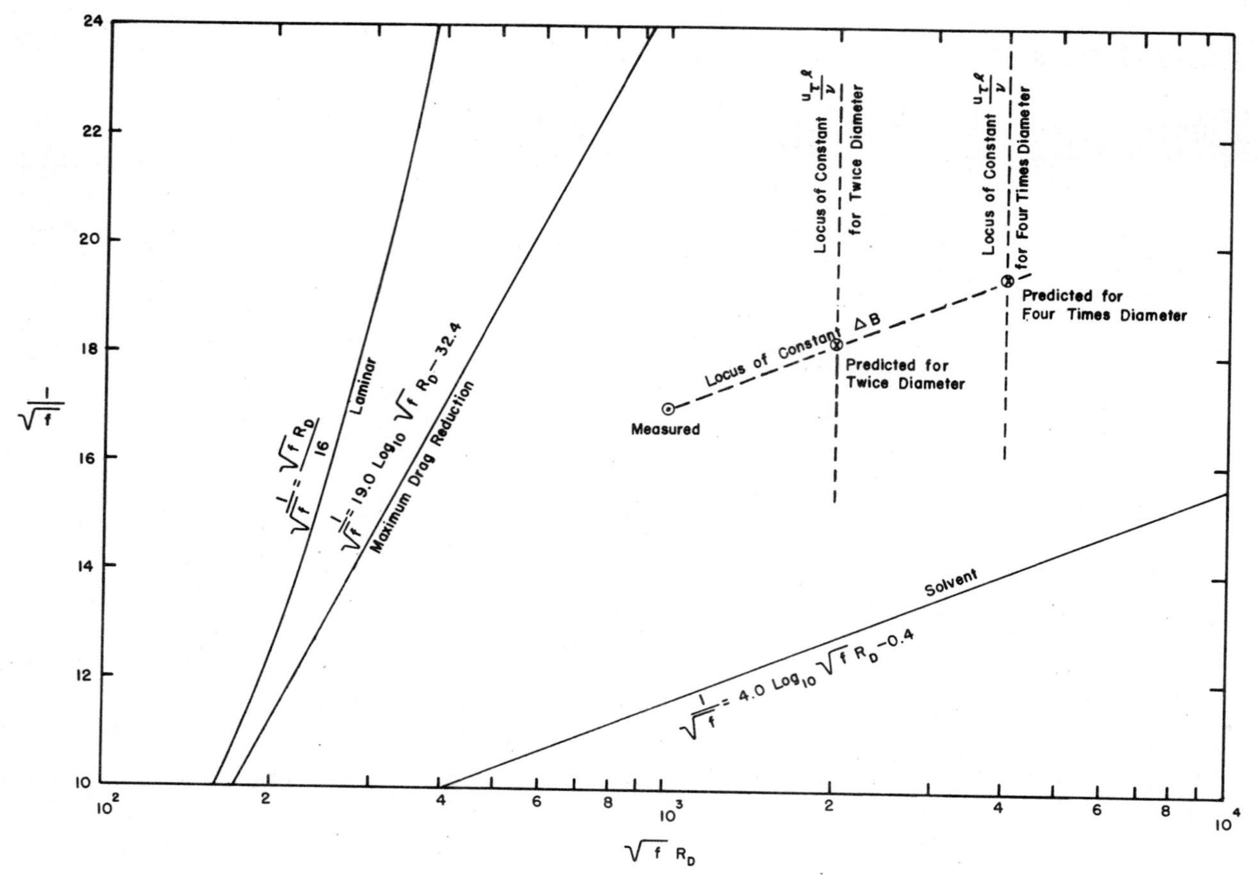

Fig.2 Scale-up method of two loci.

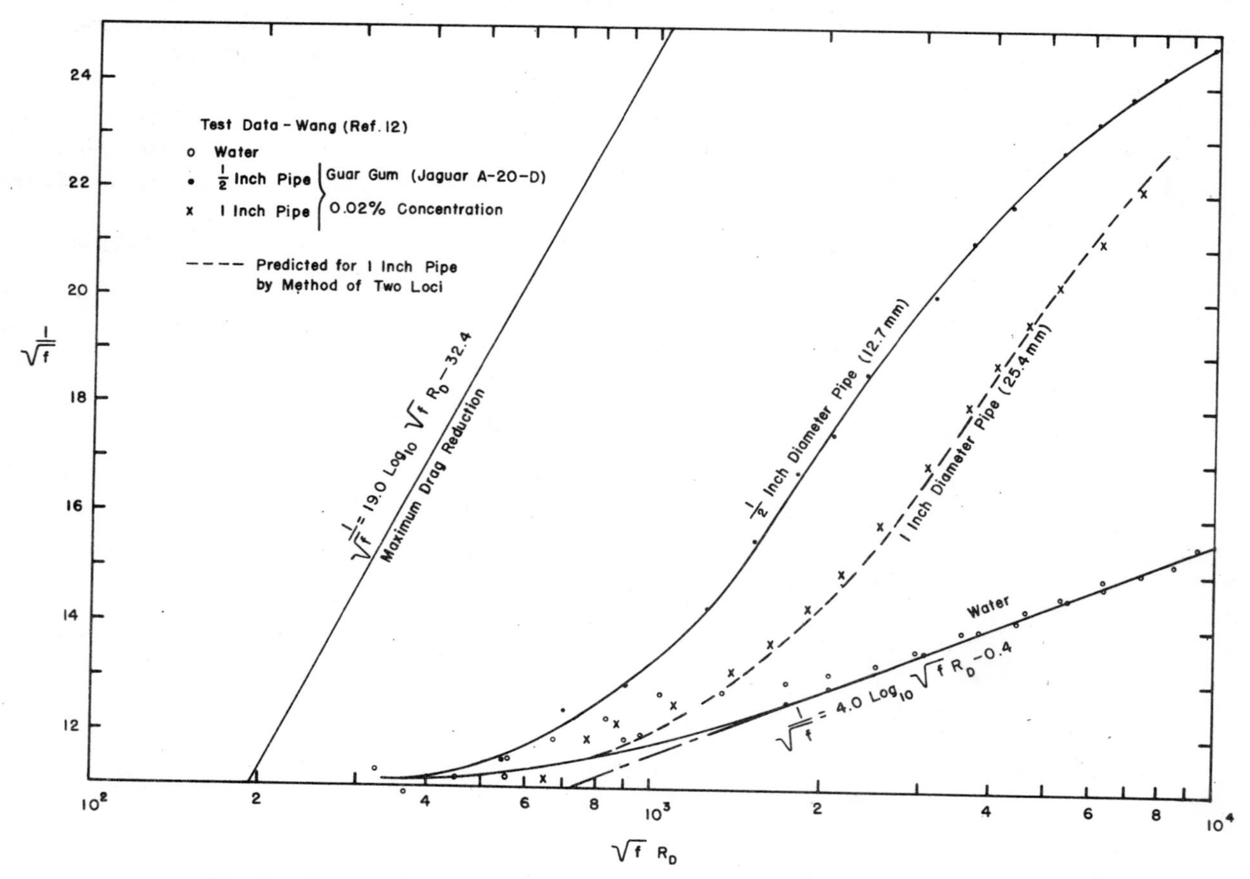

Fig.3 Comparison of scale-up predictions with test data.

Second International Conference on

Drag Reduction

August 31st - September 2nd, 1977

PAPER B2

PRESSURE DROP REDUCTION IN LARGE INDUSTRIAL DUCTS BY MACROMOLECULAR ADDITIVES

Part 1: POLYMER EFFICIENCY

J. P. De Loof, B. de Lagarde, M. Petry, A. Simon

Bertin & Cie, France.

Summary

A calculation method of pressure drop reduction in large pipes is presented here, based on the Virk's velocity profile model. The value of the shift of the logarithmic law is deduced by means of small pipe flow laboratory measurements (4 to 10 mm internal diameter) using guar gum as a non degradable polymer in our test conditions. According to Huang's experiments, it is verified that this shift is only dependent of wall friction velocity and polymer concentration, and not of the pipe diameter. This polymer flow characteristic is used to compute the pressure drop reduction in larger pipes. Comparison between experiments and calculations are made in 50, 100 and 200 mm internal diameter pipes, with mean flow velocities up to 6 m/s. Agreement is very good.

We conclude that efficiency of even a high molecular weight polymer such as Polyox 301 is not very important in large pipes, but may be of economical interest if polymer degradation is negligible.

Held at St. John's College, Cambridge, England.
Organised and sponsored by BHRA Fluid Engineering, Cranfield, Bedford, MK43 0AJ.

NOMENCLATURE

Cf	:	FANNING's friction coefficient $= \frac{\tau}{\rho V^2/2}$
D	:	Pipe diameter
J	:	Linear pressure drop $= \frac{\Delta P}{L}$
L	:	Pipe length
Q	:	Liquid flow rate
R	:	Pipe radius
$R\Delta P$	:	Pressure drop reduction (°/₀)
Re	:	REYNOLDS number $= \frac{VD}{\nu}$
S	:	Pipe cross section
$\bar{u}$	:	Local mean velocity at distant y from the wall
u^+	:	Adimensional number $= \frac{\bar{u}}{U^*}$
U^*	:	Friction velocity $= \sqrt{\tau/\rho}$
Uc	:	Fluid velocity at the pipe centerline
V	:	Mean flow velocity = Q/S
y	:	Distance from the wall
y^+	:	Adimensional number $= \frac{y\,U^*}{\nu}$
α	:	Characteristic parameter of the polymer-solvent system
ΔP	:	Pressure drop value
ΔB	:	Characteristic value of the polymer-solvent system $= \alpha \log \frac{U^*}{U^*_{cr}}$
ρ	:	Mass density of the fluid
τ	:	Wall shear stress
ν	:	Kinematic viscosity

SUBSCRIPTS

p	:	Polymer-solvent system
s	:	Solvent alone
cr	:	Onset of drag reduction ($\Delta B = 0$)

I N T R O D U C T I O N

It is well known that turbulent drag can be drastically reduced by adding small amounts of a high molecular weight polymer.

Strategical and economical implications of this phenomenon seemed attractive enough to explain the tremendous infatuation for fundamental or applied research in friction reduction (see the synthesis of J.W. Hoyt - ref. 26, 27).

The only industrial realizations having succeeded nowadays must be considered as strategical applications, i.e. those for which economical aspect can be regarded as secondary with respect to the aim that has to be reached. A typical example is the flow rate increase in fire fighting hoses leading to a more efficient intervention and consequently to a decrease in material or human losses (we resolutly eliminate military applications in this paper).

Nevertheless, we must notice that applications having a strictly economical feature have not yet been achieved, whereas they were highly promising. It seems that one of the main reasons, other than engineering difficulties, is that estimating the profit is not very easy.

According to the title of this paper, we will speak here only about duct applications.

In order to estimate the savings, one has to know :

- . the pressure drop reduction versus polymer concentration and flow parameters,
- . the life duration of the polymer in the flow,
- . the eventual modification of pumps efficiency.

Having all this information, engineers would be able to outline a first technical and economical trade off, putting on one side of the balance the cost of polymer and on the other side the savings which are for exemple a decrease of the energy consumption or a reduction of the pipe diameters, i.e. of their costs.

Of course, for a more complete estimation, it would be necessary to take into account other parameters such as the cost of polymer injection unit or the cost of the transportation of the products into the injection unit,without neglecting some important aspects as fluid chemical modifications introduced with adding polymers.

One now better understands the perplexity of technical project managers to whom is proposed a miraculous process without the means of estimating the profits.

With assistance of DGRST (Délégation Générale à la Recherche Scientifique et Technique), BERTIN & Cie has undertaken some studies to get intellectual and material potentialities needed for technical and economical trade off establishment.

In this paper, we report our research results concerned only with the pressure drop reduction calculations in the large industrial ducts. Another study on polymer degradation in large pipes has been undertaken for a few month in our laboratories and the results will be published later.

1. SHORT LITERATURE REVIEW

At the beginning of polymer flow studies, experiments had been made indicating that pressure drop reduction was less important in large pipes than in small pipes for same Reynolds number and polymer concentration.

Extrapolation of the results obtained in small pipes led to pessimistic estimates that have not been verified during later experiments in large industrial ducts.

Researches have then been oriented in two complementary ways : one of fundamental research working on the "diameter effect"and adjustement of calculation methods (Ref 7, 8, 11, 13, 14, 15, 16, 22, 28, 32), and one of applied research wishing to obtain concrete results at full scale as quickly as possible (Ref 1, 4, 5, 6, 29, 30, 31).

Unfortunately, experiments of both topics have the most often been achieved in so different conditions (pipe diameter, polymer used, ...) that they generally do not overlap and that, consequently, it has not been possible to test calculation methods on large diameter pipes.

Our aim is precisely to fill this gap.

2. VELOCITY PROFILES IN TURBULENT FLOW OF DILUTE SOLUTIONS OF DRAG REDUCING POLYMERS

Numerous measurements in turbulent flow of dilute solutions of drag reducing polymers have been made by many investigators. More or less sophisticated models have been established in order to describe the velocity profiles inside a pipe or a boundary layer, namely :

- the four layer model of Virk (Ref 11),
- the continuous model of Poreh (Ref 22).

We have prefered to use Virk's model that gives a more realistic representation of the velocity profile modification and which is the best known to date.

This model has been very well described in a paper of T.T. Huang (Ref 8) who used it in order to establish similarity laws. We frequently refer to this paper.

ADIMENSIONAL PARAMETERS

$$u^+ = \frac{\bar{u}}{U^*}$$

$$y^+ = \frac{y\,U^*}{\nu}$$

y : normal distance from the wall

$\bar{u}$: local mean velocity at distance y

ν : kinematic viscosity

U^* : friction velocity

VIRK'S MODEL

In his model, VIRK distinguishes four layers (Fig 1 and 2).

Viscous sublayer

In this layer, transverse velocity fluctuations are cancelled on account of the presence of the wall. Viscous effects prevail like as in laminar flow and we may write:

$$u^+ = y^+ \qquad (1)$$

The generally accepted edge for this layer is located at $y^+ \simeq 10.8$. Thickness of this layer is unaffected by the inside polymer.

Strongly interactive layer

This layer is highly affected by the polymer and is characterized by a KARMAN'S constant value smaller than in a Newtonian flow.

Equation of velocity profile in this layer is $u^+ = 30 \log y^+ - 20.2$ (2)
Without polymer (Newtonian flow), this layer vanishes.

Weakly interactive layer

This layer is always present, but shifted away from the wall according to the expansion of the strongly interactive layer.

Its equation is the same as in the Newtonian flow (especially the same KARMAN'S constant), taking into account the value ΔB of the parallel upward shift in a semi-logarithmic plot :

$$u^+ = 5.62 \log y^+ + 5 + \Delta B \quad (3)$$

ΔB is a function of flow characteristics as it will be specified thereafter.

For Newtonian fluids, $\Delta B = 0$.

Outer wake region

This region is not affected by the polymer. Consequently, velocity profile equation is the same as in the case of Newtonian flow.

It is expressed by the velocity defect law :

$$\frac{Uc - \bar{u}}{U^*} = - 5.62 \log \frac{y}{R} + \frac{\Omega}{0.41}\left[1 + \cos(\Pi \frac{y}{R})\right] \quad (4)$$

Uc : velocity at the pipe centerline
Ω : COLE's wake parameter
= 0,2 for internal flows
R : radius of the pipe

ΔB characterisation

Let us consider the parameter :

$$\Delta V^+ = \left(\frac{Vp}{U^*} - \frac{Vs}{U^*}\right) \text{ at constant } \frac{U^*}{\nu}$$

V : mean flow velocity

Subscripts p and s represent the polymer-solvent system and solvent alone, respectively.

This parameter represents the difference between mean flow velocities in a same pipe for two fluids having the same viscosity and the same friction velocity, one of them containing polymer and the other none.

It may also be written as :

$$\Delta V^+ = \frac{1}{S}\left(\frac{Qp}{U^*} - \frac{Qs}{U^*}\right)$$

with S : pipe cross section ; Q : liquid flow rate

Computation of flow rate Q is achieved by integrating local flow rates, taking into account the successive four layers velocity profiles equations.

$$\frac{Q}{U^*} = \iint \frac{\bar{u}\, ds}{U^*}$$

$$= \int_0^{2\pi}\int_0^{R} \frac{\bar{u}}{U^*}\, r\, d\theta\, dr$$

HUANG carried out this computation and demonstrated that it could be written with a good approximation : $\Delta V^+ \simeq \Delta B$, which will be justified thereafter by the good accuracy of our experimental results. Consequently, one can write :

$$\Delta B = \left(\frac{Vp}{u^*} - \frac{Vs}{u^*}\right) \text{ at constant } \frac{U^*}{\nu}$$

or, since $U^* = V\sqrt{\frac{Cf}{2}}$

$$\frac{\Delta B}{\sqrt{2}} = \left(\frac{1}{\sqrt{Cf}}\right)_p - \left(\frac{1}{\sqrt{Cf}}\right)_s \text{ at constant } \frac{u^*}{\nu}$$

Here, Cf is the FANNING's friction coefficient related to the mean flow velocity V (and not to the center line velocity Uc).

$$Cf = \frac{\tau}{\rho V^2/2}$$

τ : wall shear stress $= \rho U^{*2}$

Using the pipe REYNOLDS number $Re = \frac{VD}{\nu}$, the value of $\frac{U^*}{\nu}$ can be written as :

$$\frac{U^*}{\nu} = Re\ \frac{1}{D}\sqrt{\frac{Cf}{2}} \quad \text{where D is the pipe diameter}$$

If experiments are made in a same pipe, we write (see fig 3)

$$\frac{\Delta B}{\sqrt{2}} = \left(\frac{1}{\sqrt{Cf}}\right)_p - \left(\frac{1}{\sqrt{Cf}}\right)_s \quad \text{at constant } Re\sqrt{Cf}$$

Saturation

It is experimentally established that pressure drop reduction value increases by increasing polymer concentration. However, there is a maximum value which cannot be exceeded.

In this case, the strongly interactive layer dominates the entire linear logarithmic region : the weakly interactive layer has vanished and the outer wake region has become negligible. The model is then reduced to only two layers : the viscous sublayer, whose thickness can be neglected, and the strongly interactive layer. It is the oversaturaded domain where the maximum value of ΔB is given by : (ref 8)

$$\frac{\Delta B\ max}{\sqrt{2}} \simeq 17.15 \log\left(\frac{Du^*}{\nu}\right) - 34.58 \qquad (5)$$

Friction coefficient computation

A good approximation of the friction coefficient is given by equating equations (3) and (4) and assuming that they overlap near the pipe centerline, i.e. $\frac{y}{R} \simeq 1$. With respect to the mean flow velocity, its equation is (ref 21) :

$$\left(\frac{1}{\sqrt{Cf}}\right)_p = 4 \log\left(Re \sqrt{Cf}\right)_p - 0.4 + \frac{\Delta B}{\sqrt{2}} \qquad (6)$$

When $\Delta B = 0$, we recover the well known Prandtl's equation for Newtonian flow :

$$\left(\frac{1}{\sqrt{Cf}}\right)_s = 4 \log\left(Re \sqrt{Cf}\right)_s - 0.4 \qquad (7)$$

3. LABORATORY MEASUREMENTS

3.1. Chronology

Initially, we measured ΔB versus the parameter $\frac{U^*}{\nu}$ and the polymer concentration C, inside 4, 6, 8 and 10 mm internal diameter pipes.

Subsequently, we measured pressure drop reduction in 50, 100 and 200 mm internal diameter pipes (really : 2,4 and 8"), and compared these values with computation results obtained by using the previous law of ΔB.

3.2. Experimental set up

3.2.1. 4, 6, 8 and 10 mm i.d. pipes (fig 4)

An unpressurized tank containing three hundred liters and used for easy polymer solution preparation was located above a pressurized tank of same capacity. Transfer of the well mixed solution inside this latter was simply achieved by free fall. A bell-mouth outlet directed the flow from the pressurized tank through the pipes. Flow rate was regulated by varying the helium pressure in the holding tank (maximum value : 10 bars). B In this way , mechanical degradation was minimized.

Each pipe of 1 500 mm length was fitted with three pressure tappings respectively located at 850, 1 250 and 1450 mm from the inlet. Pressure drop measurements were carried out between the first and the second pressure tapping.

Flow rate was measured by means of a weighing barrel and a chronometer.

3.2.2. 50, 100 and 200 mm i.d. pipe (fig 5)

We used the experimental set-up of O.T.P. (The French firm Omnium technique de transport par pipeline). In fact this facility is usually used to study the transportation of suspended materials in a liquid.

Pipes of 50.8, 101.6, 203.2 mm i.d. pipes are respectively 50, 100 and 200 meters long and instrumented with regularly spaced pressure picks-off.

Flow circulation is achieved by an open centrifugal impeller, with variable rotational speed. An open tank inserted on the line just upstream of the impeller was used for preparing solutions. Flow rate measurement is carried out by temporary diversion of the fluid towards a weighing tank.

3.2.3. Polymer choice and solution preparation

We used GUAR CSAA M 200, a less effecient hydrosoluble polymer than the well-known POLYOX 301, but known as much more resistant to mechanical degradation (ref 18). Effectively, steady pressure drop reductions in O.T.P. facility were obtained even after several hours of fluid recirculation. But on the other hand, we met difficulties concerning chemical stability of the product (an enzyme would be responsible of fast chemical degradation : see ref 20) and the presence of impurities troubling the precise concentration measurement. This latest point seems to be responsible of the lack of repeatability of different investigations of pressure drop reduction versus concentration which are found in the literature .

Solution preparing was achieved in three stages :

- weighing of the amount of GUAR just necessary to obtain desirable concentration taking into account the used volume water,
- suspending GUAR in a 50 % ethylic alcohol - 50 % glycerol mixture,
- pouring of this mixture into the containing water tank with simultaneous agitation.

Viscosity measurements have given us :

$$\frac{\nu_p}{\nu_s} = 1 + 4.73.10^{-4}\ C^{1.157}$$

3.3. Results

3.3.1. Rough results in the small pipes (4, 6, 8, 10 mm i.d.)

Λ - Re values

Pressure drop coefficient Λ versus solvent REYNOLDS number Re_s are shown on fig 6 to 9.

Exact pipe diameters were determined by measuring water flow pressure drop and adjusting it on turbulent flow theoretical values.

Thus, exact values were found to be :

4.1 mm
6.15 mm
8.14 mm
10 mm

3.3.2. ΔB versus $\frac{U^*}{\nu}$ values

It has been seen that :

$$\frac{\Delta B}{\sqrt{2}} = \left(\frac{1}{\sqrt{Cf}}\right)_p - \left(\frac{1}{\sqrt{Cf}}\right)_s \quad \text{at constant } \frac{U^*}{\nu}$$

with $Cf = \Lambda/4$ and $\frac{U^*}{\nu} = \frac{1}{\sqrt{2}}\frac{1}{D} Re\sqrt{Cf}$

consequently, constant $\frac{U^*}{\nu}$ implies constant $Re\sqrt{Cf}$ at same D.

$\frac{\Delta B}{\sqrt{2}}$ was directly computed using the following method :

From Λ versus Re values, we deduced $\left(\frac{1}{\sqrt{Cf}}\right)_p$ for corresponding value of $\left(Re\sqrt{C_f}\right)_p$

as well as the value of $\frac{U^*}{\nu} = \frac{1}{D}\frac{1}{\sqrt{2}}\left(Re\sqrt{C_f}\right)_p$

Then, we computed Newtonian value of $\left(\frac{1}{\sqrt{Cf}}\right)_s$ by formula (7) for $\left(Re\sqrt{C_f}\right)_s$ = $\left(Re\sqrt{C_f}\right)_p$

And finally, we deduced $\frac{\Delta B}{\sqrt{2}}$ value from the difference between $\left(\frac{1}{\sqrt{Cf}}\right)_p$ and $\left(\frac{1}{\sqrt{Cf}}\right)_s$ (In our experiments, solvent was water).

All the experimental points have been reported on fig 10 where are also shown maximum theorotical values of $\frac{\Delta B}{\sqrt{2}}$ calculated by equation (5). We so confirm that values of ΔB versus $\frac{U^*}{\nu}$ are independant of pipe diameter. This is a very important result, because it expresses that experiments in only one pipe are sufficient to determine the equation of ΔB versus $\frac{U^*}{\nu}$

3.3.3. Equation of ΔB versus $\frac{U^*}{\nu}$

Experimental points are located on lines whose equations can be expressed as :

$$\frac{\Delta B}{\sqrt{2}} = \alpha \log\left(\frac{U^*}{\nu} \Big/ \left(\frac{U^*}{\nu}\right)_{cr}\right)$$

$\left(\frac{U^*}{\nu}\right)_{cr}$ is the critical value corresponding to $\Delta B = 0$ and related to the threshold value of pressure drop reduction establishment. α is the line slope. These two parameters are depending on the polymer concentration C. Experimental values are :

C	α	$\left(\frac{U^*}{\nu}\right)_{cr}$
75 ppm	4	$4.3.\ 10^4$
200 ppm	7.9	$3.8.\ 10^4$
500 ppm	14.7	$2.9.\ 10^4$

Power curve adjustment on the α experimental points gives (fig 11)

$$\alpha = 0.2073\ C^{0.6862}$$

and line adjustment on the $\left(\frac{U^*}{\nu}\right)_{cr}$ experimental points gives (fig 11)

$$\left(\frac{U^*}{\nu}\right)_{cr} = -\ 31.25\ C + 4.5 \times 10^4$$

with C in p.p.m.

These two formulae are valid only for $0 < C \leqslant 500$ ppm. General equation of $\frac{\Delta B}{\sqrt{2}}$ thus becomes

$$\frac{\Delta B}{\sqrt{2}} = 0.21\ C^{0.69} \log \frac{U^*/\nu}{-31.25\ C + 4.5 \times 10^4} \qquad (8)$$

$$0 < C \leqslant 500 \text{ ppm}$$

it is independant of pipe diameter.

3.3.4. Pressure drop reduction calculations in pipes

Pressure drop reduction can be calculated by using formulas (6), (7) and the equation of ΔB (8). Computation program has been established, taking into account the maximum pressure drop reduction values. The data are : polymer concentration, pipe diameter, polymer characteristic values versus concentration : α and $\left(\frac{U^*}{\nu}\right)_{cr}$, and the output is pressure drop reduction versus mean speed velocity.

3.3.5. Comparison between calculations and experimental results obtained in 50, 100 and 200 mm i.d. pipes

Values of J versus V^2 measured on O.T.P. set-up are shown on fig. 12 to 14.

J is the linear pressure drop $\frac{\Delta P}{L}$

L is the pipe length

V is the mean flow velocity

Exact internal diameter pipes are 52.5, 104.7, and 208 mm. Experimental points are not well located on KARMAN'S theorotical line. This might be due to pipe roughness.

According to VIRK's experiments (ref 24), we assumed that pressure drop reduction was not significantly affected by roughness. Pressure drop reduction $R\Delta P$ was deduced from experimental curves J (V^2).

$$R\Delta P = \frac{Jo - J}{Jo} \quad \text{at constant } V$$

Jo : water flow measurement

J : polymer flow measurement

Experimental results are compared with computation values on fig 15 to 17. Agreement is good.

4. RESULTS UTILIZATION FOR INDUSTRIAL APPLICATIONS

a) Pressure drop reductions obtained with GUAR versus pipe diameter are shown on fig 18. It can be seen that pressure drop reduction becomes less important as pipe diameter increases, leading to an asymptotic value.

We may consider two domains :

- the small diameters domain (a few millimiters) where pressure drop reduction changes very quickly with pipe diameter : according to that, pressure drop reduction would not exist any longer in large diameter pipes (but that extrapolation is not valid : see dotted line of fig 18),

- the large diameters domain (many tenth centimeters diameters) where pressure drop reduction is nearly constant with pipe diameter.

It clearly indicates that noticeable pressure drop reductions can be obtained even in pipe diameters larger than one meter.

b) Fig 15 to 17 show the evolution of pressure drop reduction at constant diameter with mean flow velocity and for different concentrations of GUAR. One notices that for current commercial velocities of about 1.5 to 2 m/s, pressure drop reduction is very low, nearly unextant. This is not surprising according to the low efficiency ot the polymer. (This polymer had been choosen only on account of his shear degradation resistance).

We have outlined on fig 19 estimated pressure drop reduction values for POLYOX 301, a high efficiency polymer but also highly shear degradable. Values of α and $\left(\frac{U^*}{\nu}\right)_{cr}$ used in ΔB equation are those obtained by FABULA (ref 25)

It can be seen that some noticeable pressure drop reduction can be obtained in a 0.5 meter diameter pipe at a flow velocity of about 2 m/s, but not as high as expected (everybody should remember that nearly eighty per cent pressure drop reduction are obtained in a 2 mm i.d. pipe at a REYNOLDS number of 20 000 and with concentration of POLYOX 301 not greater than 10 ppm).

Another representation of these results is given on fig 20 which shows the evolution of the pressure drop reduction versus polymer concentration at constant velocity and diameter.

Now, having such results and taking into account the cost of polymer, a technical project manager might try to get the best profit of the drag reduction process.

5. CONCLUSIONS

It has been shown during this study that agreement between theorotical calculations and experimental results is very good. This demonstrates that VIRK's model we have used is well representative of inside pipe velocity profile.

Alteration of this profile is made of an upward shift ΔB of the Newtonian logarithmic law (on a semi-logarithmic diagram u^+ versus log y^+) which is related to characteristic values of the polymer-solvent system alone and independant of pipe diameter.

Knowing ΔB values versus polymer concentration and versus the ratio of the friction velocity U^* to the kinematic viscosity ν, it is possible to calculate the obtainable pressure drop reduction for flows inside pipes of any diameter, by means of equations 6 and 7.

ΔB values are deduced from only one pipe laboratory experiment, out of the saturated domain. It implies the use of a sufficient diameter pipe (6 to 10 millimiters seems to be good values).

ΔB measurements must be achieved in the same environment conditions as in full scale application, particularly at the same temperature, in order to take into account the exact configuration of the polymer inside the fluid.

Knowing the efficiency of a drag reducing polymer with flow conditions, it is now possible to estimate the expected commercial profit from this drag reduction process. One has to notice that in the large ducts, flow velocity is generally low (1.5 to 2.5 m/s or a little more) and so are consequently the corresponding friction velocity value and the pressure drop reduction obtainable even with a high efficient polymer like POLYOX 301. Furthermore, if polymer degradation occurs, one can be sure that the process becomes uninteresting from the application point of view.

Consequently, it is of first importance to know the exact conditions of mechanical degradation occurence in order to consider the transportation of a fluid inside a pipe during several hundred kilometers without polymer degradation. This is precisely the object of the research that we are conducting now.

REFERENCES

1. Water Pollution Control Research : "Polymer for sewer flow Control". U.S. Federal Water Pollution Control Administration. Series WP 20.22. (August 1969).

2. Crowley P.A. et al : "Evaluation of friction reducing additives for pipe-line use". Columbia Research Corporation. US Army Mobility Equipment Research and Development Center. (October 1970).

3. Patterson G.K., Zakin J.L., Rodriguez J.M. : "Drag Reduction, polymer solutions, soap solutions and solid particle suspensions in pipe flow". Industrial and Engineering Chemistry. Vol. 61,n° 1 pp 22-30 (January 1969).

4. Forester R.H., Larson R.E., Hayden J.W. and Wetzel J.M. : "Effect of polymer addition on friction in a 10 inches diameter pipe". Engineering notes. Journal of Hydronautics. Vol. 3, No. 1, pp 59-62 (January 1969).

5. Lescarboura J.A., Culter J.D. and Wahl H.A. : "Drag reduction with a polymeric additive in Crude oil pipe-lines". Society of Petroleum Engineers Journal. Vol. 11, No. 3, pp 229-235 (September 1971).

6. Bazilevitch V. A. and Shabrin A. N.: "Use of Polymeric Additives for Reducing Hydraulic Resistance in Pipes". Fluid Mechanics Soviet Research. Vol. 1, No. 5 59-67, (1972).

7. Elata C., Lehrer J., Kahanowitz A.: "Turbulent Shear Flow of Polymer Solutions". Israël Journal of Technology. Vol. 4, n° 1 pp 87-95 (1966).

8. Huang T.T. : "Similarity laws for turbulent flow of dilute solutions of drag reducing polymers". The Physics of Fluids. Vol. 17, n° 2 pp 298-309 (February 1974).

9. Granville P.S. : "The frictional resistance and velocity similarity laws of drag reducing dilute polymer solutions". Research and development report n° 2502. Naval ship research and development center. Washington. (September 1967).

10. Poreh M.and Miloh T.:"Rotation of a disk in dilute polymer solutions". Journal of hydronautics. Vol. 5, n° 2 pp 61-65 (April 1971).

11. Virk P.S. : "An elastic sublayer model for drag reduction by dilute solutions of linear macromolecules". Journal of Fluid Mechanics. Vol. 45, Part 3 pp 417-440 (1971).

12. Kim S. and Tagori T. : "A correlation of the Toms phenomenon". University of Tokyo. Proc. Intern. Conference on drag reduction Paper B3, BHRA Fluid Engineering. Cranfield. U.K. (1974).

13. Katsibas P., Balakrishnan C., White D. and Gordon R.J. : "Drag reduction correlations". University of Florida. Proc. Intern. Conference on drag reduction Paper B2, BHRA Fluid Engineering. Cranfield. U.K. (1974).

14. Norgeot G. : "Studies on dilute solutions turbulent flow in pipes". (Etudes sur l'écoulement turbulent des solutions diluées de polymères dans les conduites). Thèse de 3ème cycle Université de Paris VI (1973) (in French).

15. Berman N.S. and George W.K. : "Time scale and molecular weight distribution contributions to dilute polymer solutions fluid mechanics". Proc. of the 1974 Heat Transfer and Fluid mechanics institute. Stanford University Press.

16. Lumley J.L. : "Drag reduction in turbulent flow by polymer additives". Jl of polymer science. Macromolecular Review. Vol. 7 pp 283-290 (1973).

17. White W.D. : "Turbulent drag reduction with polymer additives". Jl Mechanical Engineering Science. Vol. 8, n° 4 (1966).

18. Deb S.K. and Mukherjee S.N. : "Molecular weight and dimensions of guar gum from light scattering in solution". Indian J. of chemistry. Vol. 1 (October 1963).

19. Encyclopaedia of Polymer Science and Technology. Edited by H. F. Mark, Interscience (1972).

20. Pharmaceutica Acta Helvetiae : "Influence of preparing method and oxygen concentration on viscosity of GUAR GUM solutions". (Der Einfluss der Herstellungsmethode und der Wasser stoffionen Konzentration auf der Viskosität von guar Gummi schleimen) (1960) (in German).

21. Granville P.S. : "Hydrodynamic aspects of drag reduction with additives". Marine Technology Vol. 10, No. 3, pp 284-92 (July 1973).

22. Poreh M. and Dimant Y. : "Velocity distribution and friction factors in flows with drag reduction". Naval Hydrodynamic Symposium. (1974).

23. Davies J.T. : "Turbulence Phenomena". Academic Press. New York (1972).

24. Virk P.S. : "Drag reduction in rough pipes". Jl Fluid Mech. Vol. 45 No. 3 pp 417-440 (1971).

25. Fabula A.G. : "The Toms phenomenon in the turbulent flow of very dilute polymer solutions". Proceedings of the Fourth International Congress on Rheology - Part. 3 (1965).

26. Hoyt J.W. : "The effect of additives on fluid friction". Jl of Basic Engineering. Transactions ASME Vol. 94. Series D, No. 2, pp 25-28 (June 1972).

27. Hoyt J.W. : "Recent progress in polymer drag reduction". CNRS Symposium "Polymers and lubrication". Brest (France) (20-23 May 1974).

28. Piau J.M. : "Mechanics of dilute polymers solutions. Friction reduction application". (Mécanique des solutions diluées de polymères. Application à la réduction de frottement). CNRS Symposium "Polymers and Lubrication". Brest (France) (20-23 May 1974) (in French).

29. Sellin R.H.J. : "Experiments with polymer additives in a long pipe-line". BHRA International conference on drag reduction. Proceedings of BHRA International Conference on Drag Reduction, Paper G2, Cambridge (1974).

30. Tullis J.P. and Ramu K.L.V. : "Drag reduction in developing pipe flow with polymer injection". Proceedings of BHRA International Conference on Drag Reduction Paper G3, Cambridge (1974).

31. Thorne P.F. : "Drag reduction in fire fighting". Proceedings of BHRA International Conference on Drag Reduction, Paper H1, Cambridge (1974).

32. Virk P.S. : "Drag reduction fundamentals. AICHE Journal. Vol. 21, n° 4 pp 625-656 (July 1975).

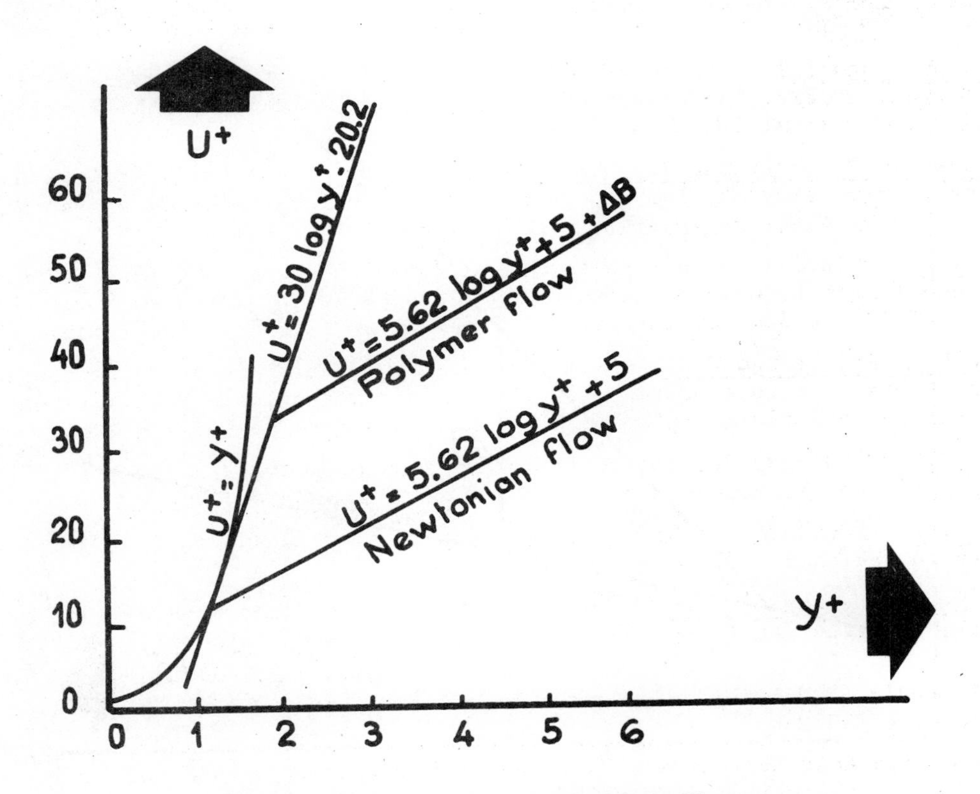

Fig.1 Internal layer velocity profiles

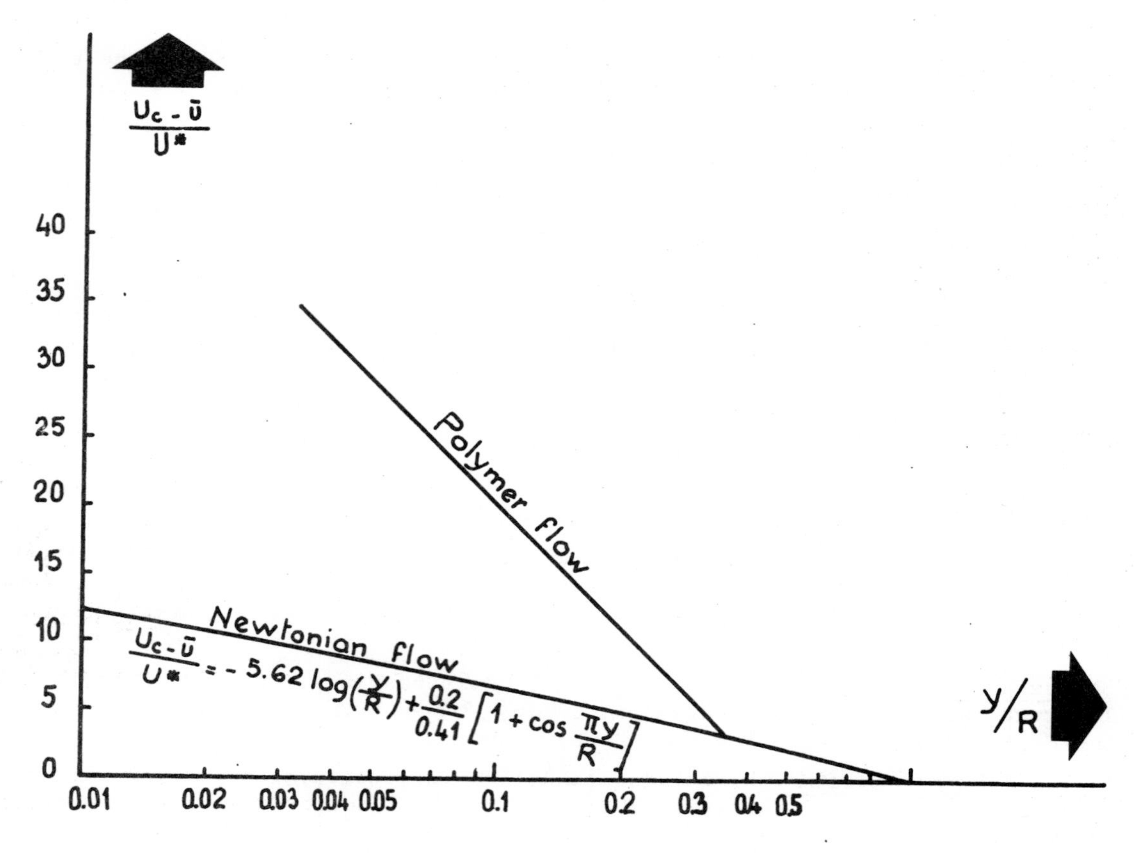

Fig. 2 Outer layer velocity profiles

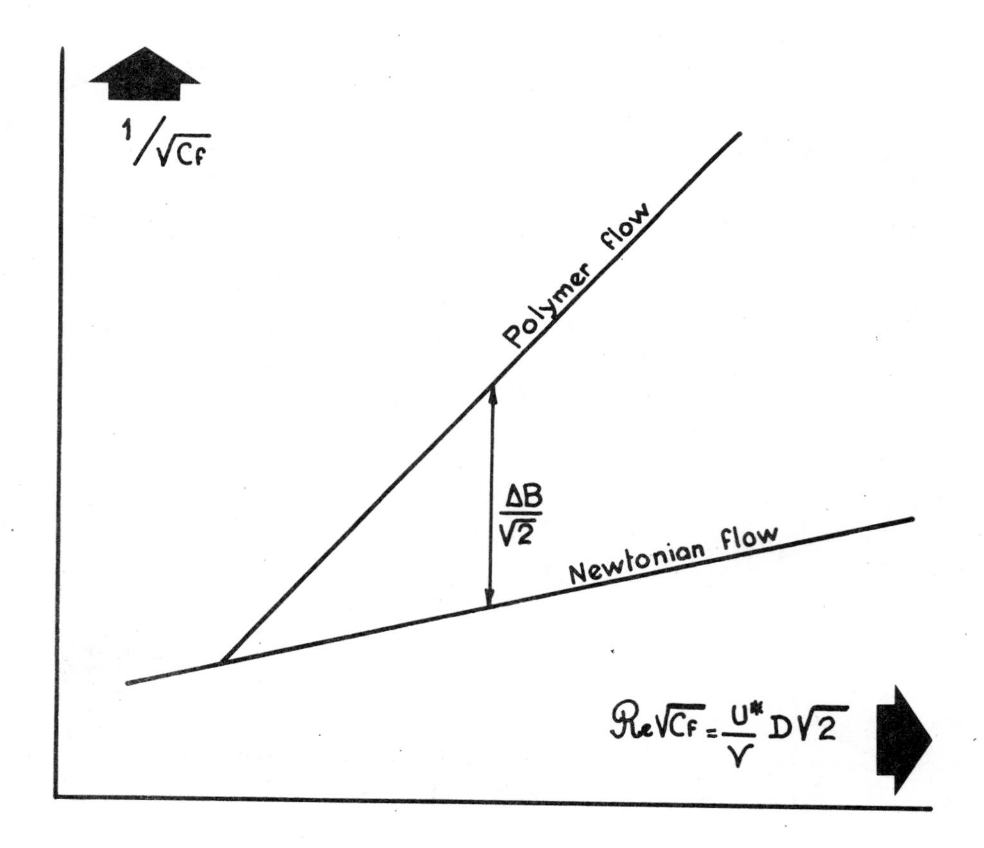

Fig. 3 Principle of ΔB measurement

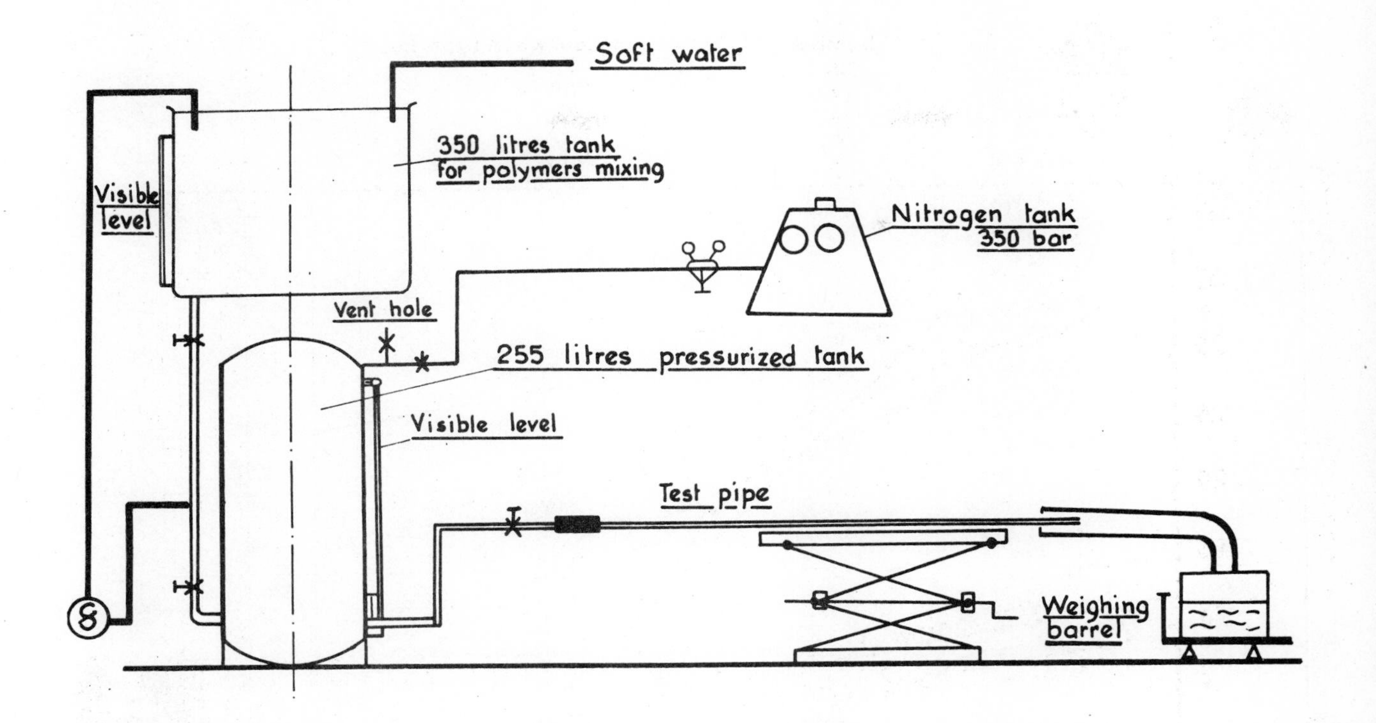

Fig.4 Experimental set up for small pipes

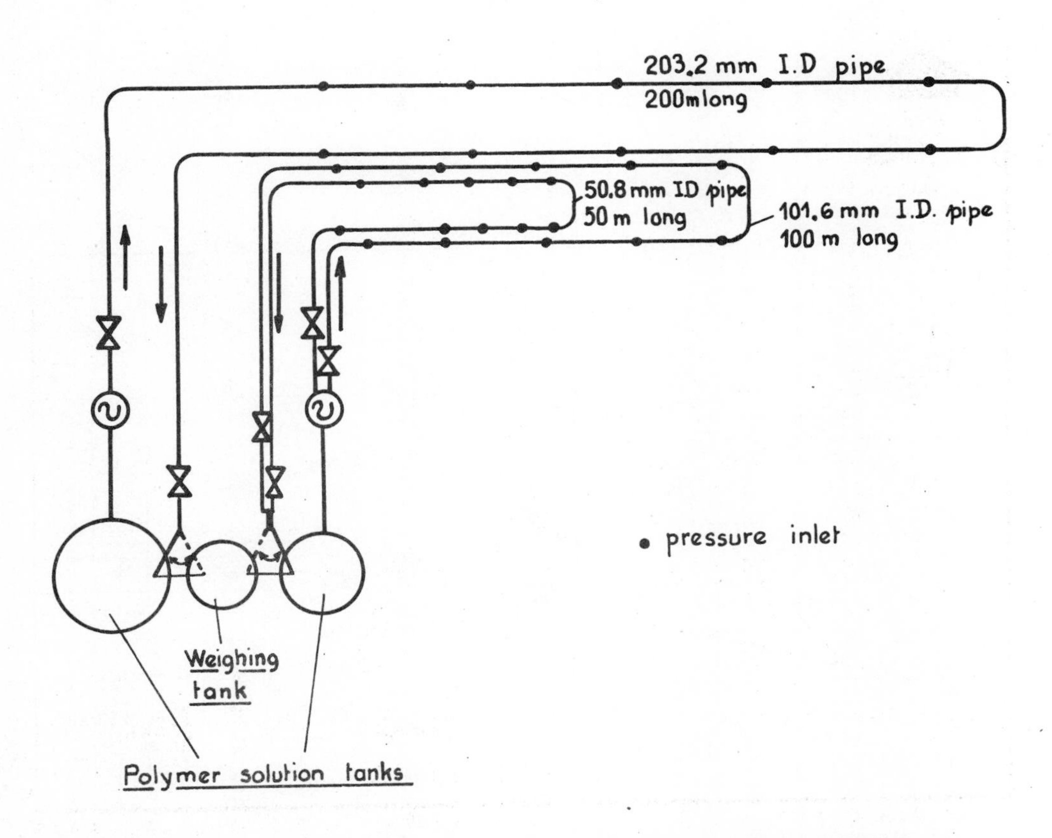

Fig.5 O.T.P. experimental installation - recirculated fluid

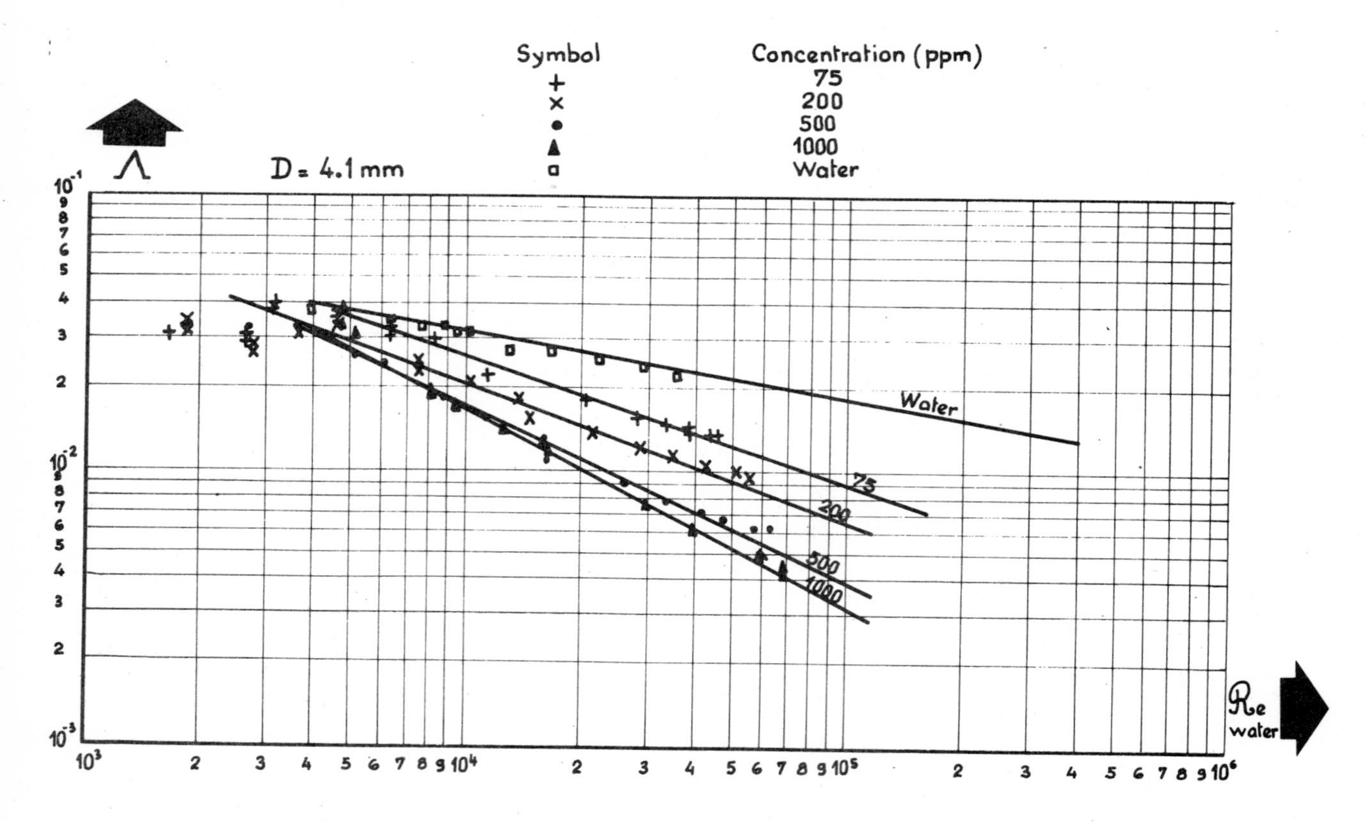

Fig.6 Pressure drop coefficient versus Reynolds number

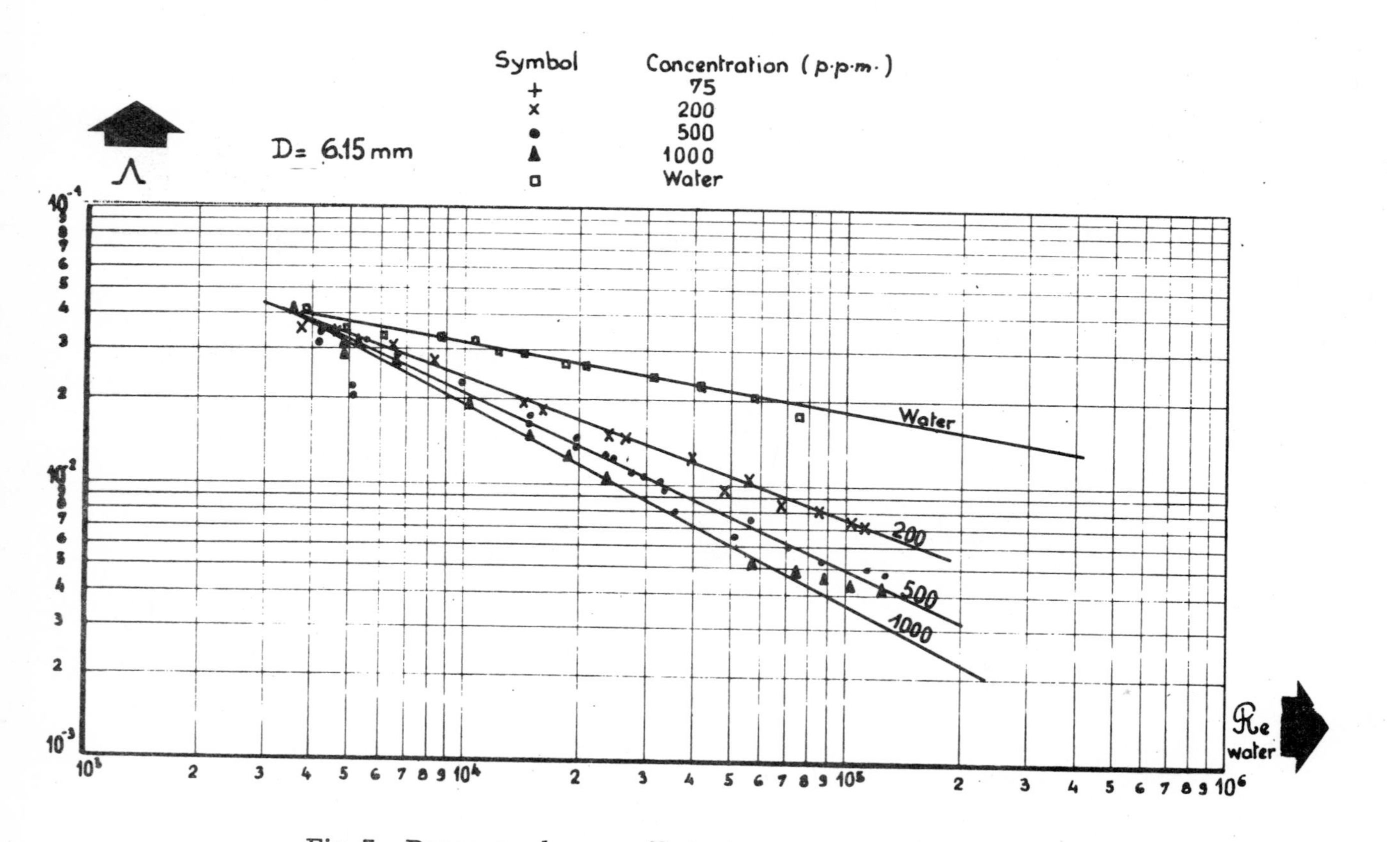

Fig.7 Pressure drop coefficient versus Reynolds number

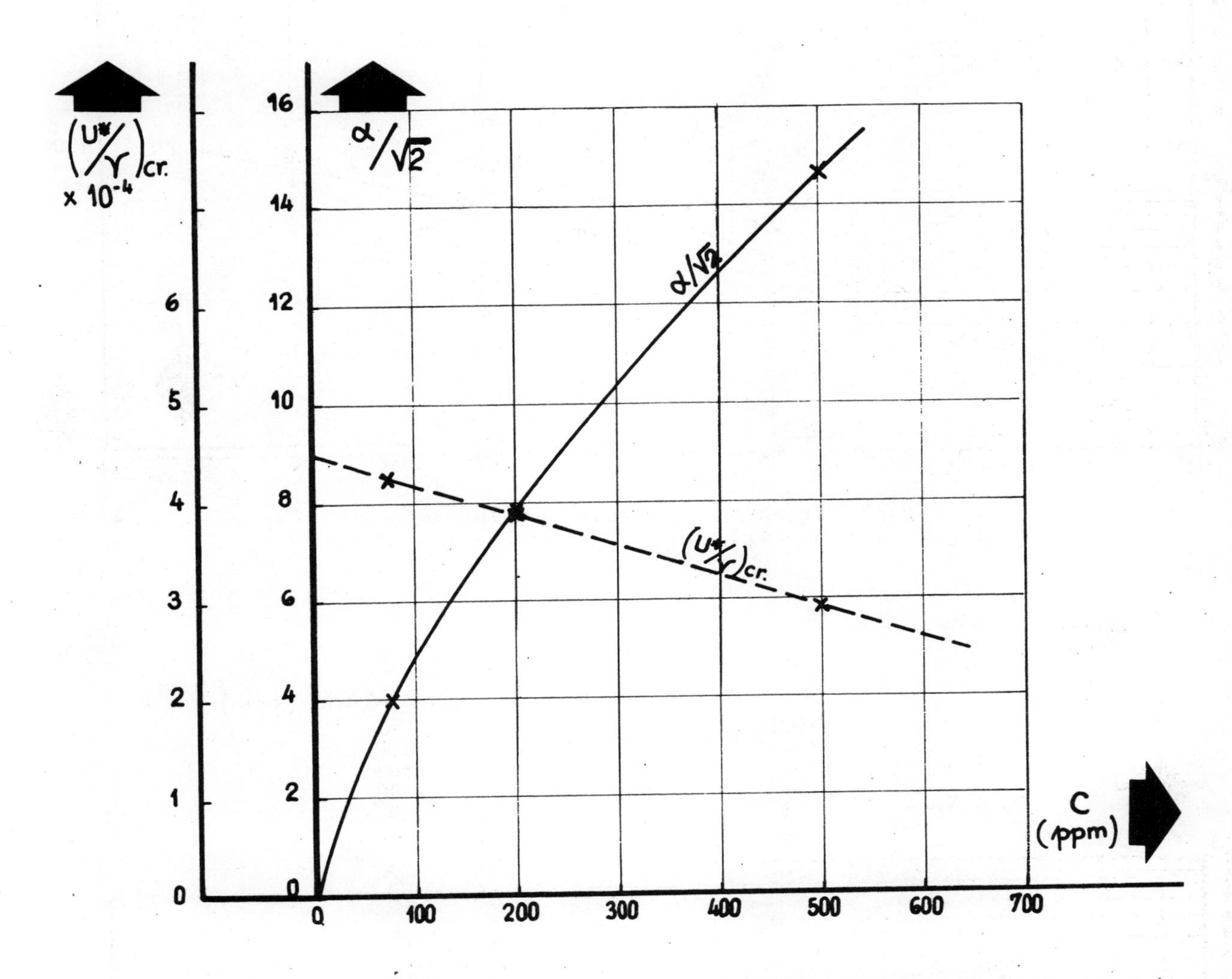

Fig.11 Values of characteristic coefficients α and $(U^*/\nu)_{cr}$ versus concentration

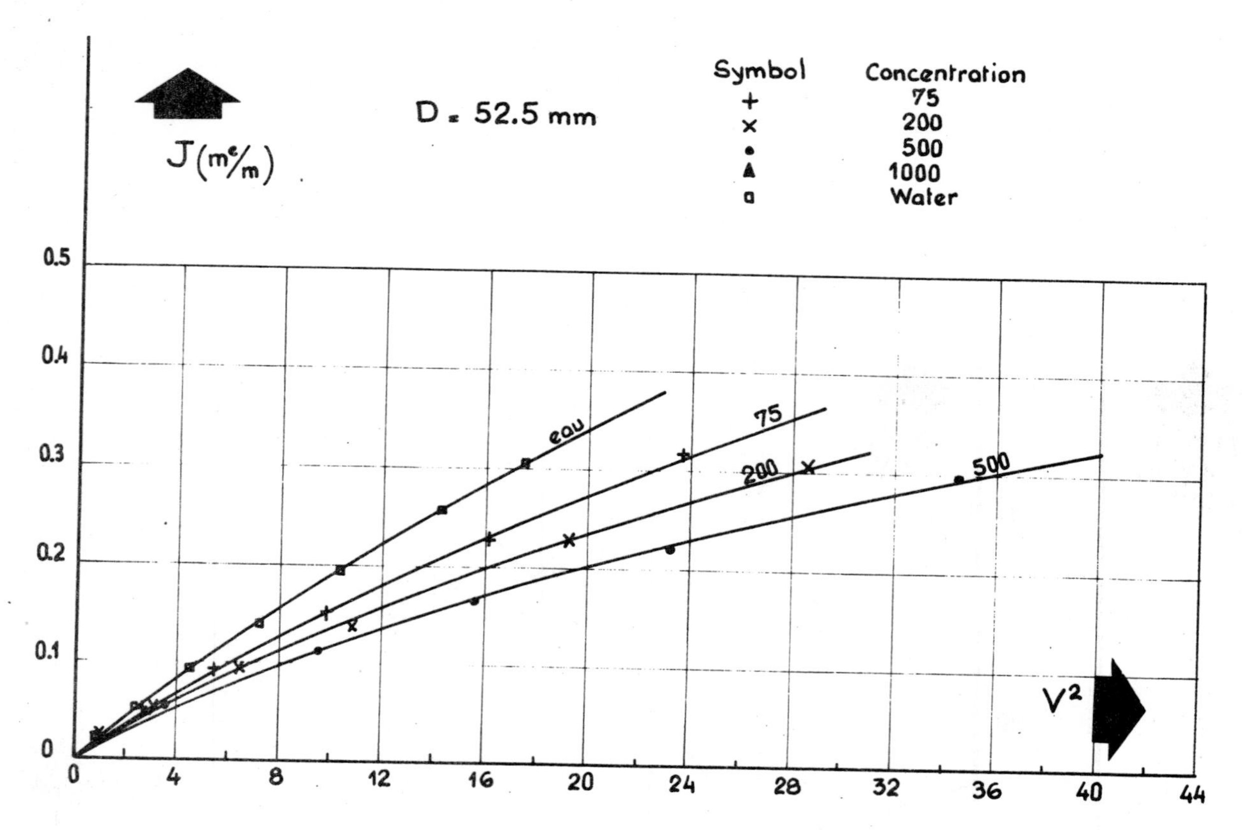

Fig. 12 Pressure drop versus mean flow velocity

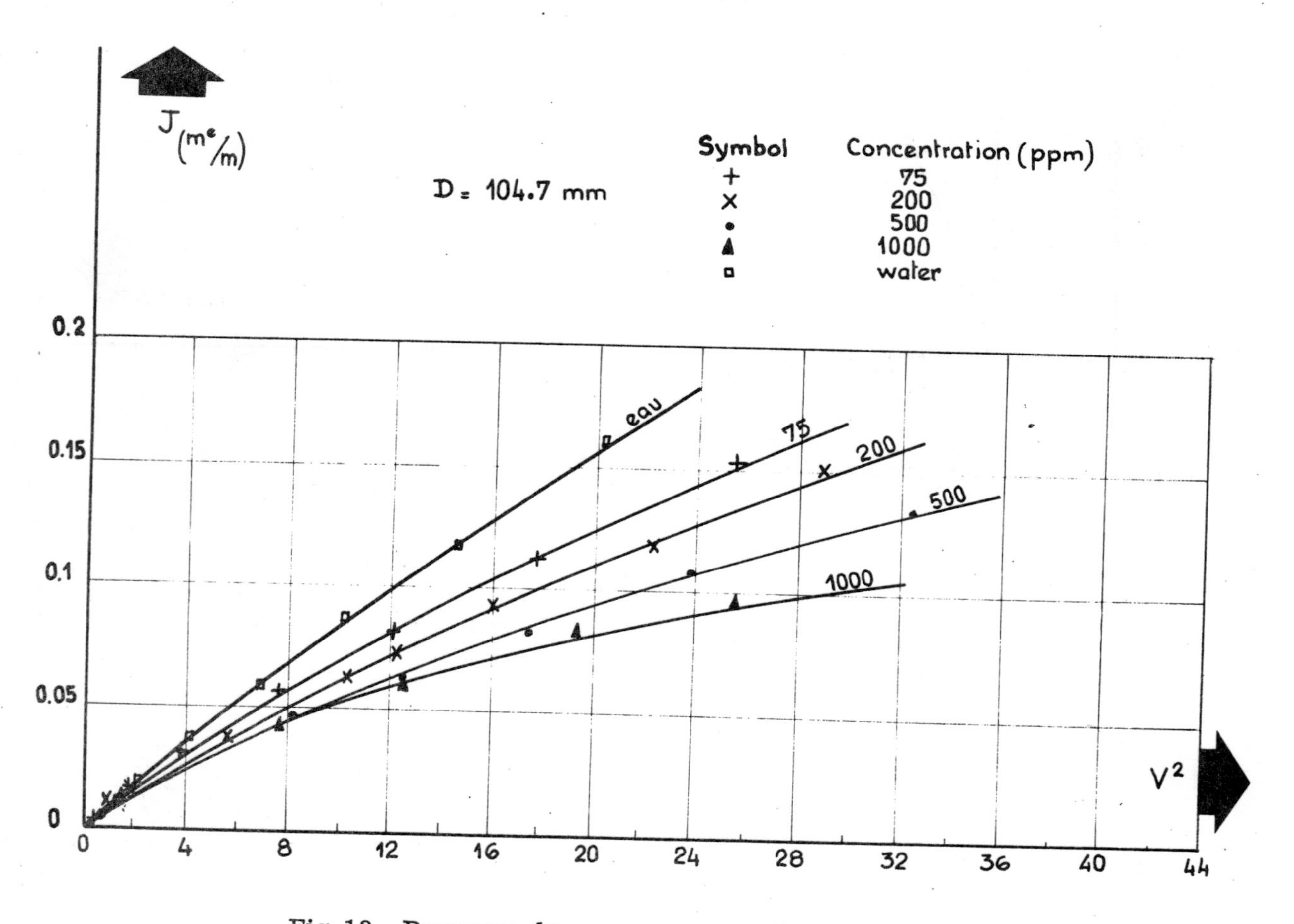

Fig. 13 Pressure drop versus mean flow velocity

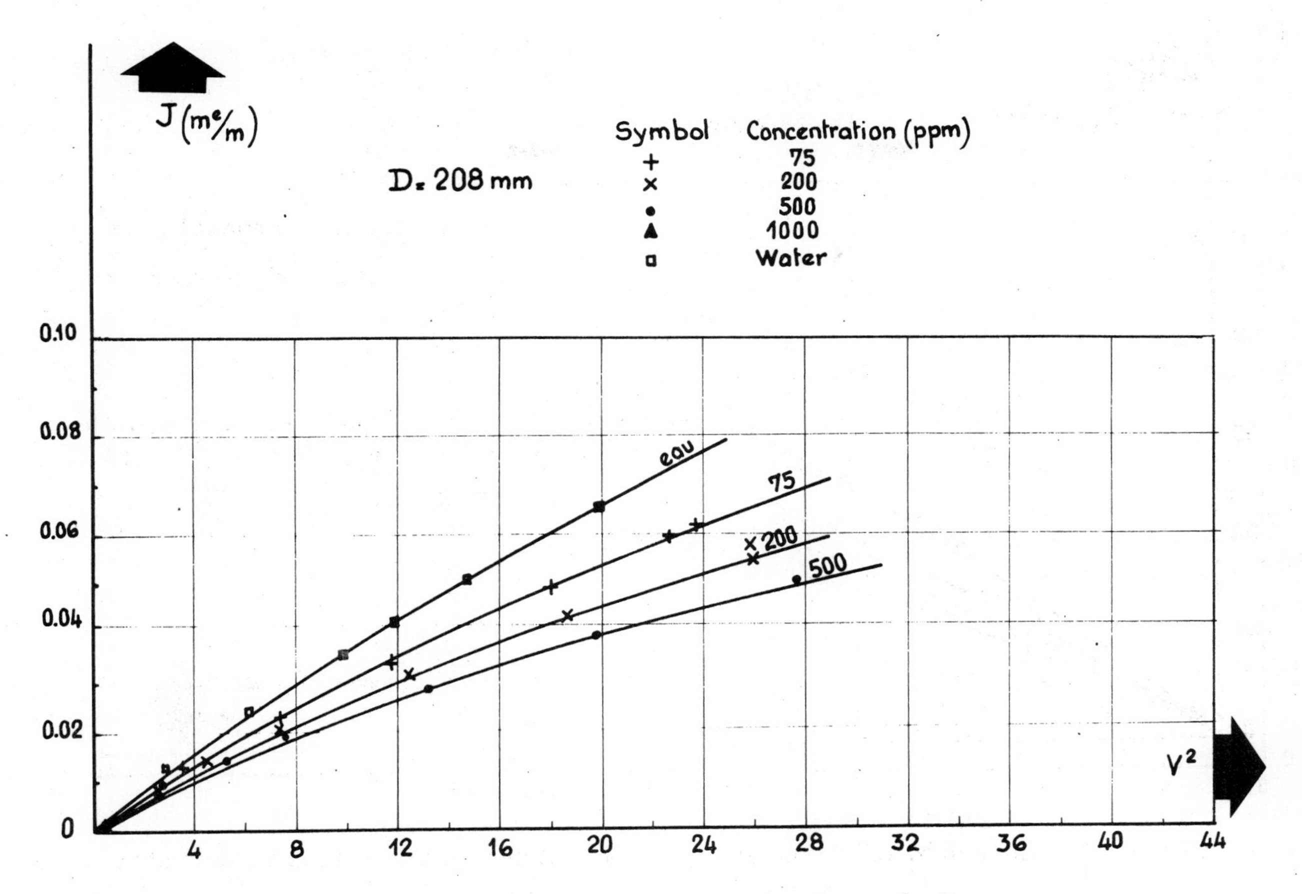

Fig.14 Pressure drop versus mean flow velocity

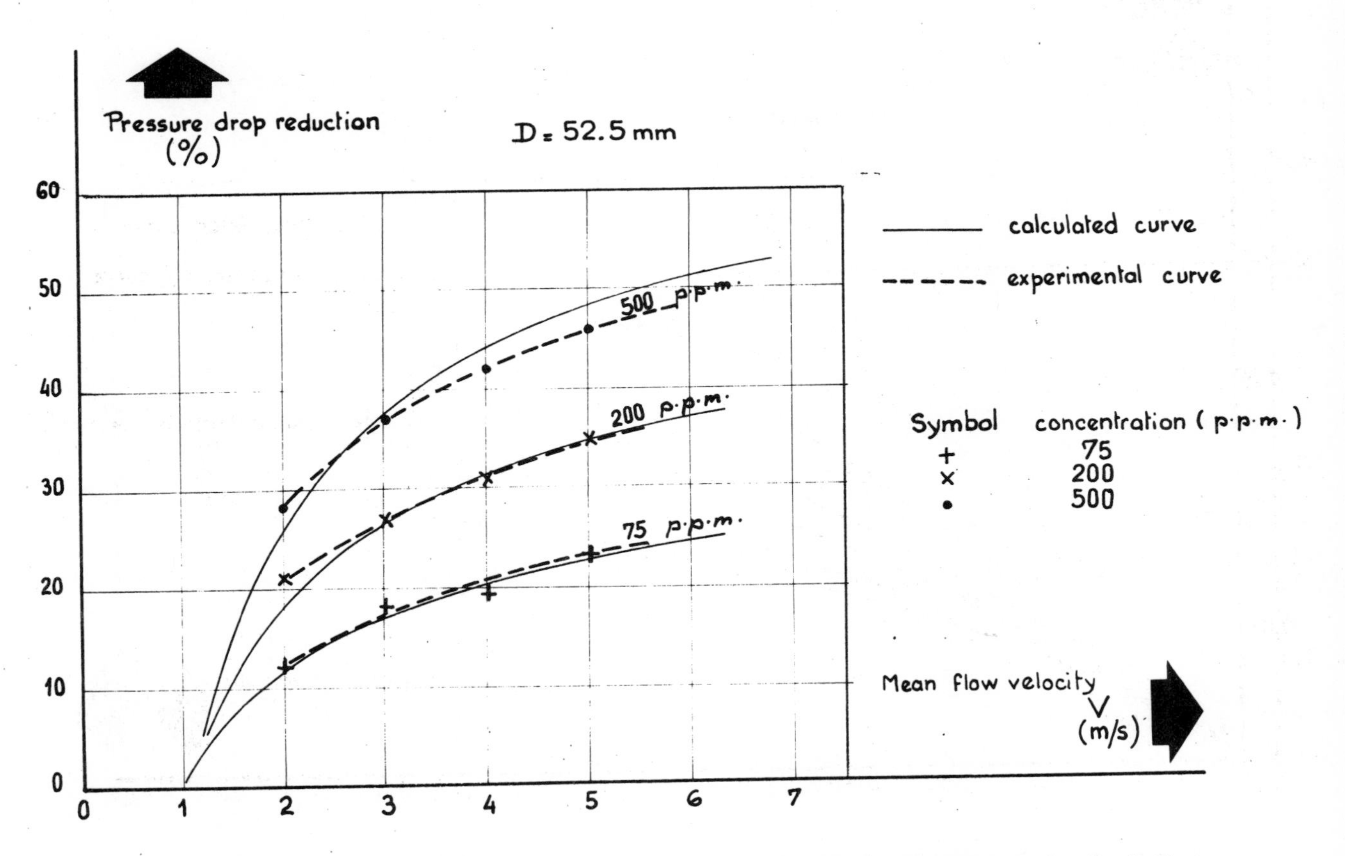

Fig.15 Pressure drop reduction - comparison between experiments and calculation

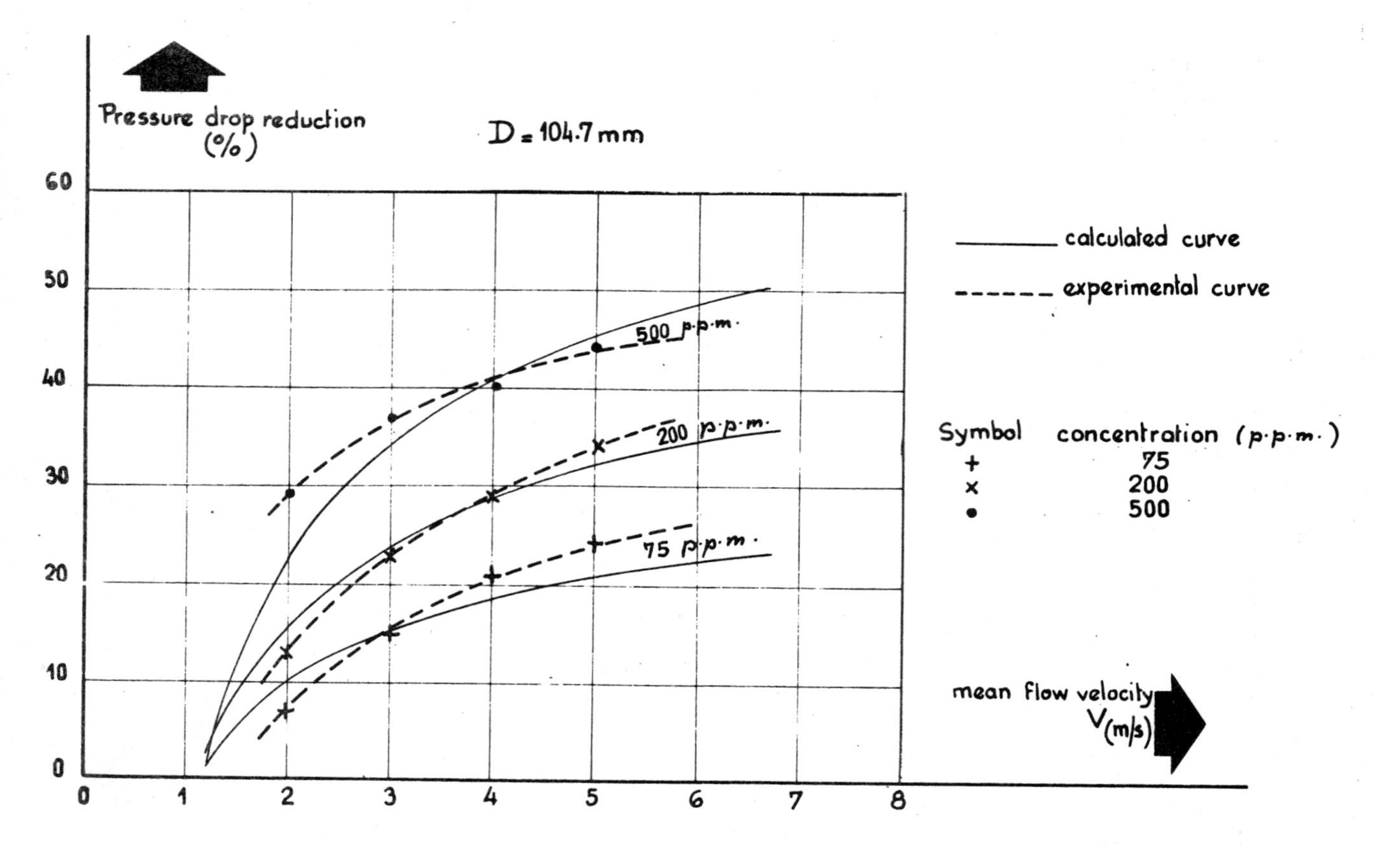

Fig.16 Pressure drop reduction - comparison between experiments and calculation

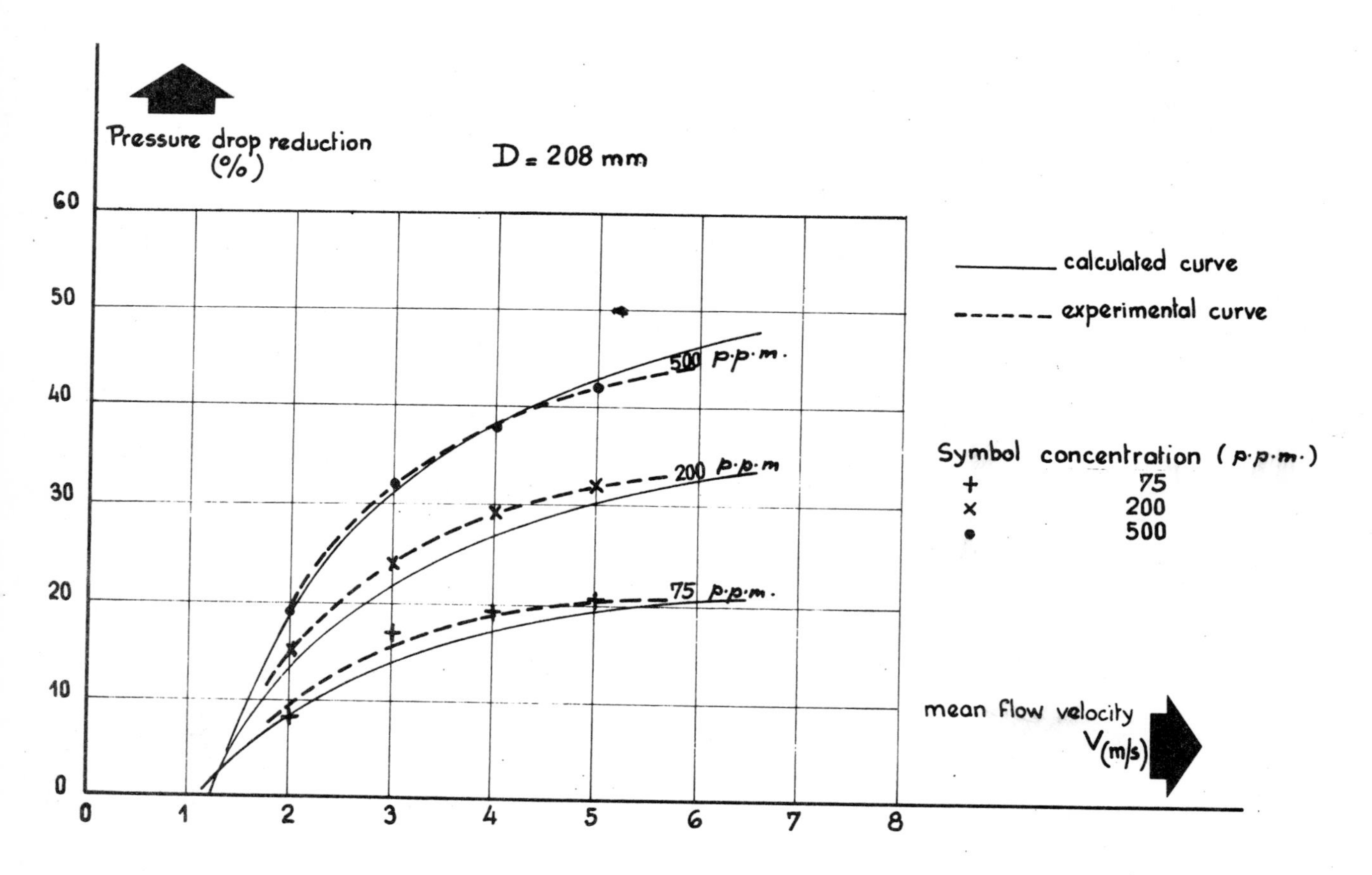

Fig.17 Pressure drop reduction - comparison between experiments and calculation

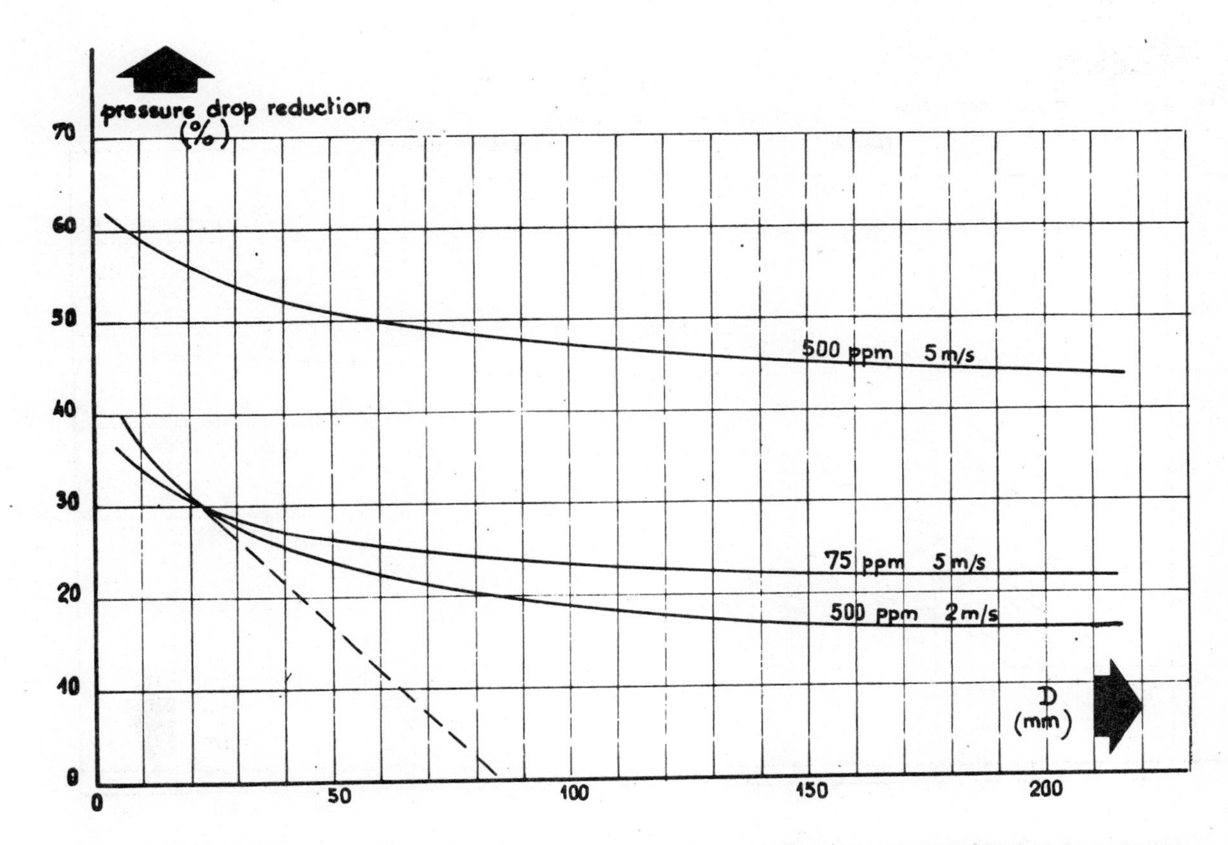

Fig.18 Diameter effect on pressure drop reduction - polymer: GUAR GUM CSAA M200

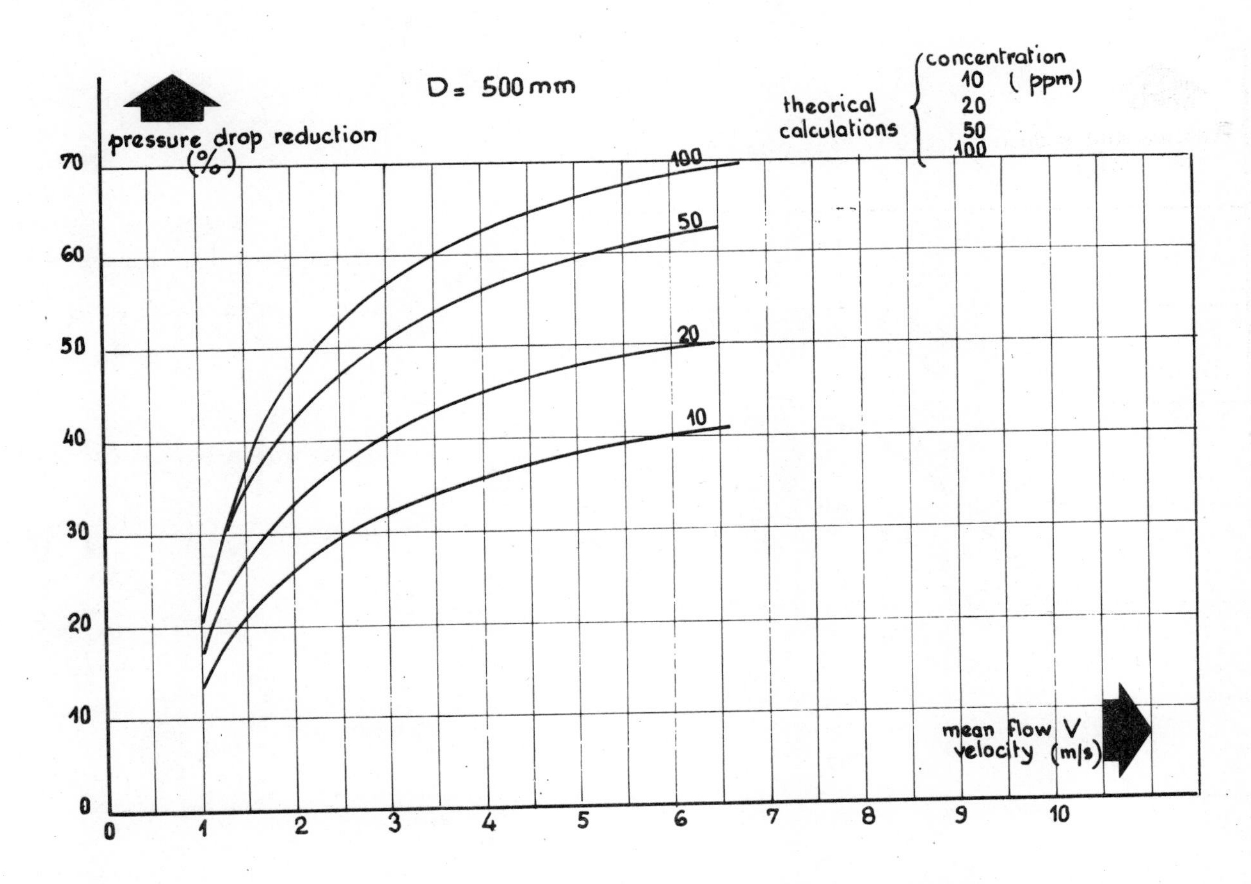

Fig.19 Calculation of Polyox 301 efficiency in large pipes

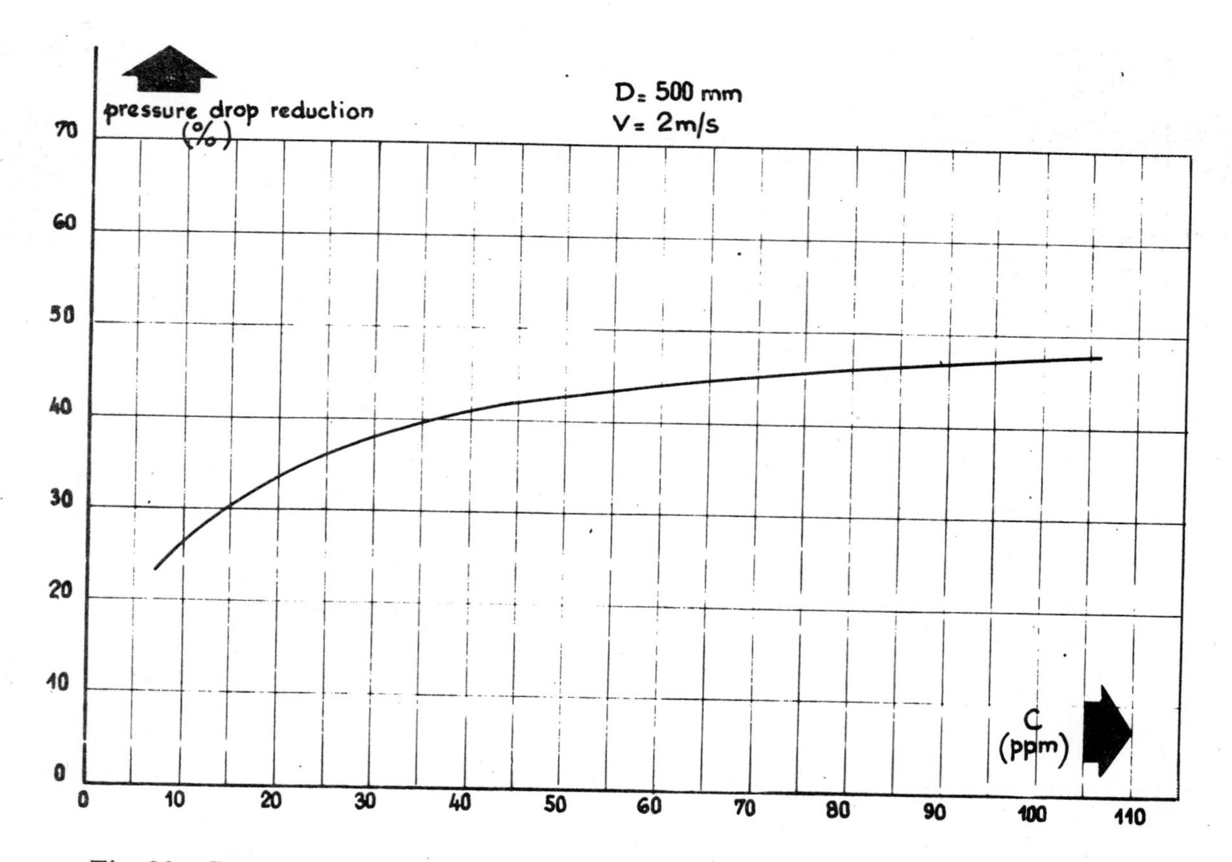

Fig. 20 Concentration effect on pressure drop reduction - Polymer: Polyox 301

Second International Conference on

Drag Reduction

August 31st - September 2nd, 1977

PAPER B3

HYDROTRANSPORT OF FLY ASH-WATER MIXTURE AND DRAG REDUCTION

J. Pollert, Ph.D., C.Eng.

Technical University of Prague, Czechoslovakia

Summary

Big investments and operating costs for pipeline systems for fly ash-water mixture in thermal power stations urge designers and workers of power stations to look for new economical ways. One of the ways is using drag reduction by polymer additives for hydrotransport of fly-ash. For that purpose three pilot plant experiments were carried out in a 0.25 m diameter steel pipeline, 4121 m in length. The basic average mixture rate of flow ranged from 125.7 to 128.1 l/s. The anoinic polymer Separan AP 273 was added to mixture in concentrations from 0.000936% to 0.00828%. Two different methods were used for dosing polymer additive into the pipeline. The economic assesment showed that using drag reduction is beneficial in cases where operation is for a limited duration.

Held at St. John's College, Cambridge, England.
Organised and sponsored by BHRA Fluid Engineering, Cranfield, Bedford, MK43 0AJ.

Nomenclature

C_v - volume concentration
g - gravitational constant
i - pipeline pressure gradient
v - average velocity
D - pipe diameter
Fr - Froude number
Q - volume flow rate
Re - Reynolds number
λ - pipe friction factor
φ - Durand function

Subscripts
m - mixture
w - water
p - polymer

Note. All concentrations are by weight.

1. Introduction

The disposal of solid waste, in this case of fly ash from power plant, is today's increasing problem. In our country this problem is even more urgent in consenquence of the use of coal with low calorific value with high ash content. Usually hydrotransport of the fly ash-water mixture is used from the power plants to the basins of ash-disposal. However the number of convenient exhausted mines which are often used and ought to be situated nearly to the power plant is lowering. New localities which must be found out are less economical because of long distances incuring severe increase of capital and operating costs for pipeline and intermediate booster stations especially.

One of the possibility how to overpower this problem is the use of friction-reducing polymers. So far the high efficiency of friction-reducing polymers was proved for water and various lengths of pipelines. But we are still short of larger practical use. This may be due to impossibility of further utilization of water after polymers. On the other side in some cases the pollution of water by polymers is not detrimental and could be useful. The fly-ash water mixture hydrotransport is the right example.

With this in mind a programme of drag reduction experiments was initiated at the Thermal Power Plant in Tisová in the Autumn of 1976.

2. Pipeline and measurements

All of the experiments were carried out on the part of the pipeline from the intermediate booster station Antonín to the basin of ash-disposal Bohemia as shown in Fig.1. A 0,25 m diameter steel pipeline, 4121 m in length was used. The mixture was pumped from a bunker through centrifugal pumps Sigma 250 NBA directly into the pipeline. A difference of elevation between the axis of centrifugal pumps and outlet of the pipeline was 7,5 m, the outlet being lower.

During all of the experiments the pressure drop, volume flow rate of mixture and mixture level inside the bunker were measured. The pressure gauge Prema (range 0 - 981 kPa, sensitivity-2,5 %) was fixed to the pipe at the intermediate booster station and volume measurement of flow rate was carried out at the outlet to the basin of ash-disposal, Fig.9. For that purpose the cylindrical tank of known volume of 1,85 m^3 was used. The mixture level inside the bunker measured with 5 cm accuracy by means of float appeared to be relevant because of decreasing its level after dosing polymer thus obtaining different positive displacement height.

From each experiment four samples were taken at the outlet thus obtained values of pH, density of mixture, volume concentration and size distribution of solids.

3. Polymer injection

The dosing system employed was as follows. To prepare polymer water solution a storage and mixing tank by the volume of 4,35 m^3 was used. The powder polymer was poured into the storage and mixing tank simultaneously with water and uniform concentration of the polymer solution was ensured by a diffused-air unit. The polymer solution being withdrawn from the bottom of the storage tank passed through a volume controlled pump into a 50 mm in diameter, 8,5 m in length steel pipe as shown in Fig.10. Feeding of the polymer solution into the main pipeline was possible to rule by a control valve. The resultant concentration of the suspension was calculated according to the rates of flof of the fly ash-water mixture and the polymer solution level drop in the storage tank in time.

During experiments two ways of dosing were employed:

a) dosing of the polymer solution from the storage tank

b) powder polymer pouring directly into the bunker of the fly ash-water mixture.

The use of the anionic polyacrylamide Separan AP 273 was at first tested in laboratory. Various concentration tests for digested sludges have been carried out, Pollert (Ref.1) showed great drag reductions. Pollert attributed a mutual interaction between sludge flocs and the additive the great significance. The use of Separan AP 273 for the fly ash-water mixture showed to be possible and efficient,

4. Technical relations applied

Certain disadvantage during measurement was that all of the experiments were carried out under full Thermal Power Plant in Tisová operation and pipeline system as well. After polymer injection some time was needed to achieve steady pressure and flow rates values as shown, e.g. in Fig.3. These changes could not be measured for only fly ash-water mixture because of a constant throughput of the installed centrifugal pumps. For determination a drag reduction $\Delta\lambda$ in a pipeline, short range experimental data for only fly ash-water mixture had to been extended and calculated beyond measured values to cover full experimental range.

Friction factor λ at all experimental points (with and without additive respectively) was calculated from Darcy-Weisbach formula

$$i_m = \lambda_m \frac{1}{D} \frac{v^2}{2g} \qquad (1)$$

Average velocities used in Eq.1 and those for extrapolating experimental points for only fly ash-water mixture were the same. It was possible, assuming that the friction factor for heterogeneous suspensions can be expressed

$$\lambda_m = \lambda_w + \Delta\lambda_m \qquad (2)$$

The friction factor λ_w was calculated from Konakov expression

$$\lambda_w = (1.8 \log Re - 1.5)^{-2} \qquad (3)$$

with respect to the temperature of flowing mixture.

For a supplementary friction factor $\Delta\lambda_m$ for a heterogeneous fly ash-water mixture, Durand expression was used

$$\Delta\lambda_m = \lambda_w \cdot C_v \cdot \varphi \qquad (4)$$

where

$$\varphi = \frac{i_m - i_w}{i_w \cdot C_w} \qquad (5)$$

Durand and Condolios (Ref.2), Bonnington (Ref.3), Newitt (Ref.4) and others introduced different methods for calculation of Durand's function for different particles. Most of investigators suggested common expresion

$$\varphi = K \cdot (Fr \cdot \sqrt{C_\varphi})^{-m} \qquad (6)$$

Considering that expression Eq.6 need not be for all material functions explicitly satisfied most of the authors recommended a calculation of function φ according to polynom (Ref.5)

$$\varphi = a_o + a_1 Fr + a_2 Fr^2 \qquad (7)$$

Note. Kupka, Hrbek and Janalík (Ref.5) introduced values of coefficients a_i for fly ash with particle size d (mm) - $0,1 < d < 1,5$. Our fly ash size felt into this range.

For calculation of a maximum drag reduction following expression was used:

$$\Delta\lambda_{max.} = \frac{\lambda_m - \lambda_p}{\lambda_p} \cdot 100$$

5. Experimental results

a) For experiment No.1 concentrated storage solution of polymer was prepared - 0,875 % Separan AP 273. During the experiment revolutions of dosing pump were three times changed. The change of revolutions corresponded with concentrations - 0,000936 %; 0,00395 % and 0,00828 % respectively. Weight concentrations of fly ash was during the experiment on an average 4 % and total volume discharge of mixture Q_m = = 127,2 l/s.

Fig.3 shows pressure and mixture volume rate of flow - time relationship with indication time of dosing. It will be seen that steady values of rates of flow have been achieved whereas to get this for pressure some longer time would be needed. For that purpose the storage tank with concentrated polymer solutions was too small. With regard to the highest concentrations of Separan AP 273 0,00828 % in this experiment maximum drag reduction $\Delta\lambda$ = 32,0 % was achieved, Fig.4.

b) During experiment No.2 dosing was cut out of operation (blown fuse for dosing pump) for 10 minutes. The dose of Separan AP 273 was decrased for only 0,00549 %, the concentration of fly ash was 4,35 % and mixture volume rate of flow Q_m = 125,7 l/s. Both rates of flow and pressure became stable. The maximum drag reduction achieved in this experiment was $\Delta\lambda$= 26,4 %.

c) For experiment No.3 simple pouring of the powder Separan AP 273 into the bunker was employed. Resultant concentration of polymer was

0,00423 %.Mixture volume rate of flow was Q_m = 128,1 l/s for the concentration of fly ash 4,16 %.The dosing of polymer had to be interrupted before finishing this experiment because the mixture level at the bunker had become critical.As seen from Fig.5 and 6 rates of flow achieved were stable while the pressure was still decreasing.The maximum drag reduction obtained was $\Delta\lambda$ = 27,6 %.

In Table 1 results and parameters from all of the experiments can be found.

6.Economic assesment

The main purpose of our experiments was to prove the possibility of using the drag - reducing polymers for the fly ash-water mixture conveyd by a long pipeline.One of the serious problems for larger practical use of this method is in our country a high cost of the polymers which must be imported.A drawback of multiplied used and whence degradation (quick loosing of friction-reducing ability) in closed systems and a lost of additive in open ones without any chance to use it again,seem also to be relevant.

As shown in Fig.7 the maximum drag reduction - concentration of additive relationship indicates a significance of the additive concentration for the economic assesment.The great increase of the additive concentration is not accompanied correspondingly by the increase of the maximum drag reduction.

The economic assesment concerned the whole pipeline from the Power Plant in Tisová to the basin of ash-disposal Bohemia 6,57 km in length.For quick filling of the basin of ash-disposal (within two years),the reconstruction of the pipeline 0,35 m in diameter instead of 0,25 m in diameter is intended.Because of larger diameter and thus decreasing of friction losses,the intermediate booster station Antonín could be put out of operation.

Using of friction-reducing polymer Separan AP 273,it would be possible to ensure full operation of the pipeline without the intermediate booster station Antonín and even without designed reconstruction of the pipeline because of increasing the rates of flow by 10 %.

In Table 2 pressure losses calculated for 10 % increased rates of flow from experimental data are shown.Additional increase of the operating costs due to the use of Separan AP 273 is much lesser than the capital costs for the pipeline reconstruction.It is worth calculating for two years when the filling of the basin of ash-disposal is to be finished.

From the obtained data the maximum drag reduction value 32 % was taken account for calculating of possible extending of the pipelines a 0,25 and 0,35 m in diameter (the pressure losses being the same as measured).Fig.8 indicates a possibility of construction of the pipelines for fly ash-water mixture over longer distances without intermediate booster station to be needed.

7.Summary of experiments and conclusions

a)Three pilot experiments of hydraulic transport of fly ash in steel pipeline (0,25 m in diameter,4121 in length)were carried out with drag-reducing polymer additive.

b)Polyacrylamide Separan AP 273 the concentration of which ranged from 0,000936 % - 0,00828 % was used.

c)Fly ash concentration by weight ranged 3,89 - 4,35 %.

d)The maximum drag reduction obtained was 32,0 %.

e) The economical advantage of the use of drag reduction was proved to be relevant in this case for time limited period.

f) Comparing Sellin's experiments (Ref.7), D = 0,203 m; l = 4190 m, lower maximum drag reduction values were obtained. This seems to be due to the presence of suspended solid particles. If different efficiency of Polyox WSR 301 used by Sellin and Separan AP 273 is relevant, this could not be compared, Polyox WSR 301 not being available for us. But according to (Ref.6) the efficiency ought to be similar.

g) From two different methods of dosing polymer employed, simple pouring of powder Separan AP 273 into the bunker of mixture appeared to be satisfactorily efficient.

Acknowledgements

The author wishes to thank Power Plant in **Tisová** for making available experimental facilities at the intermediate booster station Antonín and Dow Chemical, Co. -Rephachem Praha for providing Separan AP 273 and a dosing pump.

References

1. Pollert, J.: "Toms' effect and sludge suspensions". (Tomsův jev a kalové suspenze). Ph.D. thesis, Technical University in Prague, (February 1976), (In Czech)

2. Durand, R., Condolios, E.: "The Hydraulic Transport of Coal", Proc. of Colloq. on Hydraulic Transport of Coal, London (1954).

3. Bonnington, S.T.: "Estimation of Pipe Friction Involved in Pumping Solid Material", BHRA, TN 708, (Dec. 1961).

4. Newitt, D.M., Richardson, J.F., Abbot, M., Turtle, R.B.: "Hydraulic Conveying of Solids in Horizontal Pipes", Trans. Instr. Chem. Eng., 33:93-113, (1955).

5. Kupka, F., Hrbek, J., Janalík, J.: "Hydraulic transport in pipelines". (Hydraulická doprava potrubím). SNTL, Praha (1970), (In Czech)

6. Kolář, V.: "Non-newtonian Flow". (Nenewtonské proudění). Script, Technical University in Prague, (1975), (In Czech)

7. Sellin, R.H.J.: "Experiments with Polymer Additives in a Long Pipelines". Drag Reduction -1 st Intern. Conference Cambridge, England (1974).

Table 1

Pipeline: Antonín - Bohemia D = 250 mm; l = 4121 m				
	Experiment No.			
	1	2	3	4
Volume flow rate Q_m (l/s)	127,2	127,9	125,7	128,1
Concentration of fly-ash C_v (%)	4,0	3,89	4,35	4,16
Density of mixture ρ_m (kg/m^3)	1013,2	1014,1	1020,1	1017,1
Average temperature of mixture in pipeline (^{o}C)	27	27	28	26
Concentration of Separan AP 273 by weight (%)	0,000936 0,00395 0,00828	0,00737	0,00549	0,00423
Max. drag reduction $\Delta \lambda_{max.}$ (%)	15,7 24,3 32,0	24,7	26,4	27,6
Concentration of polymer solution (%)	0,875	0,625	0,625	

Table 2

v_m = 2,938 m/s Q_m = 144 l/s		Separan AP 273 0,002 % ; $\Delta\lambda$= 20 %	
D (mm)		250	350
		λ = 0,01287	λ = 0,01287
h_z - without additive (m)	l (km) 5 6,57 10 20	113,2 148,8 226,5 452,9	80,9 106,3 161,8 323,5
		λ = 0,01015	
h_z - with additive (m)	l (km) 5 6,57 10 20	89,3 117,3 178,6 357,2	
Cost of additive ā 50 Kčs/1 kg/year		3 153 600	
Cost of pipeline l = 6,57 km			6 570 000
Total costs whole reconstruction of pipe l = 6,57 km			13 000 000

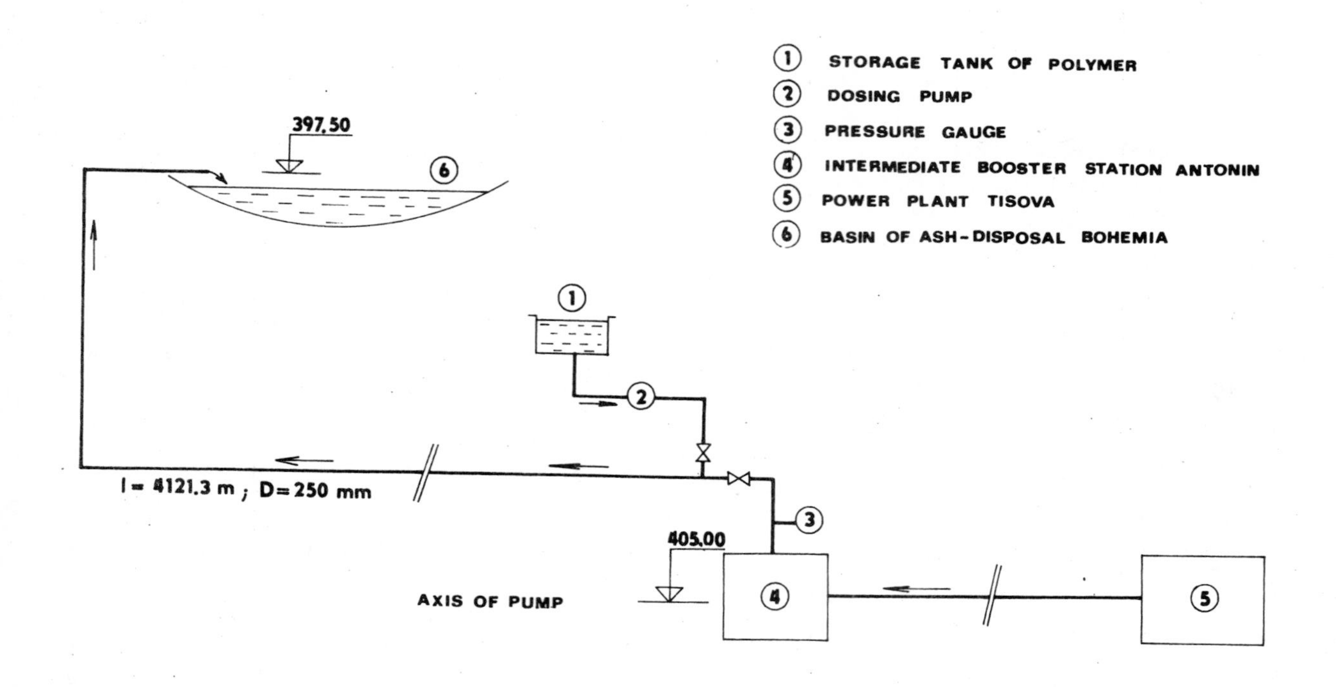

Fig. 1. Diagram of pipeline and dosing system

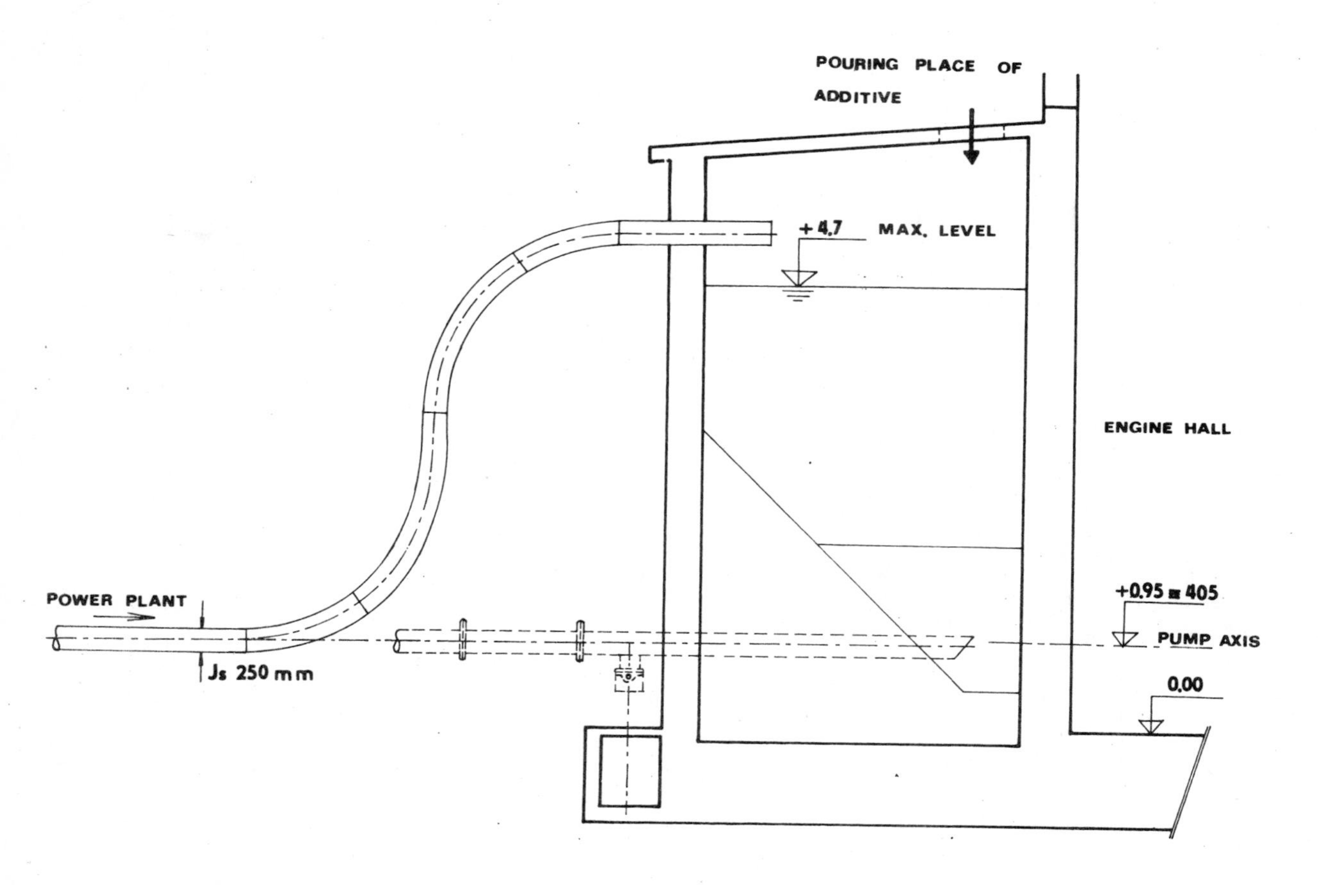

Fig. 2. Cross section view of intermediate booster station at Antonin

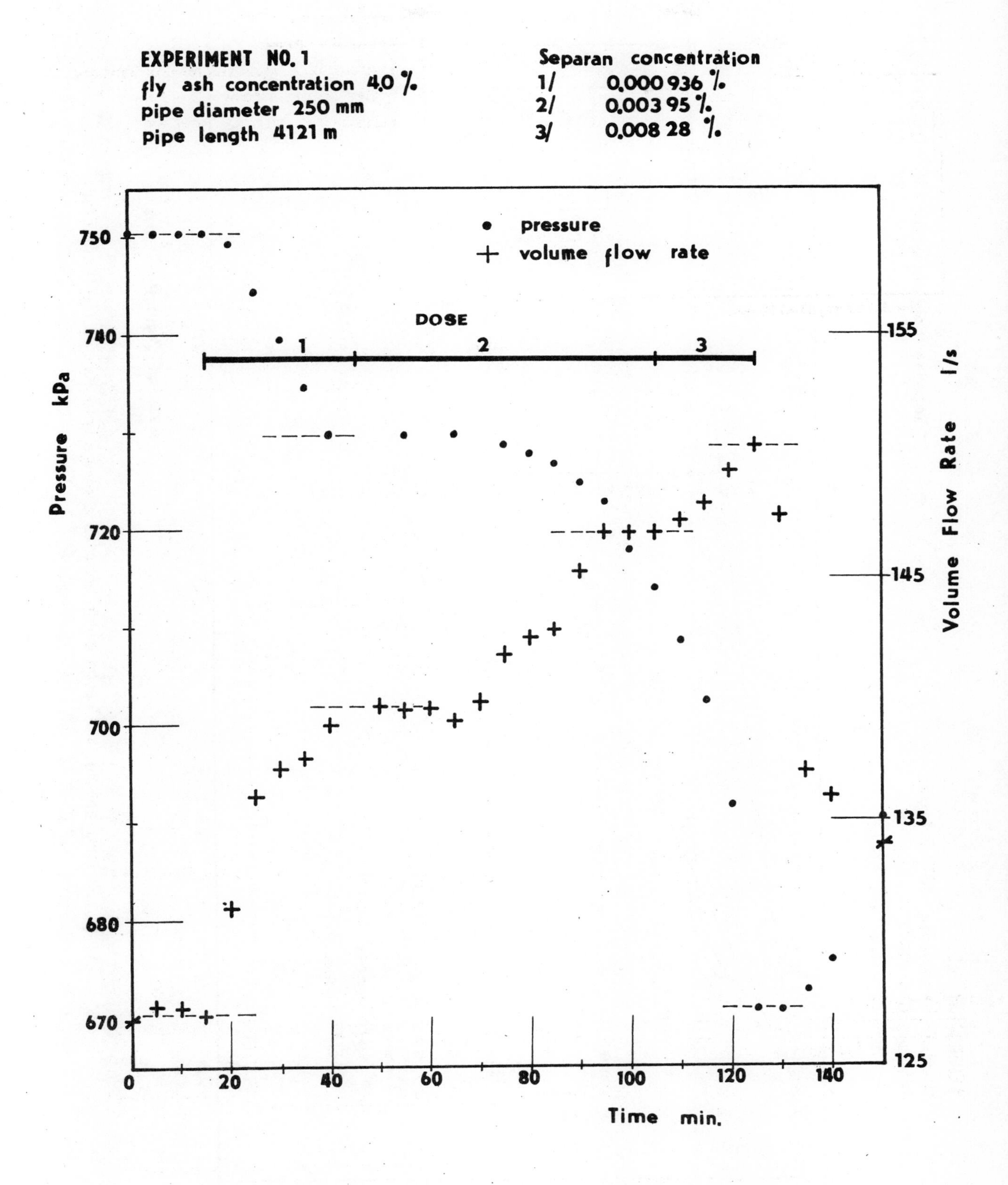

Fig. 3. Experiment No. 1 - volume flow rate and pressure gauge record

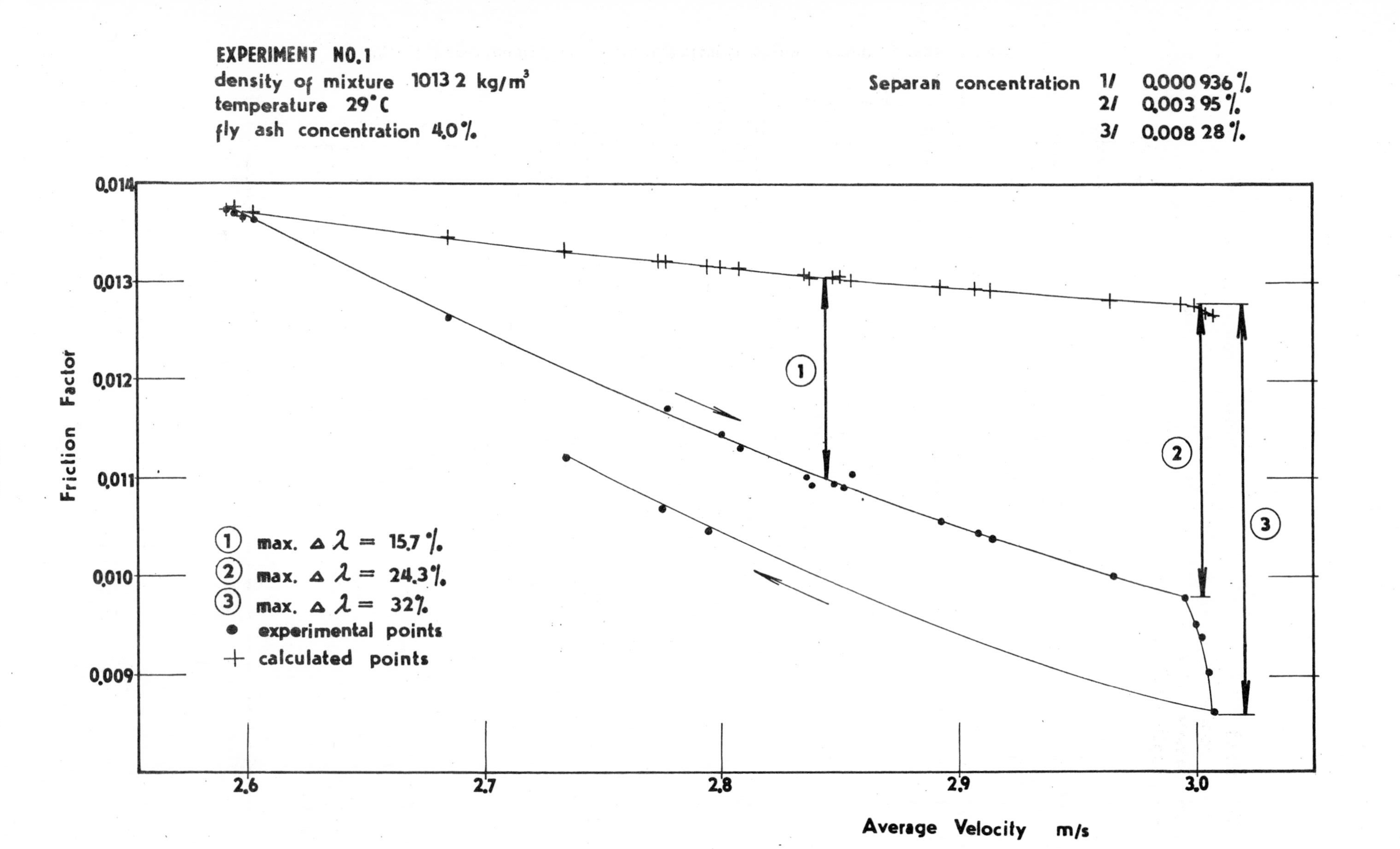

Fig. 4. Experiment No. 1 - relationship between average velocity and friction factor

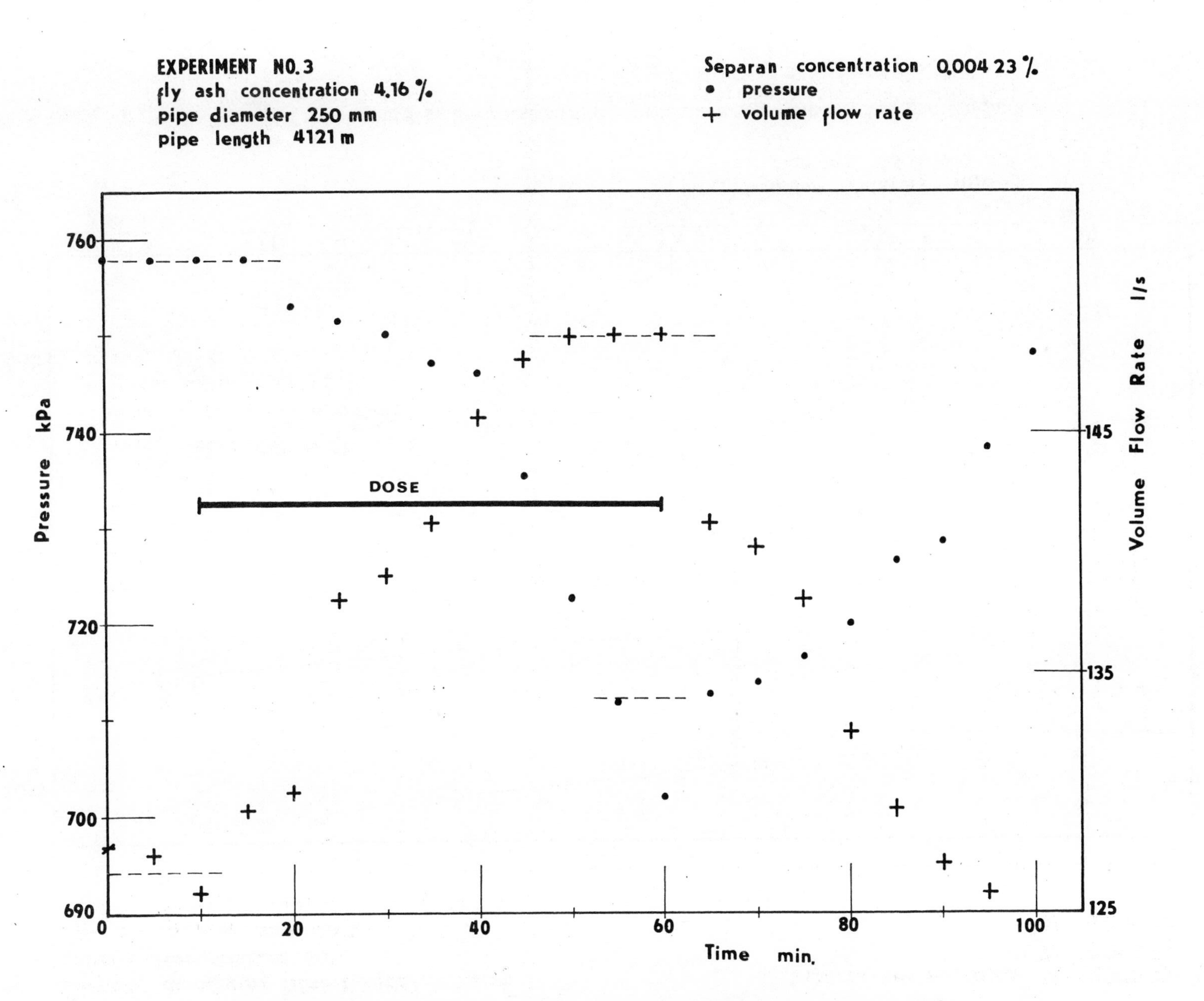

Fig. 5. Experiment No. 3 - volume flow rate and pressure gauge record

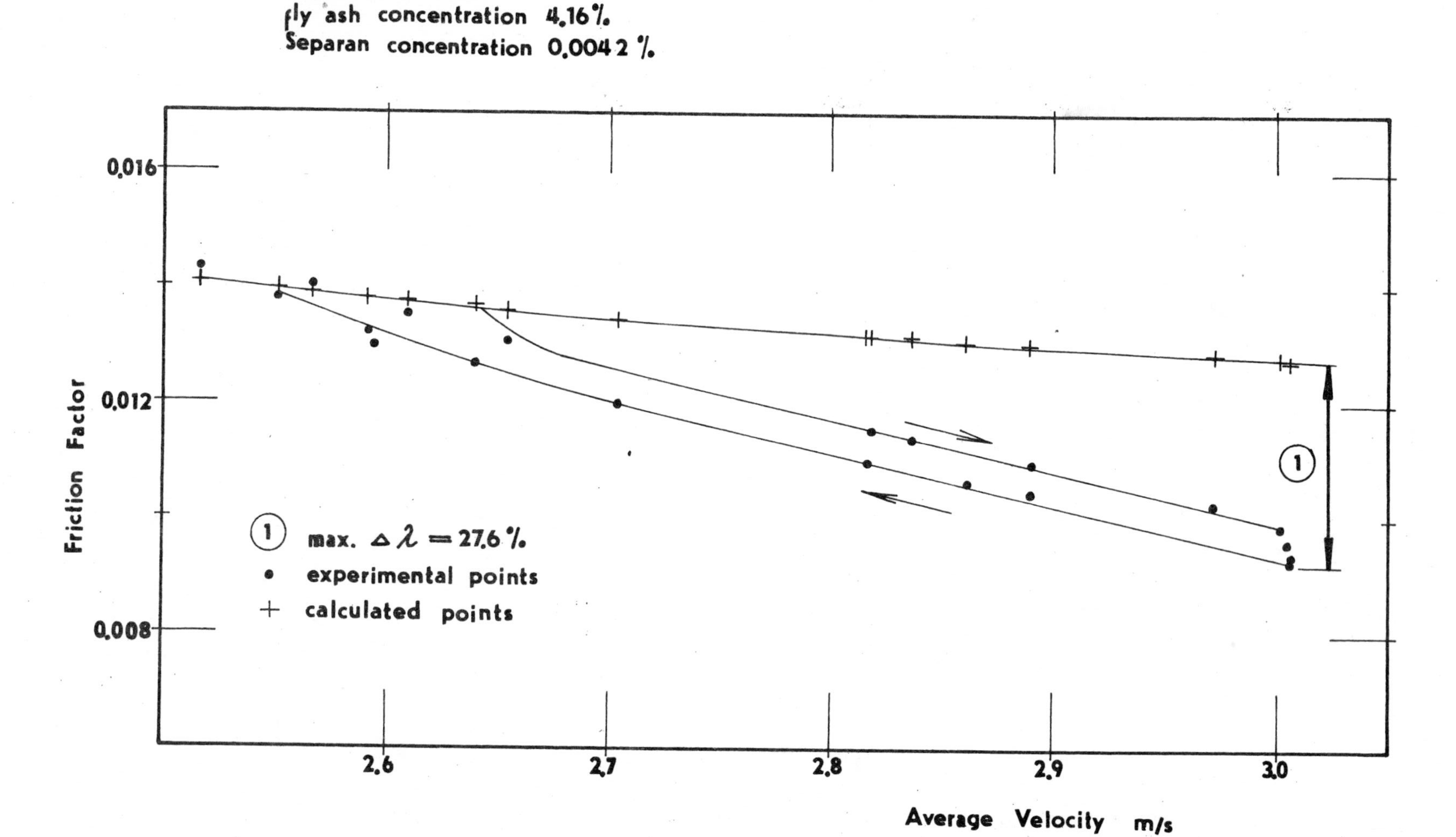

Fig. 6. Experiment No. 3 - relationship between average velocity and friction factor

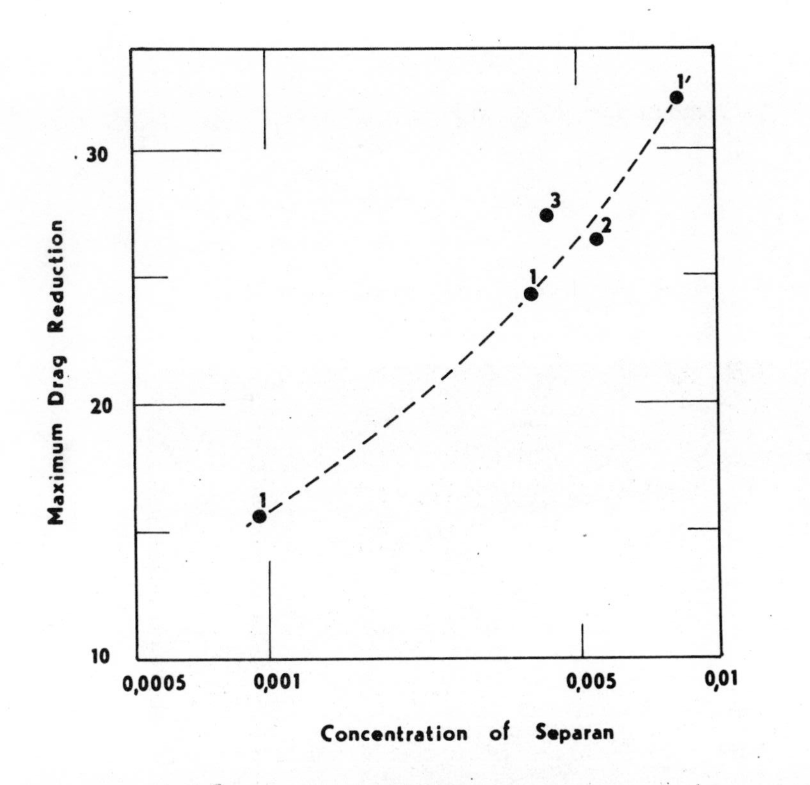

Fig. 7. Relationship between concentration of Separan AP 273 and maximum drag reduction from experiments

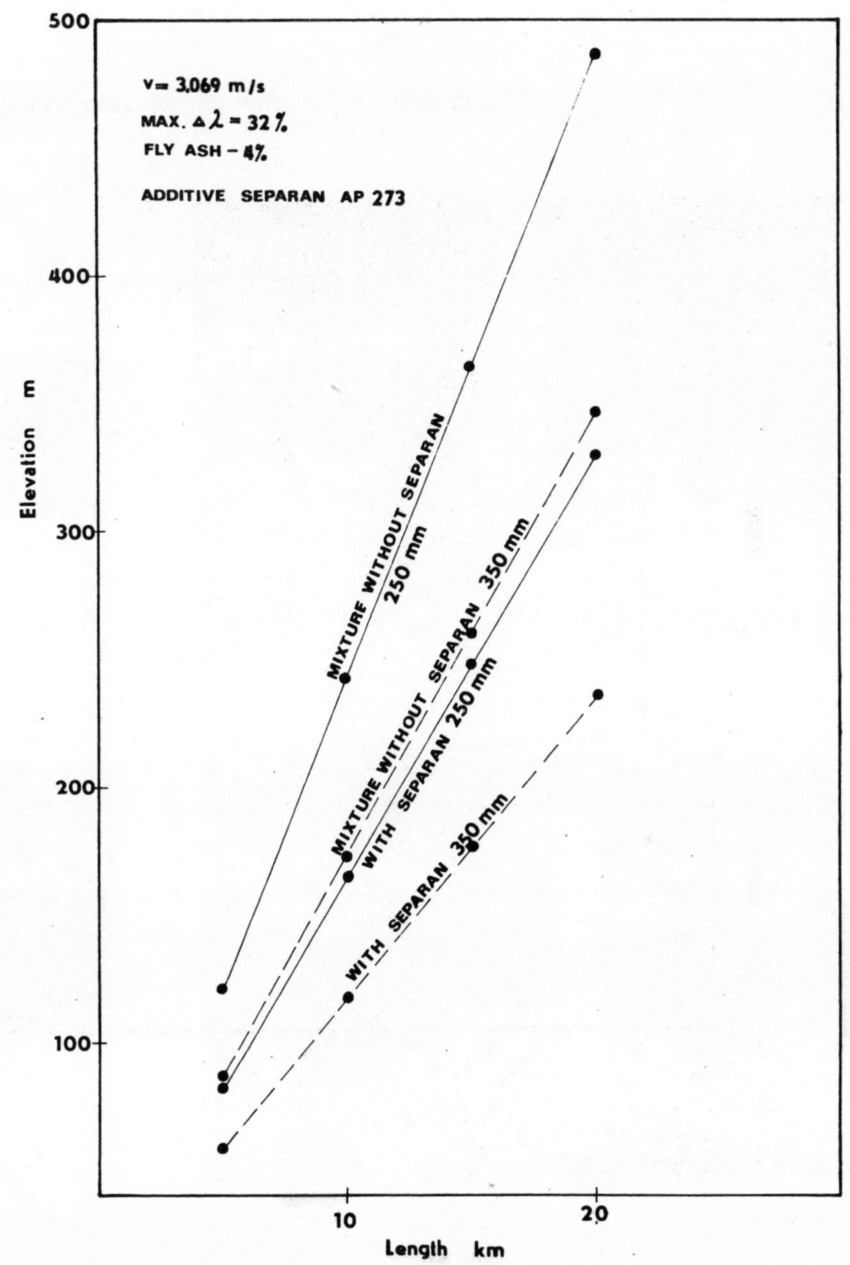

Fig. 8. Change of elevation due to adding Separan AP 273 ▶

Fig. 9. Volume measurement of flow rate

Fig. 10. Dosing pump

Second International Conference on

Drag Reduction

August 31st - September 2nd, 1977

PAPER B4

FEASIBILITY OF UTILIZATION OF DRAG REDUCING ADDITIVES IN HYDRAULIC MACHINES

E. Bilgen, Ph.D. and P. Vasseur, Ph.D.

Ecole Polytechnique de Montrêal, Canada

Summary

The feasibility of using friction reducing additives in hydraulic machines and hydraulic power stations is studied and economical aspects are discussed. Either using friction reducing agents premixed in the entire circulating water system or injecting into selected components are considered. The components considered were penstock, turbine spiral case, band and crown chambers, wicket gates, runner, band and crown seal regions and draft tube. The theoretical findings are compared with experimental results obtained in either system. The first was a Francis type pump-turbine arranged as a pump working with a friction reducing additive premixed in the entire circulating water supply. The second was a especially designed model turbine with high friction losses in selected components. Friction losses were measured with injection of additives into various components and compared to those measured before and after injection. This enabled to measure and evaluate quite accurately the power gains due to injection. It was found that the theoretical predictions compared well with the experimental results. The scale effect and step-up for prototypes are also discussed.

Held at St. John's College, Cambridge, England.

Organised and sponsored by BHRA Fluid Engineering, Cranfield, Bedford, MK43 0AJ.

NOMENCLATURE

a	disk or cylindrical surface radius (m)
b	height of cylindrical surface (m)
C_M	moment coefficient
d_h	hydraulic diameter (m)
D	drag reduction; characteristic diameter
F	turbulent drag reduction factor
L	characteristic length (m)
M	moment exerted by a disk or cylindrical surface (Nm)
N	power (W)
P	pressure head (N/m^2)
q	polymer injection rate
Q	volume flow rate (m^3/s)
Re	$= \rho \bar{u} d_h/\mu$ pipe Reynolds number in Equations 1 to 3
Re	$= \rho\omega a^2/\mu$ disk Reynolds number in Equations 8 to 13
Re	$= \rho\omega at/\mu$ Couette Reynolds number in Equations 17 to 19
S	disk clearance (m)
t	gap width, radial clearance (m)
$\bar{u}$	mean fluid velocity (m/s)
W	power (W)
Δ	increment or difference
η	efficiency
λ	coefficient of resistance in pipes
μ	dynamic viscosity (kg/m.s)
π	= 3.14159
ρ	mass density (kg/m^3)
ω	angular velocity (s^{-1})

SUBSCRIPT

c	cylinder
ℓ	laminar flow
m	model
o	water, solvent
p	prototype
pv	turbulent flow
1	pipe flow
2	disk flow
3	Couette flow

INTRODUCTION

The use of polymer additives and small particles such as fibres in turbulent flow of water or other solvents is known to reduce the friction in various geometries. When and if these additives are used in an existing flow system, its flow capacity or the efficiency of transportation can generally be increased. On the other hand, in a new flow system, the initial investment can be reduced by reducing the size for the same output. The possible areas of application would be liquid and solid transport in pipe lines (Ref. 1), fire-fighting systems (Ref. 2), torpedoes and ships (Refs. 3 and 4), and rotating surfaces of hydraulic machines (Ref. 5). Any specific application of drag reducing additives requires information about the suitable additives and their optimum utilization conditions such as concentration, shear stability, geometrical and dynamic conditions as well as mixing injection techniques (Ref. 6) and sealing of machine components (Ref. 7). In this study, the feasibility of using such additives in hydraulic machines and hydraulic power stations is considered. The additives can be used as premixed in the entire circulating water system such as a pump-turbine system. In this case, an optimum concentration for maximum drag reduction must be maintained since due to shear and biological degradations of additives, actual concentration required may be higher than the optimum concentration with fresh additives; the concentration may change also due to gain of fresh water and losses. In a closed system such as pump-turbine installation, an important parameter would be protection of water life and antipollution measures. An other possibility is to inject the concentrated solutions or suspension of drag reducing additives into selected components of an hydraulic power station. The selected components may be penstock, turbine spiral case, band and crown chambers, wicket gates, runner, band and crown seal regions and draft tube. In that case, if a recovery of additives cannot be achieved, the additives will mix into the downstream water and eventually pollute the rivers, lake and sea. Fortunately, the concentration required to obtain a drag reduction are usually low for the long chain polymers and these when mixed in the working fluids, will result in small concentration, usually less than the maximum concentration permitted by the Food and drug administrations and utilized in the food and drug industry. However, this aspect should be carefully studied before any full scale application.

In the following section, the maximum drag reduction conditions are considered theoretically. Then, the experimental results obtained in either system are discussed in the light of theoretical findings.

THEORETICAL

In order to apply the drag reducing additives in hydraulic machines and hydraulic power stations in general, information about pipe flow, Couette and disk flows of dilute polymer solutions is required.

The maximum drag reduction is defined as the relative reduction of drag or friction in water flow by additives:

$$D = \frac{\text{reduction of drag or friction}}{\text{total drag or friction in water flow system}}$$

Pipe flow - the maximum drag reduction or ultimate assymptote in pipe flow is given as a logarithmic law by Virk et al. (Ref. 8). However, for the present purpose, a simpler power law relation will be preferred (Ref. 9). Such a relation was,

$$\lambda = 2.260\ \mathrm{Re}^{-0.560} \tag{1}$$

This is compared to the power law for water flow,

$$\lambda_o = 0.316\ \mathrm{Re}^{-0.25} \tag{2}$$

The drag reduction can be calculated as,

$$D_1 = \frac{\lambda_o - \lambda}{\lambda_o} = 1 - 7.15\ Re^{-0.31} \tag{3}$$

In the estimate of drag reduction in penstock, casing, wicket gates and runner, equation (3) can be used. The pressure drop is,

$$\Delta P = \lambda \frac{L}{d_h} \frac{1}{2} \rho \bar{u}^2 \tag{4}$$

hence, the relative pressure drop , or pressure gain becomes,

$$\frac{\Delta P}{P} = \frac{\lambda_o - \lambda}{\lambda_o} = D_1 \tag{5}$$

The power loss due to friction losses is calculated as,

$$N = Q.P \tag{6}$$

from equations (5) and (6), the relative power gain would be,

$$\frac{\Delta N}{N} = \frac{\Delta P}{P} = \frac{\lambda_o - \lambda}{\lambda_o} = D_1 \tag{7}$$

Disk flow - the turbulent flow characteristics of drag reducing fluids around an enclosed disk were studied earlier both theoretically and experimentally and the following expression was found (Ref. 10):

$$C_M = 0.360\ Re^{-0.359} \tag{8}$$

By considering the effect of clearance ratio, (S/a), in flow with separate boundary layers, the moment coefficient was,

$$C_M = 0.548\ (S/a)^{0.06188}\ Re^{-0.359} \tag{9}$$

For water flow around an enclosed disk, similar expressions were (Refs. 11 and 12),

$$C_{Mo} = 0.0622\ Re^{-0.20} \tag{10}$$

and

$$C_{Mo} = 0.102\ (S/a)^{1/10}\ Re^{-0.20} \tag{11}$$

The drag reduction can be evaluated as,

$$D_2 = \frac{C_{Mo} - C_M}{C_{Mo}} = 1 - 5.788\ Re^{-0.159} \tag{12}$$

and

$$D_2 = 1 - 5.373\ (S/a)^{-0.0381}\ Re^{-0.159} \tag{13}$$

The frictional moment upon a rotating disk is determined from:

$$2M = C_M \frac{1}{2} \rho \omega^2 a^5 \quad (14)$$

and the power is

$$N = \omega M \quad (15)$$

Hence, using equations (12), (13) and (14), (15), the relative power gain would be,

$$\frac{\Delta N}{N} = 1 - \frac{C_M}{C_{Mo}} = D_2 \quad (16)$$

The estimate of drag reduction in disk portions of shroud spaces can be estimated using equations (12), (13) and the relative power gain with equation (16).

Couette flow - Couette flow of drag reducing fluids was studied earlier (Ref. 13). The following expressions apply:

$$C_M = 0.9528 \, (t/a)^{0.50} \, Re^{-0.45} \quad (17)$$

For the same conditions, the Couette flow of water was also studied (Ref. 14) and the following expression applies:

$$C_{Mo} = 0.065 \, (t/a)^{0.3} \, Re^{-0.20} \quad (18)$$

The drag reduction can be calculated as,

$$D_3 = 1 - 14.65 \, (t/a)^{0.20} \, (\rho \omega a t/\mu)^{-0.25} \quad (19)$$

or by using disk Reynolds number, $Re = \rho \omega a^2/\mu$,

$$D_3 = 1 - 14.65 \, (t/a)^{-0.05} \, (\rho \omega a^2/\mu)^{-0.25} \quad (19)'$$

The drag reduction on cylindrical rotating surfaces can be estimated by using (19) or (19)´.

Similarly, the frictional moment upon a rotating cylindrical surface is determined as,

$$M_c = C_M \, \pi \, \rho \, \omega^2 \, b \, a^4 \quad (20)$$

and the power is calculated with equation (15). Hence, by using (19),(19)´, (20) and (15), the relative power gain would be,

$$\frac{\Delta N}{N} = 1 - \frac{C_M}{C_{Mo}} = D_3 \quad (22)$$

The hydraulic machine efficiency - the efficiency of the machine, pump or turbine, is defined as the ratio of the power measured to theoretical hydraulic power.

$$\eta = \frac{W}{Q \cdot H} \quad (23)$$

EXPERIMENTAL*

i.- Test with premixed additives - tests were performed using friction reducing additives Polyox Coagulant premixed in the entire circulating water system. The first test was with clear water then by circulating 40 and 120 w.p.p.m. Polyox Coagulant through the system while executing the test procedure for measuring of a model Francis pump-turbine. The capacity of the system was 158,000 kg of water. The concentrated solution of polymer was injected into the system and distributed uniformly by circulating the water at about 0.23 m^3/s which circulated the complete water quantity in the system in about 12 minutes. The total time of injection was approximately 3 hours and the amount of Polyox 6,320 g and 18,960 g which equal 40 and 120 w.p.p.m. respectively. The flow rate was measured using a weir and a venturi, both studied earlier (Ref. 15). The duration of each test was approximately 4 hours. On completion of Polyox test, the system was recirculated for the next 14 hours. The system was then drained, washed out and refilled with water and spot checks taken confirmed that no residual effects existed from the additive. Testing was again resumed with the pump set at the same testing speed. The throat diameter was 21.6 cm; the wicket gate was set at 16° and the pump was run at 900 rev/min. The control methods of actual concentration were as described in Appendix "A".

The results are presented on Figure 1 as efficiency versus volume flow rate and pressure head developed versus volume flow rate.

ii.- Injection of concentrated solutions - tests were performed with injection of two kinds of polymers into various components of a model turbine. The polymers were Polyox Coagulant and Jaguar 508. Polyox Coagulant with an approximate molecular weight of 5×10^6 was prepared and injected at 1000 w.p.p.m. while Jaguar 508 with 2×10^5 molecular weight was at 5000 w.p.p.m. The selected components were band chamber, crown chamber, wicket gates and runner, band and crown seal regions. As with all model tests on completely homologous runners the maximum recoverable shroud chamber friction represents a very small part of the turbine output, this being both difficult and costly to measure accurately. In this model, the distributor height was reduced by the ratio 1: 16.35 hence increasing the ratio of shroud friction to approximately 15% of turbine output: consequently the efficiency of the model turbine was approximately reduced to 50%.

The water was supplied continuously from a 10 cm diameter city main water line via a 10 cm diameter hose, and the water at the outlet of the turbine was discharged into the sewer; so that the water working through the model turbine had been kept clean throughout the test period. A centrifugal pump of 40 HP supplied water to the main body of pipework; the head and the flow required by the turbine were regulated by a by-pass valve.

The water passed through the casing and the runner and discharged to the sewer via vertical draft tube. These passages were reduced from homologous sizes to handle the lower flow determined by the small distributor height and runner passages. As a result, the actual throat diameter was less than the calculated, being 12.70 cm diameter but the throat diameter for all calculations was assumed to be 35.56 cm.

The 49.47 cm outside diameter model runner had seal, flinger and runner outside diameter clearance, also shroud spaces homologous to a prototype machine to establish homologous behaviour. The runner was close coupled to a 20 HP squirrel cage motor, which, when running above synchronous speed, operated as a field excited generator, the output of which may be measured in a Watt hour meter. The water required at the floating seal was taken from a connection at the entrance of the casing, passed through a filter and then through separate band and crown measuring orifices and regulating valves and injected to the floating seal position through a manifold.

---

* Experimental studies were carried out in the Hydraulic Laboratory, Dominion Engineering Works Ltd., Montreal, Quebec, Canada.

The polymeric solutions 1000 w.p.p.m. for PEO and 5000 w.p.p.m. for Jaguar 508 were prepared in a 200 ℓ drum and then were drained into the pressurized blow down from which the polymeric solutions required for injection were supplied to band and crown shroud injection points. The injection rate was measured by weighing the blow down tank for every test.

Transparent ring/covers were provided at band and crown for observation of shroud chambers at various stages of injection. These also allowed the water ring at the runner outside diameter to be observed, when the test of the components other than the shroud chambers were performed with full aeration.

Tappings were provided at inside and outside diameters of flinger, also at outside diameter of runner. These allowed measurement of pumping conditions within the turbine shroud spaces and were displayed on water and mercury manometers situated on the gauge board.

The optimum seal water flow rate was determined by using of salt method described in Appendix "A". The conductivities of the solutions in the shroud spaces, at runner outer diameter and in seal for various seal water flow rate were measured and the corresponding concentrations were plotted: the seal flow rate at maximum concentration was set during the tests. The polymer concentrations in the chambers were controlled as described in Appendix "A".

Estimated accuracies were ± 1% for main flow rate, ± 0.5% for total head, ± 1.5% for water power, ± 0.4% for break horsepower and ± 1.9% for the efficiency.

The measured and calculated values as power gain versus polymer injection rate are shown in Figures 2 - 4.

DISCUSSION

Based on the equations presented, power gains in various components of the model turbine are computed and shown in Table 1. The components considered were band and crown chambers with and without cross flow of water, runner, runner + wicket gates, band seal, crown seal + configuration. The penstock, spiral casing and draft tube are not computed and shown since the power gains are straightforward and would be as much as in round conduits.

Figure 1 shows that by using drag reducing fluids in the entire working fluid such as in a pump-turbine installation, a considerable power gain can be achieved. The gain with the pump-turbine model is more than 4 efficiency points with 120 w.p.p.m. solution. An other important finding is that working with fully mixed solutions, the best efficiency point is shifted to higher volume flow rate values and the efficiency curves are flatter at higher volume flow rates. This may be important for two reasons: first, peak demands can be met easier and demand variations do not affect the machine performance; second, using drag reducing additives becomes more economical at peak demands since the power is scarce at that time.

The theoretical water resistance and possible power gains with additives for various components are shown in Table 1. Figures 2, 3 and 4 show the experimental results which may be compared with the theoretical predictions. It can be seen that there is good agreement at high injection rates in Figures 2 and 3. On the other hand, there is quite a discrepancy in case of wicket gates + runner and seal regions (both tested with aerated shroud spaces). The predicted power gain for the wicket gates (0.617 kW) + runner (0.598 kW) are too high, the reason for which may be too much dilution. Actual power gain is in the order of 0.40 kW. The computed power gains for band and crown seals are about 50% lower than the measured values, perhaps due to additional drag reduction in part of the shroud spaces and in the draft tube.

Scale effect and step-up - Reynolds numbers in prototype concerning the friction losses are several times those of model. Consequently, the dilution of polymers will probably take place in a shorter time, but entanglement and degradation if any will be more due to intensive turbulence.

In view of the discussions concerning the polymer consumption, it should be noted that to estimate the step-up is quite difficult as it may be very much dependent on geometry of the components and hydraulic conditions. Therefore, the probable ways to step-up may be:

i.- the polymer consumption may be dependent on the total surface of components in which case the polymer consumption can be expressed as:

$$q \sim D^2$$

ii.- it may be dependent on the volume of components, in which case the polymer consumption becomes:

$$q \sim D^3$$

iii.- finally, the polymer consumption may depend on the leakage rate or unit flow rate. In this case, for similar machines, the unit polymer consumption will be the same. Consequently, the polymer consumption for similar machine can be expressed as:

$$q \sim D^2 \sqrt{P}$$

The last seems to be the most reasonable way for step-up. The polymer consumption can be related to the power increase obtained by injecting 1 kg of polymer since the power gain per kg of polymer versus injection rate should be similar in similar machines. Figure 5 which is prepared from Figures 2 and 3 shows this relation for a 45 MW prototype machine. Theoretically, the ratio $(D^2\sqrt{P})_p/(D^2\sqrt{P})_m$ should be calculated by taking the characteristic diameter as seal dimensions at the runner outer diameter. In this way, the polymer consumption will only be dependent on the leakage rate which is very much related to shear rate rather than the purely geometrical similarity.

The required investment for injection of polymers and control is very small, hence the energy cost will depend on the polymer consumption. It can be seen that the energy cost at very small injection rates will be in the order of 1¢/kWh, which may be economical at peak periods of demand.

CONCLUSIONS

By using the theoretical and experimental results published earlier and experimental results obtained with models turbines, the power gains due to utilization of drag reducing additives in flow systems, are analysed. It appears that the use of friction reducing additives in hydraulic machines and hydraulic power stations is quite feasible. The additional investment is negligible. The power gain may be several hundred kW per kg of polymer at very low injection rates hence, it may be economical at peak demands.

REFERENCES

1. Mysels, R.J., U.S. Patent Z 492 173, (Dec. 27, 1949).
2. Fabula, A.G., "Fire Fighting Benefits of Polymeric Friction Reduction", Trans. ASME, J. Basic Engineering, pp. 453-455, (Sept. 1971).
3. Crawford, H.R., U.S. Patent 3230919, (Jan. 25, 1966).
4. Gray W.O. and Hilliard, B.A., U.S. Patent 3 289 623, (Dec. 6, 1966).
5. Sproule, R.S., U.S. Patent 3 398 696, (Aug. 27, 1968).
6. Bilgen, E. and Ackungor, A.C., Can. Patent 985492, (March 16, 1976).

7. Bilgen, E., On the Feasibility of Straight, Labyrinth and Screw Seals in Hydraulic Machines, Proc. 7th ICFS, Paper A2, pp. 31-42, Organized by BHRA, University Nottingham, (Sept. 1975).

8. Virk, P.S., Mickley, H.S. and Smith, K.A., "The Ultimate Asymptote Mean Flow Structure in Tom's Phenomenon", Trans. ASME, J. Appl. Mech., pp. 488-493, (June 1970).

9. Bilgen, E., "On the Stability of Viscoelastic Flow Due to a Rotating Disk", Chem. Engineering Progress, Vol. 67, No. 111, pp. 74-84, (1971).

10. Bilgen, E., and Vasseur, P., "On the Friction Reducing Non-Newtonian Flow Around an Enclosed Disk", Trans. ASME, J. Appl. Mech., Vol. 41-E, No. 1, pp. 45-50, (March 1974).

11. Schultz-Grunow, F., "Der Reibungswiederstand Rotierender Scheiben in Gehausen", Z.A.M.M., Vol. 15-4, pp. 191-204, (July, 1935).

12. Daily, J.W. and Nece, R.E., "Chamber Dimension Effects of Induced Flow and Frictional Resistance of Enclosed Rotating Disks", Trans. ASME, J. Basic Engineering, Vol. 82-D, pp. 217-232, (March 1960).

13. Bilgen, E., and Boulos, R., "Turbulent Flow of Drag Reducing Fluids Between Concentric Rotating Cylinders", Trans. CSME, Vol. 1, No. 1, pp. 25-30, (March 1972).

14. Bilgen, E., and Boulos, R., "Functional Dependence of Torque Coefficient of Coaxial Cylinders on Gap Width and Reynolds Numbers", Trans. ASME, J. Basic Engineering, Vol. 95-I, No. 1, pp. 122-126, (March 1973).

15. Bilgen, E., "Effect of Dilute Polymer Solutions on Discharge Coefficients of Fluid Meters", ASME Paper 70-FE-38, pp. 8, (March 1970).

16. Stratta, J.J., Private Communication, (July 25, 1967).

TABLE 1

Turbine Component	Water Resistance kW	Power Gain kW
Band chamber	1.442	0.791
Band chamber + seal + runner O.D.	1.742	0.991
Crown chamber + seal + runner O.D.	1.114	0.520
Crown chamber + seal + crown configuration	1.379	0.623
Wicket gate + runner	1.603	1.216
Band seal	0.471	0.228
Crown seal + crown configuration	0.635	0.268

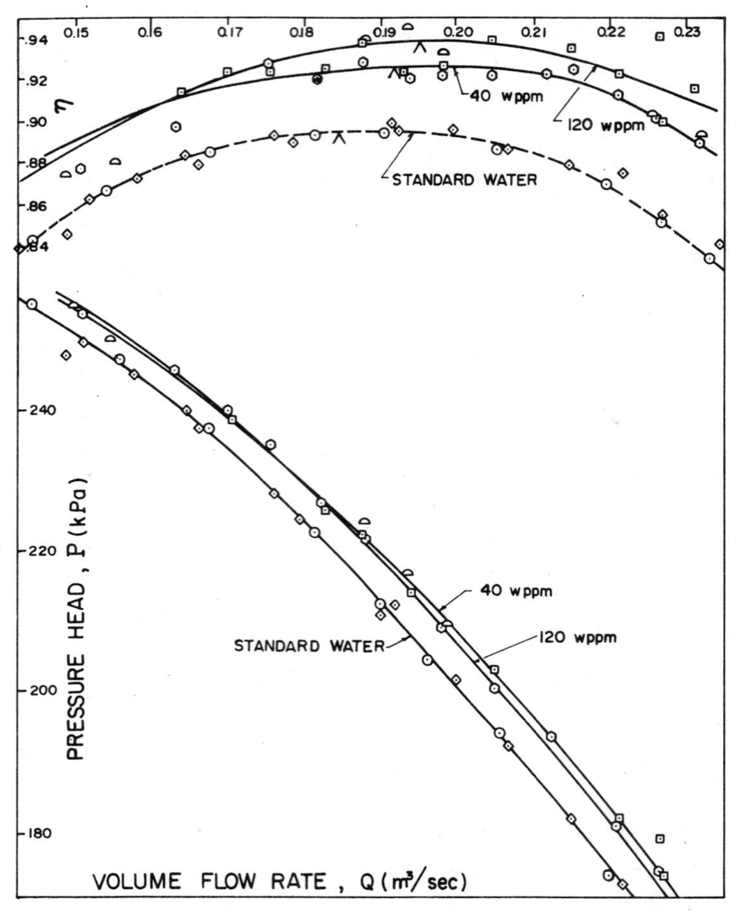

Fig. 1. Experimental results with a Francis pump-turbine working as a pump

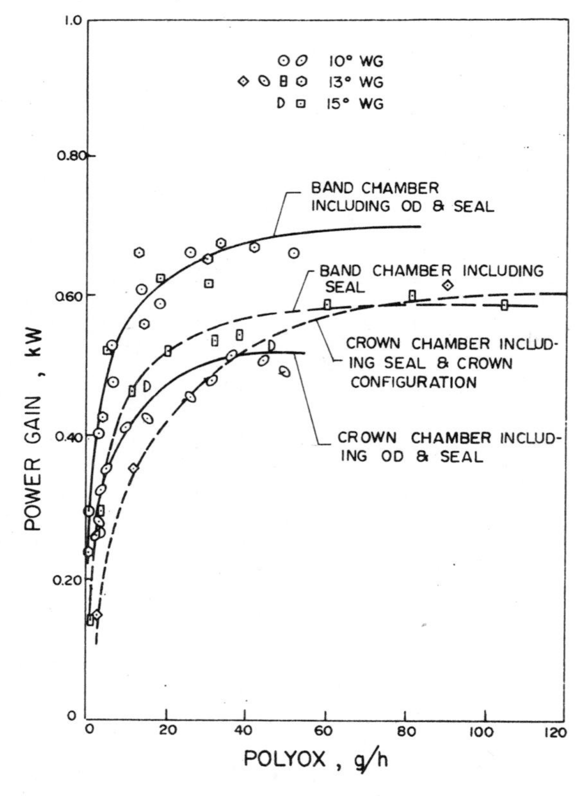

Fig. 2. Experimental power gains with Polyox injection in a model turbine. Turbine components: band and crown chambers.

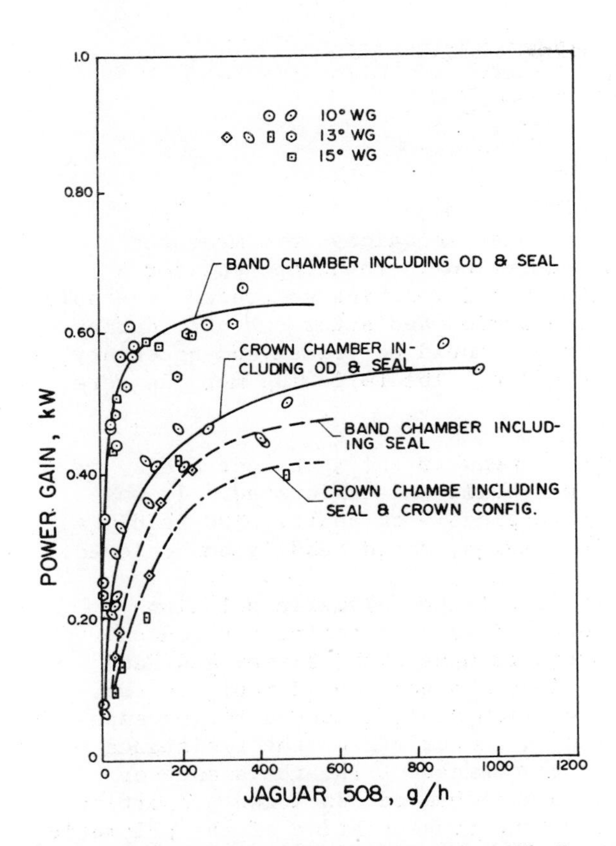

Fig. 3. Experimental power gains with Jaguar injection in a model turbine. Turbine components: band and crown chambers.

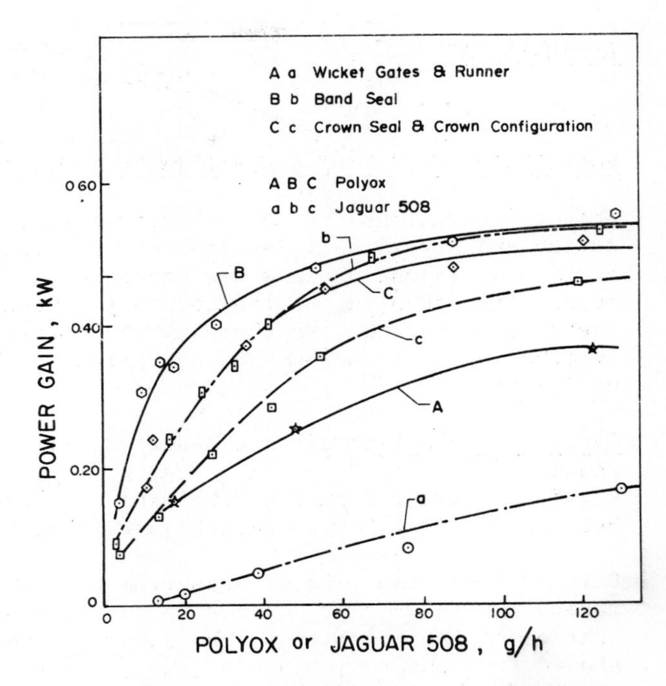

Fig. 4. Experimental power gains with polymer injection into the components of wicket gates, runner and seals.

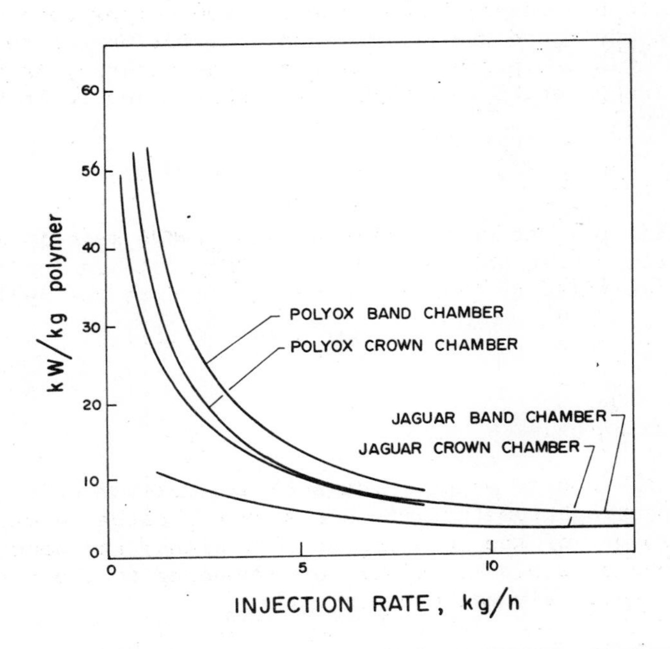

Fig. 5. Power gain - polymer consumption relation for step-up to a prototype turbine.

APPENDIX "A"

POLYMER CONCENTRATION CONTROL INSTRUMENTS

The concentrations and the effectiveness of the polymer solutions were measured during and after the injection period. The concentration in the component for a particular injection rate was continuously checked until reaching saturated or stable conditions, then during the constant concentration period and after cut off, during the dilution period, so that any concentration change could be traced and necessary instruction for readings and operations could be given. The following methods were utilized:

Capillary tube method: a compact capillary tube to measure the turbulent drag reduction factor, $F = (\lambda_{pv} - \lambda) / (\lambda_{pv} - \lambda_{\ell})$, was built and calibrated. It was a 2.57 mm inside diameter tube with two pressure tappings 5 cm apart. The readings between Re = 4,000 to 20,000 just in transition region, could readily be achieved.

Salt method: the idea of measuring the conductivity of the polymeric solutions together with another conductive additive has been utilized to define the concentration of the solutions. The laboratory measurements made with polymer and NaCl showed that the resistance of the polymeric solutions changes considerably in the range of interest and on the other hand for one particular NaCl concentration especially above 20 w.p.p.m., the existance of polymer does not effect the resistance of the solution of (polymer + NaCl), i.e. either the change of the resistance of polymer solutions or the high conductivity of an other chemical in known proportion with the polymer may be utilized to relate to the true concentration of the polymeric solutions. During the model test, the method by using the conductive chemical in a known proportion with the friction reducing polymers has been utilized. A Beckman conductivity meter and two inline cells together with an automatic temperature compensator were used to measure the conductivities during the calibration and test period.

Light scattering method: to determine trace amounts of Polyox in water, the absorbance of the solution can be measured. This method is based upon a patented procedure of Union Carbide (Ref. 16). Best results were obtained when the absorbance values were kept in the 0.1 to 0.3 range; hence, the optimum Polyox concentration range was 20 to 40 w.p.p.m. It was found that the standard deviation for ten test results at the 84 w.p.p.m. level was 3 w.p.p.m.; in general, the accuracy in determining the concentration over a range of 25 to 400 w.p.p.m. Polyox should be within 10% relative of the true value.

Reagents used:

a) Poly (acrylic acid) reagent which forms and insoluble association product with Polyox. The particle size was controlled by Tergitol Non-ionic NPX and the suspension was stabilized by gum arabic which prevents the agglomeration of the particles.

b) Source water.

c) Polyox Water soluble resin.

Then the absorbances of (source water + reagent) and (Polyox solution + reagent) were measured through a spectrophotometer, the net sample absorbance was obtained by subtracting the absorbance of the solution (colour blank) and source water (reagent blank). The corresponding w.p.p.m. Polyox corresponding to the net sample absorbance was read from a previously calibration curve.

For this purpose a Beckman Spectrophotometer Calorimeter "Spectronic 20" was used. This instrument can read in optical density or % transmittance; it uses diffraction grating and can work in wave length range of 340 to 650 μm with a band width of 20 μm.

Second International Conference on

Drag Reduction

August 31st - September 2nd, 1977

THE STUDY OF DRAG REDUCTION USING NARROW FRACTIONS OF POLYOX

N.S. Berman and J. Yuen,

Arizona State University, U.S.A.

Summary

Polydisperse samples of Polyethylene oxide are generally used in drag reduction experiments leading to results influenced by the molecular weight distribution. One way to obtain a narrow distribution is to use gel fractionation. In this work a 50 mm diameter by 500 mm column was packed with a Agarose gel (Bio-Gel A-50m), and relatively concentrated solutions of Polyox grades N-80, WSR 205 and WSR 301 were eluded through the column. The fractionated samples were analyzed for amount of polymer, instrinsic viscosity and diluted for drag reduction tests. Flow rate and pressure drop were measured in two different size tubes for several fractions of the original solutions and the results in terms of friction factor curves were examined for the dependence of drag reduction on concentration and molecular weight. Correlations of the slope increment of the drag reduction trajectories with concentration to the two-thirds power and intrinsic viscosity were obtained. These results suggest that if the molecules uncoil and align in high strain low vorticity areas of the turbulent flow they do not expand completely. Such a limit due to molecular interactions explains why increased expansion in strain fields with longer persistance times cannot be observed for these molecules.

Held at St. John's College, Cambridge, England.
Organised and sponsored by BHRA Fluid Engineering, Cranfield, Bedford, MK43 0AJ.

NOMENCLATURE

B	-	constant in logarithmic law
c	-	concentration
c_x	-	concentration to fill the volume
f	-	friction factor, $8u_\tau^2/U^2$
f_s	-	friction factor of solvent
$\overline{k}$	-	elongational rate
ℓ	-	length
N	-	number of mass points in a molecule
M	-	weight average molecular weight
R	-	Reynolds number, $D\overline{U}/\nu$
T_1	-	Rouse terminal relaxation time
u_τ	-	shear velocity, $\sqrt{\tau/\rho}$
$u_{\tau o}$	-	onset shear velocity
δ	-	slope increment
$[\eta]$	-	intrinsic viscosity
ν	-	kinematic viscosity
ν_{eff}	-	effective viscosity of elongated molecules
ρ	-	density
τ	-	shear stress

INTRODUCTION

An understanding of drag reduction depends upon a knowledge of the contribution of a single molecule to the phenomenon. Reviews of the current state-of-the-art have recently been presented by Virk (Ref. 1) and Little *et al*. (Ref. 2). Unfortunately all of the experiments cited in these reviews used polydisperse polymer solutions with unknown molecular weight distributions. The concentration, molecular weight and other molecular properties attributed to these samples are not representative of the molecules effective in drag reduction. In this work we report pipe flow drag reduction experiments for narrow fractions of two commercial Polyox grades. The results are used to further the understanding of the behavior of individual molecules.

MOLECULAR MECHANISM FRAMEWORK

One previous study using fractionated Polyox has been conducted (Ref. 3). This study and the discussion by Little (Ref. 2) compared drag reduction at constant Reynolds numbers based on the bulk velocity. Virk (Ref. 1) on the other hand has developed a scheme of correlation which is independent of Reynolds number and which leads to much more information on the individual molecules. Virk's analysis begins from the observation that on Prandtl-von Kármán coordinates ($1/\sqrt{f}$ vs. $\log R\sqrt{f}$) most drag reduction trajectories can be represented by straight lines.

$$1/\sqrt{f} = a \log R\sqrt{f} + b \tag{1}$$

The Newtonian line is

$$1/\sqrt{f_s} = 2 \log R\sqrt{f} - 0.8 \tag{2}$$

Then

$$\begin{aligned} 1/\sqrt{f} - 1/\sqrt{f_s} &= (a - 2) \log R\sqrt{f} + b - 0.8 \\ &= \delta \log [R\sqrt{f}/(R\sqrt{f})_o] \end{aligned} \tag{3}$$

where δ is termed the slope increment and $(R\sqrt{f})_o$ is the "onset" value found from the intersection of eqs. 1 and 2. Equation 3 is also the displacement of the pipe core log profile, ΔB.

Virk (Ref. 1) finds that the slope increment, δ, is proportional to concentration to the one-half power and molecular weight to the first power for polymers with the same backbone units. Such a correlation using δ alone avoids the problem with the proportion of molecules actually involved in drag reduction only if the same sample is used for all experiments. Therefore we can assume the concentration dependence has some significance, but the molecular weight which depends on different samples does not. Hunston (Ref. 4) and Berman (Ref. 5) have shown that drag reduction is dominated by small amounts of the highest molecular weights. In the initial studies leading to this work (Ref. 5), 2% of the highest molecular weight tail of a sample of Polyox N-80 gave essentially the same drag reduction as the unfractionated material.

If we propose a relationship between drag reduction, or more specifically ΔB, and the viscosity increase of random coiled molecules in elongational flows as suggested by Lumley (Ref. 6), we can proceed to develop a molecular theory of drag reduction for small amounts of drag reduction.

$$\text{Let } \Delta B = A \frac{\nu_{eff} - \nu}{\nu} \tag{4}$$

Then from Hassager (Ref. 7) for a bead rod molecular model subjected to pure strain an asymptotic solution for high strain is

$$\Delta B = A\ c\ [\eta]\ N\ \{1 - 0.8/T_1\ \bar{k} - \ldots\} \tag{5}$$

where T_1 is the Rouse Terminal Relaxation time of the molecule and $\bar{k}$ is the elongational rate. When eq. 5 is plotted on Prandtl-von Kármán coordinates, a straight-line portion can be found. Then if T_1 is replaced by an onset time scale and $\bar{k}$ by the strain rate in turbulence the result is similar to eq. 3:

$$\Delta B = A'\ c\ [\eta]\ N \log (u_\tau/u_{\tau o})\ . \tag{6}$$

Comparing eq. 6 with eqs. 5 and 3,

$$0.8/T_1 \sim \nu/u_{\tau o}^2 \tag{7}$$

and

$$\delta = A'\ c\ [\eta]\ N \tag{8}$$

This analysis assumes that the molecules act independently until they are fully elongated. We can speculate as to the result of molecular interactions by referring to Batchelor's treatment of elongated particles in nondilute suspensions (Ref. 8). Little change would be expected if the particles could extend fully and act independently. If not, we might expect that N would be replaced by an effective extensional length ratio. Such a ratio is proportional to $c^{-1/3}$ when the particles have just enough room to rotate. The molecular weight dependence essentially disappears except for the intrinsic viscosity. We therefore expect

$$\delta \propto c^{2/3}\ [\eta] \tag{9}$$

when the molecules can extend enough to touch but no farther. Overlapping molecules will lower the exponent of the concentration to the 1/2 power.

To determine the agreement with the molecular model, we have measured δ for fractions of Polyox WSR 205 and 301 for various molecular weights and concentrations. The data show striking agreement with eq. 9 and suggest that the speculative analysis presented above has some merit. In an actual turbulent flow both vorticity and strain are present and the problem is much more complex than the pure strain analysis. We assume that the polymer stretching occurs in vorticity free areas and the persistence time is long enough to stretch the molecules.

EXPERIMENTAL

Fractions of Polyox WSR 205 and 301 were separated by gel chromatography and then diluted for pipe flow tests. Initially 5000 ppm solutions of Polyox 205 and 500 ppm solutions of Polyox 301 were prepared and left to stand for 48 hours. A 50 mm diam. by 500 mm long column was prepared with 1000 ml of gel supported by a coarse glass frit. Bio Rad Laboratories Bio Gel A was used in the column with Type 50m, a 2% Agarose gel, for Polyox 205 and Type 150m, a 1% Agarose gel, for Polyox 301. Approximately 50 grams of the concentrated polymer solution were added to the top of the column and allowed to flow into the gel while removing liquid from the bottom. In order to minimize mixing, the column effluent was closed when the top level reached a preset mark and the gel was allowed to swell up to the mark. Then, the bottom stopcock was opened and water was added slowly to the top of the column to maintain a constant level. During operation the gel contracted and the pressure drop increased when the highest molecular weights were flowing through the glass frit. Samples containing measurable amounts of polymer solution were collected after the first 300 ml, representing the free volume of the column, had eluded. For each run approximately 10 samples of volume 25-50 ml and then larger ones were collected.

The samples were tested for specific viscosity and polymer concentration in dilutions so that the concentration was between 5 and 25 ppm by weight. Concentration measurements were according to Warren Spring Laboratory Method Sheet No. 75. Light transmission at 650 nm after addition of fresh Reinecke salt reagent was compared with a reference. The percent transmission was calibrated in parts per million and new calibration curves were run periodically. The viscosity was measured in a Cannon-Fenske Number 50 viscometer. To obtain the distribution curves in Fig. 1 all samples were tested as soon as possible after collection. For the pipe flow tests part of the sample was tested for concentration and viscosity and the other part was diluted for pipe flow experiments. Only selected samples could be tested so that all pipe flow runs could be completed on the same day as the fractionation.

The pipe flow apparatus was essentially the equipment described by George *et al.* (Ref. 9). For runs in a 5.54 mm diam tube five liters of solution were required. When runs in both a 14.94 mm and the 5.54 mm diam pipes were made, 25 liters of solution were prepared. The dilutions were made immediately after the fraction was collected from the column. Pipe flow runs were made after a 30-minute period of gentle mixing. Pressure drop and flow rate measurements were recorded over a Reynolds number range from 2000 to 40,000 in the smaller pipe and up to 100,000 in the larger pipe. These measurements were made in an unsteady apparatus and at least two overlapping regimes were used to obtain the drag reduction trajectories. Only one set of pressure taps approximately 1.5 m apart were used, and the entrance to both pipes contained a center trip and roughened walls. Solutions were used only once in the blow down system.

EXPERIMENTAL RESULTS

Drag reduction trajectories are shown in Figs. 2 and 3 for unfractionated Polyox 205 and 301 and for typical fractions. The unfractionated curves are for the original 50 grams diluted to 5 liters. Two series of fractionation runs represent the majority of the data examined in the following discussion. The location of the fractions on the distribution is shown in Fig. 4. These distributions differ from Fig. 1 for several reasons: non-uniformity of the original batch, mixing differences, and degradation in the column. It is necessary to characterize all samples tested from each fractionation or to obtain data on adjacent fractions. In order to obtain large enough samples to test in pipe flow the columns were somewhat overloaded leading to overlapping of the molecular weights in adjoining samples. However, the fractionated samples were considerably narrower in molecular weight distribution than the unfractionated samples and contained essentially no low molecular weight material.

The fractionated samples could be characterized by the slope of the drag reduction trajectory, δ, and by the maximum value of ΔB. Results of measurements using a 5.54 mm tube are given in Table 1. In addition, several experiments were run using a 14.94 mm tube. Two sets of curves comparing the two size pipes are shown in Fig. 5. These experiments were run for concentrations between 0.3 and 0.7 ppm and intrinsic viscosities from 4 to 16. The trajectories for the large pipe fell into two classes illustrated in Fig. 5. For the lower values of $c[\eta]$ the two pipes are identical when scaled with u_τ, but for the higher values the smaller pipe gives greater drag reduction.

DISCUSSION

The slope increment results are plotted in Figs. 6 and 7. Slopes of 1/2 for the higher concentrations and 2/3 for the lower fit the data. Another possible empirical correlation shown in Fig. 7 leads to the equation

$$\delta = a\, c\, [\eta]/(1 + b\sqrt{c}) \qquad (10)$$

where the least squares fit for a is 1.88×10^4 and for b, $5.24\times10^3 cm^3/g$, when c is in g/cm^3. Although eq. 10 can be extrapolated to zero concentration giving an intrinsic drag reduction, Fig. 6 indicates that none of the experiments involved isolated individual molecules. Concentrations below one ppm were necessary to find the slope of 2/3 which would indicate the molecules extend until they run into another molecule.

If we use a length of molecule, ℓ, such that one-half of this length defines a volume of rotation, the concentration required to take up the entire volume is approximately

$$c_x \sim 2 \times 10^{-23} \, M/\ell^3 \tag{11}$$

For Polyox we can take ℓ as the product of the length of a chain unit, b_0, and the number of extensions, n. If the molecule can be fully extended, n is the ratio of molecular weight to the molecular weight of the chain unit. In the experiments of this work concentrations are at least a factor of 20 greater than c_x for full extension. This result assumes that the measurements for the lowest molecular weights are accurate but in fact these results are most questionable. The large relative amounts of very low molecular weight material could have contaminated these samples (205 A-6-9) giving a concentration dependence to the 2/3 power for all the data.

Another indication that the effect of single molecules cannot be isolated comes from eq. 10. Comparing eq. 4 with measured turbulent velocity profiles leads to a proportionality factor on the order of 50. Then the constant "a" in eq. 10 can be compared with 50N. The number of structural units, N, is about 10^5, and 50N is much larger than "a." We conclude that extrapolation of the data is meaningless in terms of intrinsic drag reduction. If the increase in extensional viscosity is related to drag reduction, observation of the effect of the extension of a single molecule would require much lower concentrations than we have used. The result from Fig. 6 is that $\delta = 31 \, [\eta] \, c^{2/3}$, and the correlation with the extensional viscosity for independent molecules fails.

Another part of the model, however, shows some agreement. The shape of the curve given by that part of eq. 5 in the brackets fits the low concentration data very well as shown in Fig. 8. It is significant that a similar curve will fit Pitot tube error data as shown by Berman and George (Ref. 10). The curves in Fig. 8 were obtained by matching plots of eq. 5 with different maximum values of ΔB to the data. For the low concentrations of narrow fractions the slope increment by a visual fit of a straight line is twice the maximum ΔB for the plots of eq. 5. The experimental data listed in Table 1 also show agreement between the slope increment and maximum ΔB. The elongational viscosity analysis for stretched molecules given by Hassager (Ref. 7) shows that the ratio of stretched to the unstretched viscosity can be almost independent of N for small amounts of stretching and for large N. Then the viscosity should be linearly proportional to concentration. It appears that the evidence points to another limit to stretching to fit the experimental two-thirds power dependence. For Polyox perhaps the molecules expand until they stick to another head to tail.

The above analysis and Hassager's asymptotic solution apply after the molecules begin stretching. In fact, a minimum of a few percent stretching is necessary to use the equations. The concept of "onset" is difficult to apply because the curves are dependent on concentration, intrinsic viscosity and a molecular time scale. No analysis of onset is possible from the data. Also we have ignored the effect of degradation. Tests of the intrinsic viscosity after flow through the pipe show some degradation mostly occurring in the entrance. It appears that degradation is more severe in the larger pipe and for the larger molecules. The reduced drag reduction in the larger pipe is probably a result of increased degradation. Fig. 5 shows that there are conditions for which the large pipe is equivalent to the small one as well as conditions for which there is considerable difference. Almost all the experiments showed a decrease in drag reduction at the highest shear rates. If we extrapolate back to lower shear rates, a slight increase in slope increment would be expected. Such a change, however, would only bring the correlation shown in Fig. 6 closer to the 2/3 slope. If only a small proportion of the molecules are stretched and these only degrade, the true concentration dependence and the relationship to slope increment would be difficult to deduce from pipe flow data. This is certainly the picture for polydisperse polymers as shown previously (Ref. 5) but such a model for narrow fractions would correspond to a higher slope increment than observed.

The only consistent interpretation of these data is that the presence of a slope increment is due to molecular stretching which is inhibited in a manner dependent upon the effective volume of the molecules filling the entire volume. If the effective volume is the circumscribed sphere of the extended molecule, extensions 100 times less than complete uncoiling are involved in this work. It is not clear whether molecular interactions are necessary for drag reduction or not. These experiments and all others at higher concentrations apparently have always included molecular interactions when random coiling polymers were used. When molecular interactions are absent, we would expect to find a linear concentration dependence and some relationship to the persistence time of large eddies (Ref. 11). Trends in these directions are observed for a partially extended polyelectrolyte (Ref. 12) and for a rodlike molecule DNA (Ref. 13). No slope increment is observed for the molecules which are elongated from the beginning leading to a constant ΔB. The concentration dependence approaches a linear value and also ΔB is proportional to the large eddy scale.

There is some evidence indicating that additional expansion is responsible for the increased drag reduction in the larger pipe but for the present discussion it is sufficient to note that such an effect is not observed for Polyox. Thus the persistence time of the large eddies is not an independent factor for the random coiling polymers suggesting again that a molecular interaction limit is responsible.

CONCLUSIONS

Pipe flow drag reduction experiments using narrow fractions of Polyox show that drag reduction correlates with $c^{2/3}[\eta]$. For concentrations less than one ppm the results suggest that the slope increment on Prandtl-von Kármán coordinates is due to molecular stretching and that molecular interactions are always present for observed drag reduction of Polyox. Concentrations of the order of 0.01 ppm, or less for the highest molecular weights in typical Polyox distributions, are required to see the effect of individual molecules. Many aspects of the drag reduction trajectories correlate well with a model based on the increased elongational viscosity of stretched molecules in high strain rate fields.

ACKNOWLEDGEMENTS

The authors acknowledge the support of the National Science Foundation and the donors of the Petroleum Research Fund administered by the American Chemical Society.

REFERENCES

1. Virk, P.S.: "Drag reduction fundamentals." AIChE Journal, 21, 4 pp. 625-655. (July, 1975).

2. Little, R. C., et al.: "The drag reduction phenomenon. Observed characteristics, improved agents, and proposed mechanisms." I & EC Fundamentals, 14, 4 pp. 283-295. (November, 1975).

3. Morgan, D. T. G. and Pike, E. W.: "Influence of molecular weight upon drag reduction by polymers." Rheol. Acta, 11, 2 pp. 179-184. (February, 1972).

4. Hunston, D. L. and Reischman, M. M.: "The role of polydispersity in the mechanism of drag reduction." Phys. Fluids, 18, 12 pp. 1626-1630. (December, 1975).

5. Berman, N. S.: "Drag reduction of the highest molecular weight fractions of polyethylene oxide." Phys. Fluids. 20, 5 pp. 715-718 (May, 1977).

6. Lumley, J. L.: "Drag reduction in turbulent flow by polymer additives." J. Polymer Science: Macromolecular Reviews, 7, pp. 263-290. (1973).

7. Hassager, O.: "Kinetic theory and rheology of bead rod models for macromolecular solutions. 1. Equilibrium and steady flow properties." J. of Chem. Phys., 60, 5 pp. 2111-2124. (March 1, 1974).

8. Batchelor, G. K.: "The stress generated in a non-dilute suspension of elongated particles by pure-straining motion." J. Fluid Mech., 46, 4 pp. 813-829. (1971).

9. George, W. K., Gurney, G. B., and Berman, N. S.: "Technique for rapid friction factor fluid characterization." J. Hydronautics, 9, 1 pp. 36-38. (January, 1975).

10. Berman, N. S. and George, W. K.: "Time scale and molecular weight distribution contributions to dilute polymer solution fluid mechanics." Proc. 1974 Heat Transfer and Fluid Mechanics Institute edited by Davis, L. R. and Wilson, R. E., pp. 348-364, Stanford University Press. (1974).

11. Lumley, J. L.: "Two phase and non-Newtonian flows," Chap. 7 pp. 290-324 in "Turbulence" edited by Bradshaw, P. Springer-Verlag. (1976).

12. Berman, N. S.: "Flow time scales and drag reduction." IUTAM Symposium on Structure of Turbulence and Drag Reduction. Washington, D.C. (1976). To be published in Phys. Fluids.

13. Elihu, S.: "Drag reduction and turbulent production using dilute DNA solutions." M.S. Thesis, Arizona State University. (December, 1976).

Table 1

Tabulation of Experimental Results

Run	c, ppm	[η], cm³/g	δ	ΔB max	Mx10⁶
205 A - 2	0.58	2250	5	2.2	7.0
3	1.55	1620	6.3	3.1	4.6
4	2.02	1300	5.7	2.7	3.5
5	2.11	920	4.5	2.2	2.2
6	2.92	560	3.2	1.6	1.2
8	3.28	370	2.1	1.5	0.7
9	3.49	310	1.9	1.4	0.55
205 C - 3	0.3	1390	2.1	1	3.8
4	0.39	1070	1.8	1	2.7
5	0.48	860	1.6	.8	2.0
8	0.95	460	1.4	.7	0.9
301 A - 1	0.07	3300	2	1	11.4
2	0.19	2560	2.9	1.2	8.3
3	0.21	2450	2.9	1.2	7.8
4	0.24	2370	3.0	1.2	7.5
6	0.25	1740	2.3	1.2	5.0
8	0.44	1500	2.7	1.3	4.2

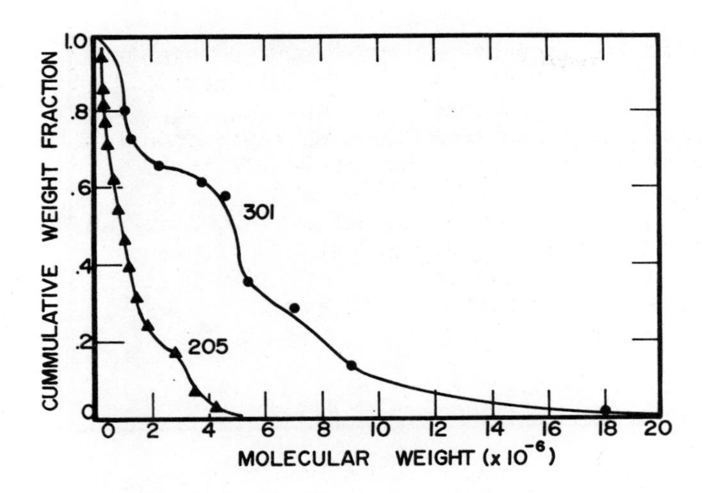

Fig. 1. Molecular weight distribution for Polyox WSR grades

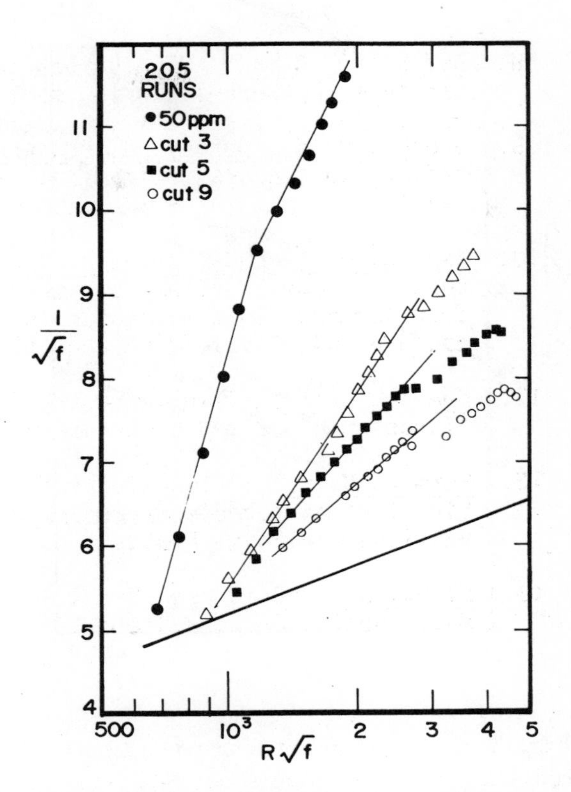

Fig. 2. Representative drag reduction trajectories for Polyox WSR 205 fractions and original solution

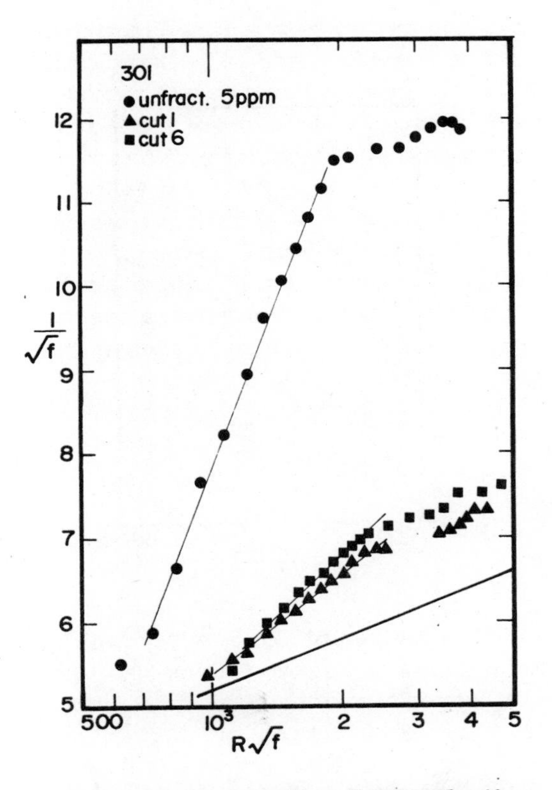

Fig. 3 Representative drag reduction trajectories for Polyox WSR 301 fractions and original solution

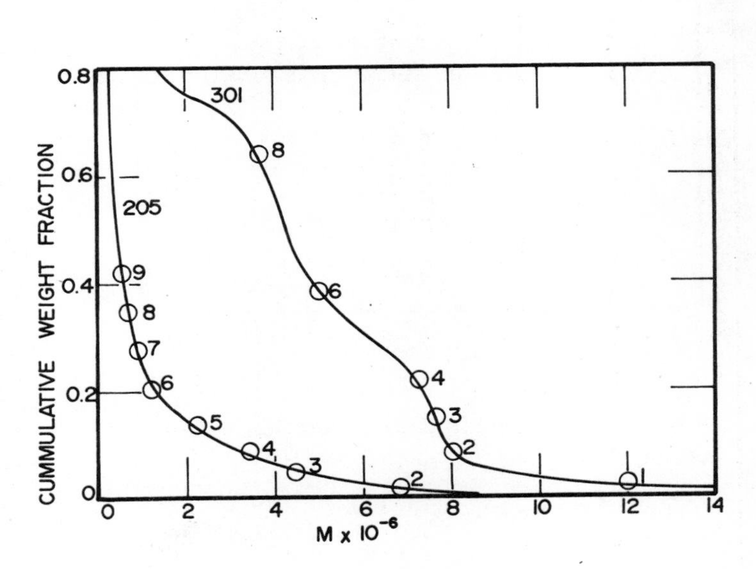

Fig. 4. Location of fractions used in drag reduction experiments on distribution curves

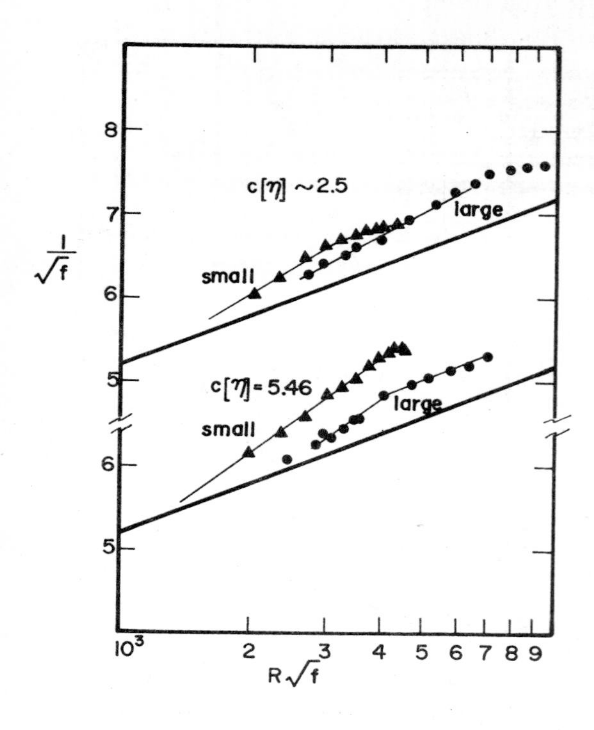

Fig. 5. Comparison of drag reduction trajectories in two pipe sizes

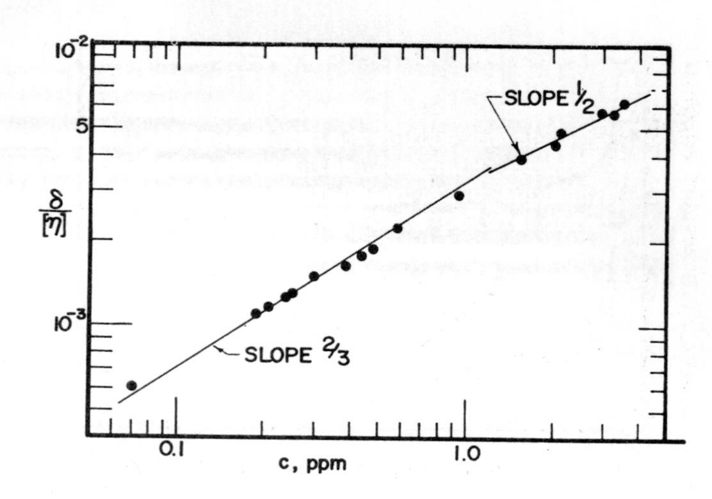

Fig. 6. Slope increment correlation with concentration and intrinsic viscosity

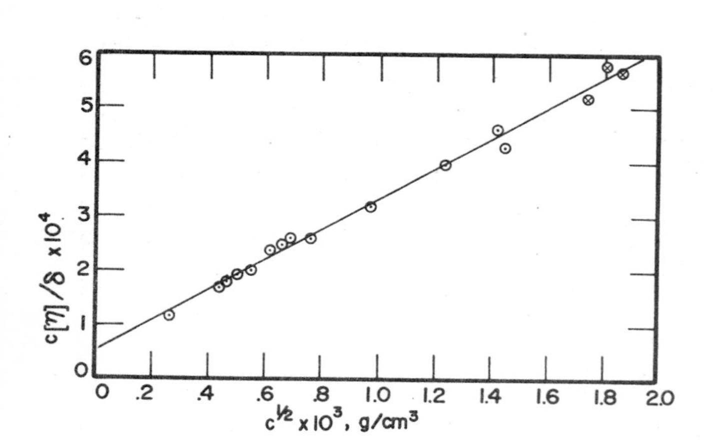

Fig. 7. Slope increment correlation based on Eq (10)

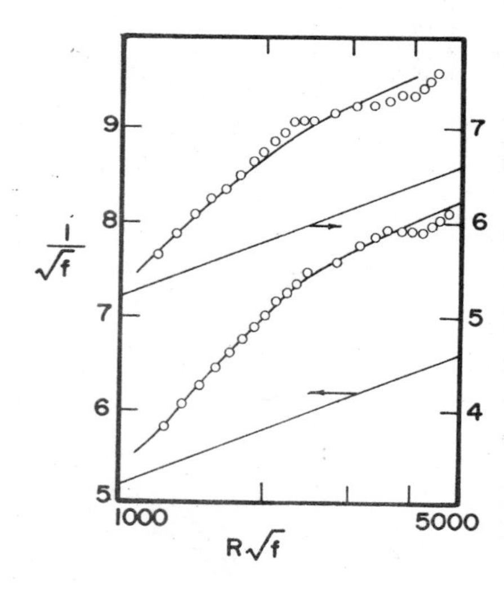

Fig. 8. Comparison of Eq (5) with drag reduction data for run two Polyox 301 fraction (upper curve) and run six Polyox 205 fraction (lower curve)

Second International Conference on

Drag Reduction

August 31st - September 2nd, 1977

PAPER C2

DRAG REDUCTION MEASUREMENTS WITH POLY (ACRYLIC ACID) UNDER DIFFERENT SOLVENT pH AND SALT CONDITIONS

R. H. J. Sellin, B. Sc., Ph. D., M. I. C. E.

E. J. Loeffler, B. Sc., Ph. D.

University of Bristol, U.K.

Summary

A turbulent flow rheometer has been developed for characterising drag reducing polymer solutions. In this a 18 ml liquid sample is tested to evaluate Reynolds number and pipe friction coefficient. The sample is driven through a 1 mm stainless steel pipe 830 mm in length by nitrogen gas at pressures up to 1.10 MNm^{-2} achieving a maximum Reybolds number of 17,000 at a shear velocity of 0.56 ms^{-1}. It has been found possible to test in this way 150 samples per hour and this has enabled experimental programmes to be carried out involving a wide range of polymer and solvent variables. So far preliminary assessments have been made of 45 polymer grades supplied by 5 different manufacturers. A more detailed study has been made of the poly (acrylic acid), Veriscol S25, in which the effect of solvent pH and dissolved salts has been investigated. It was found that the bi-valent metal ion magnesium, added as magnesium chloride, caused a sharp fall in the drag reducing effectiveness of the PAA molecules at a concentration two orders of magnitude less concentrated (10^{-2}M) than that reported in the literature for sodium chloride (a mono-valent ion) which suggested that the bi-valent ions are able to exercise a strong effect on the polymer molecule configuration in solution.

Held at St. John's College, Cambridge, England.

Organised and sponsored by BHRA Fluid Engineering, Cranfield, Bedford, MK43 0AJ.

NOMENCLATURE

c	polymer solution concentration
d	test pipe diameter
g	gravitational acceleration
h_f	friction head loss
L	test pipe length
M	molarity of salt solution
Re	solvent based Reynolds number
$\bar{v}$	mean velocity in test pipe
wppm	parts per million by weight
Δp	gas pressure in test vessel (gauge)
λ	pipe friction coefficient
ρ	fluid density
Ψ	drag reduction

1. INTRODUCTION

As part of a programme to test the drag reduction performance of the large number of water soluble polymers now available commercially a turbulent flow rheometer has been developed and built which has some unusual and useful features. Engineers interested in the drag reduction phenomenon have always been searching for advantageous applications and one which meets many prerequisites is its occasional use in storm water or foul sewers to avoid spillage into homes or natural watercourses. In sewers the nature of the drag reduction environment will be very variable and it is of the utmost importance to determine the reaction of possible polymers to dissolved and suspended impurities of all types. The turbulent flow rheometer described here is the third of a series of instruments developed over the last nine years and is well suited to investigate aqueous solutions of a large number of polymers and a wide range of solvent impurities. So far 45 different commercial grades of polymer obtained from 5 manufacturers have been tested in distilled water while more recently five polymers have been tested in solutions having pH values adjusted on the acid side of neutral and also with different concentrations of sodium chloride present. One - a poly(acrylic acid) - has also been tested with different concentrations of magnesium chloride, to examine the effect of a bivalent metal ion at differing pH values on the drag reduction behaviour of a polymer which was known to be sensitive to the pH value of the solvent.

2. TEST APPARATUS

This turbulent flow rheometer has been designed for fast comparative assessment of the drag reduction potential of a wide range of polymer solutions under laboratory conditions. The variables considered are polymer type and grade, polymer concentration, pressure gradient, pH value of the solution and salt ion type and concentration. The derived values are flow mean velocity, flow Reynolds number, wall shear stress, shear velocity, drag reduction and pipe friction coefficient. Because of the potentially large number of tests that result from the selected variable values an apparatus was required that would have a short test time and quick and simple refilling. The system adopted is shown in outline in Figure 1 while the details of the flow unit are shown in Figure 2. The apparatus is charged with 18 ml of the required solution which it discharges under nitrogen pressure through a drawn stainless steel tube of internal diameter 1.205 mm and length 850 mm. The range of variables obtainable with this apparatus is set out in Table 1 for both drag reducing and non-drag reducing liquids. Normally each sample is tested only once and, if the solutions have been prepared in advance, it is possible to test at least 150 samples per hour.

To check for inconsistencies three samples are tested under identical conditions and the time values obtained on the digital timer are averaged. An analysis of 50 such values gives a standard deviation of 0.2% but this figure tends to be higher in the transition region between laminar and turbulent flow. From this it will be appreciated that the 150 readings obtained in an hour yield data for 50 experimental points and work is conveniently organised in sessions of 450 samples tested in a day including the preparation of the solutions.

It will be seen from Figure 1 and 2 that the test cylinder is filled through a side port communicating with the bottom of the cylinder. It is necessary to close the end of the test pipe while filling is in progress to stop air entering. Any air finding its way into the test cylinder will collect below the piston and be exhausted at the bottom of the stroke after the conclusion of the flow time measurement on the sample. The sample is expelled by pressurising the cylinder above the piston with nitrogen gas; the gas control and pressure measurement arrangements are shown in Figure 1. During the test stroke the piston is a passive follower of the falling liquid level in the cylinder, the pressure difference across the piston corresponding only to the friction force in the piston and piston rod system. Attached to the piston rod is a timing cam, having axial symmetry, which operates a microswitch through a roller contact. The cam is so positioned that only 45% of its travel is timed, the first 40% being allowed for the establishment of steady flow conditions and the final 15% for the exhaustion of any air or froth remaining in the cylinder. The jacket between the stainless steel cylinder liner and the brass outer body is provided with a temperature regulated water supply but this is only used when testing samples at other than ambient temperature. It will be noticed that the stainless steel test tube is sealed and retained by a rubber gland and a gland nut. This enables the tube to be removed or changed very easily.

The pipe friction coefficient λ (sometimes called f') is defined by the equation

$$h_f = \frac{\lambda L \bar{v}^2}{2gd} \qquad 1$$

where h_f is the pipe friction head loss over a pipe length L. As it is difficult to make satisfactory pressure tappings in a 1 mm tube the total pressure drop Δp between the test cylinder and atmospheric pressure is used and therefore equation 1 has to be modified to allow for (a) the kinetic energy head of the pipe flow, based on the mean velocity $\bar{v}$ and (b) the sharp-edged pipe entrance loss, taken to be $0.5\ \bar{v}^2/2g$. By rearranging equation 1 it can be seen that λ is proportional to d^5, all other parameters remaining unchanged. A 1% error in measuring d will therefore lead to errors slightly in excess of 5% in all inferred λ values. A short length of the 1 mm nominal diameter tube used in this apparatus was cut off before installation, its ends cleaned and its diameter measured optically using a Nikon profile projector. All measurements were found to lie within the range ±1% of the average.

Expressing these terms as pressures enables the full equation to be written

$$\frac{\Delta p}{\rho} = \frac{\bar{v}^2}{2}\left\{\frac{\lambda L}{d} + 1.5\right\} \qquad 2$$

although it should be noted that it is not possible, using this single cylinder pressure value, to make any allowance for the non-uniform flow conditions in the entrance length of the pipe over which the boundary layers are attaining equilibrium. The behaviour of the apparatus is checked from time to time using distilled water and Figure 3 shows the results of such a run. It will be seen that the test points start slightly above the laminar flow transition and that at the highest Re values the points lie above the Prandtl - von Karman line for smooth turbulent flow. This latter tendency is a property of the particular pipe used in the test series described here. These pipes are readily replaced and measurements made using distilled water in a new pipe fitted after the conclusion of the Versical S25 tests do not show this rise in λ values in the higher Re range. Accordingly Figure 3 has been based on data obtained with the pipe used for Figures 5,6 and 7 but the rising λ values above Re = 6000 have not been taken into account in calculating the Ψ values for the polymer solutions.

Figure 4 shows λ - Re data plotted for three concentrations 5, 10 and 20 wppm of freshly prepared poly(ethlyene oxide), Polyox WSR - 301, dissolved in distilled water. It will be noticed that the small pipe diameter leads to high wall shear stress values at a particular Re value and this results in the "onset" point for drag reduction occurring at an Re value below the laminar-turbulent transition as observed on many occasions before. Another point also apparent in Liaw's data (Ref. 3) is the manner in which the majority of data points obtained with an efficient drag reducer, such as WSR - 301, lie below Virk's maximum drag reduction assymptote (MDRA) but this line is retained here as it constitutes a useful reference.

Other single pass turbulent flow rheometers described in the literature, both gas driven and mechanical displacement type, appear to use much larger fluid samples. Parker and Hedley's system (Ref. 8) uses the smallest noted volume of 0.5ℓ. Apart from the question of speed some methods of obtaining close molecular weight fractions from specially or commercially prepared polymer samples give only small volumes of dilute solution which must be tested without delay. Against this and the rapidity with which data can be obtained must be set the obvious disadvantages associated with capillary tube apparatus using small samples. This derives from the difficulty of making clean intermediate pressure tappings and hence avoiding the entrance length in making pressure drop measurements.

3 DRAG REDUCTION BEHAVIOUR OF POLY(ACRYLIC ACID)

Many polyelectrolytes make good flocculants and are used as such to assist sedimentation in water treatment processes. For this reason it might be advantageous to use a polyelectrolyte as a drag reducing agent in sewer applications in preference to a less surface-active non-ionic polymer. We therefore considered it worthwhile to test, in association with dissolved impurities that might represent the significant conditions found in a sewer, a polyelectrolyte which is known to be a good drag reducer in distilled water. Poly(acrylic acid), PAA, was selected because of the extensive and controversial information already published on it and because of its generally high drag reducing performance.

Considering first the effect of solvent pH value, Hand and Williams (Ref. 5) obtained their PAA development samples from the Dow Chemical Co. and in equipment essentially the same as ours obtained very good drag reduction in the pH range 1-3 using a 25 wppm solution. Outside this pH range they obtain very little drag reduction with this solution. In our experiments we obtained over 70% drag reduction with an unmodified pH value (5.7) and as the pH was lowered to 3 by the addition of HCl this figure was reduced (see Figure 5). This trend is in agreement with results published by Parker and Hedley (Ref.8), who used a 6 mm pipe although the pH values selected do not coincide. Our tests, and those of Parker and Hedley, were made using Versicol S25 supplied by Allied Colloids of Bradford, Yorkshire. More detailed results using Versicol S25 solutions in a 4.5 mm gravity flow apparatus were published by Balakrishnan and Gordon (Ref.2). Although the recorded drag reduction levels are only about half those that we obtained at the same concentration, the effect of pH variation produced identical trends. The low drag reduction obtained by Balakrishnan and Gordon may be due to the low wall shear stress inherent in their apparatus.

The earliest published results that we have found in which the combined effect of pH change and salt content (NaCl) is investigated are those of Banijamali et al (Ref.9). Here the by now familiar trends are observed. Before the salt is added, solutions having pH=4 and above give good drag reduction while pH=2 results are poor. However, the addition of NaCl to bring the pH=4 solution up to 0.5M strength leads to a decrease in its drag reducing achievement and brings it much into line with the values for the pH=2 solution. The pH=7 and 10 solutions are not affected even by 1.0M NaCl content. Our results, Figure 6, cover the pH range 5.8 - 3 and NaCl content up to only 10^{-2}M or 1/50 the concentration used by Banijamali. At this concentration of sodium ions Figure 6 shows that the salt has no inhibiting effect on the drag reduction performance of a 10 wppm solution of S25 although the pH changes have their usual effect.

We then repeated this experiment using a bi-valent ion (magnesium) instead of the monovalent sodium and Figure 7 now shows a very different situation. At a 10^{-2}M solution of $MgCl_2$ the Ψ value achieved with the unadjusted pH solution drops off from about 65% to 14%; at 10^{-1}M $MgCl_2$ it rises slightly to 18%. Figure 7 shows Ψ values obtained at Re = 6500 but at a higher nitrogen pressure producing Re = 7500 the corresponding values of Ψ were 6% and 10%, an even more marked drop. The Ψ values for the solution in which the pH value had been adjusted to 5.0 fell from 64% to 48% at 10^{-2}M salt content and then to 19% at 10^{-1}M. It appears as if the process of pH adjustment has in some way blanketed the drag reduction inhibitory action of the Mg ions present at 10^{-2}M strength but not at 10^{-1}M. There are also some less marked effects at pH 4 and 3.

4 DISCUSSION

Because of the agreement of our poly(acrylic acid) results with those of all previous authors, except Hand and Williams, under conditions of changing pH value of the solution we believe that the extended form of the molecule produces better drag reduction. Our use of a bi-valent ion (Mg) has been found to produce a large inhibitory effect on drag reduction in these solutions at pH values about 5 at a 10^{-2}M strength solution, a strength at which Na ions seem to have no effect. This suggests that the bi-valent ions are able to change the molecular arrangement of the poly(acrylic acid) perhaps by reducing the electrostatic repulsion between the ion-atmospheres surrounding each charge group in the chain and so reducing the extension of the large molecules in a way which the mono-valent Na ion is unable to do at this molarity.

It is our intention to repeat the experiments with bi-valent ions obtaining more data points than are shown here in Figure 7. It would also be of value to examine the effect of Ca and Al ions on the drag reduction behaviour of poly(acrylic acid) over a range of pH values at least as wide as that explored here. Until these results become available those shown in Figure 7 should be taken as provisional but they do suggest however that there is some significant reaction taking place between the Mg ions and the poly(acrylic acid) solutions which is dependent on pH value and which would justify caution in the selection of drag reducing additives for use with waste water, especially in the case of polyelectrolytes. The identification and concentration measurement of metal ions in waste water from different sources is obviously necessary before the full consequence of these results can be seen in this context.

5 ACKNOWLEDGEMENTS

We would like to acknowledge the assistance we have received from Dr. Brian Vincent over the course of many discussions about the molecular structure of water soluble polymers. Thanks are also due to Dr. Eric Dunlop for helpful comments about these experiments.

REFERENCES

1. Sellin R.H.J. "Increasing sewer capacity by polymer dosing". Proceedings Institution of Civil Engineers 63, part 2, pp. 49-67 (March 1977)

2. Balakrishnan C. and Gordon R.J. "Influence of molecular conformation and intermolecular interactions on turbulent drag reduction". J. of Ap. Polymer Sc., 19, pp. 909-913 (1975)

3. Liaw G.C., Zakin J.L. and Patterson G.K. "Effects of molecular characteristics of polymers on drag reduction". A.I.Ch.E. Journal 17, 2, pp. 391-397 (March 1971)

4. Parker C.A. and Hedley A.H. "A structural basis for drag reducing agents". J. of Ap. Polymer Sc., 18, pp. 3403-3421 (1974)

5. Hand J.H. and Williams M.C. "Effect of secondary polymer structure on the drag reducing phenomenon". J. of Ap. Polymer Sc., 13, pp. 2499-2503 (1969)

6. Frommer M.A., Feder-Lavy A. and Kraus M.A. "Dependence of the drag reduction efficiency of polyelectrolytes on their conformation in aqueous solutions". J. of Colloid and Interface Sc., 48, 1, pp. 165-169 (July 1974)

7. Little R.C. and Wiegard M. "Drag reduction and structural turbulence in flowing Polyox solutions." J. of Ap. Polymer Sc., 14, pp. 409-419 (1970)

8. Parker C.A. and Hedley A.H. "Drag reduction and molecular structure". Nature Physical Science 236, pp. 61-62 (March 27, 1972)

9. Banijamali S.H., Merrill E.W., Smith K.A. and Peebles L.H. "Turbulent drag reduction by polyacrylic acid". A.I.Ch.E. Journal 20, 4, pp. 824-826 (July 1974)

Table 1 - Turbulent flow rheometer - limiting parameters

* 15 wppm solution Polyox WSR - 301

	(water) Minimum	(water) Maximum	drag reducing* Maximum
Nitrogen pressure, MNm^{-2}	0.15	1.10	1.10
Recorded flow time S	2.910	1.313	0.629
Mean velocity ms^{-1}	3.61	7.99	16.68
Reynolds number	3250	7200	17174
Pipe friction coefficient (λ)	0.0276	0.0392	0.0072
Wall shear stress Nm^{-2}	45	315	250.8
Shear velocity ms^{-1}	0.21	0.56	0.50
Drag reduction (Ψ)%	-	-	77.1

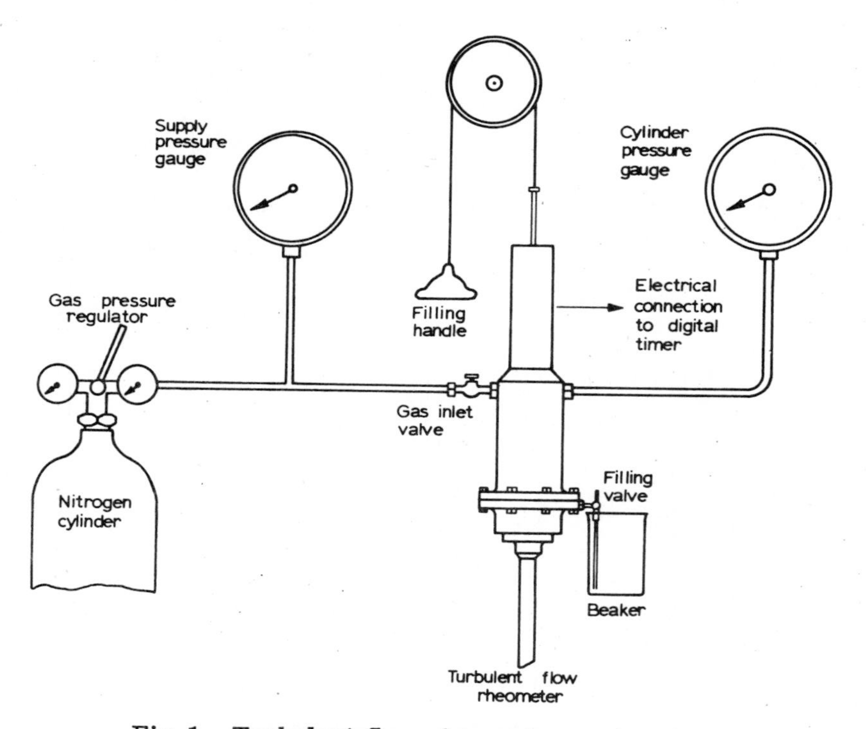

Fig.1 Turbulent flow rheometer system

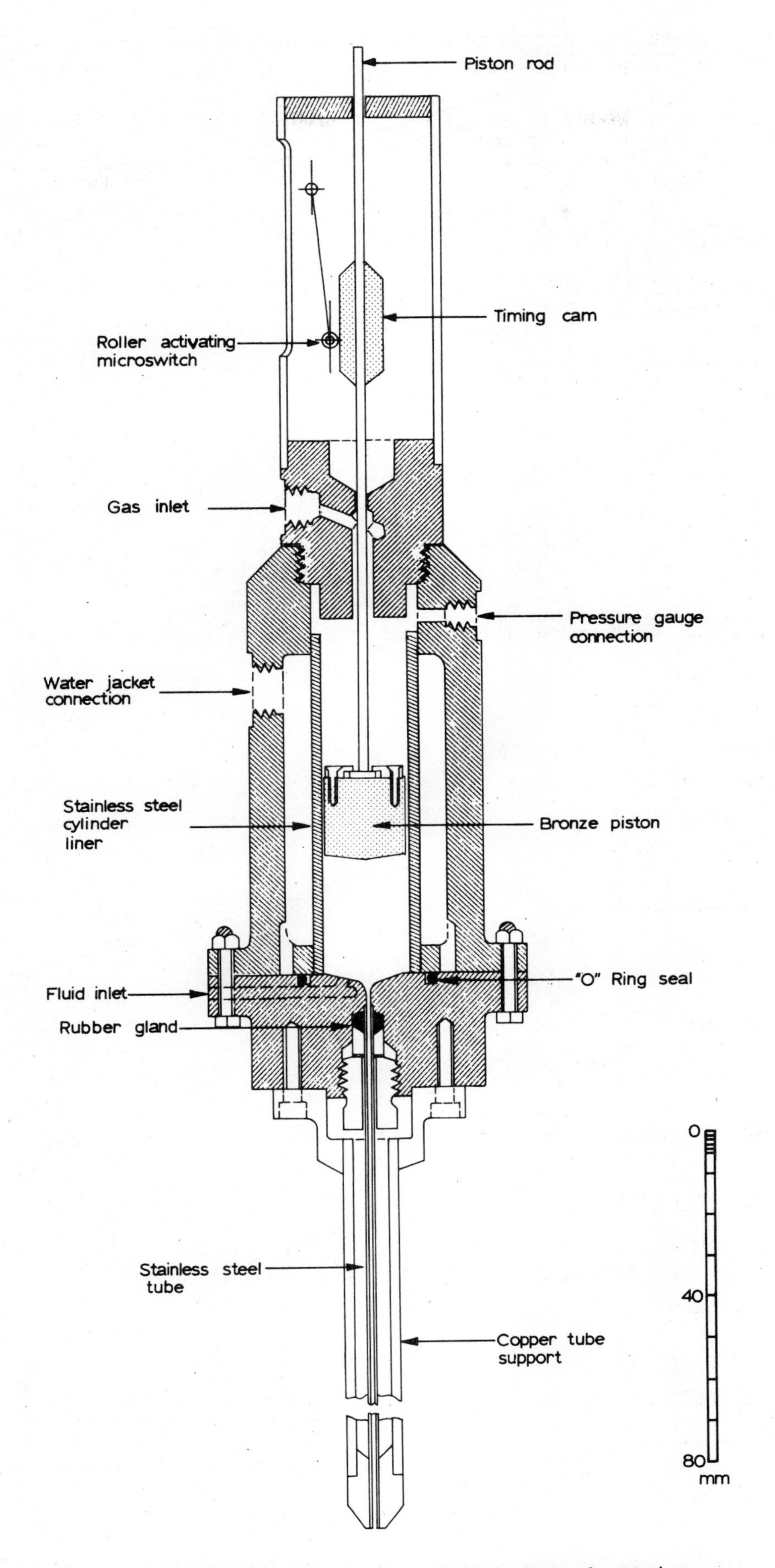

Fig. 2 Details of flow unit of turbulent flow rheometer

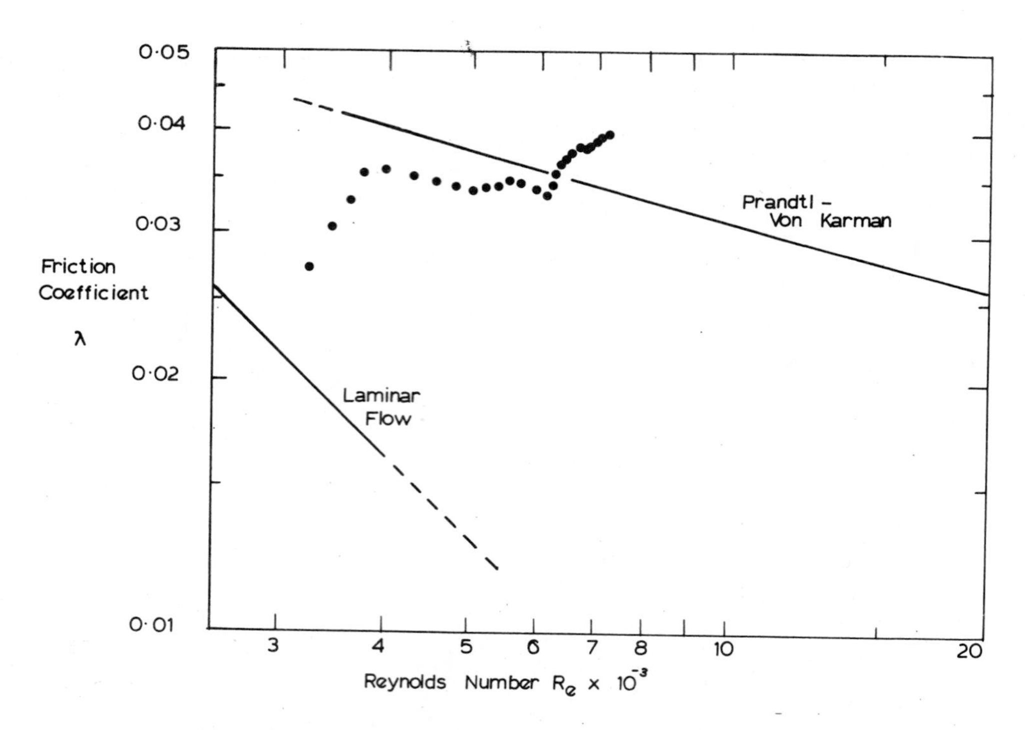

Fig.3 Friction data for distilled water in turbulent flow rheometer

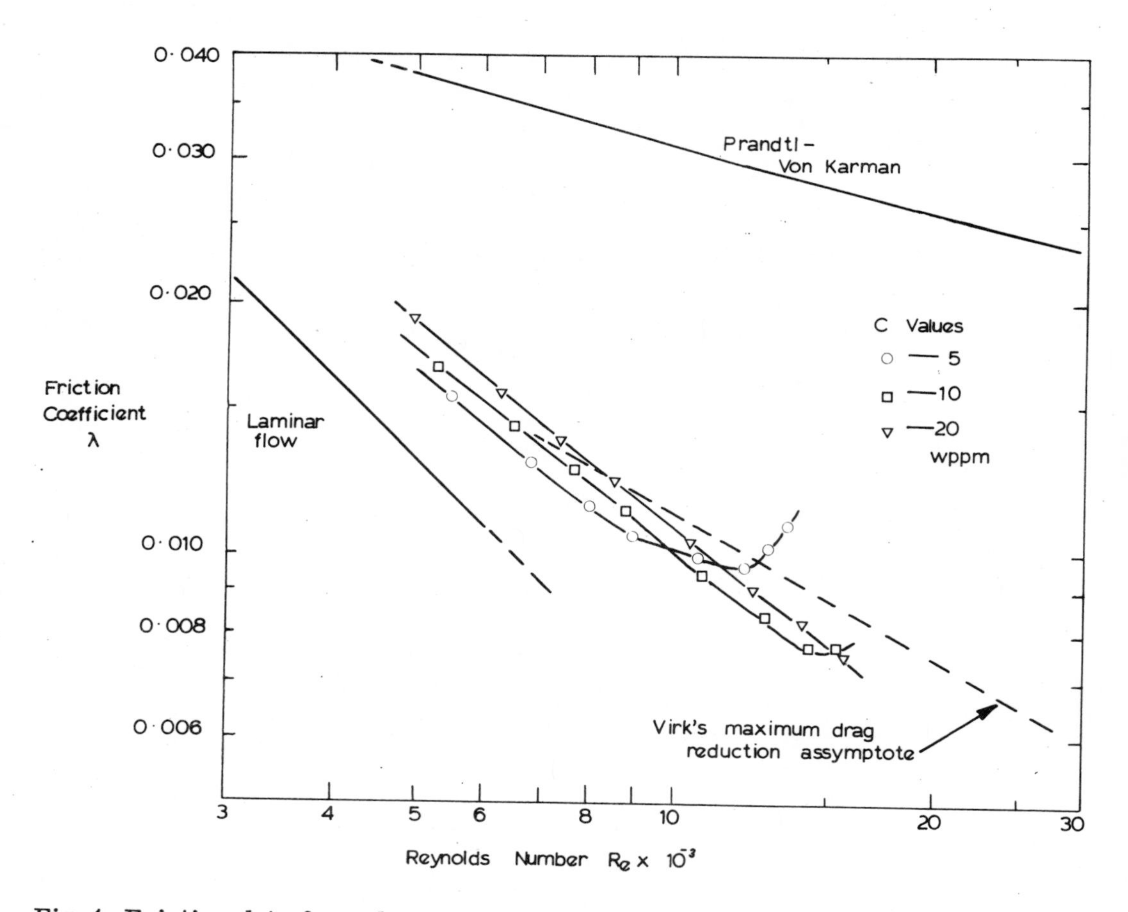

Fig.4 Friction data for poly (ethylene oxide) polyox WSR-301 in distilled water

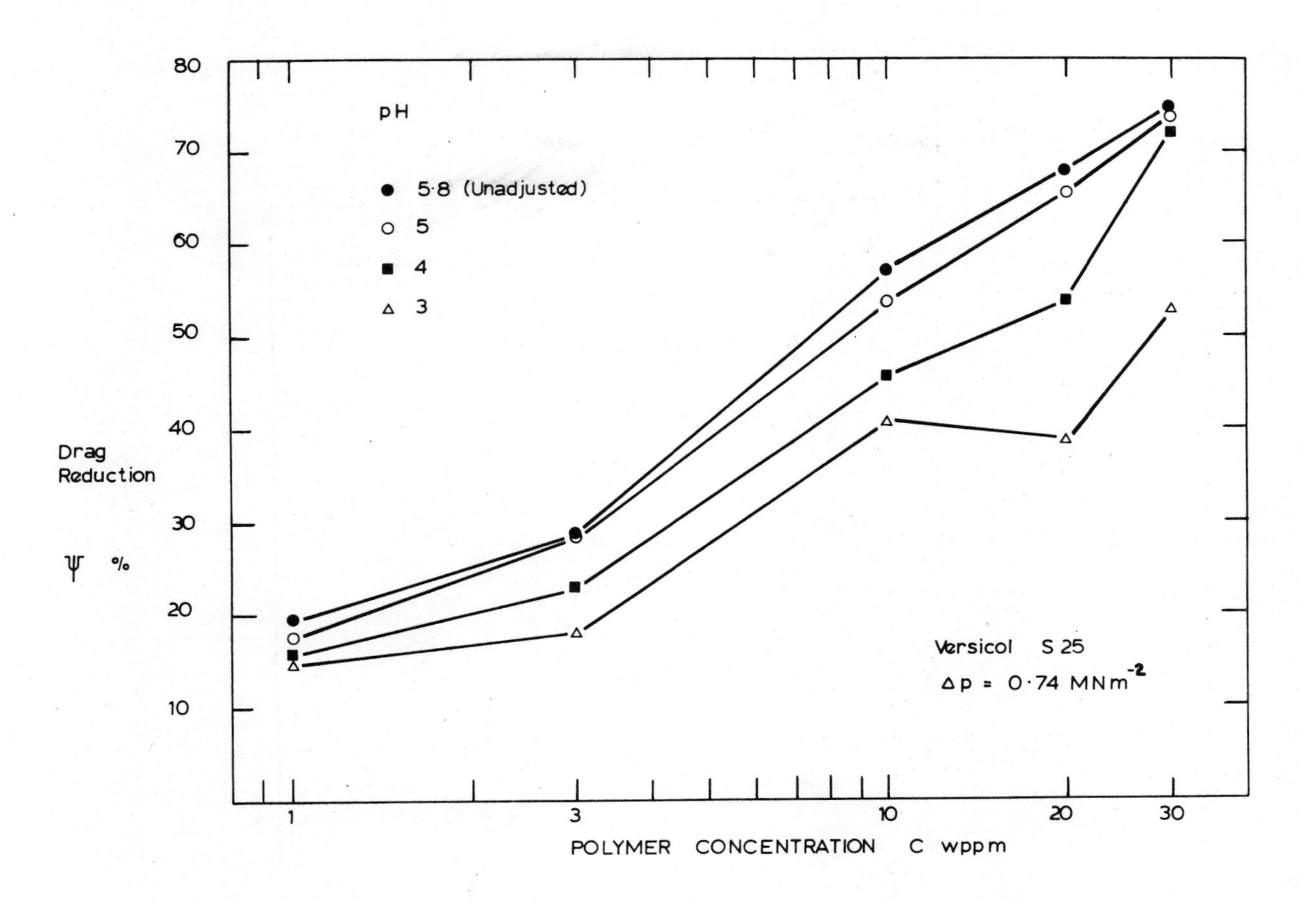

Fig.5 Effect of solution pH on drag reduction of poly (acrylic acid)

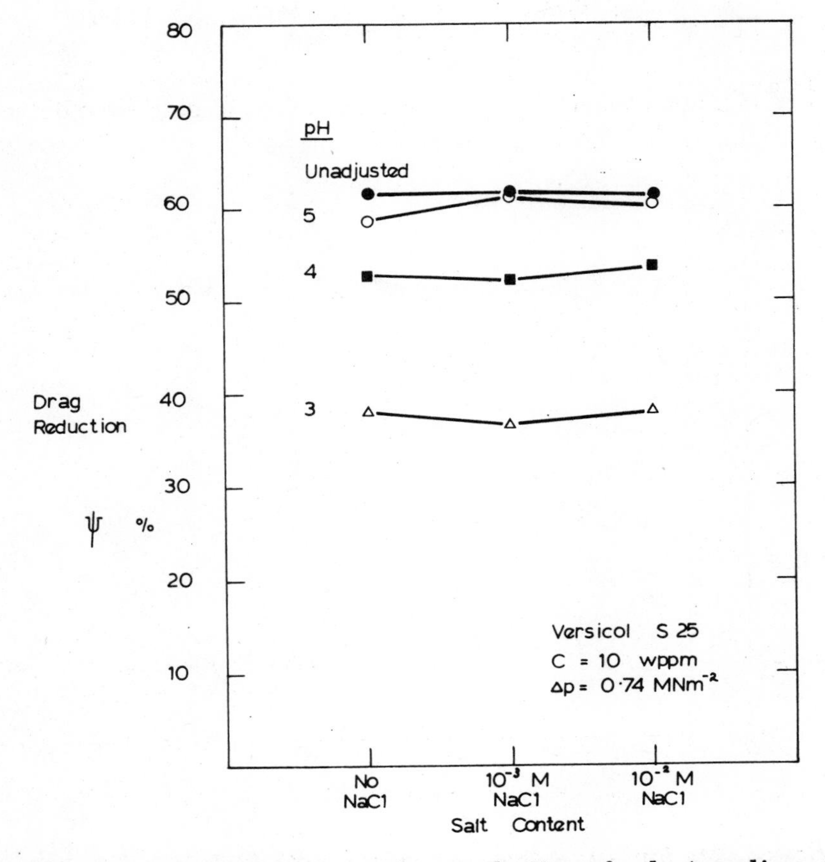

Fig.6 Effect of Na ion concentration on drag reduction of poly (acrylic acid) solutions

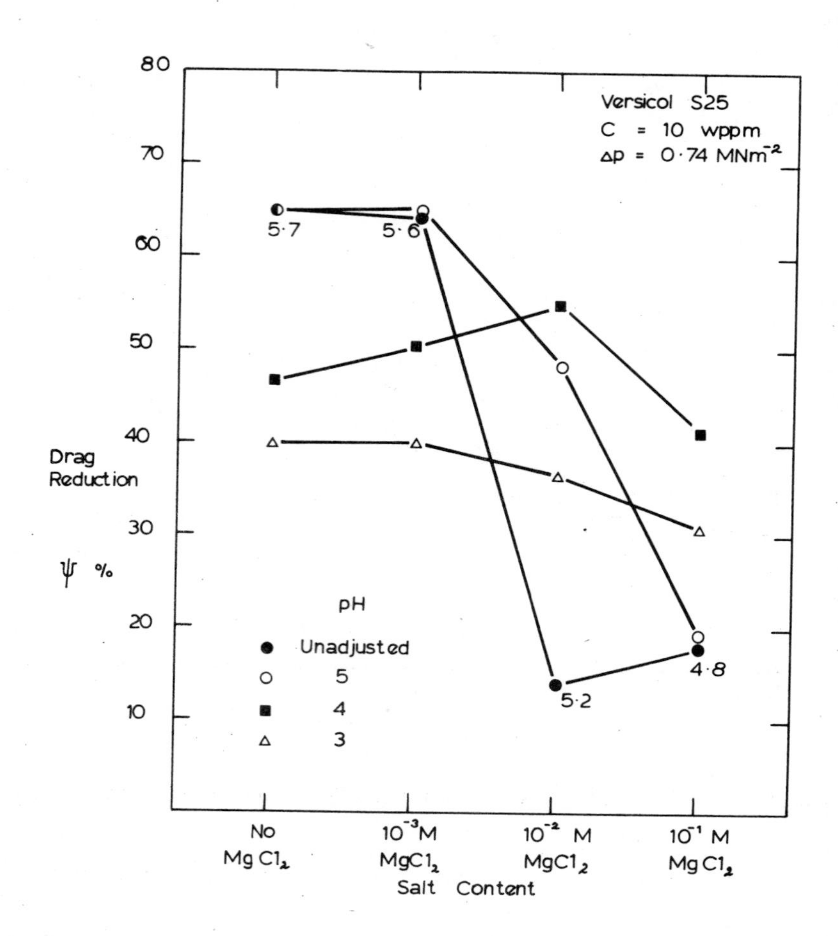

Fig.7 Effect of Mg. ion concentration on drag reduction of poly (acyrlic acid) solutions

Second International Conference on

Drag Reduction

August 31st - September 2nd, 1977

PAPER C3

DRAG REDUCTION AND ROUGHNESS BUILD-UP BY TRO-375 POLYACRYLAMIDE SOLUTION

M. Poreh

Technion-Israel Institute of Technology, Israel

and

J.P. Tullis and J.A. Hooper,

Colorado State University, U.S.A.

Summary

An experimental study of pipe flows with TRO-375 polyacrylamide solutions manufactured by Calgon is reported. The analysis of the data suggests that this drag reducing polymer increases the effective roughness of the pipe. As a result, the drag reducing ability of this polymer is attenuated. In some cases, where a small amount of this polymer is used or when the solution is degraded, a drag increase can be expected due to the use of the polymers. Friction measurements with plain water, before and after the use of this polymer solution, show that the additional roughness it produced is not easily washed away by flowing water.

Held at St. John's College, Cambridge, England.
Organised and sponsored by BHRA Fluid Engineering, Cranfield, Bedford, MK43 0AJ.

Nomenclature

A, B	- universal constant in the logarithmic law
D	- pipe diameter
f	- friction coefficient
$\overline{K}$	- average size of roughness
u	- local mean velocity
V^*	- shear velocity
V^*_{crit}	- critical shear velocity
V^{*+}_{crit}	- dimensionless parameter = $V^*_{crit} D/\nu$
y	- distance from the wall
α	- concentration dependent parameter
δ	- nominal thickness of viscous sublayer
Δu^+	- dimensionless parameter describing the upward shift of the log profile in the law of the wall formulation
ν	- kinematic viscosity

INTRODUCTION

During a study of the effectiveness of polymer additives as cavitation suppressants (Ref. 1) the drag reducing effectiveness of several polymers was tested. Several sets of data obtained in tests with dilute solutions of the polyacrylamide TRO-375 manufactured by the Calgon Corporation were not readily explainable. A steady change of the value of the friction coefficient was observed during the beginning of each new series of tests (which was initially called a "Monday morning effect.") In subsequent tests with very dilute solutions, or degraded solutions, friction factors above those recorded for plain water were recorded. These results have promoted a careful experimental investigation of the drag reducing properties of this polymer which have led the authors to a conclusion that this particular polymer modifies the shape of the inner surface of the pipe and thus increases its effective hydraulic roughness.

EXPERIMENTAL PROCEDURE AND RESULTS

Polymer solutions were premixed in a 1100 gallon tank using a dry powder aspirator. Water was supplied from a large reservoir. The water was clear of any suspended material and did not contain any chemical additives.

The solution was gently mixed and stored at least 12 hours but not more than three days before testing. Just prior to testing the solution was mixed again for approximately one hour.

An 8-in. pipe led from the tank to a variable-speed, positive displacement Moyno pump. The pump discharged the solution through a 3-in. flexible hose to a long 1 5/8-in. galvanized steel pipe in which the friction measurements were made.

The friction coefficients were determined by measuring the pressure drop along the last 15 ft. of the pipe. At a distance of 5 ft. upstream the test section an 0.62-in. orifice could be mounted. The effect of the orifice was to drastically reduce the effectiveness of the polymer. We shall refer to the solutions which have passed once through the orifice as degraded solutions.

The degraded solutions, which were collected downstream the test section, were stored in a second tank. Their drag reducing characteristics were determined later by passing them again in the same pipe without the orifice. No shear degradation occured in this second run.

To collect data on the change of the friction coefficients with time the following experiment was performed. The pipe was carefully cleaned with a stiff bristle nylon brush and the friction factor for water was measured. The measured friction factor was not time dependent and indicated an effective pipe roughness of $k/D = .00175$.

A fresh solution of TRO-375 was then pumped through the pipe without the orifice. Several manometer readings were taken as soon as the pump speed, and thus the discharge became constant. A typical set of measurements is given in Fig. 1. The data show that the friction factor increased with time. Subsequent tests with similar solutions indicated that the increase in the friction coefficient after 15 minutes was very small and it is thus assumed that an equilibrium state was reached at approximately that time. However, the phenomenon could be reproduced by cleaning the pipe with a nylon brush.

All the measurements of the friction factor reported herein, were taken after an equilibrium was reached. The data is presented in Figs. 2 to 4, which shows the variation of the Darcy-Weisbach friction coefficients with the Reynolds number, for different solutions.

ANALYSIS OF THE EXPERIMENTAL DATA

A possible explanation of the observed increase in the drag coefficient after each cleaning of the pipe is that the polymer solution changes the effective roughness of the pipe. It should be noted that this effect was not observed in earlier experiments with poly (ethylene oxide) (WSR 301).

Analysis of the friction measurements with polymer additives was therefore made in order to examine whether this data supports the above intuitive conclusion.

Previous studies (Refs. 2, 3, 4) have already indicated that drag reduction by polymer additives is drastically reduced by pipe roughness. The effect of the roughness in flows without polymers can be expressed as a function of the relative size of the roughness elements $\overline{K}$ to the thickness of the viscous sublayer δ. When $\overline{K}/\delta << 1$ the pipe may be considered to be hydraulically smooth. When $\overline{K} >> \delta$ the flow becomes independent of the viscosity of the fluid, and at the same time drag reduction disappears.

A simple semi-empirical model which attempts to describe the gross features of the effect of roughness on drag reduction has been offered by Poreh (Ref. 2). The model is based on the assumption that the diminishing effect of the drag reducing polymer is proportional to the diminishing role of the viscosity at large values of $\overline{K}/\delta$. This role has been expressed by a function P $(\overline{K}/\delta)$, which varies from 0 to 1. The function P was slightly modified in a later work (Ref. 5) following the discussion of the original paper. In this work the modified function was used. (The reader is referred to Refs. 2 and 5 for details.)

To calculate the friction factor of a polymer solution in a rough pipe using the above model, it is required to determine the behavior of the solution in a smooth pipe. It is assumed in the model that the effect of the polymers in a smooth pipe can be described by an upward shift of the log profile (Refs. 6, 7)

$$u/V^* = A \log (yV^*/\nu) + B + \Delta u^+$$

where

$$\Delta u^+ = \alpha \log (V^*/V^*_{crit}).$$

The parameter α is a function of the concentration, whereas, the critical shear velocity V^*_{crit} is primarily a function of the molecular weight. The equivalent roughness of the pipe K/D is determined using pure water. No additional coefficients are needed for estimating the behavior of the solution in a rough pipe using the model. However, it is usually required to account for the existence of nonuniform roughness by assuming that the roughness is made of elements of at least two sizes: $K_1 = \overline{K}/Z$ and $K_2 = \overline{K} \cdot Z$. A value of $Z = 2$ had been found to give good results and was also used in this study.

Although the model of Poreh is expected to give only a very approximate description of the effect of roughness on drag reduction it was decided to examine whether it can be used to describe the experimental data.

The friction factors measured in flows of fresh and degraded WSR 301 solutions are shown in Fig. 2. The figure also shows calculated friction factor curves using the model of Poreh. One sees from this figure that the experimental data can be fairly well described by the model using the relative roughness, $\overline{K}/D = 0.00175$, which had been determined in water flows, and a common value of V^{*+}_{crit} for all the fresh solutions ($V^+ = V^*D/\nu$). The behavior of the degraded 15 ppm solution can be described by the same value of α used for the 15 ppm fresh solution, which is consistent with the assumption that α is a function of the concentration, but here a different V^{*+}_{crit} had to be used to account for the decrease of the molecular weight of the degraded solution.

None of the calculated curves with $\overline{K}/D = 0.00175$ appeared to satisfactorily match the measured data. The observed minimum in the f versus Re number curve shifted to lower Reynolds numbers and this change can be described only by increasing the effective roughness of the pipe up to values of $\overline{K}/D$ around 0.003.

The measurements using a degraded TRO-375 10 ppm solution are described in Fig. 4. The friction factor curve seem to be described fairly well by the model using the earlier values of $\alpha = 15.5$ used earlier for the 10 ppm fresh solution, and $K/D = 0.003$, but with a larger value of V^*_{crit}. Finally, a degraded 2 ppm solution which had been passed through an orifice at a very high Reynolds number was tested. Previous experiments with WSR-301 suggested that this solution should lose all of its drag reducing capacity. Indeed that had happened but, as shown in Fig. 1, the measured friction factors were even higher than the original values of the friction factors measured in water flow and matched the calculated curve for water ($\alpha = 0$) in a pipe with $\bar{K}/D = 0.003$. Following this surprising result the pipe friction factors for water were remeasured. The friction factors obtained also matched the $\bar{K}/D = 0.003$ curve. However, after the pipe had been thoroughly cleaned with a nylon brush, the friction factors for water returned to the original values which correspond to relative roughness of $\bar{K}/D = 0.00175$.

DISCUSSION

It was concluded from these experiments that during the work with TRO-375 solutions the effective roughness of the pipe increased from solutions of TRO-375 which had been made before this study revealed many more records where the measured friction factor was higher than the original values for water in the cleaned pipe.

No visual changes in the shape of the surface of the pipe could be detected. However, this is not surprising since the increased roughness was only two thousandths of an inch. We have also observed that the roughness build-up in flows of fresh TRO-375 solutions did not wash out readily in water.

Although the change in the effective roughness does not seem to be large, its effect on drag reduction is not negligible at all. Figure 3 shows, for example, the effect of the increased roughness for a 3 ppmw polymer solution calculated by the theoretical model with $V^{*+}_{crit} = 300$ and $\alpha = 7$. One sees from this figure that at a Reynolds number of $3x10^5$ a value of approximately 45% drag reduction in a pipe with $\bar{K}/D = 0.00175$ is expected, whereas a value of only 20% is expected for the increased roughness of $\bar{K}/D = 0.03$.

ACKNOWLEDGEMENTS

Partial support of this work has been provided by The Hydro Machinery Laboratory Colorado State University, Ft. Collins, USA and by the U.S.-Israel Binational Science Foundation (BSF), Jerusalem Israel.

REFERENCES

(1) Hooper, John A. "Pipe-orifice flow with polymer additives", M.Sc. Thesis, Dept. Civil Engineering, Colorado State University, Ft. Collins, Colo. (1976).

(2) Poreh, M. "Flow of dilute solutions in rough pipes". J. of Hydronautics, 4, 4, pp. 151-155 (1970).

(3) Spangler, J.G. "Studies of viscous drag reduction with polymers including turbulence measurements and roughness effects", in "Viscous Drag Reduction" edited by C.S. Wells, Plonum Press, (1969).

(4) Brandt, H. McDonald, A.T. and Boyle, F.W. "Turbulent skin friction of dilute polymer solutions in rough pipes." in "Viscous Drag Reduction" edited by C.S. Wells. Plenum Press (1969).

(5) Poreh, M. "Reply by Author to A.G. Fabula and D.M. Nelson" Journal of Hydronautics, 5, 4, pp. 155-156 (1971).

(6) Myer, W.A. "A correlation of the frictional characteristics for a turbulent flow of dilute non-Newtonian fluids in pipes" A.I. Ch. E. Journal, 12, 3, (1966).

(7) Elata, C., Lehrer, J., and Kahanovitz, A. "Turbulent shear flow of polymer solutions." Israel J. of Tech., 4, p.87 (1966).

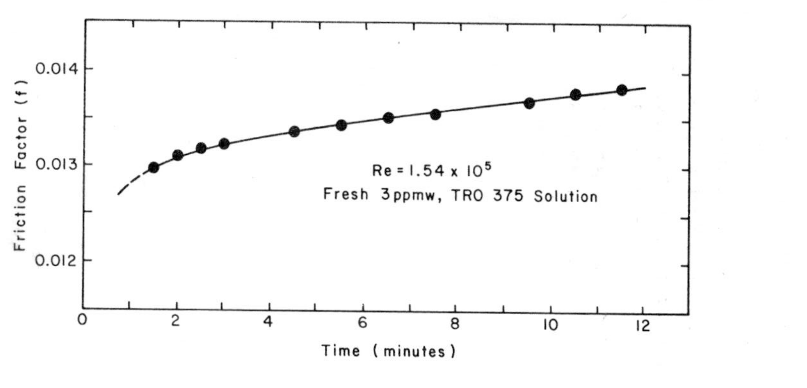

Fig.1 Change of friction factors in fresh TRO-375 solutions

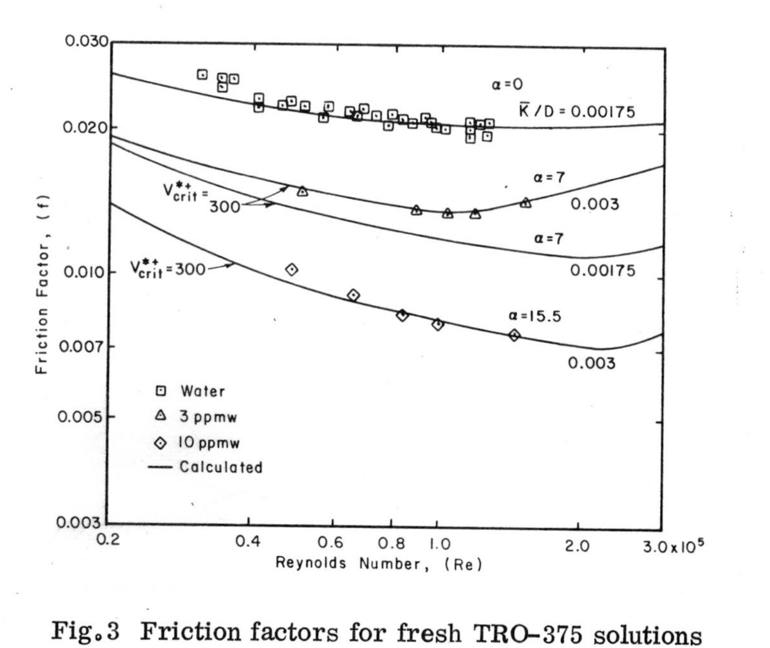

Fig.3 Friction factors for fresh TRO-375 solutions

Fig.2 Friction factors for WSR-301 solutions

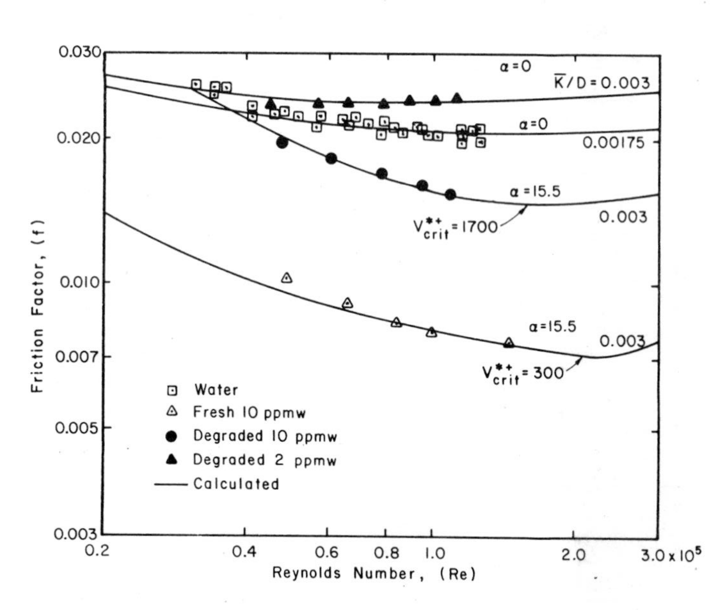

Fig.4 Friction factors for TRO-375 solutions

Second International Conference on

Drag Reduction

August 31st - September 2nd, 1977

PAPER C4

POTASSIUM POLYPHOSPHATE AND ITS DRAG REDUCING PROPERTIES

D.L. Hunston and R.C. Little,

Naval Research Laboratory, U.S.A.

Summary

Potassium polyphosphate is the first soluble totally-inorganic polymer that has been shown to reduce turbulent drag. Moreover, polyphosphates are unique in that they can be prepared with a wide variety of molecular structures. The synthesis proceeds by the heat-induced dehydration of potassium phosphate and the structure of the resulting polymer, therefore, depends on the heat treatment and the composition of the starting material. This composition can be characterized by the relative amounts of potassium and phosphorous in the potassium phosphate sample, i.e., the K/P ratio. If this ratio is increased above 1.0 by adding potassium hydroxide, lower molecular weight polymers will be produced, while decreasing the ratio of phosphoric acid will cause branching during the polymerization.

In the present study 30 different polyphosphate samples were prepared and tested. Their drag reducing efficiencies exhibited a strong dependence on the synthesis conditions with the optimum polymer found to have an efficiency equivalent to the most effective material presently known. The K/P ratio for this polyphosphate was 0.95 which indicates a branched structure. Surprisingly, however, the experiments demonstrated that hydrolysis of the branch points had little effect on the drag reducing effectiveness. The results also indicated that the efficiency was improved when the ionic strength of solution was decreased or the molecular weight was increased. These results are in keeping with the approximate correlation that was observed between the viscosities of the solutions at a particular concentration and the drag reducing efficiencies of the polymers.

Held at St. John's College, Cambridge, England.

Organised and sponsored by BHRA Fluid Engineering, Cranfield, Bedford, MK43 0AJ.

Introduction

It is well-known that the addition of very small amounts of certain high molecular weight polymers to a fluid can significantly decrease the pressure gradient required to maintain a given rate of turbulent flow (Ref. 1 and 2). This phenomenon, known as drag reduction, has attracted a great deal of interest in recent years, and as a result a wide variety of research projects have been initiated. One of the main objectives in many of these programs was to develop new, more effective materials that could be employed both in fundamental studies and as candidates for use in potential practical applications. As a result many types of polymers have been investigated (Ref. 3).

Although drag reduction has been observed for suspensions of certain insoluble materials such as asbestos (Ref. 4), the majority of new materials have been soluble polymers. Even at concentrations as low as a few parts per million by weight (ppmw), the frictional drag of the solution can be as much as 80% or 90% less than that of the solvent. Until recently all the soluble drag reducing agents have been organic or organometallic materials, but in 1973 drag reduction measurements were reported for solutions of a totally inorganic polymer (Ref.5). In this work a commercial sample of potassium polyphosphate (Monsanto Chemical Co.) was dissolved in water and at concentrations less than 100 ppmw, it gave a 50% reduction in drag (compared to water). Although this result is not impressive when compared to data for organic polymers, it must be remembered that no attempt was made to optimize the experimental conditions. It is of interest therefore to determine whether a systematic study of polyphosphate polymers might lead to the development of more effective drag reducing agents.

The chemistry of both sodium and potassium polyphosphates has been investigated in detail (Ref. 6). With regard to drag reduction however, the potassium polyphosphates are of most interest because they can be prepared by a simple technique and are known to have extremely high molecular weights (Ref. 6). Their synthesis, which has been extensively studied (Ref. 7-13) proceeds through the dehydration of potassium phosphate monobasic with heat.

$$nKH_2PO_4 \xrightarrow{\Delta} HO\left[\begin{array}{c} OK \\ | \\ -P-O- \\ \| \\ O \end{array}\right]_n H$$

The primary factor in determining the molecular structure of the polymer is the number of reaction sites (-OH groups) per molecule in the starting material. This in turn is related to the ratio of the number of potassium atoms to the number of phosphorus atoms; molecules of K_2HPO_4 (K/P = 2) should terminate chain propagation whereas molecules of H_3PO_4 (K/P = 0) should serve as potential branching points. Consequently, if the starting material has an average K/P ratio close to 1.00, the reaction will tend to produce linear molecules of high molecular weight. When the ratio is decreased by adding phosphoric acid, a branched polymer should be produced. If the ratio is increased by adding potassium hydroxide, linear polymers of lower molecular weight would be expected. Since the pH of a potassium phosphate solution is a function of the K/P ratio, pH measurements are a convenient method to monitor this ratio in samples prepared for polymerization (Ref. 8).

These predictions have been examined experimentally (Ref. 6, 8 and 14), and the results indicate that the polymerization is strongly dependent on the K/P ratio when it is below 1.0 but relatively insensitive to this ratio when it is above 1.0. It is suggested that excess potassium cannot enter the potassium phosphate crystal structure and thus does not affect the polymerization (Ref. 8). Consequently, it is the lower K/P ratios that are of most interest. By varying the K/P ratio, the heating time and the heating temperature, it is possible to synthesize a series of linear polymers with average molecular weights ranging from 212 (monomer) to 1 million or more. Likewise a series of branched polymers can be produced with structures that vary from a high molecular weight polymer with a few widely spaced branches to a highly cross-linked network (Ref. 9). This ability to vary the molecular structure suggests that a systematic synthesis program should make it possible to find a molecular architecture which optimizes the drag reducing ability of these materials. The purpose of this paper is to report the results of such a study.

Experimental

To provide some data for comparison, the commercial sample from Monsanto was used for some of the initial studies in this work. It was, for example, employed in tests designed to optimize the solution preparation technique. The K/P ratio for this material was 0.95; however, no other details of the synthesis were available. The remaining polymer samples were prepared using Fisher certified potassium phosphate monobasic which had a measured pH of 4.48 in a 0.1 molar solution. This corresponds to a K/P ratio of 0.999 which is similar to what Pfansteil and Iler (Ref. 8) have found for commercial samples. In the first set of experiments four different K/P ratios were selected: 0.900, 0.950, 0.999, and 1.020. The starting materials were prepared in the manner suggested by Pfansteil and Iler. Calculated quantities of phosphoric acid (Fisher Certified 85.5%) or potassium hydroxide (Fisher Certified 85% minimum) were added to solutions containing weighed samples of the potassium phosphate and the mixtures were dried with continuous stirring so as to distribute the added component over the surface of the crystals (Ref. 8). A 120 g sample was prepared for each of the four K/P ratios. Each material was then ground in a mortar and pestle, and thoroughly mixed. To verify that the calculated K/P ratios were in fact obtained, pH measurements were made using 1 molar solutions of each material. The K/P ratios were then estimated from a standard curve that was obtained by titrating 1 molar solutions of potassium phosphate monobasic with standardized phosphoric acid and potassium hydroxide solutions (Ref. 8).

The potassium polyphosphate polymers were synthesized by placing samples of the specially prepared potassium phosphate in platinum dishes. The dishes were placed in a preheated Hotpach muffle furnace for a prescribed length of time. When removed from the furnace, the dishes were placed on a cold stone counter top and allowed to cool rapidly to room temperature. Some preliminary tests were performed where the samples were cooled even more rapidly by setting the dishes in cold water as suggested by Pfanstiel and Iler. Since this did not seem to improve the drag reduction properties of the resulting polymers, however, this method was not used for subsequent sample preparations.

In addition to the K/P ratio, two other synthesis variables were studied: the temperature and duration of the heat treatment. Three temperatures (500°C, 750°C, and 900°C) and two time periods (30 minutes and 120 minutes) were selected based on the observations of previous workers (Ref. 10-13). To prepare the polymers each of the four starting materials was divided into six 20g samples so that all six heat treatments could be performed on samples with identical K/P ratios. The syntheses were carried out using four platinum dishes so that a sample from each of the four groups could be placed in the oven at the same time thereby eliminating any differences that might result from variations in the heat treatment. This procedure resulted in a set of 24 samples which ranged from granular solids having low molecular weights to fibrous materials with very high molecular weights. Each polymer was given a designation which consisted of the heating temperature, the K/P ratio, and the heating time separated by hyphens; thus 750-0.950-30 corresponds to the polymer made by heating a potassium phosphate sample with a K/P ratio of 0.950 for 30 minutes at 750°C.

To test these samples, it is necessary to prepare solutions. The technique used for dissolving the polymer, however, is very important if maximum drag reducing properties are to be obtained. As a result a number of experiments were performed to determine the best procedure. Although potassium polyphosphate is not directly soluble in water, it will dissolve in the presence of certain solubilizing agents such as lithium chloride or sodium metaphosphate (Ref. 6). It is generally believed that the cations from these materials exchange with the potassium ions in the crystal lattice (Ref. 6). The difference in size between the two kinds of ions results in a destabilization of the lattice and therefore promotes dissolution. If the concentration of cations is too large, however, the polymer again becomes insoluble (Ref. 6). For the work reported in this paper, a 1% by weight solution was found to be a reasonable compromise and was used for most of the tests.

Experiments designed to evaluate various techniques for preparing solutions indicated that the best procedure was a low shear version of the method suggested by Van Wazer (Ref. 6). The solubilizing agent was first dissolved in distilled water to form 500 ml of a 2% by weight solution. One gram of the sample was then dispersed in 500 ml of distilled water using a Waring blender and, since the polyphosphate is insoluble, there is little or no degradation. After the sample was completely dispersed, 300 ml of the 2% solution was added and simultaneously the blender was turned off. In this way most of the sample was dissolved quickly without the use of rapid stirring. The contents of the blender were carefully washed into a flask using the remaining 200 ml of 2% solution. This mixture was allowed to stand until the polymer was completely dissolved. The time required for this varied from a few hours if stored at room temperature to 6 or 8 hours if stored in a refrigerator. The final solution contained 1000 ppmw of the sample and 1% of the solubilizing agent. The experiments were then performed by diluting these stock solutions to the desired concentrations.

For the present study one set of experiments was performed with the commercial sample to determine the stability of the polymer solutions as a function of storage time. In this case the polymer was allowed to dissolve at room temperature. For all other experiments the mixtures were stored in the refrigerator until the polymers were completely dissolved and then tested within 2 to 3 hours after removing the flask from the refrigerator. To determine the effect of different solubilizing agents, tests were performed with several materials including lithium chloride, sodium pyrophosphate, and sodium metaphosphate. These experiments showed sodium metaphosphate to be most satisfactory and thus all subsequent solutions were prepared using this material. The role of metaphosphate concentration was examined by performing many of the experiments in two different ways. In one case the stock solution was diluted with distilled water and in the second case the dilutions were made with a 1% sodium metaphosphate solution.

The drag reducing properties of each polymer were measured by determining the flow behavior in a capillary tube (internal diameter 0.1575 cm) using an automated flow rheometer (Ref. 15). This device records the wall shear stress as a function of flow rate between Reynolds numbers, Re, of 1,000 and 12,000. The percentage drag reduction, DR, was calculated at Re = 8,000 using the definition

$$DR = 100[1 - (\tau_p/\tau_w)] \qquad (1)$$

where τ_w and τ_p are the wall shear stresses in water and solution respectively (1% of the solubilizing agent does not appreciably alter the wall shear stress of water in the turbulent flow regime). Since the laminar to turbulent transition for this apparatus occurs at a Re of about 3,500, the data were measured for fully developed turbulent flow. The apparatus (Ref. 15) is a single pass system and thus fresh solution was used for each test to minimize the chance for flow-induced degradation. With the commercial sample, however, a second experiment was also performed. In this case the same volume of solution was measured repeatedly to give drag reduction as a function of the number of passes through the apparatus. When the data are displayed in graphical form, the loss of drag reduction with increasing exposure to the flow provides information as to how sensitive the molecules are to the hydrodynamic forces present in the apparatus (Ref. 16).

For each of the 24 polyphosphate samples, a complete characterization consisted of drag reduction measurements at a minimum of five different polymer concentrations using both water dilutions and 1% metaphosphate dilutions. Based on these results, a second set of polyphosphates was prepared in an effort to further identify the optimum K/P ratio. Four 20g samples (K/P ratios of 0.925, 0.940, 0.960, and 0.975) were prepared by the same procedure described above except that only the 2 hour, 750°C heat treatment was used. The drag reducing properties of these materials were also characterized at Re = 8,000.

In addition to drag reduction measurements two other types of experiments were performed. First the viscosity of selected samples was measured using a Cannon 4-bulb viscometer (50-S400) in a water bath controlled to 25.00 ± 0.01°C with a Tronac Model 40 procession temperature controller. This viscometer provides measurements at four different average shear rates ranging from under 60 sec^{-1} to over 500 sec^{-1}. For the viscosity measurements, all dilutions were made using 1% sodium metaphosphate. The second set of experiments that was performed with these samples involved an attempt to purify the polymers by washing with distilled water (Ref. 6 and 8). Small quantities of material were taken from selected samples of the synthesized polymer. In each case the powder was placed in a mortar and pestle, cold water was added, and the resulting suspension was ground for a short time. More cold water was then added and the mixture was filtered to remove the polymer. Since the high molecular weight polyphosphate is insoluble in water, this procedure can remove soluble impurities and some low molecular weight materials. Each of these samples was then tested for drag reduction so the effects of washing could be judged.

Results

It is believed that the stability and drag reducing efficiency of polyphosphates depend on a number of factors. These include the solution preparation technique, the nature and concentration of the solubilizing agent, the K/P ratio, the heat treatment, and the presence of impurities or low molecular weight materials. Based on the results obtained in this work, the effect of each of these factors can be analyzed. The efficiencies for most of the polymer samples can be measured by fitting the drag reduction vs concentration data to the empirical equation first proposed by Virk (Ref. 17) and later adapted by Little (Ref. 18).

$$\frac{DR}{DR_{max}} = \frac{c}{[c] + c} \qquad (2)$$

The concentration in ppmw is represented by c while DR_{max} and [c] are empirical constants which characterize each polymer. The meaning of these constants can be illustrated as follows. The maximum drag reduction, DR_{max}, is a theoretical upper limit for the amount of drag reduction that can be obtained with a given sample at high concentrations. The maximum drag reduction divided by the intrinsic concentration, $DR_{max}/[c]$, is a measure of the efficiency of a polymer since it corresponds to the drag reduction per unit concentration at infinite dilution (Ref. 19).

$$\lim_{c \to o} [DR/c] = \lim_{c \to o} [DR_{max}/([c]) + c] = DR_{max}/[c] \qquad (3)$$

The values of these constants can be determined from the slope and intercept in a plot of (c/DR) vs c since it will be linear if the data conform to eq. (2).

By using $DR_{max}/[c]$ as a measure of efficiency, it is possible to consider the effect of the solution preparation technique and the solubilizing agent. In the original work with polyphosphate (Ref. 5), the potential degradation of the polymer during solution preparation was minimized by exposing the solution to high speed mixing in a blender for only a few seconds. Even this brief exposure to high shear stresses, however, provides some opportunity for degradation. Consequently, the technique employed for the present work eliminated this step. The importance of this can be illustrated by comparing the efficiencies measured for solutions of the commercial sample prepared by the two methods. With the original technique values of 1.0 $ppmw^{-1}$ were obtained for $DR_{max}/[c]$. With the new procedure however, the efficiency of this sample was increased by an order of magnitude to 10.6 $ppmw^{-1}$ (all solutions contained 1% sodium metaphosphate).

Another important factor in determining the efficiency and stability of the solution is the nature and concentration of the solubilizing agent. As expected the solutions exhibited the behavior typical of polyelectrolytes in that the efficiency decreased as the ionic strength of the solution increased. At a given value of ionic strength however, the efficiency showed only moderate variations for different solubilizing agents. The stability of the solutions, on the other hand, was much more dependent on the nature of the solubilizing agent. Fig. 1, for example, shows how the efficiency of the commercial sample changed as a function of the length of time that a 1000 ppmw solution was stored at room temperature (all dilutions made with distilled water). The Figure compares data for solutions made using two different solubilizing agents: 1% sodium metaphosphate and 0.1N lithium chloride. Despite a considerable difference in ionic strength, the initial values for the efficiencies were quite similar. The solution containing lithium chloride, however, showed a very rapid loss of efficiency with time. The solution made with sodium metaphosphate, on the other hand, exhibited very little change even after 315 hours of storage. The metaphosphate then offers a significant advantage and is therefore used in subsequent tests.

The reason for the stability of the polymer-metaphosphate solution was not studied in detail; however, it is worth noting that the degradation proceeds through hydrolysis. Pfanstiel and Iler (Ref. 8) indicate the rate of hydrolysis increases dramatically when the pH is lowered below 6.0 which suggests acid catalysis (Ref. 14 and 20). In a 1% metaphosphate solution, the buffering action maintains a pH of 6.7 while in the 1% lithium chloride solution the absorption of carbon dioxide gives a pH less than 6.0. Consequently, this buffering action may contribute to the stability. In addition, a number of authors (Ref. 14,21 and 22) have found that the rate of hydrolysis for branch points is 10^3 to 10^5 times the rate for hydrolysis of the backbone. For example, Strauss, Smith, and Wineman (Ref. 21) observed that even in the absence of flow most branch points degraded within a few hours while the backbone was stable for periods of 120 hours or longer. Since it is generally believed that linear molecules are the most effective drag reducers (Ref. 1), the data in Fig. 1 are consistent with previous degradation studies.

Although the proper choice of solubilizing agent gives a solution with a very low degradation rate for the commercial samples, other polyphosphates exhibit somewhat higher degradation rates. As a result no solution was stored at room temperature for more than a few hours or in a refrigerator for longer than overnight. Polyphosphate solutions also exhibit degradation when exposed to shear stresses. Fig. 2 shows the flow-induced degradation curve for a 10 ppmw solution of the commercial sample. This result which is similar to what is seen for other polymers (Ref. 16) indicates why all of the other experiments reported in this work were made using fresh solution for each test.

By employing the optimum solution preparation technique and the best solubilizing agent for all samples, it was possible to study the effect of K/P ratio and heat treatment. The measured efficiencies for the 24 synthesized samples are listed in Table I. It was found that the behavior of all the polymers except the most effective ($DR_{max}/[c] > 30$ $ppmw^{-1}$) could be described by eq. (2). Although eq. (2) has been found to describe the behavior of a great many polymers, its failure to fit the data for the most effective polyphosphates is not really surprising since several other experiments with highly efficient polyelectrolytes have shown similar results (Ref. 23). The factors which produce this unusual behavior are undoubtedly the same as those which give rise to the complex variation of viscosity with concentration for polyelectrolytes (Ref. 24). Since only two polyphosphates show large deviations from eq. (2), approximate values of $DR_{max}/[c]$ are given in Table I for these materials so that comparisons can be made.

By using the results shown in Table I, it is now possible to examine four of the parameters which control the drag reducing properties of polyphosphates. First it is clear that all of the polymers show the expected increase in efficiency when the ionic strength is decreased (water dilution). Second, the temperature of the heat treatment is obviously an important factor. Potassium phosphate has a melting point (Ref. 9) of 813°C and thus the samples prepared at 900° were synthesized as liquids. Since these materials showed little or no drag reduction, the data suggests that in general higher temperatures are beneficial so long as 800°C is not exceeded. The third variable to be considered is the duration of the heating time. Table I indicates that up to a certain point a longer heating time is desirable. When the combination of high temperature and low K/P ratio is used, however, less time is needed to produce a polymer having maximum drag reducing efficiency. Previous studies (Ref. 10-13) have also indicated that, in the initial stages of the reaction, polymerization predominates and thus longer heating times promote higher molecular weights. When long heating times are used, however, degradation can become a problem and lower molecular weights are obtained. The results shown in Table I illustrate this behavior.

The fourth and most interesting parameter studied in these tests is the K/P ratio. Table I indicates that when K/P > 1.0 very little drag reduction is obtained. As the ratio is lowered below 1.0, the effectiveness increases to a maximum and then declines. This behavior can best be illustrated in a three dimensional graph where the efficiency is shown as a function of K/P ratio and heat treatment temperature. Fig. 3, for example, shows the data for samples made with the 120 minute heat treatment and tested using distilled water dilutions. This data indicate that a sample with maximum efficiency should be obtained at a K/P ratio somewhere near 0.950. Moreover, it should have an efficiency of at least 105 $ppmw^{-1}$ which is two orders of magnitude better than that found in the original tests with the commercial sample. This value is also substantially larger than the efficiencies observed for some of the highly effective organic materials such as polyethylene oxide. For example, the most widely known drag reducing agent is probably Union Carbide's Polyox WSR-301. Tests with a particular sample of this material indicated that it had a molecular weight (Ref. 16) of 3.3×10^6 which corresponds to an efficiency of about 36 $ppmw^{-1}$ in the flow rheometer used here (Ref. 18). The best polyphosphates are clearly superior to this material which can be represented in Fig. 3 as a plane. In fact, only the most effective polyethylene oxide polymer (a particular sample of Union Carbide's Coagulant with a molecular weight of 7.0×10^6) has exhibited an efficiency, 106 $ppmw^{-1}$, which is comparable to the polyphosphates (Ref. 18).

Based on the results shown in Table I, the second set of polyphosphates was synthesized in an effort to more precisely define the best polymerization conditions. A two hour, 750°C heat treatment was used with K/P ratios of 0.925, 0.940, 0.960, and 0.975 since this appears to be the area of most interest. As expected all of these samples were extremely effective drag reducing agents and exhibited strong polyelectrolyte behavior. Consequently, the data could not be described by eq. (2). As an alternative, the results were compared by plotting curves of drag reduction vs K/P ratio at a series of different concentrations (Fig. 4). These graphs indicate that the optimum ratio is between 0.945 and 0.965. There is, however, a considerable range of K/P values where the polymers have drag reducing efficiencies close to the maximum.

In a final effort to optimize the efficiencies, selected samples were washed in cool water to remove low molecular weight materials which do not contribute to drag reduction. Flow measurements were made before and after washing to determine its effect. In general these measurements indicated that although significant improvement was obtained with some of the less effective samples, the most efficient polymers were not appreciably altered (Table II). These data are similar to the results of previous studies (Ref. 6 and 14) which found that washing produced the largest increases in viscosity for samples having K/P ratios greater than 1.0. It appears, therefore, that washing can expand the range of K/P ratios that give highly effective polymers, but the optimum ratio is not altered.

The preceding results indicate that the most efficient polymer is obtained when the starting material contains about one potential branch point in every twenty monomer molecules. The synthesized polymer therefore will have a large number of branches along the backbone. As indicated earlier however, the degradation rates indicate that most of the branch points were hydrolyzed before measurements were made and thus it is quite likely that the data here correspond to essentially linear polymers. This suggests that a comparison between drag reduction and viscosity might be instructive. Curves reported in the literature for the viscosity of polyphosphates as a function of K/P ratio show a marked resemblance to the drag reduction data given in Fig. 3. Van Wazer (Ref. 6) has observed a broad peak with a maximum at K/P $\approx$ 0.9 while Pfanstiel and Iler (Ref. 8) have reported a sharper peak with a maximum at K/P = 1.000. These variations arise from several factors. First, very small differences in the composition of the starting material can produce significant changes in the resulting polymer even though the synthesis conditions are identical (Ref. 25). Second, the branch points are degrading during the experiment and thus the viscosity changes with time. Van Wazer (Ref. 6), for example, suggests that samples having K/P ratios below 1.0 lose much of their viscosity after one day of storage at room temperature. In the present work the drag reduction curves exhibit a flow behavior which is right between the two literature results for viscosity. As a result it was of interest to experimentally measure the viscosities for the samples used here.

As expected, the samples showed the large shear rate dependence characteristic of polyelectrolytes. By using a 4-bulb viscometer, it is possible to evaluate the flow properties of each solution at 4 different average shear rates. Fig. 5 gives some typical results for one of the more efficient polymer samples, 750-0.975-120. It is of interest to note that the characteristic shear rate for the onset of non-Newtonian behavior is very low. Similar observations have been made previously for other materials that show strong drag reducing properties. Darby (Ref. 26) has studied the viscosity of a polyacrylamide sample in detail and noted that non-Newtonian behavior can be observed at shear rates as low as 10^{-2} sec^{-1}. By using data such as that shown in Fig. 5, it is possible to estimate values of viscosity for the polyphosphate samples at a specific average shear rate, 70 sec^{-1}. Since the polyelectrolyte effect made extrapolation to zero concentration very difficult, different samples were compared at a fixed concentration, 50 ppmw. Fig. 6 shows a plot of viscosity vs K/P ratio for the series of samples that was prepared using the 750°C, 2 hour heat treatment.

For the viscosity measurements degradation was more of a factor than it was in the drag reduction tests. With some samples the viscosity decreased with time at the relatively rapid rate which has been associated with hydrolysis of branch points. For the same samples the drag reduction showed no change with time. This suggests that either the drag reduction was unaffected by hydrolysis at branch points or the high stress levels present in the turbulent flow rheometer degraded the branch points before the measurements were made. The second explanation seems more likely. This result however adds some uncertainty to the comparison between drag reduction and viscosity. Nevertheless, if Figs. 3 and 6 are compared, the correlation is surprisingly good. Polyphosphates then show the same type of approximate correlation that has been observed for other polymers (Ref. 27). One particular result, however, indicates that caution is necessary. The commercial sample had a viscosity much higher than would be expected based on a comparison with other samples of similar drag reducing efficiencies. Patterson and Abernathy (Ref. 28) have also observed a lack of correlation between viscosity and drag reduction when they compared degraded and undegraded samples of polyethylene oxide. They suggested that the correlation fails because the molecular weight distributions for the degraded samples were significantly different than those for fresh samples. Since the synthesis conditions for the commercial polyphosphate sample was different from those for the polymers prepared in this work, the molecular weight distributions could be quite different and this could be responsible for the differences in flow behavior. In general, however, it appears that a correlation between drag reducing efficiency and viscosity can be obtained for polyphosphate polymers when all the samples are prepared by a similar synthesis technique.

Conclusions

The results reported here clearly demonstrate that good drag reduction can be obtained at extremely low concentrations using an inorganic polymer, potassium polyphosphate. By carefully controlling the synthesis conditions and solution preparation technique, it is possible to get polyphosphate solutions that exhibit a drag reduction effect as large as that obtained with solutions of the most effective polymers presently known. In solution the polyphosphates degraded when exposed to high shear stresses or long storage times. The rate of degradation is similar to that reported for hydrolysis of the polymer backbone. When a series of polyphosphates having different molecular structures are synthesized by similar polymerization techniques, an approximate correlation is observed between drag reducing efficiency and viscosity. This suggests that the optimum structure for a polyphosphate drag reducing agent is a flexible, linear molecule with extremely high molecular weight.

Moreover, the maximum effect is obtained when the ionic strength in the solution is low so that the polyelectrolyte effect makes the molecules highly extended. This result is consistent with the conclusions of similar studies (Ref. 1) with other polymers.

Acknowledgements

The authors wish to thank Mr. Robert Proodian for his help with the experimental aspects of this work and Dr. E. J. Griffith of the Monsanto Chemical Company for providing the commercial potassium polyphosphate sample. Presentation of this work was made possible by a grant from the Office of Naval Research.

References

1. Lumley, J. L.: "Drag Reduction by Additives". Ann. Rev. Fluid Mech., 1, pp. 367-384. (1969).

2. Hoyt, J. W.: "The Effect of Additives on Fluid Friction". Amer. Soc. Mech. Eng. Trans., D 94, pp. 258-285. (June, 1972).

3. Hoyt, J. W.: "Drag-Reduction Effectiveness of Polymer Solutions in the Turbulent-Flow Rheometer: A Catalog". Polymer Letters, 9, pp. 851-862. (November, 1971).

4. Peyser, P.: "The Drag Reduction of Chrysotile Asbestos Dispersions". J. Appl. Polym. Sci., 17, pp. 421-431. (February, 1973).

5. Hunston, D. L., Griffith, J. R. and Little, R. C: "Drag Reducing Properties of Polyphosphates". Nature Physical Science, 245, 148, pp. 140-141. (October, 1973).

6. Van Wazer, J. R: "Phosphorus and its Compounds". Vol. 1, chap. 10. Interscience. (1966).

7. Strauss, U. P. and Wineman, P. L: "Molecular Dimensions and Interactions of Long-Chain Polyphosphates in Sodium Bromide Solution". J. Amer. Chem. Soc., 80, pp. 2366-2371. (May, 1958).

8. Pfanstiel, R. and Iler, R. K: "Potassium Metaphosphate: Molecular Weight, Viscosity Behavior and Rate of Hydrolysis of Non-cross-linked Polymer". J. Amer. Chem. Soc., 74, pp. 6059-6064. (December, 1952).

9. Van Wazer, J. R: "Phosphorus and its Compounds". Vol. I. chap. 12. Interscience. (1966).

10. Malmgren, H., Acta Chem. Scand., 6, p. 1 (1952).

11. Lamm, O. and Malmgren, H: "Dispersity Measurements on the High Polymer Metaphosphate of Tammann". (Dispersitatmessungen an einen hochpolymeren Metaphosphat nach Tammann). Z. Anorg, u. Allgem. Chem., 245, pp. 103-120. (1940).

12. Malmgren, H. and Lamm, O: "Dispersity Measurements on High Molecular Weight Potassium Metaphosphates". (Dispersitatmessungen an hochmolekularen Kaliummetaphosphaten). Z. Anorg. u. Allgem. Chem., 252, pp.256-271. (1944).

13. Malmgren, H. Acta Chem. Scand., 2, p. 147. (1948).

14. Osterheld, R. K. and Audrieth, L. F: "Polymerization and Depolymerization Phenomena in Phosphate-Metaphosphate Systems at Higher Temperatures. Condensation Reactions Involving the Potassium Hydrogen Orthophosphates". J. Phys. Chem., 56, pp. 38-42. (January, 1952).

15. Little, R. C. and Wiegard, M: "Drag Reduction and Structural Turbulence in Flowing Polyox Solutions". J. Appl. Polym. Sci., 14, pp. 409-419. (February, 1970).

16. Ting, R. Y. and Little, R. C: "Characterization of Drag Reduction and Degradation Effects in the Turbulent Pipe Flow of Dilute Polymer Solutions". J. Appl. Polym. Sci., 17, pp. 3345-3356. (November, 1973).

17. Virk, P. S., Merrill, E. W., Mickley, H. S., Smith, K. A. and Mollo-Christensen, E. L: "The Toms Phenomenon: Turbulent Pipe Flow of Dilute Polymer Solutions". J. Fluid Mech., 30, pp. 305-328. (November, 1967).

18. Little, R. C: "Drag Reduction in Capillary Tubes as a Function of Concentration and Molecular Weight". J. Colloid Interfac. Sci., 37, pp. 811-818. (December, 1971).

19. Hunston, D. L.: "Effects of Molecular Weight Distribution in Drag Reduction and Shear Degradation". J. Polym. Sci., 14, pp. 713-727. (March, 1976).

20. Van Wazer, J. R: "Phosphorus and Its Compounds". Vol. I. p. 452. Interscience. (1966).

21. Strauss, U. P., Smith, E. H. and Wineman, P. L: "Polyphosphates as Polyelectrolytes. I. Light Scattering and Viscosity of Sodium Polyphosphates in Electrolyte Solutions". J. Amer. Chem. Soc., 75, pp. 3935-3940. (August, 1953).

22. Van Wazer, J. R: "Phosphorus and Its Compounds". Vol. I. p. 437 and p. 761. Interscience. (1966).

23. Hunston, D. L. and Ting, R. Y.: Unpublished results.

24. Armstrong, R. W. and Strauss, U. P: "Encyclopedia of Polymer Science and Technology: Polyelectrolytes". Vol. 10. pp. 781-861. Interscience. (1969).

25. Griffith, E. J: Personal communication. (1975).

26. Darby, R: "Transient and Steady State Rheological Properties of Very Dilute Drag Reducing Polymer Solutions". Trans. Soc. Rheol., 14, 2, pp. 185-212. (1970).

27. Little, R. C., Hansen, R. J., Hunston, D. L., Kim, O. K., Patterson, R. L. and Ting, R. Y: "The Drag Reduction Phenomenon. Observed Characteristics, Improved Agents, and Proposed Mechanisms". I and EC Fundamentals, 14, pp. 283-295. (November, 1975).

28. Paterson, R. W. and Abernathy, F. H: "Transition to Turbulence in Pipe Flow for Water and Dilute Solutions of Polyethylene Oxide". J. Fluid Mech., 51, pp.177-185. (January, 1972).

Table I: Drag Reducing Efficiencies of Polyphosphate Samples

Heat Treatment Temperature (°C)	Heat Treatment Heating Time (min.)	Efficiency[a] at Indicated K/P Ratio 0.900	0.950	0.999	1.020
500	30	16.0(13.8)	16.6(10.7)	1.4(0.9)	0(0)[b]
500	120	23.2(19.0)	26.7(23.3)	4.0(3.0)	0(0)
750	30	20.2(17.5)	33.3(27.0)	12.5(11.1)	0.1(0)
750	120	5.5(4.7)	105(83.3)	20.0(17.3)	0.1(0)
900	30	0.9(0.4)	17.6(10.9)	0(0)	0(0)
900	120	0.6(0.3)	5.8(3.3)	0.1(0)	0(0)

[a] The drag reducing efficiencies given in parentheses refer to dilution with 1% metaphosphate; the other values refer to aqueous dilution.

[b] Zero efficiency indicates no drag reduction is detectable at concentrations up to 500 ppmw.

Table II: Effects of Washing

Sample Designation	Drag Reducing Efficiency[a] Before Washing	After Washing
750-0.900-120	5.5	10.5
750-0.950-120	105	101
750-0.999-120	20	31
750-1.020-120	0.1	10

[a] All dilutions made with distilled water.

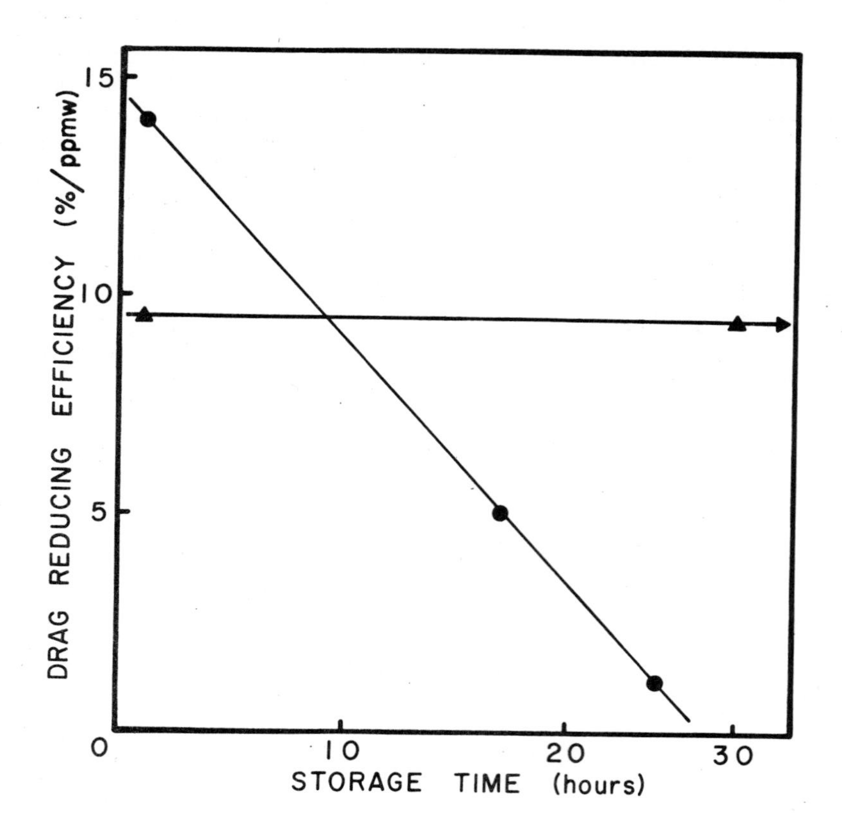

Fig. 1: The influence of room temperature storage time on drag reducing efficiency is shown for two polyphosphate solutions made with different solubilizing agents: 0.1N lithium chloride, ●; and 1% sodium metaphosphate, ▲. With the metaphosphate solution very little change is observed even after 120 hours.

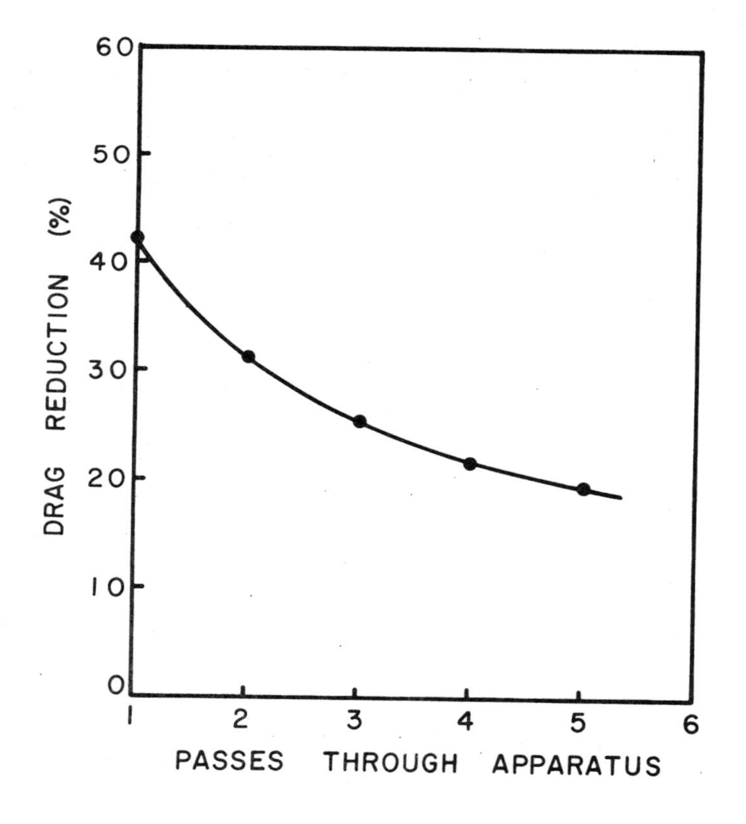

Fig. 2: A 10 ppmw polyphosphate solution shows significant degradation when repeated measurements are made on the same solution.

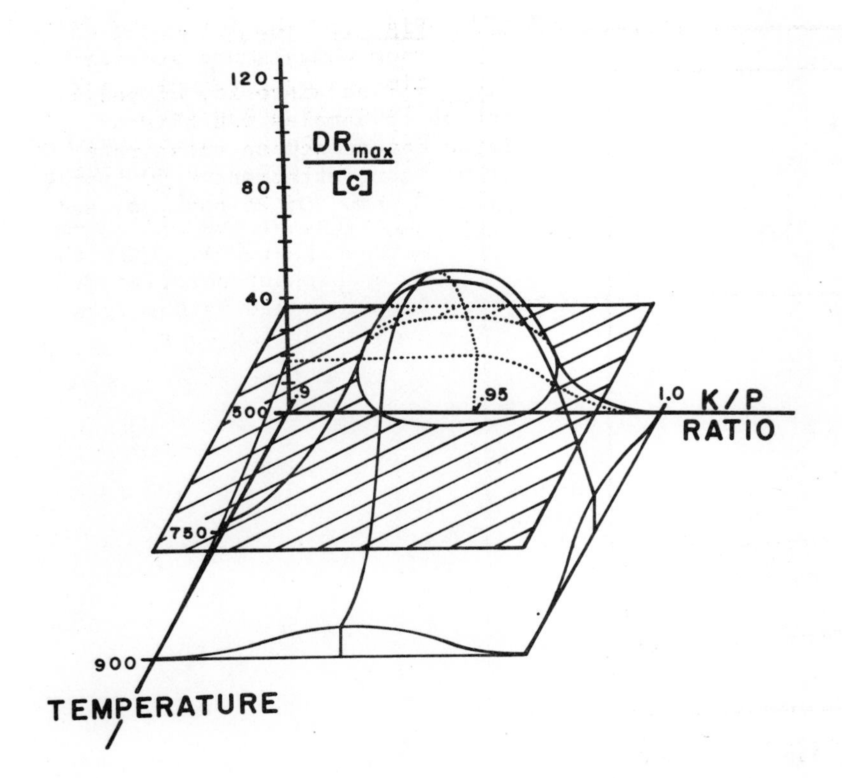

Fig. 3: The drag reducing efficiency of polyphosphate polymers exhibits large variations with changes in synthesis temperature and K/P ratio. For comparison the efficiency for a particular sample of polyethylene oxide is shown as a plane. Although the maximum efficiency is at K/P ratios near 0.95, the data in Fig. 4 indicates that the surface is skewed toward higher K/P ratios.

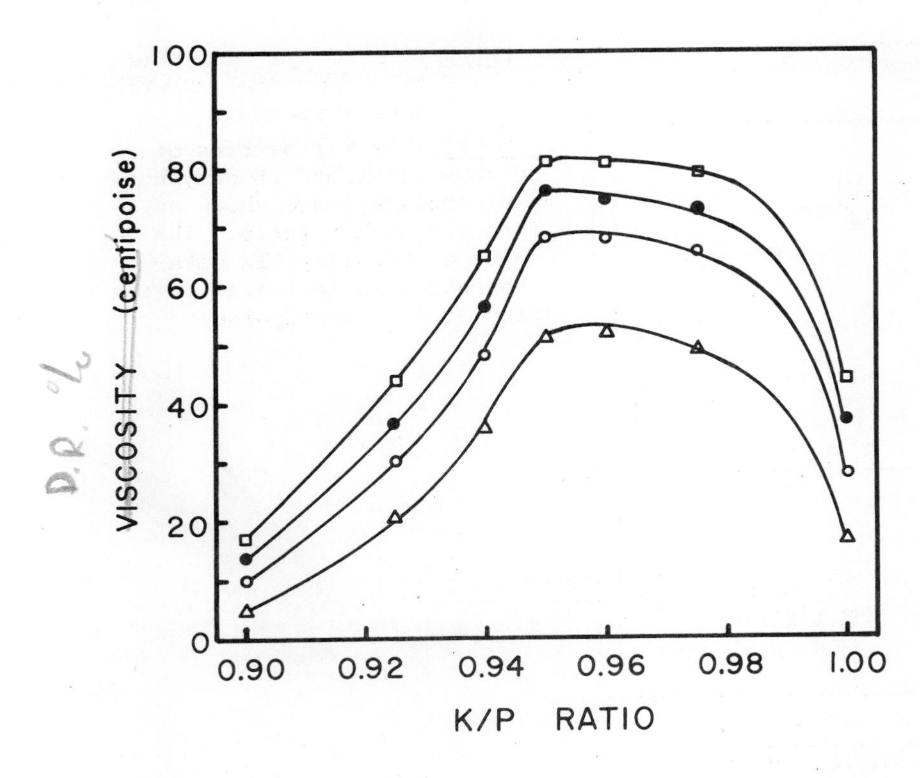

Fig. 4: Drag reduction is plotted as a function of K/P ratio for a series of polyphosphate samples at various concentrations: 1 ppmw, Δ; 2 ppmw, O; 3 ppmw, ●; and 4 ppmw, □ .

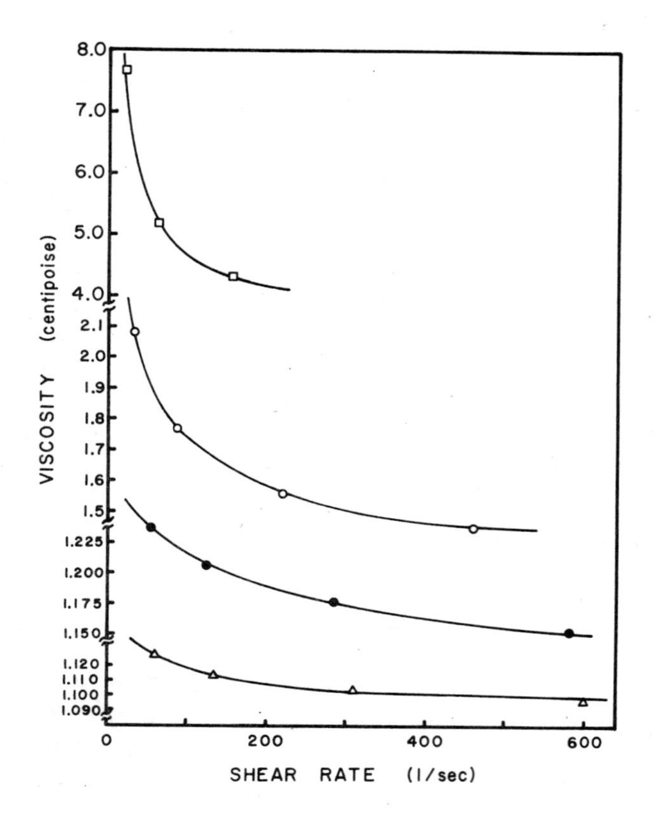

Fig. 5: The viscosity of polyphosphate samples exhibits a large dependence on shear rate at all concentrations: 500 ppmw, □ ; 50 ppmw, O; 25 ppmw, ●; and 12.5 ppmw, Δ.

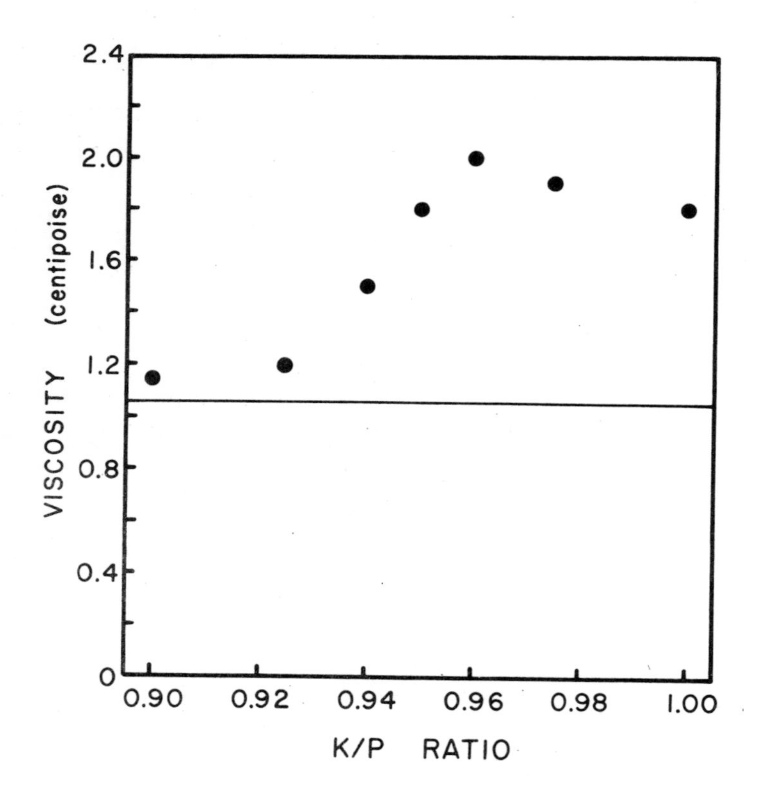

Fig. 6: The viscosities of 50 ppmw solutions of polyphosphate polymers are shown as a function of K/P ratio. The solvent viscosity (1% metaphosphate) is indicated by the line at 1.05 centipoise.

Second International Conference on

Drag Reduction

August 31st - September 2nd, 1977

PAPER C5

MECHANICAL DEGRADATION AND DRAG REDUCING EFFICIENCY OF DILUTE SOLUTIONS OF POLYSTYRENE

J.L. Zakin* and D.L. Hunston,

Naval Research Laboratory, U.S.A.

Summary

Mechanical degradation of high polymers in dilute solution is a major hindrance to their use in practical drag reduction applications. Unfortunately, experimental data on mechanical degradation are sparse and the available results on the effects of solvent nature and polymer concentration are inconclusive. A major problem has been the difficulty of following changes in molecular weight at low concentrations. In the present study this problem was overcome by using drag reduction measurements as well as gel permeation chromatography (GPC) to monitor molecular weight changes. When possible, GPC results and drag reducing efficiency were measured on the same polymer solutions.

Three narrow molecular weight distribution polystyrene samples, with molecular weights of 2, 4, and 7 million were studied. Solutions containing from 100 to 1500 parts per million by weight of polymer were characterized for drag reducing behaviour and then degraded using a high speed mixer. The extent of degradation was observed as a function of mixing time. In addition to studying the effects of concentration and molecular weight on the kinetics of degradation, the role of solvent nature was also investigated. Measurements were made both in toluene which is a good solvent and in a toluene-isooctane mixture which is a poor solvent.

The experimental results indicate that the rate of mechanical degradation increases with increasing molecular weight. The effect of increasing molecular weight is also to increase the polymer's maximum drag reducing ability with the largest increases occurring at the lower molecular weights. Thus, above some high value of molecular weight, no practical drag reducing advantage would be obtained by further increase in molecular weight.

In poor solvents the degradation rate increases as concentration is lowered while in poor solvents there is little concentration effect. At all of the concentrations studied, however, the degradation was more rapid in the poor solvent than in the good solvent. Moreoever, the drag reducing effeciency and mechanical stability are higher in good solvents, practical systems should be selected with good solvency as an important consideration.

*On leave from University of Missouri-Rolla

Held at St. John's College, Cambridge, England.
Organised and sponsored by BHRA Fluid Engineering, Cranfield, Bedford, MK43 0AJ.

INTRODUCTION

The practical use of high polymers as drag reducing agents in turbulent flows has been hampered by the sensitivity of the most effective molecules to mechanical degradation. Mechanical degradation refers to the chemical process in which the activation energy for scission of the polymer chain is provided, at least in part, by mechanical action on the polymer. When sufficient energy is concentrated in the polymer chain, bond rupture occurs. Degradation reduces the effectiveness of the polymer because of the strong dependence of drag reduction on molecular weight (1).

A paper by Casale, Porter, and Johnson (2) and more recent reports by Culter (3) and by Yu (4) review much of the literature on mechanical degradation of high polymers. Most of the studies, however, have been conducted using concentrated solutions or melts so there is only a limited literature on mechanical degradation in the dilute solutions that are of interest in drag reduction. Consequently, the present investigation was designed to study the effects of polymer molecular weight, polymer concentration and solvent nature on the degradation of well-characterized polystyrene samples at low concentration. Drag reduction measurements were performed on these samples so that the trade off between high drag reducing efficiency and high resistance to mechanical degradation could be investigated. When possible, gel permeation chromatography (GPC) measurements of molecular weight distributions (MWD) were made on the same samples.

Literature Review

In the vast literature describing mechanochemistry, a few papers are of particular pertinence to this investigation and will be reviewed here. The earliest theory on the mechanism of mechanical degradation of high polymers was offered by Frenkel. He considered a single polymer molecule in a shear field with the molecule stretched in the direction of flow. The chain ends were not really affected only the central portion of the chain was stretched. As the shear field increased, the stress on the central chain bonds increased until chain rupture occurred.

Bueche (5) also considered a randomly coiled polymer chain in a shear field but calculated that it could not be stretched enough to cause breakage since the shear gradient would lead to rotation of the molecules. If entanglements occurred, however, he suggested that they would increase the force on the molecule with the greatest force occurring near the center of the chain. Because the force depends on molecular size, Bueche's theory predicts that the higher molecular weight molecules will degrade more rapidly, degradation rate increasing as the square of molecular weight. This should result in a narrowing of all but very narrow molecular weight distributions.

Other authors have proposed that degradation might occur randomly at any position on the polymer chain. Boyd and Lin (6) showed theoretically that the MWD would tend toward the "most probable" distribution ($M_w/M_n=2$) as increasing amounts of degradation took place randomly on the molecular chains (random scission).

The most controlled degradation experiments have been carried out under laminar flow conditions in concentric cylinder apparatuses. However, the difficulty in maintaining the narrow clearances that are required to obtain the high shear stresses desired for degradation studies has led many investigators to use capillary tubes (laminar flow for viscous solutions, laminar flow in tiny tubes for low viscosity solutions or turbulent flows for low viscosity solutions in larger tubes) or to use high speed mixers.

Abdel-Alim and Hamielec (7) using 0.4% and 0.7% polyacrylamide solutions and Porter, et al (8) using 9.6 volume percent polyisobutene in n-hexadecane and 9.7 volume percent polyisobutene in 1,2,4-trichlorobenzene showed by GPC analysis that for severe laminar shearing conditions (concentric cylinder apparatus) significant amounts of very low (on the order of 1000) molecular weight chains were produced. The broadening of the molecular weight distribution to $M_w/M_n \approx 2$ led Porter, et al to conclude that Bueche's theory was incorrect in predicting the locations of the breaks, and they suggested that the degradation was random. However, the more rapid degradation of high molecular weight polymer was confirmed. At the lower concentrations studied in Ref 7, the results were in general agreement with Bueche's midpoint break theory. The formation of very short chains at very high shear stresses was explained as breaking of the polymer chains before they had time to extend appreciably. The absence of short chains in low shear stress degradation experiments suggested that the molecules had been extended somewhat before bond scission. Abdel-Alim and Hamielec (7) also observed a maximum molecular weight that could exist without degradation at a given shear stress. This indicates that there is a limiting molecular weight below which degradation does not occur for any particular set of shear conditions.

Harrington and Zimm (9) studied the degradation of dilute (0.03 - 0.06% weight/volume) polystyrene solutions in toluene and in methyl ethyl ketone in laminar capillary flow. They analyzed their data by determining an average force, favg, on the molecule:

$$f_{avg} = G(\eta - \eta_o)/nz \quad (1)$$

where η is solution viscosity, η_o is solvent viscosity, G is shear rate, n is number of molecules/volume, and z is average of the projection of molecular end-to-end lengths on the surface normal to the velocity gradient. Both $(\eta-\eta_o)/n$ and z are molecular weight dependent. Their ratio is approximately proportional to molecular weight. Harrington and Zimm determined the minimum force for degradation by extrapolating their results to zero rate of degradation and found that the critical force was lower in a fair solvent (methyl ethyl ketone) than in a good solvent (toluene).

Patterson, et al (10) degraded polyisobutylene solutions in toluene and cyclohexane by pumping them through a one-inch recirculation system in turbulent flow. While the viscosity average molecular weight in a toluene solution was reduced by only 20%, the drag reducing ability of the solution was reduced by 56%. A GPC analysis of the degraded polymer showed that the higher molecular weight portion was preferentially degraded.

Chang (11) degraded various polymers by pumping them in turbulent flow through the same recirculation system used by Patterson, et al (10). He observed more rapid degradation for a polyisobutylene sample in benzene (poor solvent) then in cyclohexane (good solvent) at concentrations near 0.3%. Because of the much lower intrinsic viscosity, $[\eta]$, in the poor solvent, the 0.9% polyisobutylene in the benzene had approximately the same polymer pervaded colume fraction (12) in solution ($1.08C[\eta]$) as the 0.3% polyisobutylene in cyclohexane. Nevertheless, measurements made in the poor solvent at 0.9% polyisobutylene concentration still showed more rapid degradation.

Finally, Rodriguez and Winding (13) studied degradation of polyisobutylene in a high speed stirrer in the concentration range of 0.05 - 0.16 g/dl. They concluded that the rate of chain scission could be extrapolated to zero at zero concentration presumably because entanglements could not be formed. However, Nakano and Minoura (14), studying polystyrene and poly (methyl methacrylate) degradation in several solvents at solution concentrations of 0.04 to 2% weight/volume, observed a concentration dependence below 0.5% weight/volume; the rate of degradation increased with decreasing concentration. At higher concentration, no effect of concentration was observed.

These literature results raised a number of important questions concerning the degradation process at the very low concentrations used in drag reduction: (1) Does the mechanism favor scission near the center of the chain or random bond breakage, (2) are small molecules (molecular weights on the order of 1000) produced during degradation, and (3) what is the effect of concentration, molecular weight, and molecular weight distribution and solvent type? The present study addresses these questions.

Experimental

Three narrow molecular weight polystyrene samples 2M, 4M, and 7M whose properties are shown in Table 1 were studied in toluene, which is a good solvent, and in a mixture (58% toluene - 42% isooctane) which is a poor solvent. The 7 million molecular weight polystyrene sample begins to precipitate if isooctane is added to this mixture at 25°C indicating that it is close to theta point conditions. In addition to these six polymer-solvent combinations, two other samples were evaluated. These samples were prepared by mixing the 7M and 2M polymers - one mixture contained 13% of 7M while the other contained 42% 7M. Solutions were prepared using toluene and the results of tests with these samples provided information on the effects of polydispersity.

All of the samples were characterized for drag reducing properties by running experiments at a series of different concentrations. Solutions at selected concentrations were then degraded by high speed stirring. At low concentrations it is difficult to make the molecular weight measurements necessary to monitor the degradation process. Changes in drag reduction effectiveness, however, are very sensitive to changes in polymer molecular weight and thus can provide a measure of degradation. At higher concentrations (600 ppm or above), molecular weights were also obtained by GPC analysis. The drag reduction was measured in a capillary tube (0.158 cm in diameter) using an automated flow apparatus (15). This device records the wall shear stress, τ_w, as a function of flow rate and thus the percent drag reduction is 100 $(1-\tau_{wp}/\tau_{ws})$ where τ_{wp} and τ_{ws} are the values of τ_w for the solution and solvent respectively. These measurements can be made at concentrations as low as a few parts per million by weight and can detect small changes in molecular weight that are difficult to monitor by other techniques such as viscometry (1,10).

The GPC measurements were made at the University of Akron by Mr. Richard Seegar using a Waters Associates Model 502 fitted with both a differential refractometer and a UV(254nm) detector. To cover a wide range of molecular weights, and particularly to check for the presence of very low molecular weight polymer, a set of 9 columns with the following porosity designations (A^o), 10^7, 10^6, 10^6, 5×10^5, 1×10^5, 5×10^4, 1×10^4, 5×10^3, 2000-700, were joined together. The large number of columns caused diffusion effects which are manifested as broader molecular weight distributions or high values of M_w/M_n or M_{z+1}/M_z. For example, the GPC analysis with this set of columns gave $M_w/M_n = 1.4$ for 4M, whereas earlier analysis with a smaller set gave a value of 1.1. However, the results obtained with these columns were quite useful for making comparisons and for determining trends in the data.

Degradation experiments were performed using a Virtis homogenizer at 3000 rpm. Sample temperatures were held at $25^oC \pm 3^oC$ during degradation. At selected time intervals during the degradation, samples were taken and diluted to between 100 and 250 ppm for drag reduction measurements. The diluted concentrations were chosen to obtain convenient drag reduction levels in the apparatus used. The ratio of the drag reduction at time, t, to the drag reduction of the fresh solution (both measured at the same flow rates) was determined and plotted as a function of t. For toluene the ratio was determined at a Reynolds number of 16,000 and for the mixed solvent it was determined at 17,000.

RESULTS

Drag Reduction in Good and Poor Solvents

Figure 1 shows percent drag reduction (% DR) versus concentration results for the three polystyrene samples in toluene. The polymer and solution characteristics are listed in Table 1. Similar data for the mixed solvent are shown in Figure 2. The expected increase in drag reduction effectiveness with increasing molecular weight is observed. It is also of considerable interest to note that at the higher concentrations ($\approx$500 ppm-2M, $\approx$100 ppm - 4M, $\approx$30 ppm - 7M), drag reduction in the poor solvent is two-thirds or more of that in the good solvent.

Effect of Molecular Weight on Degradation of Polystyrene

Figure 3 shows results for degradation under similar conditions of mechanical shear and concentration (1000 ppm) for the three polystyrene samples in toluene. Measurements of % DR, particularly at low values, are subject to considerable error and thus the curves drawn in Figures 3 and 4 are dashed at long degradation times to point up the uncertainties of the results in this region. Figure 3 shows a clear trend to more rapid relative degradation with increasing molecular weight. The degraded samples were taken at times well before equilibrium molecular weights were reached. Therefore, at all degradation times, the actual molecular weights were higher for those samples with initially higher molecular weight even though the relative amounts of degradation were greater for these samples.

GPC results on fresh and degraded samples of 970 ppm 4M in toluene are shown in Table 2. Values of M_z and M_{z+1} along with ratios of degraded to initial molecular weight and % DR/% $DR_{initial}$ are listed. At this concentration, the results in Figure 1 show that fresh 2M ($M_w = 2.4 \times 10^6$, $M_{z+1} = 4.0 \times 10^6$) gives drag reduction in toluene of about 60% of that of fresh 4M ($M_w = 4.1 \times 10^6$, $M_{z+1} = 5.6 \times 10^6$), both samples having relatively narrow molecular weight distributions. Table 3 shows GPC analysis results for the degradation of 2M in the mixed solvent at 1000 ppm. Values of M_z and M_{z+1} are shown at different degradation times along with values of % DR/%$DR_{initial}$.

Effect of Solvent Nature and Concentration on Degradation of Polystyrene

Figure 4 shows results for degradation of polystyrene 4M in good and in poor solvents. At equal absolute concentrations, 150 ppm, the rate of degradation is far greater in the poor solvent than in the good solvent. If the comparison is made at approximately equal values (0.1) of the pervaded volume fraction of the polymer molecules ($1.08C[\eta]$, see Table 1) rather than equal concentrations, the results are similar. There is little difference between the rates of degradation of polystyrene at 150 ppm and 600 ppm in the poor solvent while a significant difference between rates at different concentrations is seen in the results for the good solvent.

DISCUSSION OF RESULTS

Drag Reduction of Fresh Polymer

Figures 1 and 2 indicate that, as expected, the drag reducing efficiency increases with increasing molecular weight regardless of solvent type. The large increases shown in these Figures are consistent with the fact that the molecular weights are less than one order of magnitude greater than the minimum molecular weight, M_{min}, required to observe drag reduction with polystyrene ($M_{w,min} \sim 10^6$ for the flow conditions of these experiments). Previous studies (16) have shown that the maximum drag reducing ability of a polymer increases most rapidly in the range from M_{min} to $10M_{min}$. The effects of molecular weight distribution can also be examined by considering the 7M - 2M mixtures. The data for mixtures containing either 13% 7M or 42% 7M superimpose on the 7M results shown in Figure 1 if the concentration of the 7M component is plotted as the abscissa. The presence of concentrations of 2M which by themselves would give significant drag reduction had no measurable effect at these ratios of the two polymers. Hunston and Reischman observed similar effects for polystyrene in benzene at concentrations of 2.5% to 10% of 7M (17).

Earlier investigators observed that drag reduction levels in poor solvents were lower than in good solvents at equal concentration under similar flow conditions (18-21). The data in Figures 1 and 2 show clearly how concentration, molecular weight and solvency are related to drag reduction efficiency for polystyrene. Above some limiting concentration for each molecular weight, drag reduction in the poor solvent is about two-thirds that in the good solvent at the same concentration and flow rate (approximately equal Reynolds numbers). Concentrations required for equal percent drag reduction are three to four times greater in the poor solvent. This is roughly equal to the ratios of intrinsic viscosity in the good solvent to those in the poor solvent. Thus, for a given polymer sample, $C[\eta]$ values which are close to pervaded volume fractions at equal levels of drag reduction are roughly constant, a result similar to previous observations (21).

Degradation Rates

The loss of drag reducing effectiveness through mechanical degradation can be described for all the samples tested by the equation

$$\frac{\%DR}{\%DR_o} = \left(1 - \frac{\%DR_\infty}{\%DR_o}\right)e^{-Rt} + \frac{\%DR_\infty}{\%DR_o} \qquad (2)$$

where % DR is percent drag reduction, t is time and the subscripts oo and o refer to infinite and zero time. The constant R is a measure of the rate of decay of drag reduction effectiveness and hence of molecular weight. If DR_{oo} is approximately zero, Equation 2 reduces to

$$\frac{\%DR}{\%DR_o} = e^{-Rt} \qquad (3)$$

Values of R calculated from Equation 3 for degradation in both solvents using all three polymers at various concentrations are shown in Table 4. Since only a few measurements were taken during each run and since the values of % DR are subject to considerable uncertainty, particularly at low values of % DR, the R values can only be compared qualitatively. Nevertheless, a number of important trends can be seen.

For all of the samples dissolved in toluene, there is a clear trend to higher values of R, corresponding to larger decay rates, as the concentration is decreased. For concentrations near 1000 ppm in toluene the values of R decrease with decreasing molecular weight. At lower concentrations the molecular weight effect in toluene is not as clear since the decreasing values of R as the molecular weight is decreased may be due to increasing concentration and/or decreasing molecular weight. These lower concentration results are however all at $1.08C[\eta]\approx 0.1$ (Table 1), so it is likely that the decreased values of R are associated with lower molecular weight of the polymer.

With the 7M - 2M mixtures in toluene the decay rates are close to those which might be expected if the 7M alone were in solution. It should be noted that the decay rates are based on percent drag reduction which, as mentioned above, depends only on the 7M concentration. The presence of large amounts of 2M, therefore, appears to have little effect on the degradation rate of 7M.

For the mixed solvent, the decay constant also decreases with decreasing molecular weight. There is, however, little change in R values with concentration changes. The decrease from 5.4×10^{-2} to 4.3×10^{-2} for concentrations of 4M ranging from 1500 ppm to 150 ppm may not be significant considering the uncertainties in the data.

The results in Table 4 also indicate that the decay rate is larger in the poor solvent than in the good solvent. These data are in agreement with the previous results of Zakin and Hunston (22) who concluded that degradation is more rapid in a poor solvent because of a lower threshold molecular weight for mechanical degradation under fixed shear conditions and/or because of thermodynamic conditions in the poor solvent which favor mechanical degradation.

Since drag reduction effectiveness is dependent on the highest molecular weight portions of the MWD (10) and since the z+1 moment weights these portions very heavily, M_{z+1} results were compared with drag reduction efficiency. Table 2 lists the ratios of M_{z+1} at time t to the initial value of M_{z+1} for the degradation of 970 ppm 4M in toluene. A rough correlation is observed between these values and the drag reduction ratios at 45 and 135 seconds degradation time. The results at 405 seconds show a value for $M_{z+1}/M_{z+1\ initial}$ of 0.42 while the relative drag reduction value is only 0.16, a much larger fall-off. However, at this level of degradation, the polystyrene molecular weight is approaching a level ($M_{z+1,min} \approx 2 \times 10^6$) where there is little or no drag reduction under these flow conditions. A much poorer correspondence between $M_{z+1}/M_{z+1\ initial}$ and the drag reduction ratios is seen in Table 3 for 1000 ppm 2M in the poor solvent. This is probably due to the fact that the level of drag reduction is much lower for this sample (see Figure 2) and that its molecular weight is close to the level ($M_{z+1,min} \approx 2 \times 10^6$) at which there is little or no drag reduction under these flow conditions. Correlation of $(M_{z+1} - M_{z+1,min})/(M_{z+1\ initial} - M_{z+1,min})$ would be in better agreement with the drag reduction ratios for these mixed solvent data; however, it would worsen the correspondence of the data in Table 2. Consequently, no definitive correlation can be established with these data.

Tables 2 and 3 also indicate that the ratios, M_{z+1}/M_z do not change with increasing degradation. GPC determined ratios of Mw/Mn for 1000 ppm 4M in toluene are also constant at $1.39 \pm .04$ independent of degradation time. Because of the diffusion effects in the GPC analysis, 1.39 corresponds to an actual ratio near 1.1 and thus no significant narrowing of the MWD can be expected. For 1000 ppm 2M in the poor solvent, Mw/Mn does decrease from 1.80 to 1.48 after 270 seconds. The tendency toward narrow MWD supports the mechanism proposed by Bueche (5), i.e., bond breakage near the midpoint of a polymer chain is favored. These results definitely do not support a random scission mechanism.

Finally, although the GPC analysis had provision for detection and characterization products down to molecular weights of $\sim$1000, no low molecular weight material was observed. Apparently, either the shear conditions in the Couette viscometer experiments of Abdel-Alim (7) and of Porter, et al (8) were far more severe than those in the experiments reported here, or the higher concentrations and uniform shear field in their Couette viscometers caused the degradation mechanism to be different.

The rate of mechanical degradation has been shown experimentally to increase with increasing molecular weight in both good and poor solvents. Consequently, for each polymer-solvent system, there must be an optimum molecular weight above which the mechanical degradation would be so rapid for any given set of flow conditions that further increases in molecular weight have no significant practical benefit. The results also indicate that the decay rates for a given polymer are lower in the good solvent than in the poor solvent at all concentrations. Since the amount of drag reduction is also greater in a good solvent, it is advantageous on both counts to take solvency into account in selecting polymers for use in practical applications.

Conclusions:

(1) To obtain equivalent drag reduction at similar flow conditions requires 3 to 4 times the concentration of polystyrene in poor solvent than is needed in good solvent. At equal concentrations, drag reduction in the poor solvent is about two-thirds that in the good solvent.

(2) The rate, R, at which the drag reducing effectiveness of polystyrene is lost through mechanical degradation in dilute solutions, increases with increasing molecular weight in both good and poor solvents (equivalent shear conditions).

(3) In dilute poor solvent solutions, R is greater than in dilute good solvent solutions at equal molecular weights and concentrations (equivalent shear conditions).

(4) For the good solvents, R increases with decreasing polymer concentration while little concentration effect is observed in the poor solvent.

(5) The presence of large quantities of low molecular weight polymer has no "protective" effect on the degradation of high molecular weight polymer.

(6) The mechanical degradation mechanism supported by these results on dilute solutions is one of preferred degradation near the center of the polymer molecules. There is no evidence supporting a random scission mechanism.

(7) No low molecular weight degradation product such as was observed under very severe, but uniform, shear conditions and higher concentrations in Couette viscometers by Abdel-Alim and Hamielec (7) and Porter, et al (8) was observed.

Acknowledgments:

The authors wish to thank Mr. R. Patterson of the Naval Research Laboratory for his assistance during the course of this work. Support was provided in part by a grant from the Office of Naval Research and by an IPA assignment agreement between the Naval Research Laboratory and the University of Missouri - Rolla.

References

1. Paterson, R. W. and Abernathy, F. H., "Turbulent Flow Drag Reduction and Degradation with Dilute Polymer Solutions," J. Fluid Mech., 43, 689-710 (1970).

2. Casale, A., Porter, R. and Johnson, J. F., "The Mechanochemistry of High Polymers," Rubber Chem. Tech., 44, 534-577 (1971).

3. Culter, J. D., "Mechanical Degradation of Polymers in Dilute Solutions," Ph.D. Thesis, University of Missouri-Rolla, 1976.

4. Yu, F. S., "A Study of the Mechanisms of Mechanical Degradation of High Molecular Weight Polymers in Dilute Solution," Ph.D. Thesis (in preparation), University of Missouri-Rolla.

5. Bueche, F., "Mechanical Degradation of High Polymers," J. Appl. Polymer Sci., 4, 101-106 (1960).

6. Boyd, R. H. and Lin, T. P., "Theoretical Depolymerization Kinetics III, The Effect of Molecular-Weight Distribution in Degrading Polymers Undergoing Random Scission Initiation," J. Chem. Phys., 45, 778-78, (1960).

7. Abdel-Alim, A. H. and Hamielec, A. E., "Shear Degradation of Water-Soluble Polymers I, Degradation of Polyacrylamide in a High Shear Viscometer," J. Appl. Polymer Sci., 17, 3769-3778 (1973).

8. Porter, R. S., Cantow, M. J. R., and Johnson, J. F., "Polyisobutene Degradation in Laminar Flow: The Effect on Molecular Weight Distribution," J. Polymer Sci., Part C, No. 16, 1-12 (1967).

9. Harrington, R. E. and Zimm, B. H., "Degradation of Polymers by Controlled Hydrodynamic Shear," J. Phys. Chem., 69, 161-175 (1965).

10. Patterson, G. K., Hershey, H. C., Green, C. P., and Zakin, J. L., "Effect of Degradation by Pumping on Normal Stresses in Polyisobutene Solutions," Trans. Soc. Rheology, 10:2, 498-500 (1966).

11. Chang, I. C., "A Study of Degradation in the Pumping of Dilute Polymer Solutions," M.S. Thesis, University of Missouri-Rolla, 1965.

12. Simha, R. and Zakin, J. L., "Compression of Flexible Chain Molecules in Solutions," J. Chem. Phys., 33, 1791-1793 (1960).

13. Rodriguez, F. and Winding, C. C., "Mechanical Degradation of Polyisobutylene Solutions," Industrial Engineering Chemistry, 51, 1281-1284 (1959).

14. Nakano, A. and Minoura, Y., "Effects of Solvent and Concentration on Scission of Polymers with High Speed Stirring," J. Appl. Polymer Sci., 19, 2119-2130 (1975).

15. Little, R. C. and Wiegard, M., "Drag Reduction and Structural Turbulence in Flowing Polyox Solutions," J. Appl. Polymer Sci., 14, 409-419 (1970).

16. Little, R. C., "Drag Reduction in Capillary Tubes as a Function of Polymer Concentration and Molecular Weight," J. Colloid Interfac. Sci., 37, 811-818 (1971).

17. Hunston, D. L. and Reischman, M. M., "The Role of Polydispersity in the Mechanism of Drag Reduction," Phys. Fluids, 18, 1626-1629 (1975).

18. Merrill, E. W., Smith, K. A., Shin H. and Mickley, H. S., "Study of Turbulent Flows of Dilute Polymer Solutions in a Couette Viscometer," Trans. Soc. Rheology, 10:1, 335-351 (1966).

19. Hershey, H. C. and Zakin, J. L., "A Molecular Approach to Predicting the Onset of Drag Reduction in the Turbulent Flow of Dilute Polymer Solutions," Chem. Eng. Sci., 22, 1847-1857 (1967).

20. Peyser, P. and Little, R. C., "The Drag Reduction of Dilute Polymer Solutions as a Function of Solvent Power, Viscosity, and Temperature," J. Appl. Polymer Sci., 15, 2623-2637 (1971).

21. Little, R. C., "The Effect of Salt Concentration on the Drag Reduction Efficiency of Polyethylene Oxide Polymers," Nature Phys. Sci., 242, 79-80 (1973).

22. Zakin, J. L. and Hunston, D. L., "Effects of Solvent Nature on the Mechanical Degradation of High Polymer Solutions," J. Appl. Polymer Sci., in press.

Table 1 Polymer Solution Properties

Polystyrene Sample	Solvent	M_w	$\frac{M_w}{M_n}$	$[\eta]$ dl/g	Concentration ppm	Pervaded Volume Fraction (12) 1.08 C $[\eta]$
7M	toluene	7.1×10^6	1.1	10.4	1000	1.0
7M	toluene	7.1×10^6	1.1	10.4	100	.10
7M	mixed solvent	7.1×10^6	1.1	2.4	904	.20
7M	mixed solvent	7.1×10^6	1.1	2.4	452	.10
4M	toluene	4.1×10^6	1.1	7.2	970	.68
4M	toluene	4.1×10^6	1.1	7.2	150	.11
4M	mixed solvent	4.1×10^6	1.1	1.8	1500	.24
4M	mixed solvent	4.1×10^6	1.1	1.8	600	.10
4M	mixed solvent	4.1×10^6	1.1	1.8	150	.02
2M	toluene	2.4×10^6	1.2	5.0	1000	.48
2M	toluene	2.4×10^6	1.2	5.0	250	.12
2M	mixed solvent	2.4×10^6	1.2	1.4	1000	.13

Table 2 Degradation of 970 ppm Polystyrene 4M in Toluene

Time Seconds	$M_z \times 10^{-6}$	$\frac{M_z}{M_z \text{ initial}}$	$M_{z+1} \times 10^{-6}$	$\frac{M_{z+1}}{M_{z+1} \text{ initial}}$	$\frac{M_{z+1}}{M_z}$	$\frac{\% DR}{\% DR \text{ initial}}$
0	4.95	1.00	5.63	1.00	1.20	1.00
15	--	--	--	--	--	.92
45	3.13	.63	3.74	.66	1.25	.75
135	2.38	.48	2.83	.50	1.27	.51
405	2.35	.47	2.37	.42	1.22	.16

Table 3 Degradation of 1000 ppm Polystyrene 2M in Poor Solvent

Time Seconds	$M_z \times 10^{-6}$	$\frac{M_z}{M_z \text{ initial}}$	$M_{z+1} \times 10^{-6}$	$\frac{M_{z+1}}{M_{z+1} \text{ initial}}$	$\frac{M_{z+1}}{M_z}$	$\frac{\%\ DR}{\%DR \text{ initial}}$
0	3.28	1.00	4.04	1.00	1.23	1.00
15	--	--	--	--	--	.53
30	2.69	.82	3.24	.80	1.20	.47
90	2.37	.72	2.86	.71	1.20	.25
270	1.77	.54	2.07	.51	1.16	0

Table 4 Decay Constants for Equation 3

Polymer	Solvent	Concentration ppm	$R \times 10^{-2}$ sec^{-1}
7M	toluene	1000	.8
7M	toluene	100	2.0
7M	mixed solvent	904	5.6
7M	mixed solvent	452	5.5
4M	toluene	970	0.5
4M	toluene	150	1.5
4M	mixed solvent	1500	5.4
4M	mixed solvent	600	4.9
4M	mixed solvent	150	4.3
2M	toluene	1000	0.2
2M	toluene	250	0.5
2M	mixed solvent	1000	2.6
13% 7M } 87% 2M	toluene	1000	1.8
		150	4.8
42% 7M } 58% 2M	toluene	800	0.7
		150	4.0

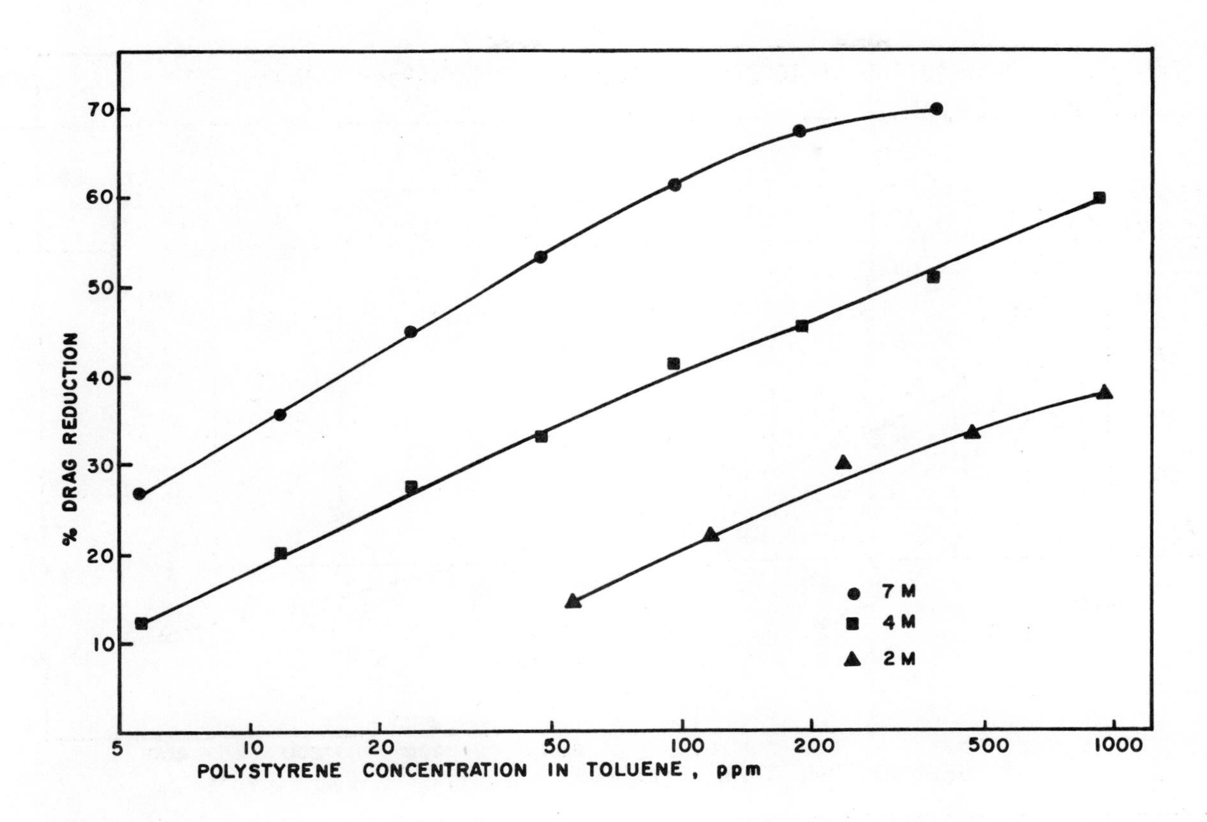

Figure 1 - Drag reduction effectiveness versus polystyrene concentration in toluene (tube diameter 0.158 cm, Reynolds number = 16,000).

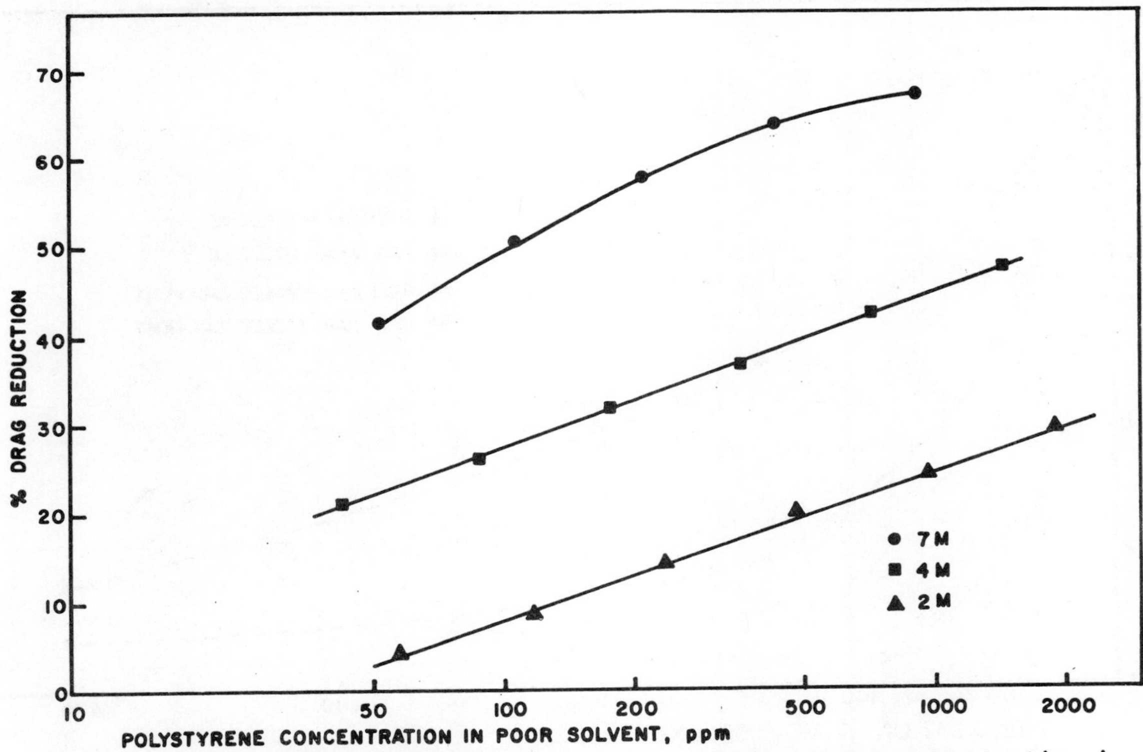

Figure 2 - Drag reduction effectiveness versus polystyrene concentration in 58% toluene 42% isooctane (tube diameter 0.158 cm, Reynolds number = 17,000)

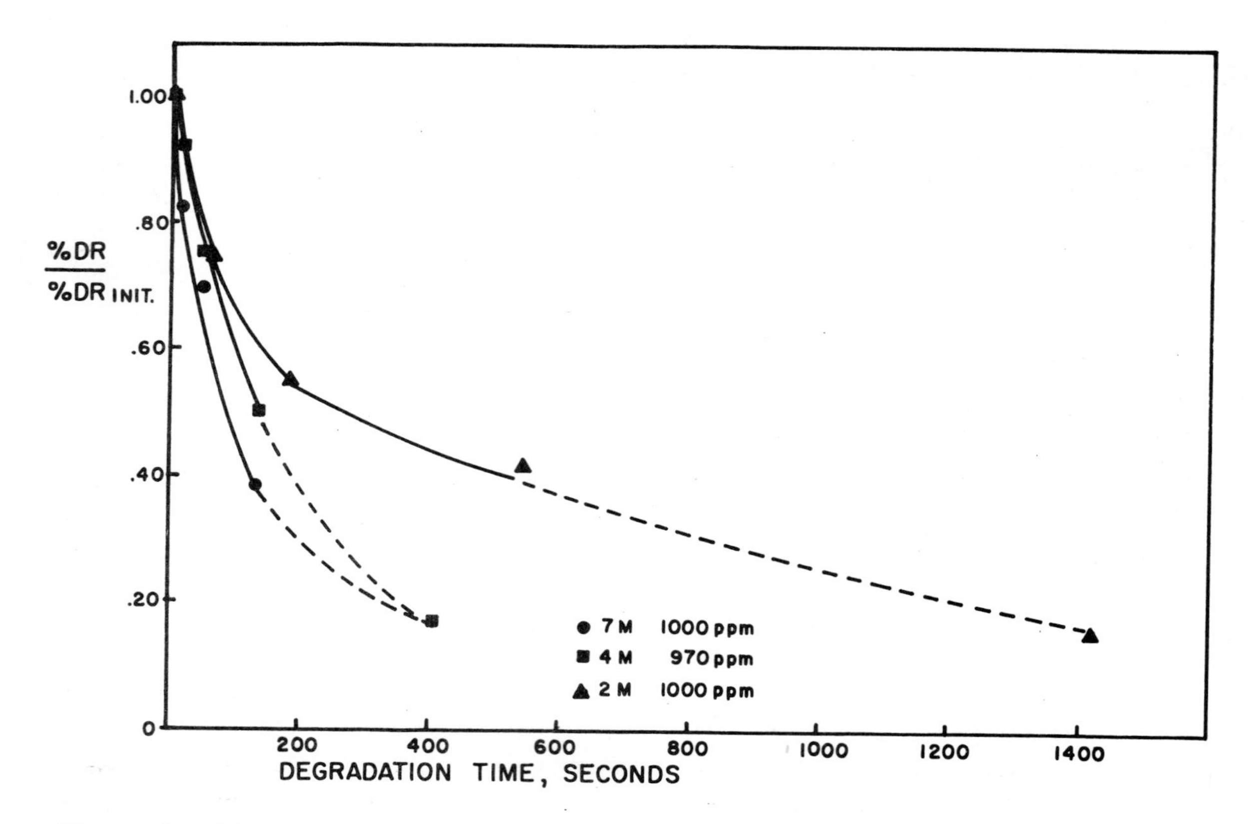

Figure 3 - Effect of molecular weight on degradation of polystyrene in toluene.

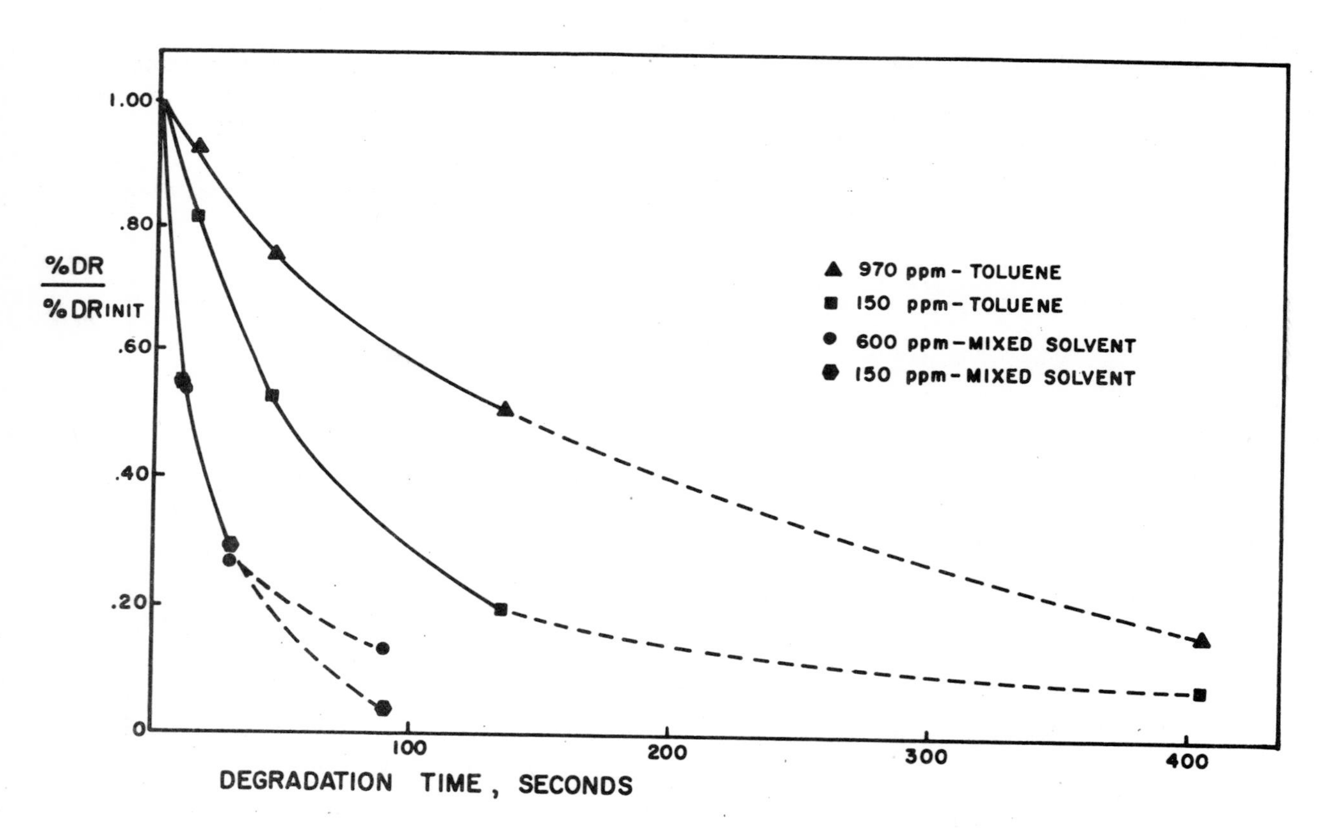

Figure 4 - Effect of concentration on degradation of 4M polystyrene in toluene and in 58% toluene - 42% isooctane.

Second International Conference on

Drag Reduction

August 31st - September 2nd, 1977

PAPER D1

SOME ASPECTS OF TWO-PHASE FLOW DRAG REDUCTION

D.R.H. Beattie, B.Sc., B.E., M.Eng.Sc.,

Australian Atomic Energy Commission, Australia

Summary

Drag reduction is characterised by coefficients in the non-dimensional velocity profile $j^+ = a \log y^+ + b$ which differ from those normally encountered. In the case of drag reduction in gas-liquid flows with no drag reducing additives, empirically obtained values of a and b suggest that only certain well-defined values exist, and that these values appear to form parts of some simple series. Similarly, only well-dêfined types of sub-layers exist. Flows are therefore classified by the triad (m, n, z) where m and n are integers from which the non-dimensional velocity profile coefficients can be calculated, and z specified the sub-layer structure.

Typical data are subdivided into groups each having a particular (m, n, z) set. For each group, a demonstration is given of the compatability of data for velocity profile, friction factor, and average void fraction with the theoretical expressions for those parameters determined by the appropriate (m, n, z) set.

The (m, n, z) sets for high pressure steam-water flows are tabulated. Compatibility with typical steam-water data is demonstrated.

Held at St. John's College, Cambridge, England.

Organised and sponsored by BHRA Fluid Engineering, Cranfield, Bedford, MK43 0AJ.

NOMENCLATURE

a_1-a_5	mixing length theory coefficients
a,b,c,d,e	sublayer types (Table 1)
f	friction factor (Table 1)
j	local mixture velocity, $u_\ell(1-\alpha)+u_g\alpha$
j^+	non-dimensional velocity (Table 1)
m,n	integers in generalised equations (13)-(16)
u	local phase velocity
y	wall distance
y^+	non-dimensional wall distance (Table 1)
z	flow direction
A	area fraction of core for separated flow
B	volume flow fraction of core for separated flow
C_o	distribution parameter
D	equivalent diameter
G	mass flux
P	pressure
Q^+	non-dimensional film flowrate in separated flow
Re	Reynolds number
S^+	non-dimensional film thickness in separated flow
V	drift velocity, u-j
We	Weber number
α	local voidage
β	volume flow ratio, $\langle j_g \rangle / \langle j \rangle$
μ	dynamic viscosity
σ	surface tension
ρ	density

Subscripts

c	value for core of separated flow
f	value for film of separated flow
F	frictional component
g	gas value
ℓ	liquid value

Operator

<--->	quantity is averaged over the flow section

1. INTRODUCTION

Pressure losses for low quality (i.e. with low gas content) gas-liquid bubble flows and for high quality gas-liquid dry-wall flows are consistent with the single-phase Colebrook friction factor equation if the relevant dimensionless groups are appropriately defined (Ref. 1). An examination of the variation of friction factor with quality for these two flow types shows that for the intermediate quality region, pressure losses are not consistent with the Colebrook equation.

Flows encountered in the intermediate region are characterised by velocity profiles which deviate from the 'universal' profile and, as such, are similar to single phase drag-reduced flows generated by small amounts of suitable additives, although additives are not necessary to stimulate two-phase drag reduction.

This paper examines various aspects of two-phase drag reduction of the type encountered without drag-reducing additives. Some preliminary aspects have been reported in Ref. 2. Since then several papers have appeared on the topic (e.g. Ref.3) but all are concerned with the effect of additives on two-phase pressure losses.

A feature which distinguishes two-phase drag reduction without additives from additive-induced drag reduction (for both single and two-phase flows) appears to be the strong suggestion in the former of eigen-value characteristics of the type often encountered in the solution of bounded field problems. This behaviour may be significant in the further interpretation of the turbulent flow field.

2. BACKGROUND TWO-PHASE FLOW EQUATIONS

Two-phase flow equations relevant to the examination of gas-liquid flow drag reduction are summarised below. Their derivation, which is outside the scope of this paper, will be given in Ref. 4. Relevant aspects of the theory are in Refs. (5-8). Where appropriate, use of the equivalent diameter renders the equations applicable to non-circular ducts.

2.1 Definition of Dimensionless Groups

As in single phase flow, the nature of the sub-layer flow near the wall determines the form of the dimensionless-group definitions. The various possible structures of the two-phase sub-layer lead to different definitions of the same basic dimensionless groups, each definition being valid for the particular sub-layer structure. Five sub-layer types are considered here. Others can be postulated, but these are either unrealistic or lead to dimensionless group definitions which are either identical, or very close, to those for the five sub-layer types considered. The sub-layer types are:

(a) A liquid sub-layer containing rigid surface bubbles. Bubble flows, some bubble/slug flows, and annular flows with entrained bubbles in the liquid film have sub-layers of this type if the sub-layer bubble surfaces are rigid. Surfactant impurities have this effect on small bubbles in liquids with polar molecules;

(b) A liquid sub-layer containing non-rigid-surface bubbles. Typical flows with this sub-layer are similar to those above, except that bubble surface movement can occur, allowing viscous dissipation within the bubble;

(c) A sub-layer in which local instantaneous shear stress alternates between values expected from calculations based on a treatment of the local flow as gas only or liquid only. 'Annular' flows consisting of a wavy liquid film with no entrained bubbles and a gas, droplet, or froth core; and some 'slug' flows, have sub-layers of this type;

(d) A gas-only sub-layer, ('dry-wall' flows) as expected with, for example, post heat-transfer crisis flows; and,

(e) A liquid sub-layer with attached wall bubbles protruding within the turbulent flow, as expected e.g. for some diabatic flows.

Dimensionless-group definitions for the five sub-layer types are summarised in Table 1.

TABLE 1. DIMENSIONLESS-GROUP DEFINITIONS FOR VARIOUS TWO-PHASE FLOW SUB-LAYER TYPES

Sub-layer Type	Pressure Loss Characteristics	Local Structure
Viscosity dependent flows (a,b,c,d)	$f = \frac{D(dP/dz)_F}{2\rho\langle j\rangle^2} = f(Re = \frac{\rho\langle j\rangle D}{\mu})$	$j^+ = \frac{j}{\langle j\rangle}\sqrt{\frac{2}{f}} = j^+(y^+ = \frac{y}{R}\frac{Re}{2}\sqrt{\frac{f}{2}})$
Surface-tension dependent flows (e)	$f = \frac{D(dP/dz)_F}{2\rho\langle j\rangle^2} = f(We = \frac{\rho\langle j\rangle^2 D}{\sigma})$	$j^+ = \frac{j}{\langle j\rangle}\sqrt{\frac{2}{f}} = j^+(y^+ = \frac{y}{R} We \frac{f}{2})$

Sub-layer Notation/ Type	Density Definition	Viscosity Definition
a. Rigid surface bubbles	$\rho = \rho_\ell(1-\beta) + \rho_g\beta$	$\mu = \mu_\ell(1+2.5\ \beta)$
b. Non-rigid surface bubbles	$\rho = \rho_\ell(1-\beta) + \rho_g\beta$	$\mu = \mu_\ell(1 + \frac{2.5\ \mu_g + \mu_\ell}{\mu_g + \mu_\ell}\beta)$
c. Wavy-gas/liquid interface	$\rho = \rho_\ell(1-\beta) + \rho_g\beta$	$\mu = \mu_\ell(1-\beta) + \mu_g\beta$
d. Dry wall	$\rho = \rho_g$	$\mu = \mu_g$
e. Attached wall bubbles	$\rho = \rho_\ell(1-\beta) + \rho_g\beta$	N.A.

2.2 Non-dimensional Velocity Profiles

Two-phase non-dimensional velocity profiles have much in common with single phase behaviour. Round tube non-dimensional velocities and wall distances as defined in Table 1 are related by

$$j^+ = a_1 + a_2 \ln y^+ \tag{1}$$

in the turbulent region. A number of such equations may be necessary for a complete description of the turbulent velocity profile (cf. the 'buffer' and 'turbulent' regions of single phase flow) and, as in single phase flow, the region adjacent to a smooth wall is assumed to be described by

$$j^+ = y^+ \quad . \tag{2}$$

2.3 Friction Factor Relations

Averaging equation 1 over the flow section leads to equations of the form

$$1/\sqrt{f} = a_3 + a_4 \log Re\sqrt{f} \tag{3}$$

for viscosity dependent flows, and

$$1/\sqrt{f} = a_5 + a_4 \log We\ f \tag{4}$$

for surface tension dependent flows.

2.4 Average Void Fraction

The behaviour of the average void fraction $<\alpha>$, and also of some other quantities, depends on whether the flow is distributed (one phase is continuous throughout the flow section) or separated (the flow section can be divided into liquid-phase continuous and gas-phase continuous regions). Note that the discussion of sub-layer types in section 2.1 indicates that, while a particular flow type implies a particular sub-layer type in some cases, evidence of a particular sub-layer is not sufficient to indicate whether a flow is distributed or separated.

2.4.1 Distributed two-phase flows

Liquid-phase continuous distributed flow void fractions are described by

$$<j_g>/<\alpha> \quad = \quad C_o <j> + V_g \tag{5}$$

and gas-phase continuous void fractions, by

$$<j_\ell>/<1-\alpha> \quad = \quad C_o <j> + V_\ell \quad . \tag{6}$$

In the above, C_o is a 'distribution parameter' given by

$$C_o \quad = \quad 1 + 0.66\, a_4 \sqrt{f} \tag{7}$$

where a_4 is the logarithmic term coefficient in equation 3 or 4 used to obtain f. The V_g and V_ℓ are dispersed phase 'drift velocities' which account for the fact that the local dispersed phase velocities differ from the local mixture velocities ($V_{g,\ell} = u_{g,\ell} - j$). Values of V_g can be found in Ref. 9 for some flows. Values for V_ℓ are at present unknown but, fortunately, can be considered to be negligible at the high velocities encountered in gas-phase continuous distributed two-phase flows.

For negligible drift velocities, equations 5 and 6 are equivalent to the simpler expressions

$$<\alpha> \quad = \quad \beta/C_o \tag{5a}$$

and

$$1-<\alpha> \quad = \quad (1-\beta)/C_o \tag{6a}$$

where β is the fraction of gas flowing by volume.

Note that one of the sub-layers a,b,c,e of section 2.1 would apply in the friction factor calculation for equations 5 and 5a; and sub-layer d, for equations 6 and 6a.

2.4.2 Separated two-phase flows

The following equations apply to annular flows with a liquid-phase continuous film and a gas-phase continuous core. The equations presumably can be adapted to inverse annular flows (gas-phase continuous film and liquid-phase continuous core) but not data have been found to confirm this.

Entrainment is allowed for by introducing core and film voidages, α_c, α_f, and gas volume flow ratios β_c, β_f, all treated as having flat profiles over the appropriate regions of the flow cross section. The area fraction of the core, A, and the volume flow fraction of the core, B, (i.e. the volume flowrate) are then given by

$$A \quad = \quad (\alpha_f - <\alpha>)/(\alpha_f - \alpha_c) \tag{8}$$

and $$B = (\beta_f - \langle\beta\rangle)/(\beta_f - \beta_c) \quad . \tag{9}$$

For no entrainment, $A \equiv \langle\alpha\rangle$ and $B \equiv \beta$. Film entrainment effects are significant only at lower voidage annular flows ($\langle\alpha\rangle \sim 0.5$) and core entrainment effects, only at higher void annular flows ($\langle\alpha\rangle \gtrsim 0.9$). Entrainment will be discussed more fully elsewhere (Ref. 8). In the present paper, entrainment is neglected since the flow conditions discussed are such that the effects of entrainment on average void values are mostly quite small.

Defining non-dimensional film volume flowrate Q^+ and film thickness S^+ by

$$Q^+ = \frac{1-B}{2(1+\sqrt{A})} \, Re \tag{10a}$$

(Viscosity dependent flows)

$$S^+ = \frac{1-\sqrt{A}}{2} \, Re\sqrt{\frac{f}{2}} \tag{11a}$$

$$Q^+ = \frac{1-B}{1+\sqrt{A}} \, We\sqrt{\frac{f}{2}} \tag{10b}$$

(Surface tension dependent flows)

and $$S^+ = (1-\sqrt{A}) \, We \, \frac{f}{2} \tag{11b}$$

then $$Q^+ = \int_{j^+=o}^{S^+} j^+ \, dy^+ \tag{12}$$

For known entrainment conditions, equation 12 may be used to calculate voidages for separated flows. Instead of $\langle\alpha\rangle$, the method, in fact, predicts $1-\langle\alpha\rangle$ which is more significant than $\langle\alpha\rangle$ at the low $1-\langle\alpha\rangle$ values normally encountered in annular flow, since apparently small errors in $\langle\alpha\rangle$ become unacceptable errors in $1-\langle\alpha\rangle$. Moreover, percentage errors in density calculations are approximately equal to percentage errors in $1-\langle\alpha\rangle$.

3. GENERALISED TWO-PHASE FLOW DRAG REDUCTION EQUATIONS

Drag reduction is characterised by mixing length coefficients which are different from those usually encountered in flow structure or pressure loss equations. Coefficient values for additive-induced drag reduction appear to form part of a continuous spectrum of values; in two-phase flows with no additives, they appear to form part of a spectrum of discrete values. An analysis of values obtained for empirical coefficients a_1 to a_4 in equations 1 and 3 indicates that nearly all velocity profile and friction factor equations for viscosity dependent two-phase flows with no additives can be generated from

$$j^+ = (\sqrt{2})^{m+5} \{\log y^+ - (0.11\,m + 0.50\,n)\} \quad , \tag{13}$$

and from the integrated form of equation (13)

$$1/\sqrt{f} = (\sqrt{2})^{m+4} \{\log Re\sqrt{f} - (0.11\,m + 0.50\,n + 1.1)\} \quad , \tag{14}$$

by appropriate integer choices for m and n.

Similarly, with the exception of a particular class of flow which is discussed later, friction factor equations for surface-tension dependent flows can all be generated from

$$1/\sqrt{f} = (\sqrt{2})^{m+4} \{\log We\ f - (0.2\ m + 0.3\ n - 1.1)\} \quad (15)$$

with the implied velocity profile expression

$$j^{+} = (\sqrt{2})^{m+5} \{\log y^{+} - (0.2\ m + 0.3\ n - 2.05)\} \ , \quad (16)$$

again with integer choices for m and n. Equations 15 and 16 are more tentative than equations 13 and 14, largely because of a relative scarcity of surface-tension dependent flow data.

As well as being relevant in friction factor and velocity profile relations, the values of m and n are relevant for specifying the average voidage characteristics of the flow, as is indicated by the theoretical expressions of section 2.4. For example, the distribution parameter, as given by equation 7, becomes

$$C_o = 1 + 2.60\ (\sqrt{2})^{m} \sqrt{f} \ . \quad (17)$$

The significance of n for distributed flow voidage is less evident, its influence on the distribution parameter being via f (equation 14). Similarly, for given m and n, the sub-layer type affects f, and hence C_o, owing to the dependence of dimensionless parameter definitions on sub-layer type (Table 1).

Reference to section 2.4.2 indicates that m,n and sub-layer type have a similar effect on separated flow void characteristics. They also have similar effects on other two-phase flow phenomena (momentum flux, choked flow, etc.).

Considering equations 13 and 14 it is worth noting that the usual smooth tube single-phase equations are generated by (m,n) = (0,-2). For some situations, the Nikuradse equation is inadequate in single phase flows. For those situations, the correct single phase equation should be used in the two-phase case for flows in which (m,n) = (0,-2). For example, equation 14, with (m,n) = (0,-2) and sub-layer d, describes smooth tube dry wall two-phase flows. The equation should be replaced by Colebrook's equation if the tube wall is rough.

It is also worth noting that equation 13 with (m,n) = (2,0) generates an equation very close to that used to describe the buffer region in single-phase flow.

3.1 Friction Regime Classification Method

At this stage it is appropriate to introduce the concept of 'friction regimes'. A friction regime is characterised by the sub-layer type as described in section 2.1 and the coefficients 'a' in equation 3 or 4.

In this paper, a notation of the form (m,n,z), in which m and n are integers and z is a letter, will respectively indicate the relevant m and n values of equations 13 to 16 and the appropriate sub-layer type described in section 2.1. The triad (m,n,z) thus specifies a friction regime.

The above convention is not sufficient to specify a flow completely. Other relevant information is required (e.g. separated or distributed flow).

4. COMPARISON WITH DATA

Many of the features encountered in the analysis of two-phase drag reduced data are of special relevance to two-phase fluid mechanics research only. In keeping with the conference theme, this section considers only drag reduction aspects of the data analysed. The detailed analysis will be reported elsewhere (Ref. 10).

4.1 Examples of Friction Factor Data Characterised by m=4

Although data with various m and n values will be discussed later, it is interesting to see the trend in a family of friction factor curves with constant m and varying n. Typical two-phase data covering the range (4,0 to 8,c) are shown in Fig. 1.

Two points about Fig. 1 are worth noting. Firstly, the wide range of Reynolds numbers (several cycles) involved in such data sets has allowed the n coefficient, 0.50, in equations 13 and 14, to be determined with some accuracy. Similarly, the range of m values encountered in the analysis of data has led to reasonably accurate assessments of other empirical parameters. Secondly, the data of Fig. 1 correspond to genuine reductions in drag losses even though much of the friction data are above values expected from the Nikuradse equation. This is a result of changes in sub-layer structure occurring in friction regimes. Consider, for example, the data corresponding to (4,4,c). The same data are shown in a different form in the second region of the data in Fig. 5a. There they are lower than would be expected from the trend shown by the lowest quality data as characterised by (0,-2,a) in which the m and n values correspond to the Nikuradse equation. In this context, drag reduction is quite evident.

In the remainder of this section pressure loss data are discussed in which there is associated flow structure and average voidage data.

4.2 Argon-water Upflow in Round Tubes

Round tube argon-water upflow data, as reported in Refs. 11 to 18, have been analysed in terms of the non-dimensioned parameters of section 2. Data cover the gas density range $10 \leqslant \rho_g \leqslant 36.1$ kg m^{-3}. Most data are for a 25 mm tube, some for a 15 mm tube.

Data were sorted into groups having the same set of (m,n,z) values. The regime map so obtained is shown in Fig. 2a. Experimental friction factors, velocity profiles and average voidages are compared with the theoretical equations, using the appropriate (m,n,z) values, in Figs. 2b, 2c and 2d. Because of the vast number of data involved, each symbol on Fig. 2a corresponds to representative data only for $\rho_g = 36.1$ kg m^{-3}, although the regime map applies for other gas densities. Other data in Figs. 2 cover all the values of gas density in the data examined. Theoretical rather than experimental values of friction factors were used for Figs. 2c and 2d.

The regime map of Fig. 2a is only approximate. Data for different axial positions indicated a dependence of regime boundaries on flow development, as expected from theoretical considerations (Ref. 19). Regime boundaries are such that shear stresses are approximately continuous across a boundary.

The velocity measuring probes apparently influenced the flow causing the regime boundaries for the velocity data to be shifted. This shift explains the existence of an additional regime type in the velocity data. A similar probe effect on steam-water data was noted in Ref. 2. It will be noted that, as is the case of the 'universal' velocity profile, when the turbulent core must be subdivided according to velocity profile behaviour, each portion of the profile is consistent with equation 13 with appropriate integer choices for m and n.

The boundary of the (2,1,e) data of Fig. 2a, marked ▼, coincides with the visually-based bubble regime boundary Ref. 14. Such coincidences of visually-based boundaries with those obtained by the present method are rare. Nevertheless, on that basis the void data for this regime were treated as being for distributed flow, (Fig. 2d(i)). The negative, rather than the normal positive, drift velocity (V) indicated by Fig. 2d(i), is flowrate dependent and therefore is consistent with the attached bubble sub-layer for these data. The distribution parameter (C_o) values of Fig. 2d(i) are consistent with the empirical value of m=2 for the data.

Although shear stress data for conditions marked ? in Fig. 2a could not be correlated by the present model, the lower void data for these conditions are consistent with a zero drift velocity, and m=0 together with the experimental value of friction factor being used in equation 17 (Fig. 2d(ii)).

Separated flow void data for each of the regimes of Fig. 2a are shown in

Fig. 2d(iii), together with the theoretical curves (equation 12) for the appropriate m and n values. Since $\frac{1-\sqrt{\alpha}}{2}$ $Re\sqrt{f/2}$ is approximately proportional to the liquid volume fraction, errors in the abscissa represent errors in calculated liquid volume fractions. Errors are larger for smaller liquid volume fractions where neglected entrainment effects are expected to be more significant.

Fig. 2 shows that friction factors, velocity profiles and average voidage values for the argon-water flows considered are compatible with the theoretical equations of section 2 when analysed in terms of the generalised equations of section 3.

4.3 Argon-Alcohol Upflow in Round Tubes

Argon-ethyl alcohol data reported in Refs. 15 to 18 which also covered the range $10 \leqslant \rho_g \leqslant 36.1$ kg m^{-3} were analysed similarly to the argon-water data (section 4.2). The results are shown in Fig. 3.

Void data for conditions marked Δ and • in Fig. 3a were omitted from the analysis since core entrainment, which is not included in the present analysis, cannot be reasonably neglected for these regimes, which border the dry wall regime (marked *).

The visually-based 'bubble flow' regime boundary for the data (Ref. 14) approximately coincides with the ▲ boundary of Fig. 3a, hence void data for the friction regime (0,-2,c) were treated as being for distributed flows (Fig. 3d(i)). Despite the visual interpretation of the flow, the sub-layer type suggests that bubble flow is unlikely and that on the contrary the flow has a froth core. Since drift velocities can be expected to be zero for froth flows, the data were analysed in terms of equation 5a, instead of equation 5, together with equation 17.

For dry wall flows the drift velocities, which are expected to be small compared with the large flow velocities encountered under these flow conditions, were neglected in the analysis of the dry wall void data (Fig. 3d(iii)).

As was the case for the argon-water data, the argon-alcohol data are compatible with the equations of sections 2 and 3, although different friction regimes were encountered for the differing flow constituents.

4.4 Separated Surface Tension Dependent Flow Data

Data analysis results reported in sections 4.2 and 4.3 confirm the consistency of average voidage data, shear stress data, and velocity profile data with the theoretical equations in sections 2 and 3 for distributed and separated viscosity dependent flows, and, except for the absence of associated velocity profile data, also for distributed surface tension dependent flows, but not for separated surface tension dependent flows.

Some separated surface tension dependent flow data for the friction regime (1,2,e) have been analysed to evaluate the compatibility of such velocity profile, void fraction, and friction factor data with theory. The data are for argon-water upflow in a seven-rod cluster, covering the gas density range $10 \leqslant \rho_g \leqslant 36.1$ kg m^{-3} (Ref. 20), and for steam-water upflow in a 17 mm tube at 7 MPa (Ref. 21). The data involved are the higher quality data from Ref. 20 and the lower quality adiabatic data from Ref. 21.

The data, resolved into appropriate non-dimensional terms, are compared with the theoretical curves in Fig. 4. The data are in reasonable agreement with the theoretical equations expected for the (1,2,e) friction regime. The discrepancy of the lower y^+ velocity data (Fig. 4b) is not significant since it is the core profile that determines the flow characteristics. In any case, the void data are consistent with a treatment in which the core profile is assumed to extrapolate to j=0 (Fig. 4c), suggesting that the lower y^+ velocity data of Fig. 4b may be in error.

The results of section 4.2 and this section indicate that, in addition to the previously noted dependence of friction regimes on flow constituents, the friction regimes are also flow geometry dependent.

5. EQUATIONS FOR HIGH PRESSURE STEAM-WATER FLOWS

An earlier paper (Ref. 19) presented recommended equations for calculating friction losses in high pressure steam-water flows. The equations were based on an analysis of approximately 10^4 sets of data, and allowed for entrance, heat transfer and geometry effects. They applied for $P \gtrsim 2$ MPa and $G \gtrsim 200$ kg $m^{-2}s^{-1}$. A U.K. comparison with a wide range of experimental friction data for adiabatic two-phase flows in round tubes (Ref. 22) showed that the Ref. 19 equations for such flows were superior to alternative equations for the mass flux and pressure conditions for which the Ref. 19 equations were developed, but were inferior to alternatives at lower pressures and when applied to non-steam-water systems. In view of the data analysis results of section 4, the latter result is not surprising. The equations of Ref. 19 incorporated the friction regime concept of the present work, but were developed before the existence of the generalised equations of section 3 was recognised.

With one exception, all the Ref. 19 equations could be made to conform to the present generalised equations with only minor alterations to their empirical coefficients. The new equations are specified by the parameters given in Table 2. A comparison with various data is shown in Fig. 5.

TABLE 2. FRICTION REGIMES FOR HIGH PRESSURE STEAM WATER SYSTEMS ($P \gtrsim 2$ MPa, $G \gtrsim 200$ kg $m^{-2}s^{-1}$)

Use in conjunction with equations of sections 2 and 3

Flow System			Friction regimes, in order of increasing quality[1]
Round Tubes	Adiabatic Flow	Settling length > 500D	(0,-2,a); (4,4,c)[2,3]; (2,0,a); (0,-2,d)
		Settling length $\lesssim$ 500D, D < 6 mm	As above[4]
		Settling length $\lesssim$ 500D, D > 6 mm	(0,-2,a); (4,4,c)[2,3]; (4,2,a); (0,-2,d)
	Diabatic Flow	Unobstructed	(0,-2,a); (3.61, 3.74, e); (0,-2,d)[5]
		Upstream obstruction, G < 1500 kg $m^{-2}s^{-1}$	As above
		Upstream obstruction G > 1500 kg $m^{-2}s^{-1}$	(0,-2,a); (4,4,c)[2,3]; (4,2,a); (0,-2,d)[5]
Annuli and rod Clusters		Spacer separation > 40D	(0,-2,a); (1,2,e); (0,-2,d)[5,6]
		Spacer separation < 40D	Variable behaviour

Notes:

1. Friction regime boundaries (≡ friction factor equation validity limits) are given by intersects of friction factor equations unless otherwise specified.

2. Replace (4,4,c) with (4,6,c) if (0,-2,a):(4,4,c) intersect occurs at $x < 0$ (D < 20 mm) or $x < 0.1$ (D > 20 mm). When (0,-2,a):(4,6,c) intersect occurs at $x < 0$, replace (4,6,c) with (4,8,c).

3. Omit the second regime when the friction equations for the first and third regimes intersect at a lower quality than the intersect for the first and second regimes.

4. Reduce regime boundary qualities by ~ 0.1.

5. Regime boundary given by appropriate critical heat flux correlation in diabatic flow case.
6. Use pre-dryout equations with D=4. Flow area/(non-dried perimeter) if only some surfaces are dry (Ref. 7).

Not all high pressure steam-water data conform with the equations to the same extent as the Fig. 5 data. Where disagreement occurs, it is usually because the friction regime differs from expectations. The data then conform to the equations for that regime. Examples are the (4,1,a) set of data in Fig. 1 and the steam-water data of section 4.4.

Confirmatory void and velocity profile data have not been presented here. Using Table 2 as a guide, virtually all existing high pressure steam water void data are for the (0,-2,a) regime. These agree with the theoretical equations for this regime provided entrainment is taken into account in separated flow voidage calculations. The little void data that could be found for other friction regimes (Ref. 3) also agree with the theory.

6. DISCUSSION

Apart from the coefficients in the generalised equations of section 3, and the evaluation of appropriate friction regimes, the present model uses no empiricism, despite the wide range of applications. Comparable published models tend to involve empiricism even for limited applications. For example, the method of Ref. 23, invokes an empirical function which has different values depending on whether it is being applied to pressure loss or voidage prediction.

It has been seen that the triad (m,n,z) characterises two-phase void values, velocity profiles, and pressure losses, all in a coherent fashion. The generalised equations of section 3, as applied to the theory of Refs. 6 and 7 also indicate that other phenomena (such as those for choked flow and heat transfer coefficient) are similarly characterised.

The generalised equations for viscosity dependent flows (equations 13 and 14) differ in some respects from those for surface-tension dependent flows (equations 15 and 16). The logarithmic term coefficients remain the same for the two flow types, but the other terms behave differently. This is not surprising; mixing length theory indicates that the logarithmic term coefficients are determined solely by the turbulent transport in the core of the flow. Such transport is expected to be identical for the two flow types. The other terms are determined by boundary conditions which apply to the turbulent core, and these are different for the two flow types. In a similar way, and for the same reasons, the logarithmic term coefficients in single phase equations are the same for smooth and roughened walls, whereas the other coefficients in the single phase equations are altered by the changed boundary conditions.

The spectrum of discrete values for friction regime parameters does not occur with all two-phase friction regimes. Exceptions tend to be encountered for low mass flowrates at extremely high voidages, of the type analysed so extensively by the Harwell group (e.g. Ref. 24), where, for example, changes in mixing length coefficients are smooth rather than discontinuous. It is postulated that the mechanism by which drag is reduced for these flows is different from that of the present work. Flow conditions are such that the liquid film at the wall is expected to be so thin as to fall within the laminar sub-layer. Such a film may act as a compliant coating.

Apart from the above exception, all two-phase friction regimes with drag reduction characteristics involve a two-region flow, with the interface between the regions lying in the vicinity of the sub-layer. Given present views that turbulence is generated in the sub-layer region, it is possible that the drag reduction is due to the interface inhibiting the turbulence transport. This appears particularly likely in the present model, whose basis (Ref. 5) suggests that, locally, turbulence is confined to one phase only, and that the turbulence-carrying phase for two-region flows is different on different sides of the interface separating the regions. The inhibition of turbulence transport, as opposed to the inhibition of turbulence generation in single-phase drag reduction (Ref. 25), is a possible cause of differences between single- and two-phase drag reduction characteristics.

Although single-phase drag reduction behaviour tends not to have the spectrum of discrete values of the type noted here, analysis of single-phase drag reduction data indicates that particular f:Re curves are preferred. Single-phase drag reduction data were not considered in the development of the generalised equations of section 3, but the preferred f:Re curves tend to agree with equation 14 with appropriate integers chosen for m and n. Examples of such single phase data are: much of the data of Ref. 26 ((m,n) = (5,1)), and of Ref. 27 ((4,0)). If half-integers are considered, Virk's initial (Ref. 28), (4.5,0) and revised (Ref. 29) (5,0.5) 'maximum drag reduction asymptotes' also conform to the generalised equation 14.

Finally, the present work provides suitable equations for three types of two-phase flow system (steam-water, argon-water and argon-alcohol). They should not be generalised to other systems, and in any case, should be treated with caution. Thus, the (2,1,e) regime of Fig. 2, marked ▼, appears to apply only for conditions when a particular wall device, used for shear stress measurements (Ref. 14) was used. Void data obtained for similar flow conditions in the absence of the wall-shear device (Ref. 18) are shown in Fig. 6. A comparison of Fig. 6 and Fig. 2d(i) suggests that the friction regime is different for the two cases.

7. CONCLUSIONS

In the application of mixing length theory to two-phase flow, the generalised dimensionless velocity profile may be characterised by two parameters, m and n. Theoretical velocity profiles, pressure drops and average voidages agree with experimental data when m and n are restricted to discrete integer values associated with a particular type of sub-layer (z), so that the triad (m,n,z) may be chosen as the basis for two-phase flow classification and prediction.

Further work is necessary to interpret low mass flux, very high voidage flows which are not compatible with the proposed classification scheme, and provide a basis from which terms in the triad (m,n,z) may be estimated for any given flow.

8. ACKNOWLEDGEMENTS

Helpful discussions with Dr. K.R. Lawther (Head, Heat Transfer Section, AAEC) and Professor J.J. Thompson (Head, School of Nuclear Engineering, UNSW) are gratefully acknowledged.

9. REFERENCES

1. Beattie, D.R.H.: Nucl. Eng. Des. 25, 395 (1973).
2. Beattie, D.R.H.: Paper D3, Int. Conf. on Drag Reduction, BHRA Fluids Engng, Cranfield, U.K. (Sept. 1974).
3. Thwaites, G.R., et al.: Chem. Eng. Sci. 31, 481 (1976).
4. Beattie, D.R.H.: Ph.D Thesis, School of Nucl. Eng., UNSW (in preparation).
5. Beattie, D.R.H.: Nucl. Eng. Des. 21, 46 (1972).
6. Beattie, D.R.H.: Paper D3, 2nd Int. Conf. on Pressure Surges, BHRA Fluids Engng, Cranfield, U.K. (Sept. 1976).
7. Beattie, D.R.H.: p.329, Proc. 2nd Australasian Conf. on Heat and Mass Transfer, University of Sydney (Feb. 1977).
8. Beattie, D.R.H.: 'Void fraction and regime transitions in two-phase flow' paper in preparation.
9. Zuber, N. and Findlay, J.A.: J. Heat Transfer, 87, 453 (1965).
10. Beattie, D.R.H.: Australian Atomic Energy Commission - report in preparation.
11. Adorni, N., et al.: CISE-R-35 (1961).
12. Adorni, N., et al.: CISE-R-41 (1961).
13. Casagrande, I., et al.: CISE-R-73 (1963).
14. Cravarolo, L., et al.: CISE-R-82 (1964).
15. Cravarolo, L., et al.: CISE-R-93 (1964).
16. Alia, P., et al.: CISE-R-105 (1965).
17. Alia, P., et al.: CISE-R-109 (1966).
18. Colombo, A., et al.: CISE-R-225 (1967).
19. Beattie, D.R.H.: ASME paper 75-WA/HT-4 (1975).
20. Alia, P., et al.: CISE-R-108 (1968).
21. Subbotin, P.L., et al.: FEI-421 (1973).
22. Clarke, C.J.T.: Engineering Sciences Data Unit - U.K. - private communication. See also ESDU Data Item 76018 (1976).

23. Chisholm, D. and Sutherland, L.A.: Paper 4, Proc. Inst. Mech. Eng. (1969-70), 184, pt 3c, 24 (1970).
24. Hewitt, G.F. and Hall Taylor, N.S.: 'Annular Two-Phase Flow', Pergamon Press (1970).
25. Achia, B.U. and Thompson, D.W.: Paper A2, Int. Conf. on Drag Reduction, BHRA Fluid Engng. Cranfield, U.K. (Sept. 1974).
26. McMillan, M.L.: C.E.P. Symposium Series, No. 111, 67, 27 (1971).
27. White, A.: Nature, 214, 585 (1967).
28. Virk, P.S., et al.: J. Fluid Mech. 30, 305 (1967).
29. Virk, P.S., et al.: J. Appl. Mech. 37, 488 (1970).
30. Gaspari, G.P., et al.: CISE-R-83 (1964).
31. Fukuda, K., et al.: SN941 74-57 (1974).
32. Harrison, R.F.: ME Thesis, School of Engineering, University of Auckland (1975).
33. Adorni, N., et al.: CISE-R-31 (1961).

CISE reports are available from Centro Informazioni Studi Esperienze Documentation Service, Casella Postale 3986-20100 Milan, Italy; FEI reports from Physical Energetics Institute, Obninsk, USSR; ESDU reports from Engineering Sciences Data Unit, 251-259 Regent St., London, WIR 7AD, UK; SN reports from O-arai Engineering Centre, Power Reactor and Nuclear Fuel Development Corporation, Ibaraki, Japan.

Symbol	Friction regime (See section 3.1)	Refs	Note: All data (except that marked ∇) are upflow
Δ	(4,0,a)	14 (Table 9)	Argon-alcohol, see Δ data of Fig. 3
o	(4,1,a)	33	Steam-water high quality adiabatic flow at 7 MPa in 8.25 mm/5.02 mm annulus
x	(4,2,a)	30	Steam-water intermediate quality, adiabatic flow at 5 and 7 MPa in a 9.18 mm tube. See Fig. 5b
•	(4,3,c)	14 (Table 9)	Argon-alcohol, see • data of Fig. 3
+	(4,4,c)	30	Steam-water lower intermediate quality, high mass flux adiabatic flow at 7 MPa in a 5 mm tube. See Fig 5a
∇	(4,6,c)	31	Steam-water lower intermediate quality adiabatic horizontal flow at 7 MPa in a 76.2 mm tube
▲	(4,8,c)	32	Steam-water flow at 2 MPa in a 200 mm tube

Fig. 1. Notation

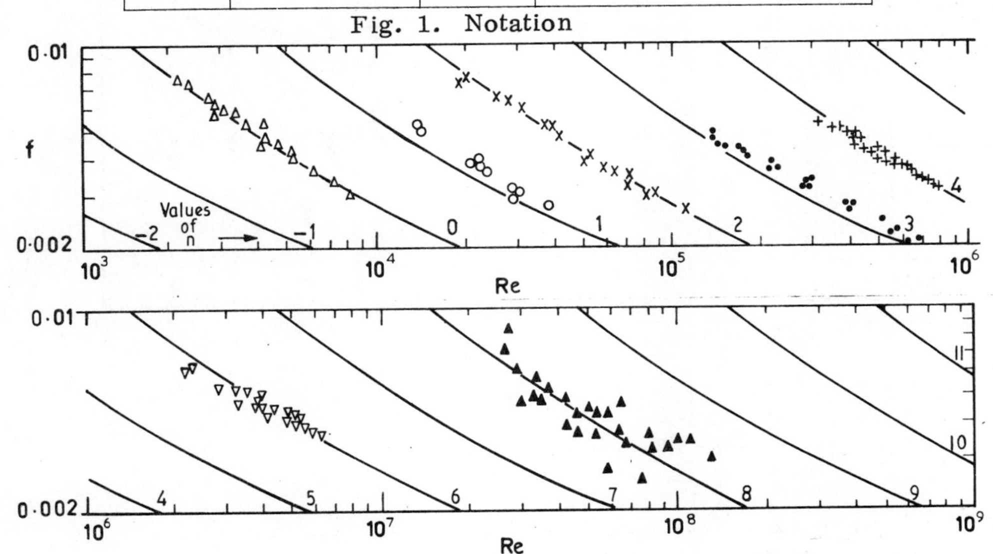

Fig. 1. Friction factors for m = 4

Symbol	Friction regime (See section 3.1)	Core region velocity profile and friction factor expression	Velocity profile between core region and wall
Δ	(1,3,c)	$j^+ = 8 \log y^+ - 12.88$ $1/\sqrt{f} = 5.66 \log Re\sqrt{f} - 15.33$	a) ? ($m \leqslant -4$) b) $j^+ = y^+$
x	(2,0,a)	$j^+ = 11.31 \log y^+ - 2.49$ $1/\sqrt{f} = 8.0 \log Re\sqrt{f} - 10.56$	$j^+ = y^+$
O	(3,1,b)	$j^+ = 16 \log y^+ - 13.28$ $1/\sqrt{f} = 11.31 \log Re\sqrt{f} - 21.84$	a) $j^+ = 2.83 \log y^+ + 2.04$ $[(m,n) = (-2,-1)]$ b) $j^+ = y^+$
+	(4,1,a)	$j^+ = 22.63 \log y^+ - 21.27$ $1/\sqrt{f} = 16 \log Re\sqrt{f} - 32.64$	a) $j^+ = 5.66 \log y^+ + 2.83$ $[(m,n) = (0,-1)]$ b) $j^+ = y^+$
∇ and ▼ (Fig. 2d(i))	(2,1,e)	$j^+ = 11.31 \log y^+ + 15.27$ $1/\sqrt{f} = 8 \log Wef + 3.20$	?
? and • (Fig. 2d(ii))	(0,?,?)	Friction data uncorrelatable with present model. The experimental void:friction-factor relation is consistent with m=0 (Fig.2d(ii))	

Fig. 2. Notation

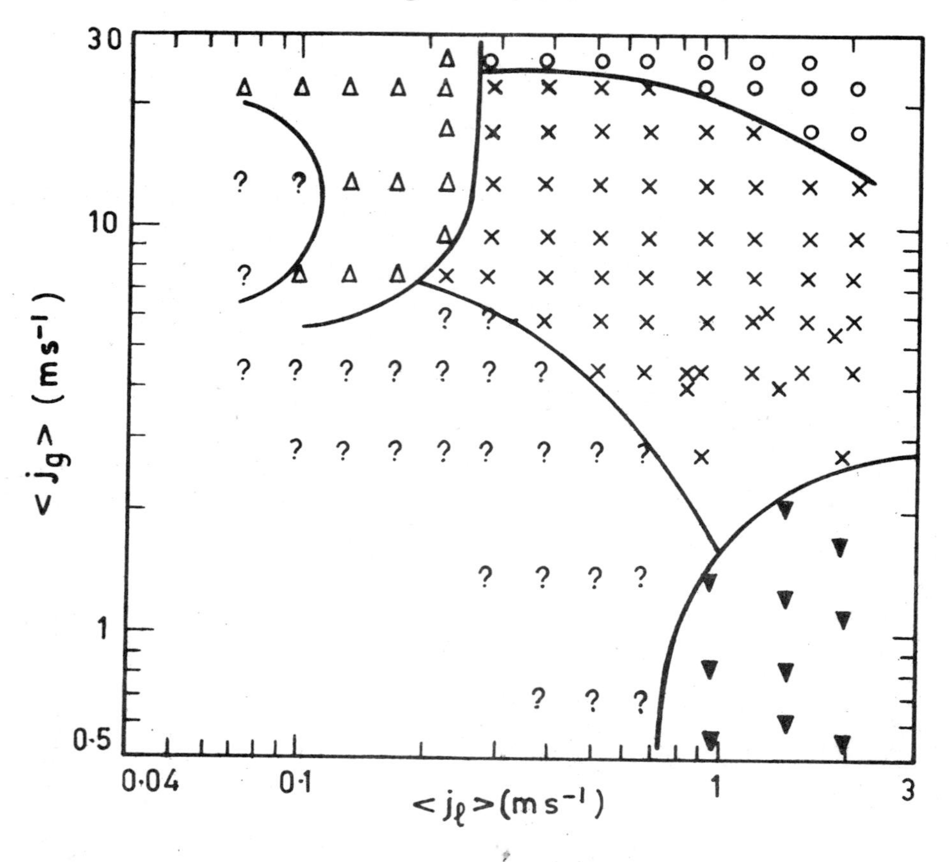

Fig. 2a. Friction regime map

Fig.2 Flow characteristics for argon-water upflow in round tubes. Data from refs. 11 to 18

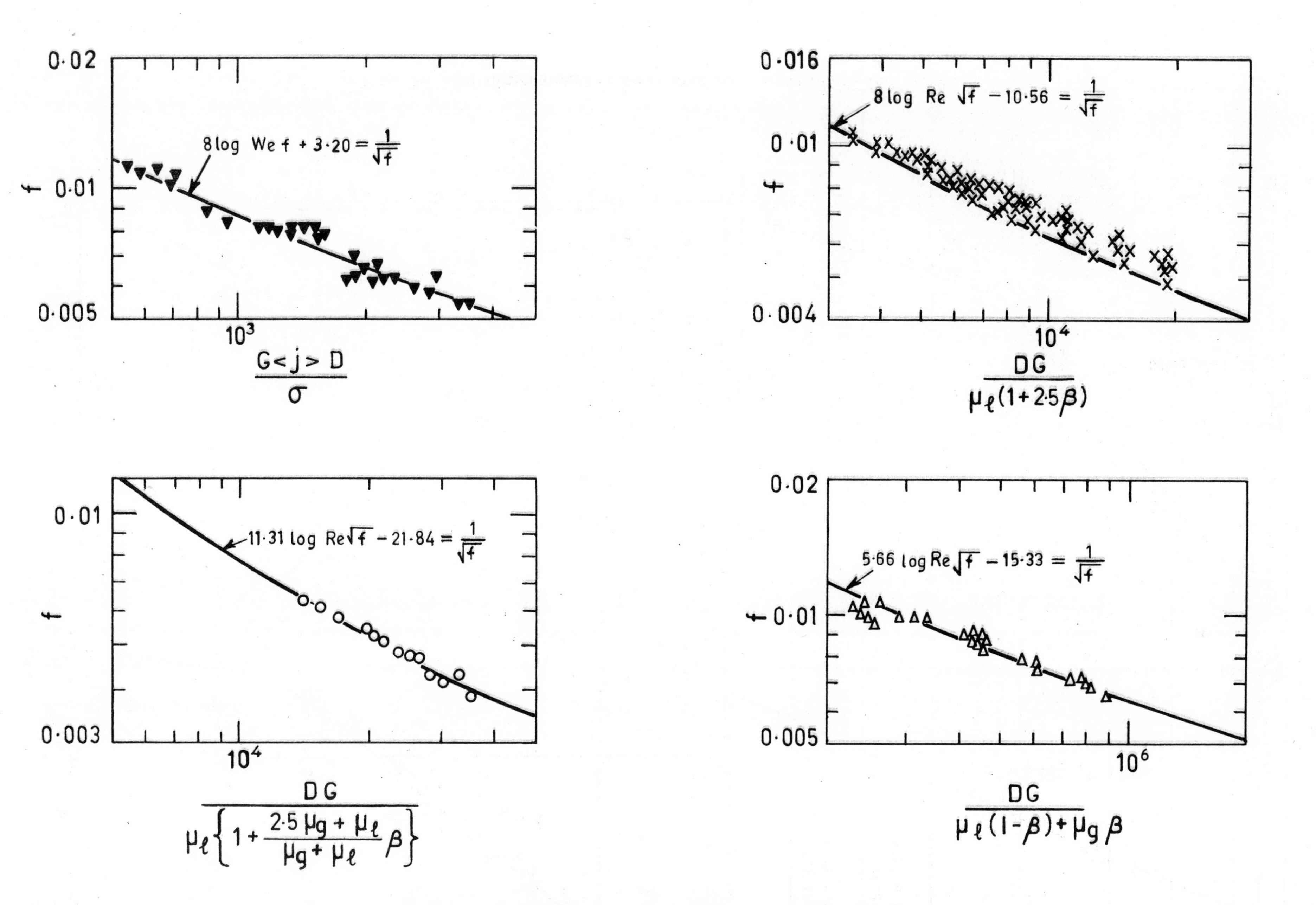

Fig. 2b. Friction factor data. Notation as in Fig. 2a.

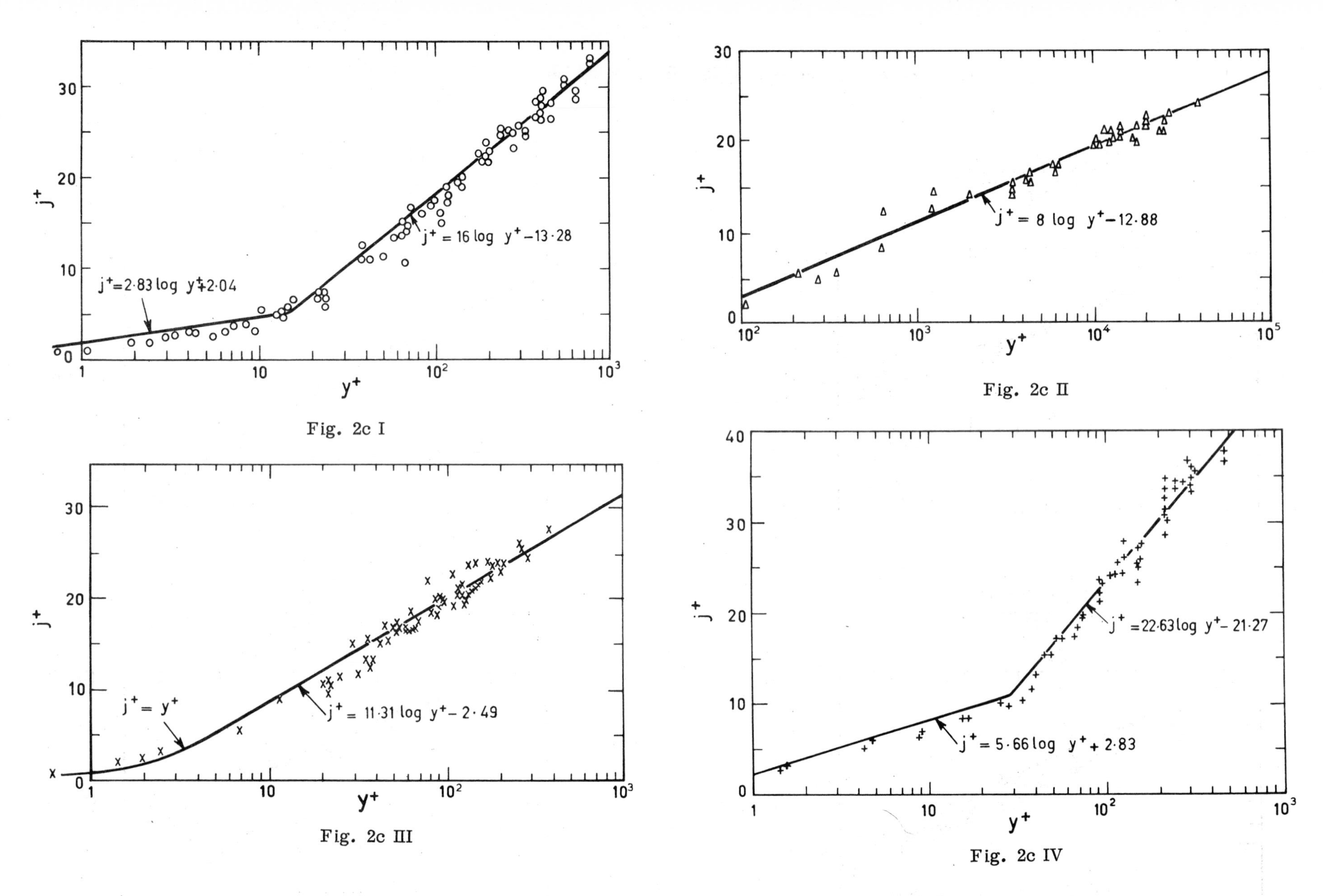

Fig. 2c. Non-dimensional velocity profiles. Notation as in Fig. 2a.

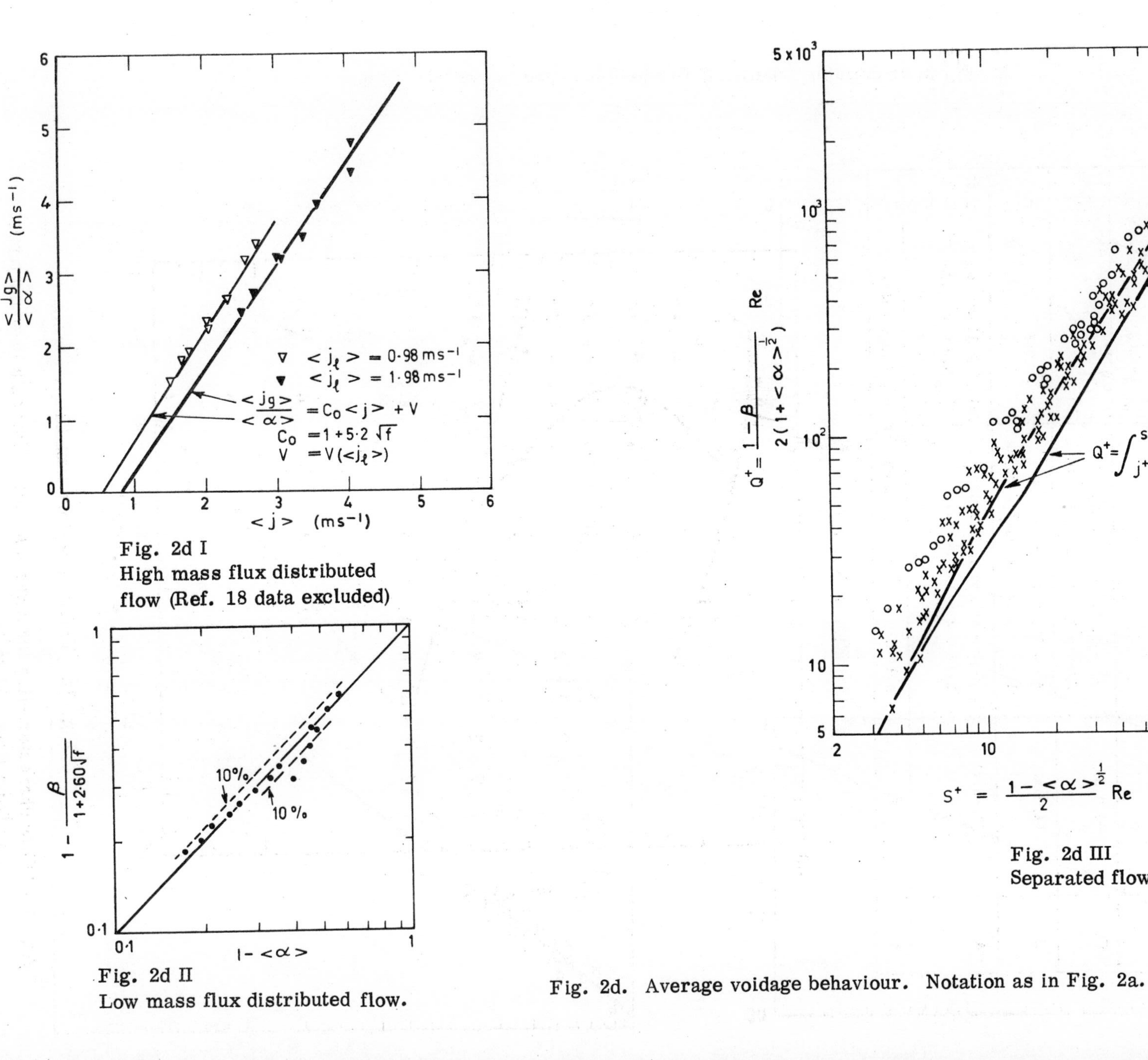

Fig. 2d I
High mass flux distributed flow (Ref. 18 data excluded)

Fig. 2d II
Low mass flux distributed flow.

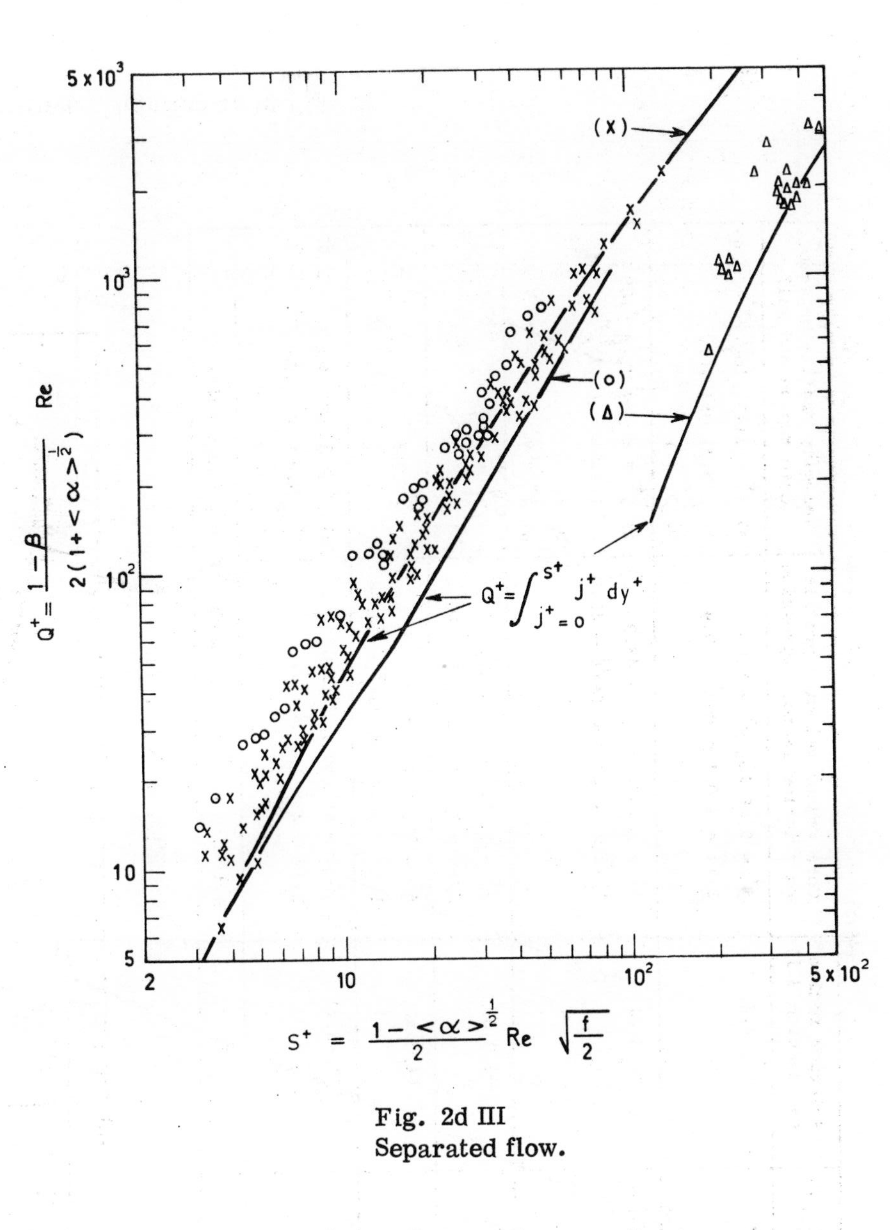

Fig. 2d III
Separated flow.

Fig. 2d. Average voidage behaviour. Notation as in Fig. 2a.

Symbol	Friction regime (See section 3.1)	Core region velocity profile and friction factor expression	Velocity profile between core region and wall
*	(0,-2,d)	$j^+ = 5.66 \log y^+ + 5.66$ $1/\sqrt{f} = 4.0 \log Re\sqrt{f} - 0.4$	a) $j^+ = 11.31 \log y^+ - 2.49$ $[(m,n) = (2,0)]$ b) $j^+ = y^+$
•	(4,3,c)	$j^+ = 22.63 \log y^+ - 43.9$ $1/\sqrt{f} = 16.00 \log Re\sqrt{f} - 48.6$	a) $j^+ = 8 \log y^+ - 8.88$ $[(m,n) = (1,2)]$ b) ? c) $j^+ = y^+$
□	(3,0,b)	$j^+ = 16 \log y^+ - 5.28$ $1/\sqrt{f} = 11.31 \log Re\sqrt{f} - 16.2$	$j^+ = y^+$
Δ	(4,0,a)	$j^+ = 22.63 \log y^+ - 9.96$ $1/\sqrt{f} = 16.0 \log Re\sqrt{f} - 24.64$	$j^+ = y^+$
▲	(0,-2,c)	$j^+ = 5.66 \log y^+ - 5.66$ $1/\sqrt{f} = 4.0 \log Re\sqrt{f} - 0.4$	a) $j^+ = 11.31 \log y^+ - 2.49$ $[(m,n) = (2,0)]$ b) $j^+ = y^+$
?	?	Friction data uncorrelatable with present model	

Fig. 3 Notation.

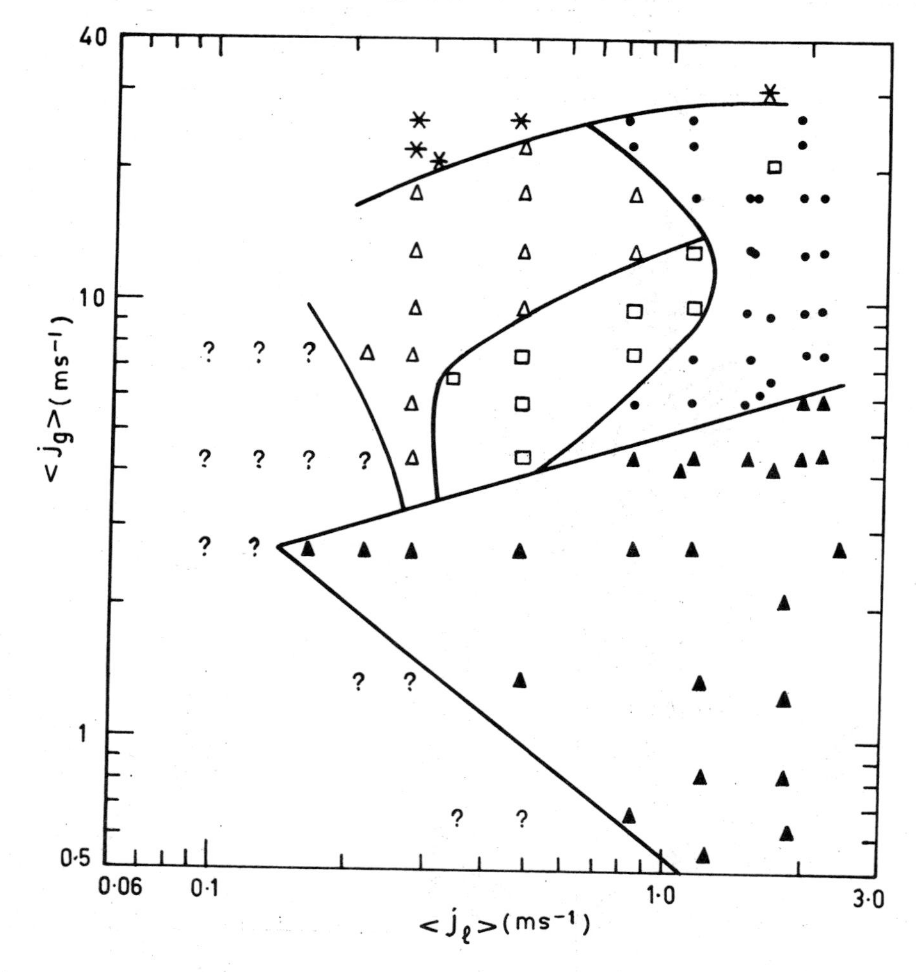

Fig. 3a. Friction regime map.

Fig. 3 Flow characteristics for argon-alcohol upflow data in round tubes. Data from Refs. 15 to 18.

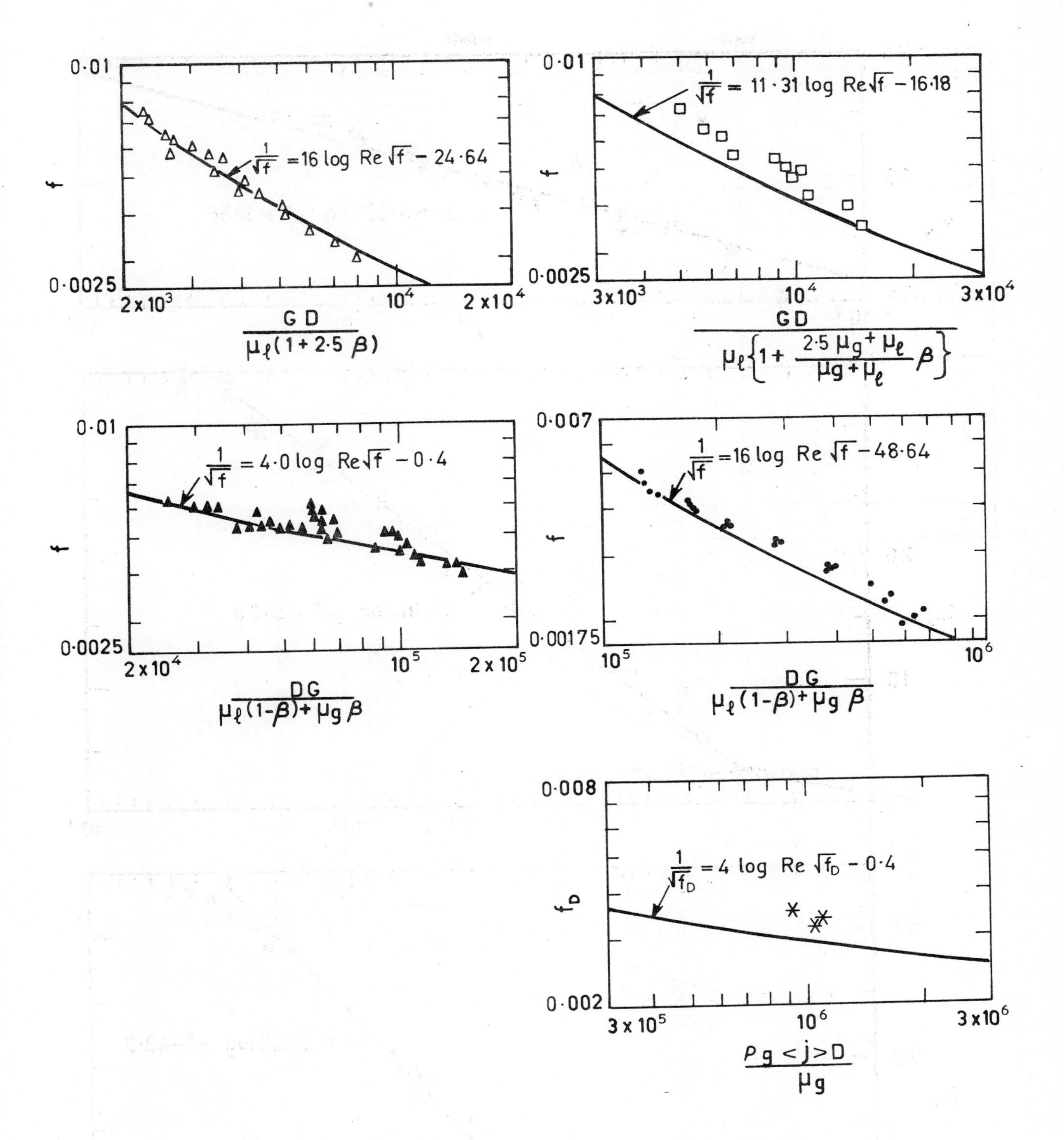

Fig. 3b. Friction factor data. Notation as in Fig. 3a.

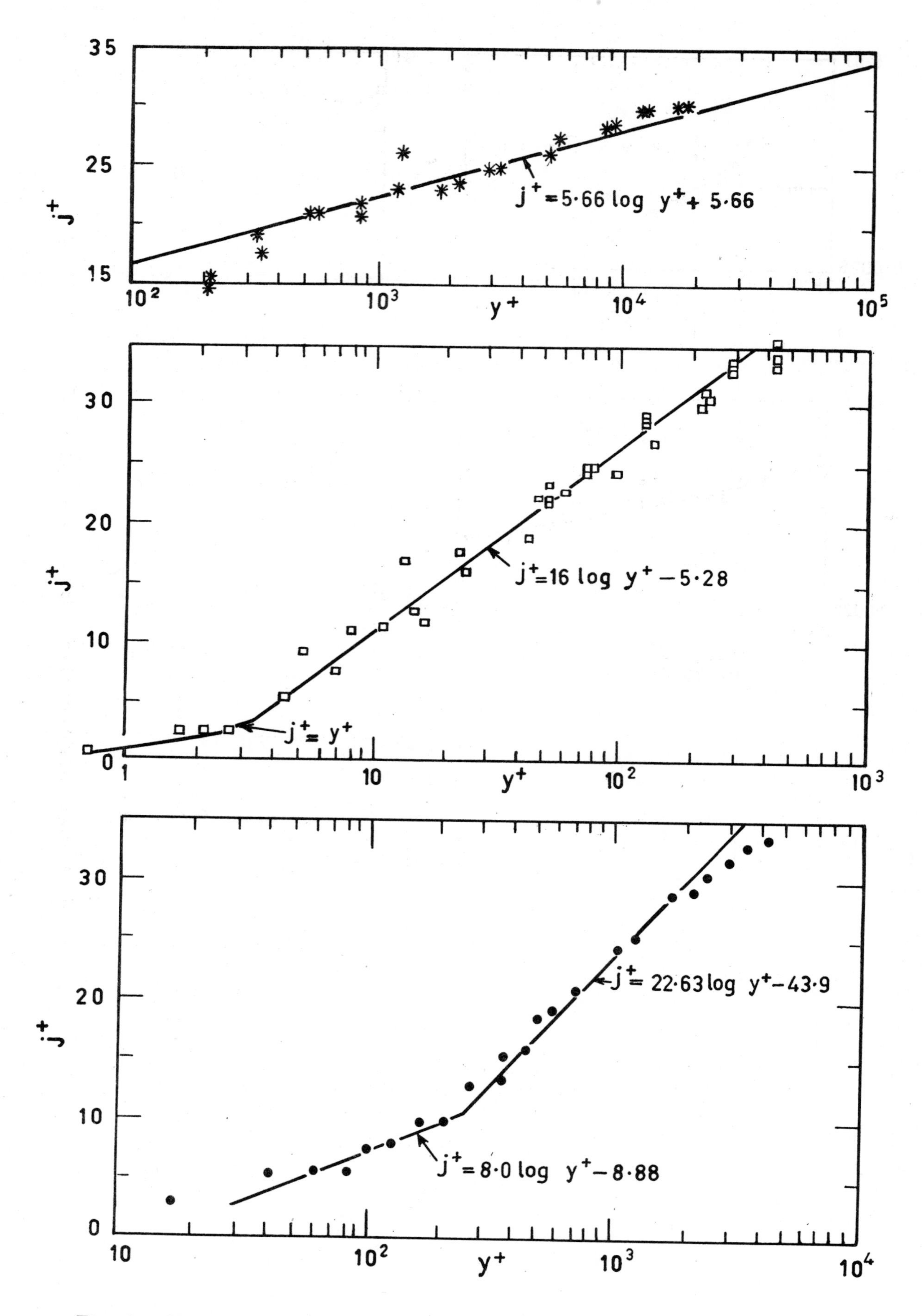

Fig. 3c. Non-dimensional velocity profiles. Notation as in Fig. 3a.

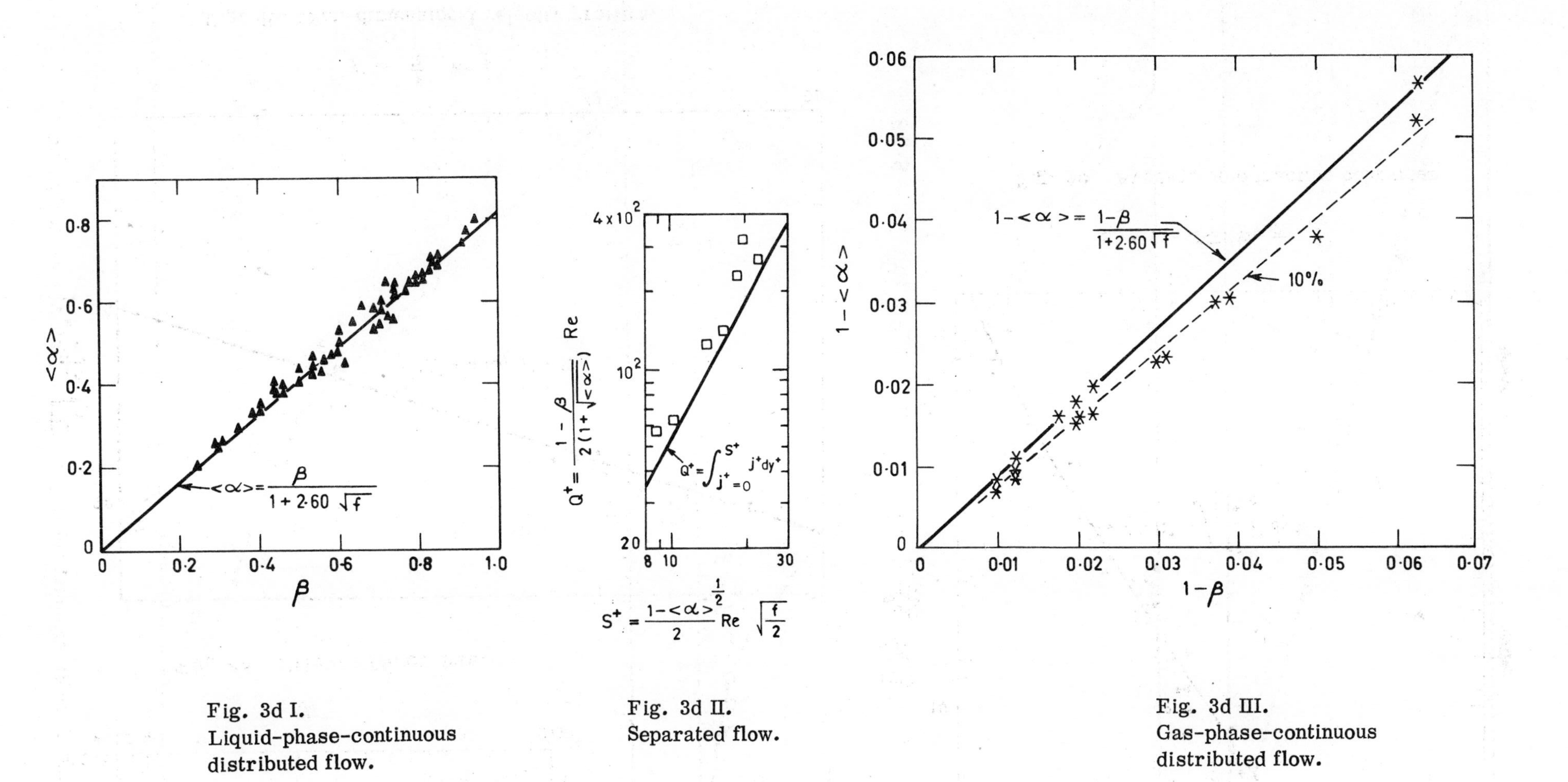

Fig. 3d I.
Liquid-phase-continuous distributed flow.

Fig. 3d II.
Separated flow.

Fig. 3d III.
Gas-phase-continuous distributed flow.

Fig. 3d. Average voidage behaviour. Notation as in Fig. 3a.

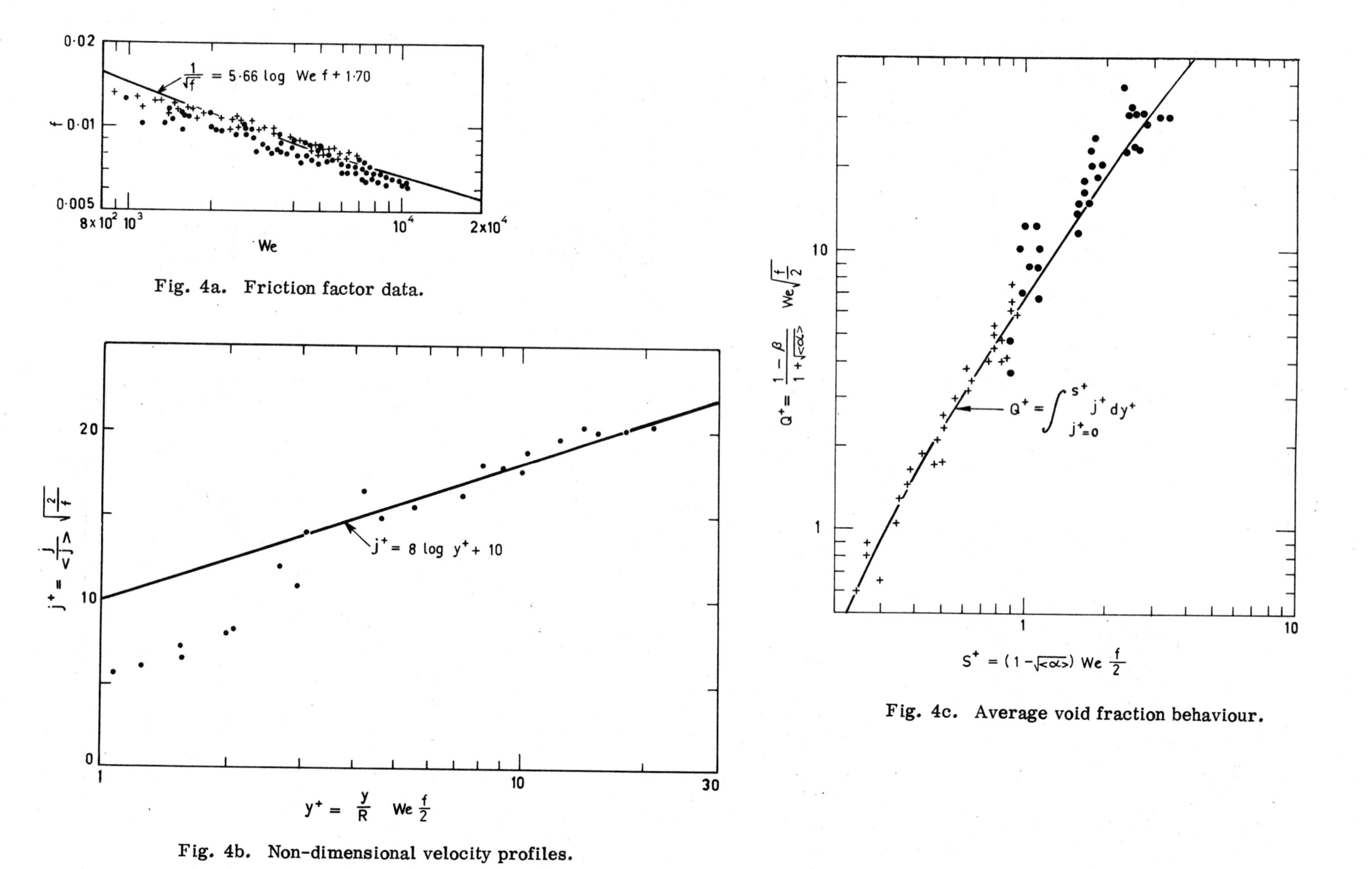

Fig. 4a. Friction factor data.

Fig. 4b. Non-dimensional velocity profiles.

Fig. 4c. Average void fraction behaviour.

Fig. 4. Flow characteristics for separated (1, 2, e) flow data. Data from Ref. 20(+) and Ref. 21 (.).

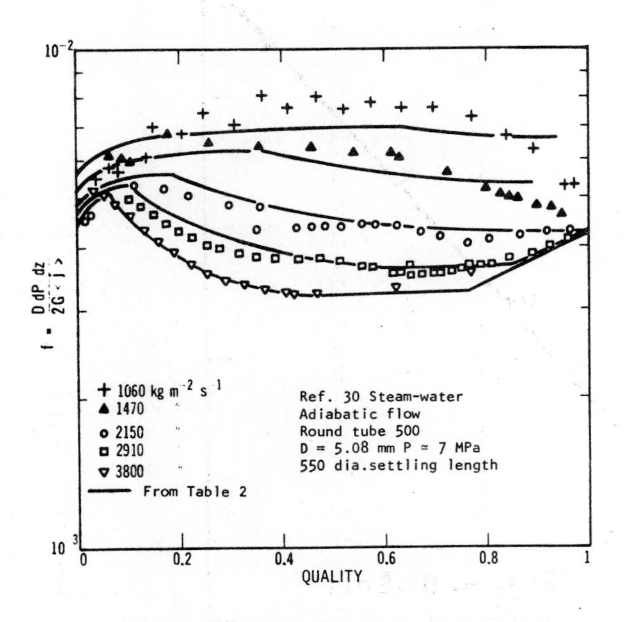

Fig. 5a. Round tube adiabatic flow with long settling length.

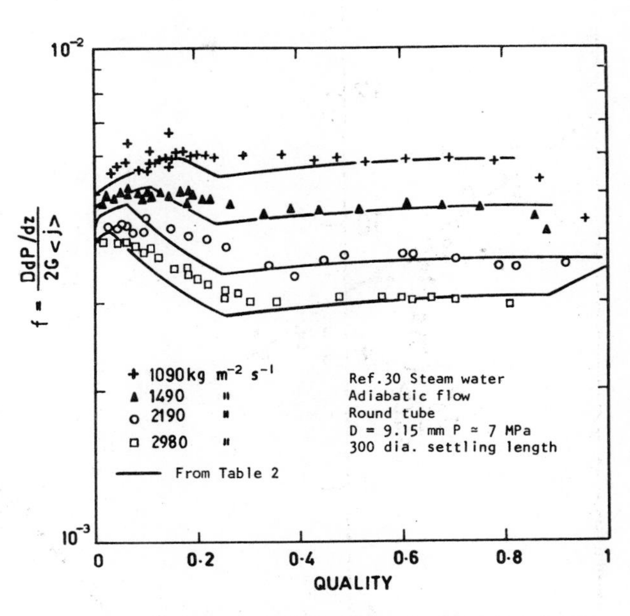

Fig. 5b. Round tube adiabatic flow with short settling length.

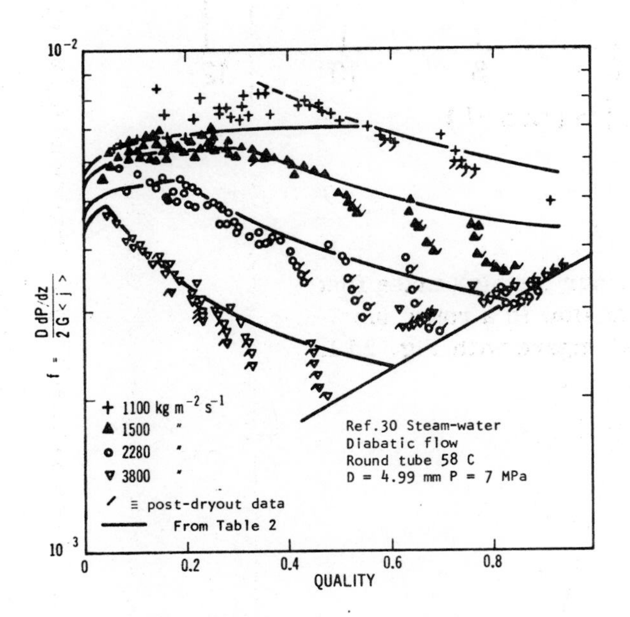

Fig. 5c. Round tube diabatic flow.

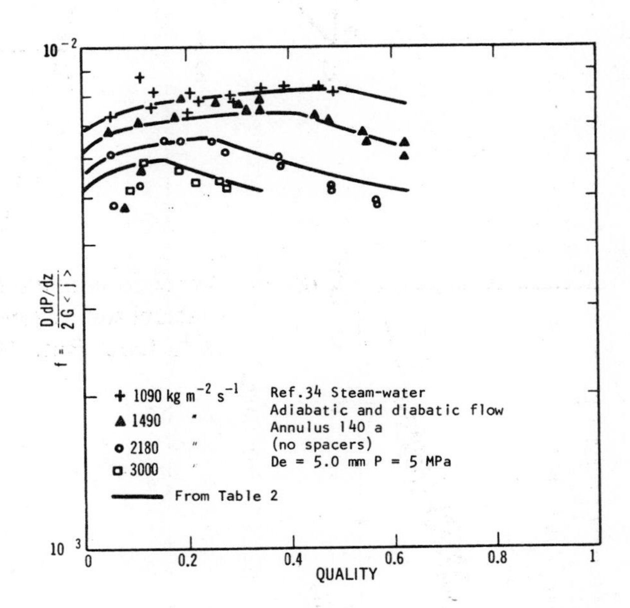

Fig. 5d. Flow in an annulus with no spacers.

Fig. 5. Friction factors for high pressure steam-water flows.

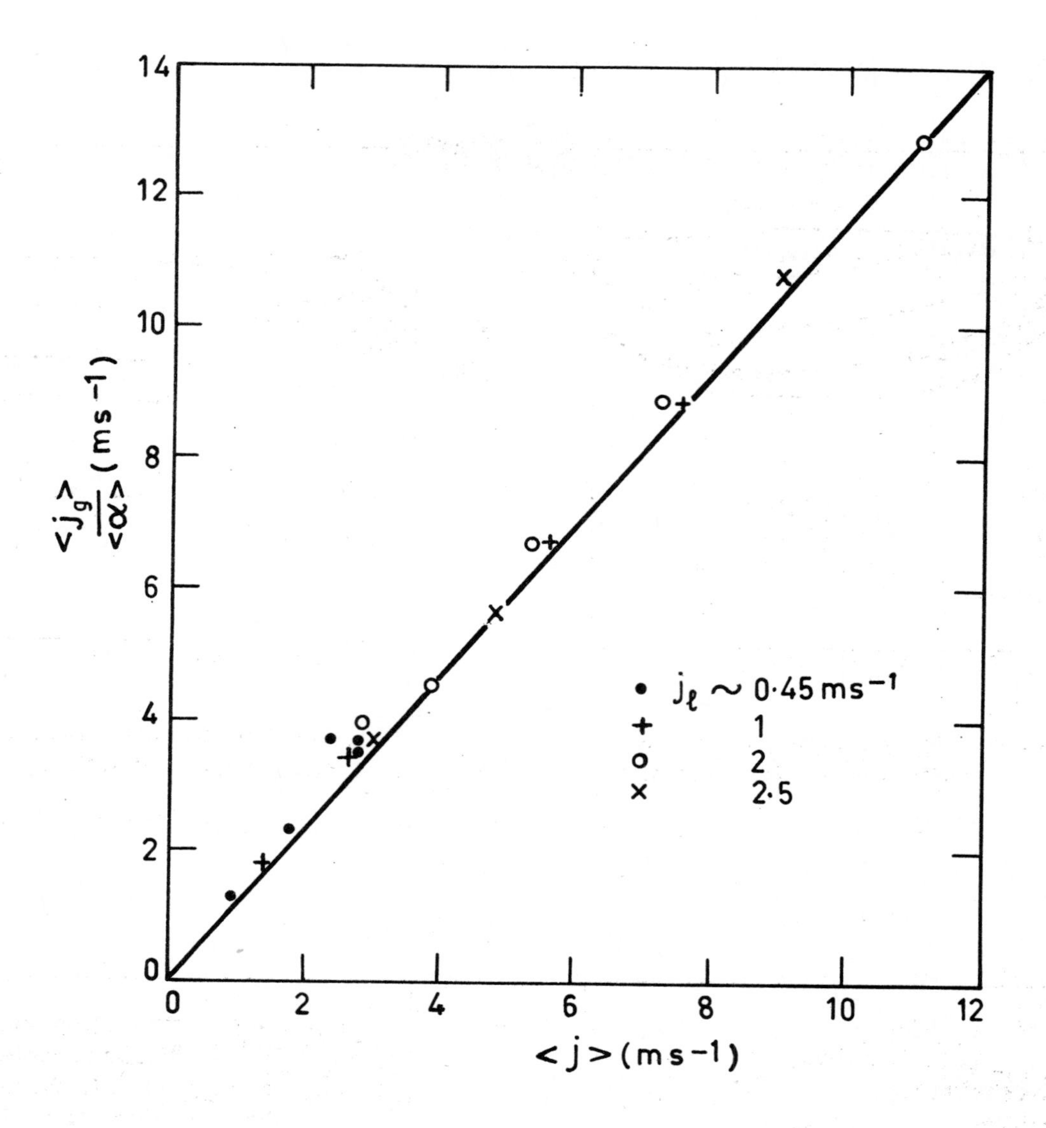

Fig. 6. Average voidage behaviour for high mass flux distributed argon-water flow in a round tube. Data from Ref. 18. Compare with Fig. 2d I.

Second International Conference on

Drag Reduction

August 31st - September 2nd, 1977

PAPER D2

GAS-SATURATION EFFECT ON NEAR-WALL TURBULENCE CHARACTERISTICS

V.G. Bogdevich, A.R. Evseev, A.G. Malyuga and G.S. Migirenko

Institute of Thermophysics, USSR

Summary

The paper presents the results of studying surface friction, gas concentration and pressure pulsations in a turbulent boundary gas-saturated layer. To exert a favourable effect directly on the turbulent generation process in the near-wall boundary layer, it is this region that should be saturated with small gas bubbles with size of about the viscous sublayer thickness. In this case the efficiency of gas saturation significantly increases.

Held at St. John's College, Cambridge, England.

Organised and sponsored by BHRA Fluid Engineering, Cranfield, Bedford, MK43 0AJ.

Nomenclature

C_f' – friction factor with flow injection;

$\bar{C}$ – gas concentration;

$\bar{C}_*$ – maximum gas concentration in boundary layer cross -section;

C_d – drag coefficient of medium;

δ – thickness of boundary layer;

δ_* – thickness of blow-off;

d – bubble diameter;

d_κ – pore diameter;

$\bar{f} = f\delta_*/U_o$ frequency;

$\bar{j} = v_g/U_o$ injection parameter;

L_n – length of porous insert;

$\bar{p}'$ – pressure pulsation;

ρ_o – liquid density;

$Re = U_o L/\nu$ Reynolds number;

σ – surface tension factor;

$Ta = \tau_w d/\sigma$ Taylor number;

τ_w – near-wall friction drag;

U_o – flow rate;

v_g – normalized gas injection flow;

$\bar{x} = x/L_n$ relative streamwise coordinate;

$\bar{y} = y/\delta$ relative distance from the stream-lined wall;

ν – kinematic viscosity factor;

Index "O" refers to pure fluid parameters.

1. Introduction

Gas injection into a liquid fluid has attracted attention for more than a century. It was Frud (1875) and Laval (1883) who first formulated this problem (Ref.1). At first it was suggested that gas injection into a near-wall flow and gas "lubricant" layer formation would ensure a considerable liquid friction reduction in the liquid friction due to the insignificant gas viscosity.

Theoretical studies of the turbulent variable density and viscosity boundary layer in terms of both the homogeneous model (Ref. 2,3) and gas-liquid interface (Ref.4) have shown that on the basis of this concept a repeated friction reduction effect can be obtained. The film flow regime, however, cannot be realized over a wide range of parameters due to the instability of the gas-liquid interface. In refs. (Ref.11,12) the calculation of the turbulent boundary layer using a homogeneous model but with the acount of viscosity of the liquid with gas bubbles, show that a 30% surface friction reduction is achieved at the limiting gas concentration on the stream-lined wall (80%). For a bubble flow regime which is realized using gas saturation of the boundary layer through porous stream-lined surfaces or electrolysis, the above friction reduction effect is limiting, provided that one takes into consideration the physical constants of the liquid-gas bubbles suspension only. Experimental studies (Ref.5,6) show the possibility to obtain a 15-20% reduction in the drag of plates and well stream-lined bodies using gas saturation of the boundary layer by gas injection through holes and electrolysis.

2. Experimental Procedure

Experimental investigations of the bubble gas saturation of the turbulent boundary layer were made on a plate and in the initial channel section.

Experiments with a plate were performed in a cavitation tube with a rectangular working section 120 x 1000 mm in size. In its middle part a plate was inserted with a rounded front edge 955 mm long and 40 x 244 mm in the cross-section. The flow velocity was equal to 4 - 11 m/s.

Studies in the initial channel section were carried out in a wind tube with an open 50 x 180 mm jet working section 1200 mm long. The flow velocity was 2 - 6 m/s.

Bubble gas saturation of the boundary layer was reached using porous inserts flush-mounted with the stream-lined surface. On the plate the porous insert was at a distance from 80 to 400 mm from the front point. In the channel it was 180 mm long and its front edge was at a distance of 365 mm from the beginning of the jet working section.

Aliminium was used as a porous material with most probable pore transverse size of 2 μm. About 95% pores had the size of about 1-3 μm. They were arranged in cross rows, the row porosity factor being 0.5 (0.005 for the sample).

Experiments were made for measuring surface friction, gas bubbles concentration, stream-lined surface pressure pulsations, gas flow rate and friction drag.

The surface friction was measured by tensometric probes with "floating" surface elements (Ref.7). The space around the floating element 25 mm in diameter was no more than 0.15 mm, and the step of its positioning was admitted of no more than 0.01 mm. The probes were precalibrated in a closed channel. The correlation between their readings and the calculated and measured pressure differential in a calibration channel showed no more than $\pm 3\%$ error.

The gas bubbles concentration was measured by probes sensitive to the medium electrical conductivity (Ref.8). Two types of probes were used - plate and double-curled with 1-1.5 mm height plates or 0.35 mm diameter, respectively. The probes were precalibrated in a device with gas injection through a porous bottom. The concentration of gas bubbles was measured in a calibration device using Co^{60} isotope, as well as according to the free liquid surface level. The measurement error was equal to ± 5%.

Pressure pulsations were measured by a probe with a sensitive piezoceramics element 1.4 mm in diameter. A special calibration device was used to determine the probe frequency characteristics by correlating its readings with the reference one.

3. Experimental Results

The principal aim of the present study was to gain an insight into the characteristics and efficiency of bubble gas saturation formed on a porous insert and to examine the influence of a liquid with gas bubbles with a downstream removal from the porous insert. To solve the first problem, probes measuring surface friction, bubble concentration and pressure pulsations were positioned directly down stream the porous insert, at a distance of 0.15 its length from the back edge. These experiments made on a plate in a cavitational tube. The second problem was solved in an open jet working section where probes measuring the surface friction and bubble concentration were positioned at distances from the back porous insert edge of 0.36, 0.64, 1.2, 1.7, 2.4 and 3.9 its length.

The results of measuring the plate surface friction are plotted in Fig.1.

From this plot it can be seen that the dependence of the friction reduction on the relative injection rate is distributed over the flow velocity. At a constant relative injection rate the friction reduction increases with the flow rate and the same reduction degree is reached at a lower value. It is characteristic that at a certain relative injection flow, depending on its velocity, the friction drag becomes minimum and with a further increase in the relative injection flow grows. This experiment has shown that the surface friction can be considerably (by many-fold) changed. In this sense the bubble gas saturation is an efficient method for the turbulent near-wall flow control.

The results of measuring the volume concentration of bubbles in the boundary layer are represented in Fig.2.

The measurements have shown the complex form of bubbles concentration profile. In the near-wall portion of the boundary layer (0.1 - 0.2) the maximum concentration is detected, which approximately conforms to the maximum RMS velocity pulsations. Its value increases with the injection flow rate and flow velocity, until the limiting value of 0.75 - 0.80 is reached. This value corresponds to the limiting sphere package. The concentration probe flush installed on a stream-lined wall shows that there are no gas bubbles in the direct vicinity of the wall. Between the maximum concentration layer and the stream-lined wall there is a bubble-free layer, which thickness several times exceeds that of a viscous sublayer.

This decisively important fact was extensively studied in (Ref.9) above the maximum concentration layer the concentration smoothly decreased due to the diffusion process in the boundary layer nucleus.

Fig.3 represents the normalized spectra of the pulsation pressure energy for the flow velocity U_0 = 10.9 m/s at various values of the relative injection flow rate. Spectrum 1 is responsible for the flow without gas saturation. The mode characterized by the initial friction reduction (C_f/C_{f0} = 0.9) is conformed to the decreasing pressure pulsations intensity, in the range of a so-called cut frequency ($\bar{f}$ = $5 \times 10^{-2} - 10^{-1}$, see spectrum 2).

The effect of gas bubbles on the spectrum of pressure pulsations is most pronounced over the range of dimensionless frequencies $5 \times 10^{-2} - 5 \times 10^{-1}$. The intensity of pressure pulsations, as compared with a single-phase flow in the frequency range mentioned, decreases almost by two orders in magnitude (see spectrum 6).

Two distinctive trends should be noted. One implies the presence of bubbles in the boundary layer to cause the increasing pressure pulsations intensity over the range of dimensionless frequencies exceeding the value 5×10^{-1}. Another manifests itself in the region of low frequencies ($f < 10^{-2}$). An increase in the pressure pulsations intensity in this region occurs at the flow regime in the boundary layer corresponding to the minimum friction drag and the bubbles concentration close to limiting.

To study the characteristic features of the bubble film cooling, when removing downstream from the gas injection region in an open jet working section, the surface friction was measured versus the relative injection rate at various flow velocities and relative in injection rates. The dependences obtained are similar to those given in Fig. 2 and are characterized by the same features. As an example, Fig. 4 represents the surface friction variation downstream a porous insert at a constant gas flow rate.

4. Discussion

In contrast to a homogeneous flow at bubble gas saturation of the boundary layer a drastic dependence is observed between the local friction and the Reynolds number (see Fig. 5).

The analysis of experimental results shows the presence of the critical Reynolds number, when no friction reduction occurs. For a porous material the critical velocity gradient proved to be equal to 1.5×10^4 s.$^{-1}$. Taking into consideration a fairly large size of bubbles generated with a porous insert at small Reynolds number, the above result can be explained by a significant disturbing pulse of bubbles with a separation from a porous surface (roughness of injection). When increasing the Reynolds number above its critical value, the gas saturation efficiency drastically grows. In this case the local friction value approaches its calculational magnitude for a laminar flow in a homogeneous flow for the same Reynolds number.

This phenomenon is likely to be caused by the formation of a stabe layer near the stream-lined wall. This layer has high concentration of small-size bubbles forming a space cell structure limiting the turbulent transfer by the amount of transport between the adjacent layers in a carrier phase. In this case the intensity of turbulent pressure pulsations represented in Fig. 3 decreases much more than it is determined by decreasing density. The portion of the flow kinetic energy transferred to the pulsation transport is decreased due to the viscous-elastic effect of a close-packed layer of small gas bubbles, and the near-wall flow is stabilized.

One of the main characteristics of a gas-liquid flow is the discrete phase concentration. The above consideration of the results of measuring the volume bubble concentration in the boundary layer section has shown that in its near-wall portion there is pronounced concentration maximum. In addition, let three more experimental results characteristic of the bubble concentration profile formation be mentioned. First, the concentration on various levels from the stream-lined wall at a given Reynolds number is proportional to the injected gas flow rate (and at a designated flow rate it is proportional to the squared flow velocity). Second, the zone of maximum concentration with decreasing flow velocity (Reynolds number) approaches the stream-lined surface. Third, the maximum bubble concentration grows with flow velocity at a constant gas flow rate.

The role of a bubble maximum concentration layer is illustrated in Fig. 6, where the degree of friction reduction is plotted as a function of the maximum bubble concentration in the near-wall boundary layer portion. The presence of the generalizing regularity for friction reduction over the investigated range of Reynolds numbers shows that the maximum bubble concentration in a boundary layer is one of the basic parameters in the turbulent flow control. As is seen from Fig.6, friction begins to decrease at about 0.15 volume concentration, accompanied by a rather close bubble package. The maximum friction reduction is reached at the maximum package of gas bubbles.

Consider the condition accounting for the minimum surface friction (see Fig. 1) or specific points where the general regularity shown in Fig.6 begins to branch. Increasing frictionat a gas flow rate exceeding its value for the minimum dependence between friction factor and flow rate, may be caused by two reasons. One implies an increase in the bubble separation diameter due to the decreasing surface friction at the point of their separation from the stream--lined wall under the effect of gas-saturation. Taking into account only the motion drag and the surface tension for the separated small bubble in a boundary layer, we obtain, as a first approximation, the expression relating the bubble separation diameter and value of surface friction and other parameters.

$$d = (\tau_w)^{-0.5} \cdot (32 \cdot C_d^{-1} \cdot d_\kappa \cdot \sigma \cdot \nu_0^2 \cdot \rho_0)^{0,25}$$

This dependence is in good agreement with the experimental results obtained for a single capillary in a boundary layer (Ref. 13).Another cause responsible for increasing friction can be explained by the possible coalescence and crushing of large bubbles formed in a shear flow with an increase in the packing density. The above phenomena require additional energy consumptions and may result in the destabilization of the boundary layer flow. These both effects are described by the Taylor criterion. The analysis has shown that the minimum surface friction for the examined range of Reynolds numbers is achieved at the same Taylor criterion equal to 0.012. The dependence of the Taylor number on the maximum bubble concentration in a boundary layer is shown in Fig.7.

The analysis of experimental results has shown that at the Reynolds numbers somewhat above critical value, the insignificant friction reduction is related to small variations in the Taylor number. Since the Taylor number characterises the mean bubble deformation in a shear flow, the above-said can be explained by the small stability margin of a liquid-gas bubble layer. Further increase in the bubble concentration under these conditions results in coalescences (Ref.10). The increasing Reynolds number leads to widening the range of the Taylor number variations. This is likely to create the conditions for stable existance of a layer of close-packed bubbles down to the limiting concentration.

The critical Taylor number equal to 0.012 under the above experimental conditions is responsible for about 50 μm bubbles. Provided that our assumptions on the bubble layer stability margin are valid, gas saturation of boundary layer through large pores (above 50 μm) should not cause a significant friction reduction. In practice, the experimental size of the pores is about 50 to 100 μm.

The analysis of the friction reduction state versus the maximum concentration, made according to experimental data in the initial channel section for the cuts at various distances downstream from the injection point, has shown a universal character of this dependence. Fig. 6 represents the results for $\overline{X} = 0.36$ cut. All experimental points are in good agreement with the results obtained for $\overline{X} = 0.15$.

Experimental results have shown that the curve of friction reduction against the dimensionless coordinate $\overline{X}$ is distributed over the relative injection rate. It can be eliminated, if the complex $C_f' \cdot C_*$ is plotted against the dimensionless coordinate $\overline{X}$. As is seen from the plot in Fig.8, all experimental points are generalized by one line. This illustrates the conclusion on the maximum bubble concentration role.

5. Conclusion

Bubble gas-saturation of the boundary layer is one of the efficient method to influence the intensity of near-wall turbulence generation.

The efficiency of the turbulent boundary layer control using bubble gas saturation under the above experimental condition is determined by two dimensionless criteria : the characteristic bubble concentration responsible for a flow region in a carrying phase, and the Taylor number governing the discrete phase behaviour in a shear flow.

References

1. Grewe P.R., Eddington W.J. The hovercraff - a new conception marin transport - "Quart. trans. Rov. inst. of Naval arch", 1960, v. 112, N 3.

2. Fedyaevsky K.K. Friction reduction by varying the physical constants of liquid and wall. (Уменьшение сопротивления трения путем изменения физических констант жидкости у стенки) Izv. AN SSSR, OTN, 1943, N 9-10. (In Russian)

3. Loitsyansky L.G. On friction drag variation by filling the boundary layer with liquids with different physical constants. (Об изменении сопротивления тел путем заполнения пограничного слоя жидкостями с другими физическими константами) OMM, 1942, t.6. (In Russian)

4. Sperrow E.M., Jonson V.K., Eccert E.R. Two-phase boundary layer and the total plate friction reduction. Trans. ASME, "Applied Mechanics", ser.E, 1962, v. 29, vyp. 2.

5. Hirata M., Nisivaki M., Torin K. Friction in a boundary layer of two-phase gas-liquid flow. Trans. Japan Eng. Mech. Soc., 1967, 33, N 254.

6. McCormick M., Bhattachoryya R. Drag reduction of a submersible hull by electrolysis - "Naval Engng.J.", 1973, v.85, 2, p.11-16.

7. Dershin, Leonard, Hallahker. Direct measurement of surface plate friction with injection. AIAA Journal, Russian translation,1967, v. 5, N 11.

8. Troitsky V.P., Kokorin Yu.V., Gutnikov V.S., Lyubimov L.I., Yuldashev V.I. "Izv. VNIIG", 1970, v. 92.

9. Dubnischev Yu.N., Evseev L.P., Sobolev V.S., Utkin E.N. The study of gas-saturated turbulent flows using the laser Doppler velocimeter. (Исследование газонасыщенных турбулентных потоков с применением лазерного допплеровского измерителя скорости) PMTE, 1975, N 1 (In Russian)

10. Shigeru H., Morimatsu O., Bull. ISME, 1971, v.14, N 75.

11. Basin A.M., Korotkin A.I., Kozlov L.F. Control of the boat boundary layer. (Управление пограничным слоем судна) "Sudostroenie", 1968. (In Russian)

12. Bogdevich V.G., Saren Yu.A. Two-phase (liquid-gas) turbulent boundary layer on a permeable plate. (Двухфазный турбулентный пограничный слой на проницаемой пластине) "Dynamics of Continuous Media",Novosibirsk, 1971, Vyp. IX.

13. Bogdevich V.G., Evseev A.R. On the influence of gas saturation on near-wall turbulence.("Studies on the boundary layer control" (О влиянии газонасыщения на пристенную турбулентность)

Institute of Thermophysics, Siberian Branch of the USSR Ac.Sci., Novosibirsk, 1976.

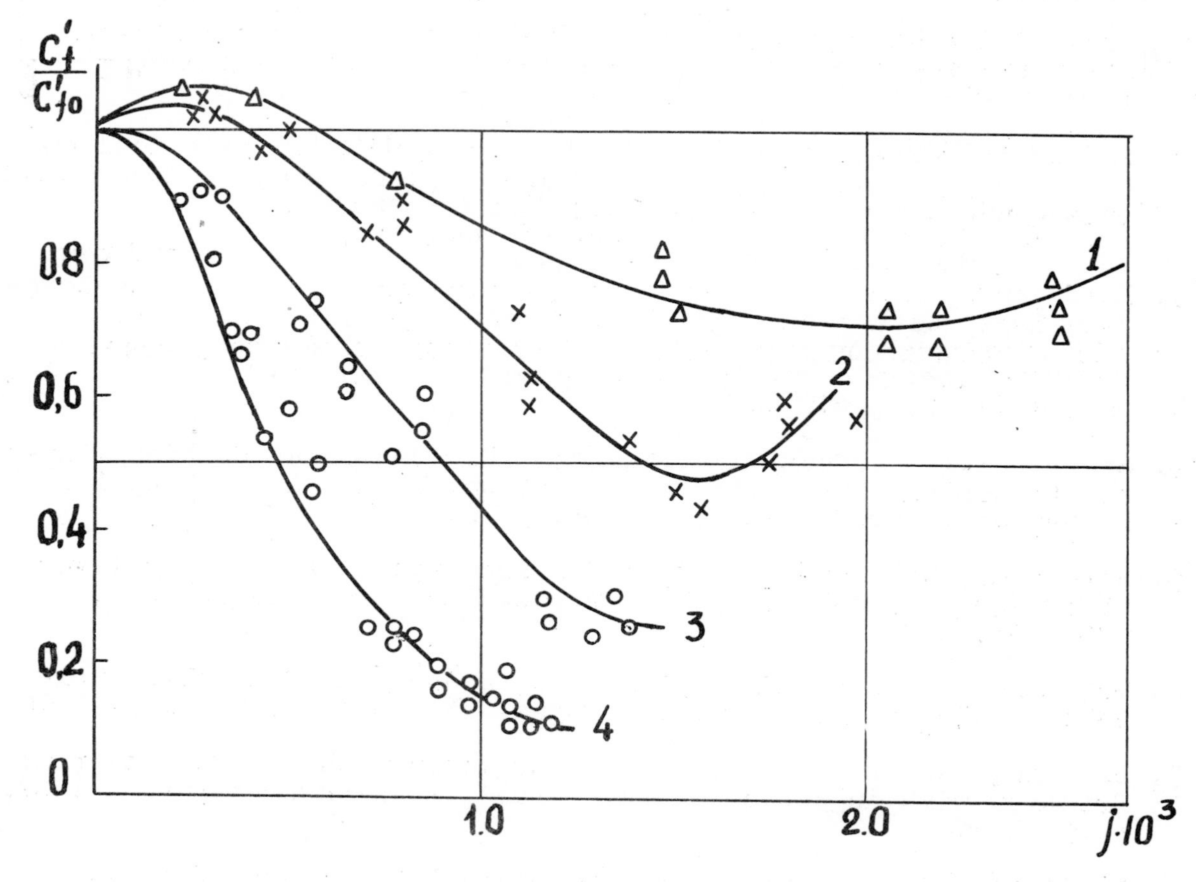

Fig. 1. Local friction variation as a function of the Reynolds number and injection parameter.

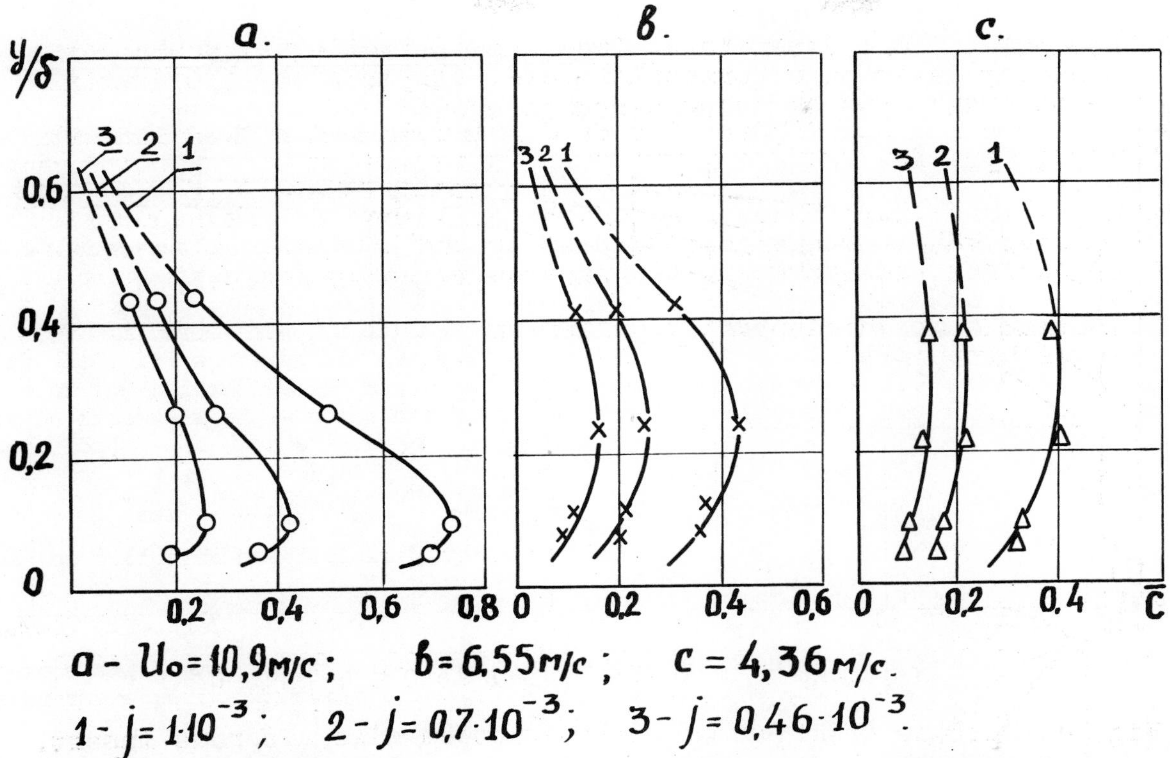

Fig. 2. Distribution of gas concentration in a boundary layer.

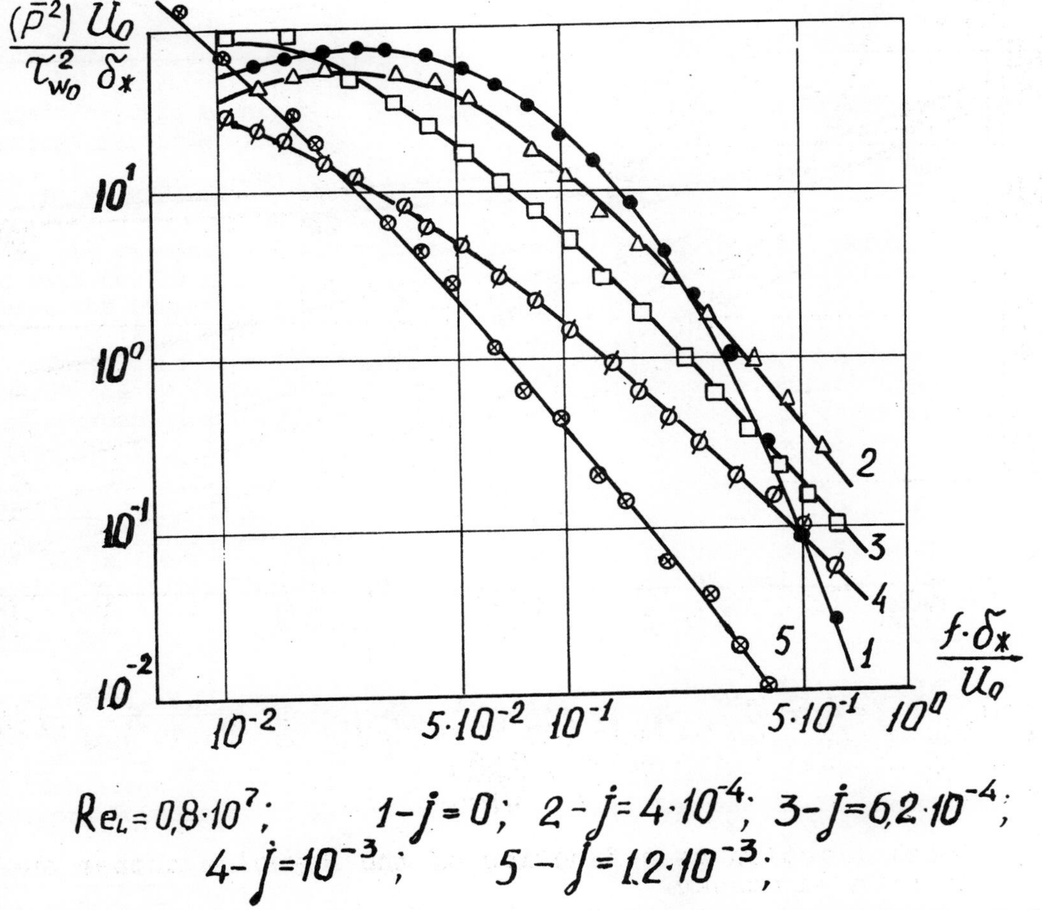

Fig. 3. Normalized spectra of pressure pulsations power at various injection parameters.

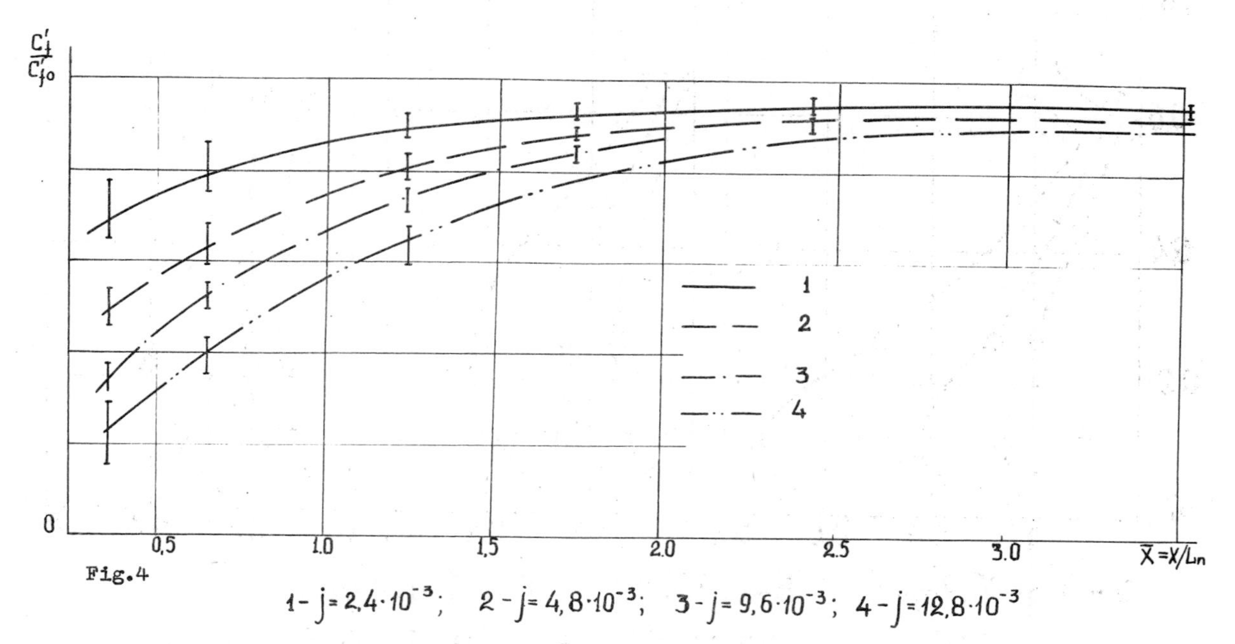

Fig. 4. Surface friction variations stream-wise a porous insert.

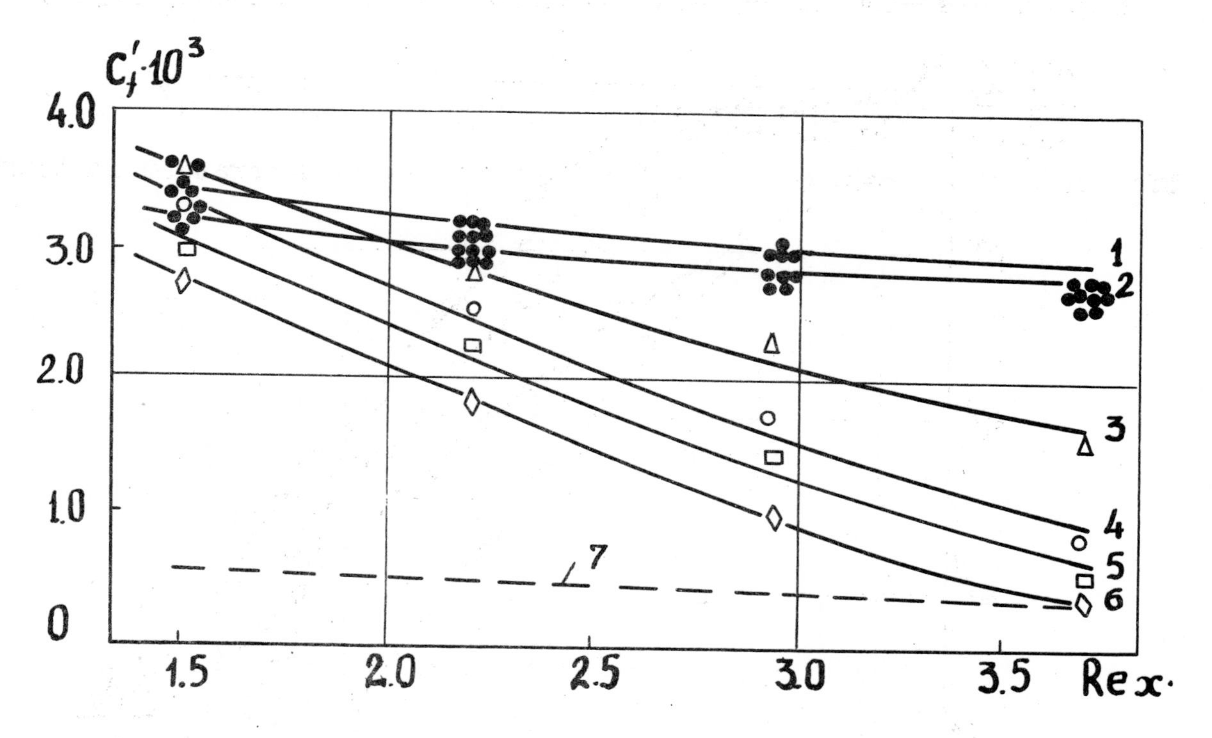

$1,2 - j = 0$; $3 - j = 4{,}6 \cdot 10^{-4}$; $4 - j = 7{,}15 \cdot 10^{-4}$; $5 - j = 0{,}98 \cdot 10^{-3}$; $6 - j = 1.16 \cdot 10^{-3}$;
$7 - C_f' = \frac{0{,}664}{\sqrt{Re_x}}$.

Fig. 5. Local friction as a function of the Reynolds number and injection parameter.

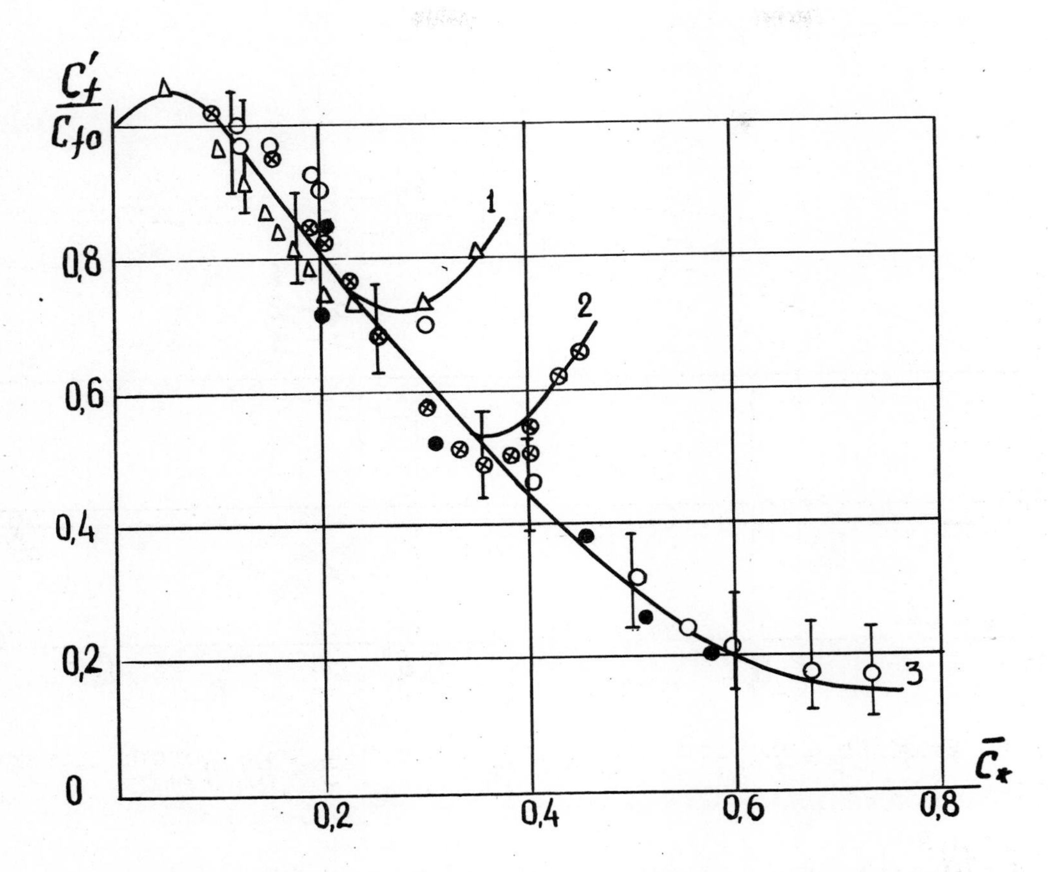

Fig. 6. Local friction versus the maximum gas concentration in boundary layer.

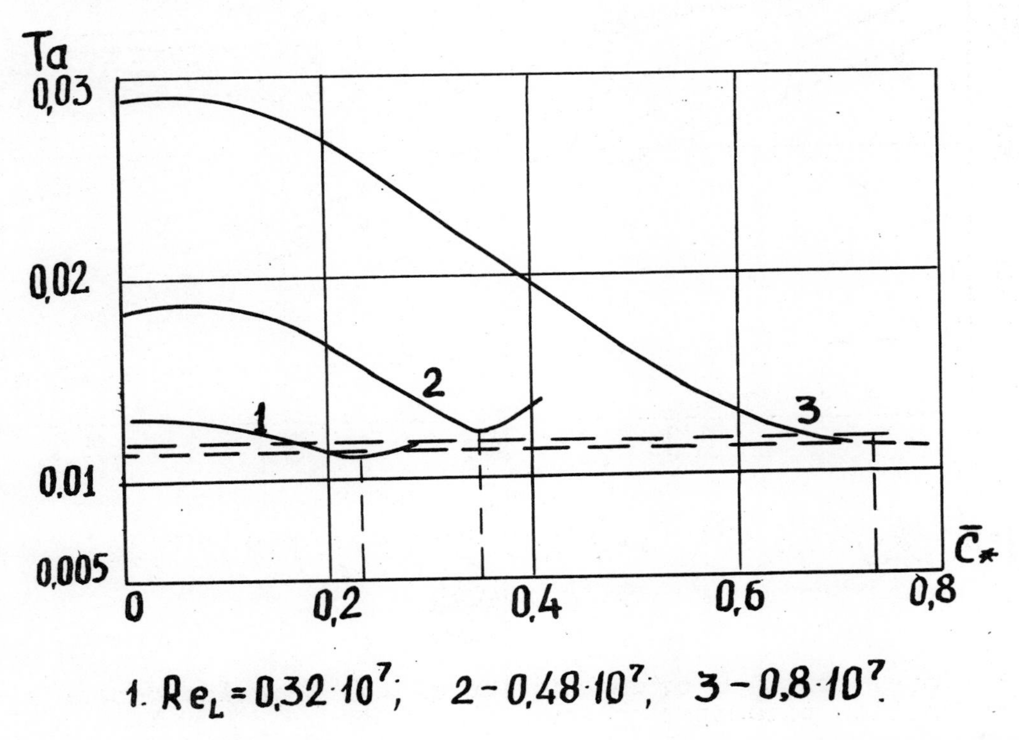

Fig. 7. Taylor number as a function of the maximum gas concentration in boundary layer.

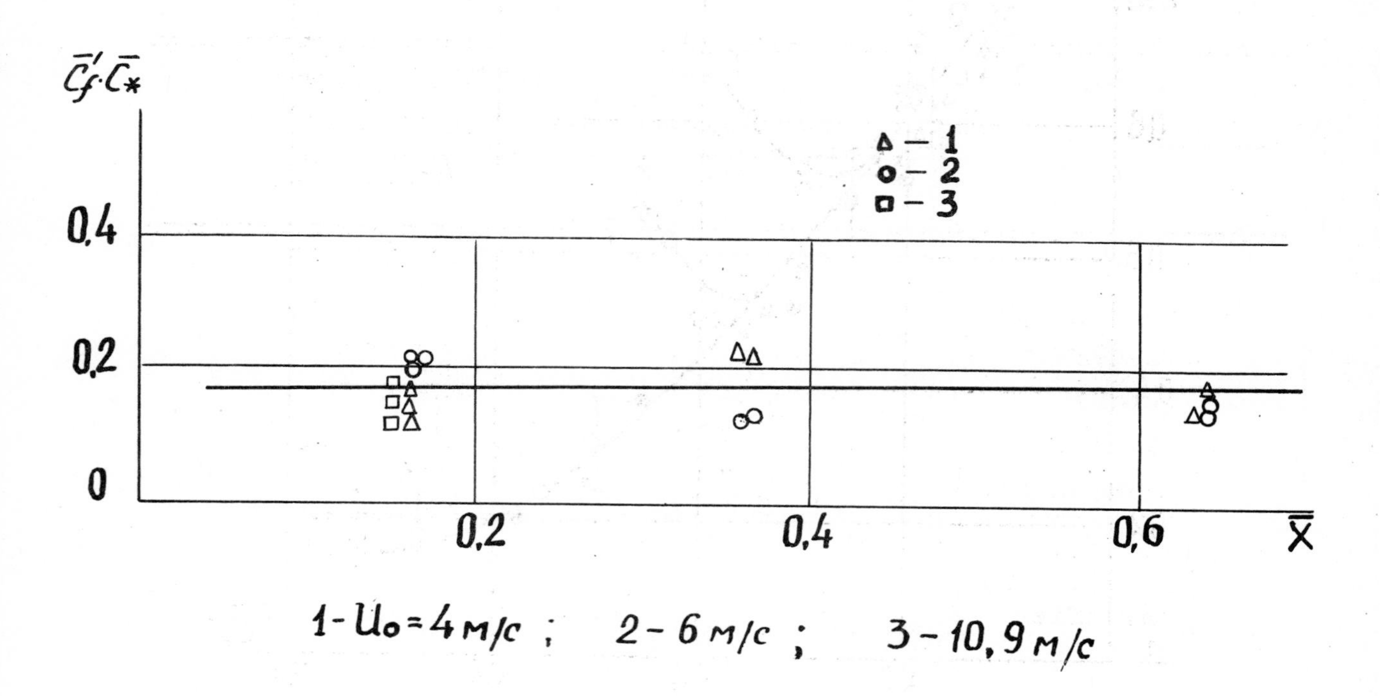

Fig. 8. Generalization of results on local friction stream-wise a prous insert.

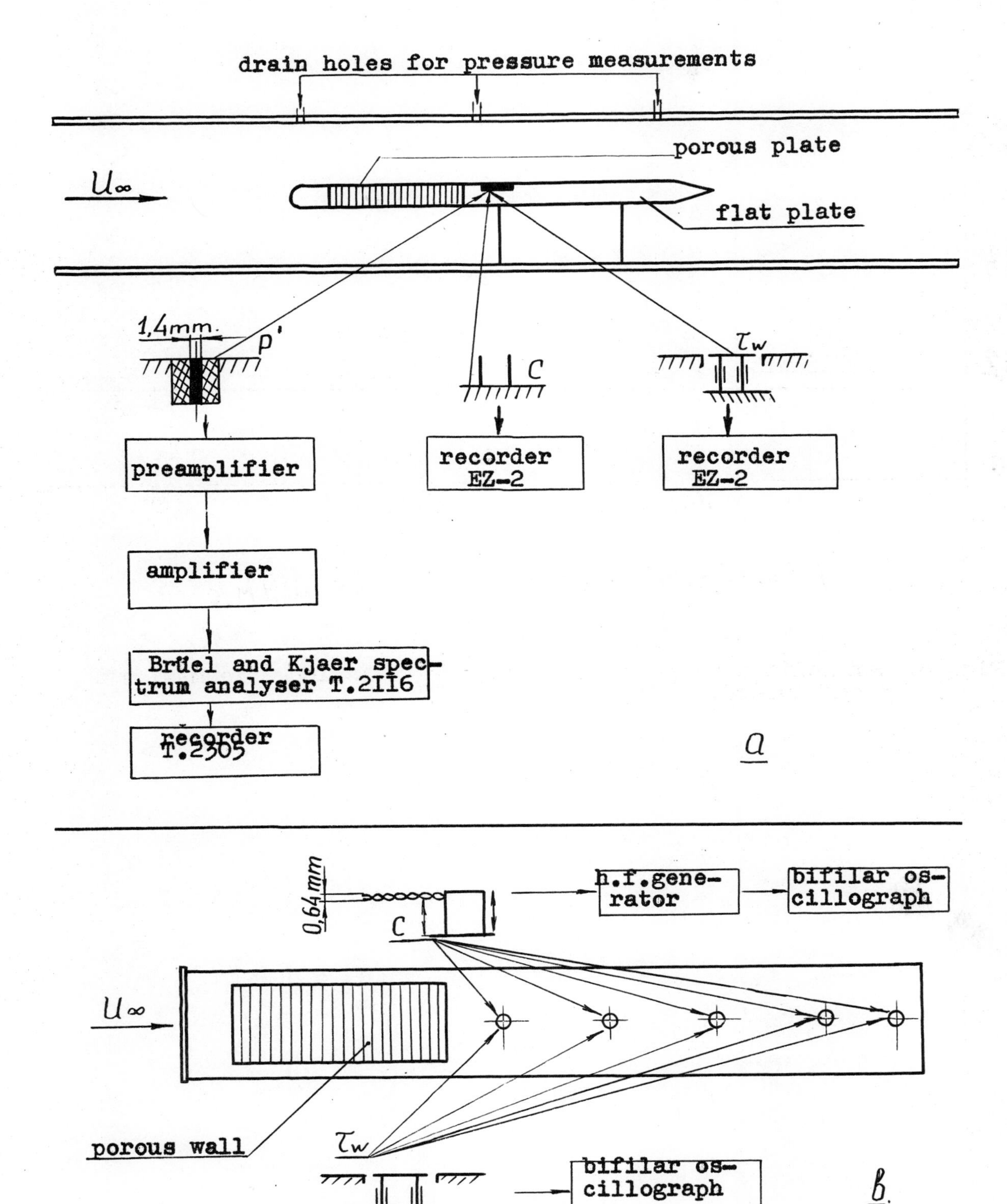

Fig. 9 Schematic diagram of installation and apparatus.

a) Test section of cavitation tube with a plain model.

b) Tube section for wall-and stream-wise friction measurements.

Second International Conference on

Drag Reduction

August 31st - September 2nd, 1977

PAPER E1

GENERAL ASPECTS OF TURBULENCE AND DRAG REDUCTION

B. Hlaváček and J. Sangster,

Ecole Polytechnique de Montreal, Canada

Summary

A new approach toward the origin of turbulence has been developed recently. In this theory, the key concept is the entropy of the flowing liquid. Any liquid in flow is characterized by a lower entropy than the liquid at rest. It can therefore be considered as more organized when flowing. Turbulence is treated as the loss of some of this organization. The theory is particularly applicable to drag reduction. The effect of a polymer in solution can be followed from very small concentrations (where it exhibits a stabilizing effect in drag reduction) up to verh high concentrations (where in contrast it has a stabilizing influence on the flow).

Through this new approach, the discrepancy between the two treatments of drag reduction represented by Astarita and Virk can be overcome. These two quite different theories (the one taking the maximum relaxation time of a polymer molecule as critical parameter, the other the size of the molecule) become complementary in our approach and are subsumed under a more general principle.

Held at St. John's College, Cambridge, England.

Organised and sponsored by BHRA Fluid Engineering, Cranfield, Bedford, MK43 0AJ.

Nomenclature

a	=	characteristic segmental dimension of a macromolecule (Eq. 16)
BR	=	bursting rate from the wall region of a flowing liquid
d	=	diameter of a polymer molecule in solution
D	=	diffusion coefficient of a polymer molecule in solution (Eq. 17)
E_{kin}	=	kinetic energy of a flowing liquid, $0.5\rho\bar{v}^2$
F	=	free energy
F_t	=	$F_v + F_{el}$
F_v	=	$- T\Delta S_v$
F_{el}	=	$- T\Delta S_{el}$
f	=	normal friction factor, τ_w/E_{kin}
f'	=	generalized friction factor, τ_w/F_t
f_o	=	segmental friction factor for a macromolecule in solution (Eq. 16)
$\mathcal{F}$	=	force of entropic origin (Eq. 11)
$g(n)$	=	F_v/E_{kin} (Eq. 5)
k	=	Boltzmann constant
K	=	constant in power law $\tau = K\dot{\gamma}^n$
m	=	mass of a moving particle
n	=	exponent in the power law
N	=	number of segments in a macromolecule (Eqs. 16 and 17)
p_{12}	=	shear stress
$(p_{11}-p_{22})$	=	first normal stress difference
Re	=	Reynolds number
ΔS	=	statistical entropy difference
ΔS_v	=	entropy of a flowing liquid deduced from its velocity profile (Eqs. 3 and 4)
ΔS_{el}	=	entropy of a flowing liquid deduced from its elastic properties (Eqs. 6 and 7)
S_R	=	recoverable elastic deformation
T	=	absolute temperature
$\underline{u}_{rms}$	=	root-mean-square velocity of thermal motion
$\bar{v}$	=	mean velocity of liquid flow
$v(r)$	=	local velocity of liquid flow
v_o	=	slip wall velocity $(\tau_w/\rho)^{\frac{1}{2}}$
V	=	volume of a flowing liquid
$\dot{V}$	=	dV/dt
$\dot{\gamma}$	=	shear rate
δ_o	=	thickness of laminar boundary sub-layer "$=\nu/v_o$".
ε	=	rate of energy dissipation in turbulence
η	=	viscosity of a liquid
λ	=	streak spacing in flow visualization experiments
λ^+	=	non-dimensional streak spacing, $\lambda v_o/\nu$
ν	=	kinematic viscosity, η/ρ
ρ	=	liquid density
τ	=	viscous stress
τ_w	=	wall stress
τ_{max}	=	maximum relaxation time of a macromolecule in solution (Eq. 16)
R	=	radius of tube

Introduction

In this work we are looking for a general description of turbulence phenomena. Other previous attempts have considered turbulence separately for Newtonian or non-Newtonian liquids, drag-reducing systems or molten polymers. In contrast, we wish to find a unifying principle which covers both turbulence and laminarity in terms of order and disorder, terms which are applicable to all systems which show onset of turbulence. This principle is distinct from interpretations of turbulence based upon the ratio of inertial and friction forces (for Newtonian and non-Newtonian liquids) or the ratio of elastic and viscous forces (for molten polymers).

This paper develops the basic ideas and equations of our approach previously sketched (Ref. 1a,b) in which the entropy of a flowing liquid is calculated from probability statistics of thermal motion.

A New Thermodynamic Aspect

Classical thermodynamics deals with systems at equilibrium, i.e. equations such as $P = f(V,T)$ for gases were developed for conditions under which there are no macroscopic changes in these variables in time. States described by a more general formulation, e.g. $P = g(V,T,\dot{V})$ are not thus treated but are considered in recently-developed non-equilibrium thermodynamics (Ref. 2).

Our theory is somewhat similar to this latter case in that we use an expression originally derived for systems at rest (the Maxwell velocity distribution of thermal motion) and adapt it to a system (flowing liquid) which is non-equilibrium in the strict sense. The Maxwell distribution for three dimensions is (Ref. 3)

$$p(\vec{u}) = (m/2\pi kT)^{3/2}\exp\{-m(u_x^2 + u_y^2 + u_z^2)/2kT\} \qquad (1)$$

for particles of mass m. If now we impose a velocity gradient on the system, the velocity distribution will be disturbed. Suppose that $\vec{A}$ is the "extra" velocity imposed, relative to the coordinate system normally associated with the center of mass of a moving and deformable body; the probability distribution p(u) will change to $p(\vec{u} + \vec{A})$. The situation for a liquid flowing through a pipe is given in Fig. 1, in which the coordinate system moves downstream with the mean velocity $\bar{v}$ parallel to the streamlines.

We have shown (Ref. 1a) that a statistical entropy difference may be calculated from the ratio of the probabilities $p(\vec{u})$ and $p(\vec{u} + \vec{A})$. If, for example, A_i of a particular microdomain "i" of liquid is oriented in the axial direction, the resultant velocity is $\Delta v_i = \bar{v} - A_i$ and the entropy is

$$-T\Delta S_i = 0.5\rho(\Delta v_i)^2 \qquad (2)$$

In the derivation of Eq. 2, it has been assumed that the liquid is of a discrete nature at the microscopic level and there are hydrodynamic units which undergo Brownian motion. The size of these microdomains may be set plausibly at the order of 100 Å, but the exact value is unimportant since it does not appear in the final expression, Eq. 2. This equation may be integrated readily over the whole volume of flowing liquid providing we know its velocity profile. The entropy thus calculated (Ref. 1a) is negative; this may be interpreted to mean that the moving liquid is more "organized" than the liquid at rest or in plug flow for which $\Delta v = 0$ and $\Delta S = 0$.

Thus for any flowing fluid we can calculate its level of entropy; all that is required is its velocity profile. This result is general and applies to both Newtonian and non-Newtonian fluids. We have interpreted the energy represented by $-T\Delta S$ as a "stored" energy associated with the build-up of a velocity profile. Thus even a Newtonian fluid can store energy in creating an organized structure, in the sense of Prigogine (Ref. 4). It is to be noted that this stored energy results from the velocity profile and not from orientation or elongation of individual particles, which is discussed in polymer rheology.

This new element is common to all fluids in motion and probably will have considerable consequences for the general rheology of fluids. We test this new approach here by considering a few simple examples in turbulence and drag reduction, and by considering both bulk and boundary layer regions.

Energies and Friction Factors in the Bulk Flowing Liquid

a) Purely Viscous Liquids

Eq. 2 may be integrated over the whole system of volume V thus

$$\sum_i T\Delta S_i = T\Delta S_V = -\rho\{\int_V (v(r) - \bar{v})^2 dV\}/2V \qquad (3)$$

and $T\Delta S_V$ is then the stored energy per unit volume associated with the formation of the velocity profile represented by $v(r)$. We have chosen (for reasons which will become apparent later) to refer to this stored energy as a kind of thermodynamic free energy F of a system under isothermal conditions:

$$F_V = -T\Delta S_V \qquad (4)$$

The functional dependence of F_V upon velocity profile may be delineated by considering, for purely viscous liquids, the flow rate and the power law exponent $n(\tau = k\dot{\gamma}^n)$. Eq. 4 may be rewritten

$$F_V = -T\Delta S_V = 0.5\rho v^2(-T\Delta S_V/0.5\rho v^2) = E_{kin}g(n) \qquad (5)$$

where E_{kin} is the bulk kinetic energy and $g(n)$ is the ratio of free to kinetic energy at the same $\bar{v}$. Of the two quantities E_{kin} and $g(n)$, only $g(n)$ depends upon n. Fig. 2 shows how the velocity profile changes with n, and Table 1 gives the values of $g(n)$ as a function of n.

An examination of Fig. 2 and Table 1 together shows that liquids exhibiting higher values of n (i.e., of more pointed or cone-like velocity profile) are clearly less stable from the point of view of stored energy. If, for example, the laminar flow of a Newtonian liquid is increased through the transition region, $g(n)$ decreases from 0.33 to 0.02 and the behaviour of F_V is shown in Fig. 3. Some of the free energy in the laminar flow has been dissipated during the transition. The turbulent liquid (represented by the Blasius profile) represents a more stable system (lower free energy) than the laminar case.

This is illustrated more completely in Fig. 4 which shows the variation of $g(n)$ with Reynolds number for water passing through the transition. There is an abrupt change at $Re \sim 2300$, but elsewhere $g(n)$ varies little. The dashed line shows the same function for 0.3% Carbopol C in water, for which $n = 0.6$. The arrow indicates the effect of adding the small amount of polymer to water in laminar flow. The flow is still laminar, but the free energy has decreased, meaning that the velocity profile has been flattened somewhat.

b) Viscoelastic Liquids

This case will be discussed in more detail in a forthcoming article (Ref. 5). The following sketch will suffice here to indicate the direction we take.

We have seen that, for purely viscous liquids, the entropy associated with a particular velocity profile is given by Eq. 3. If the liquid also possesses some elastic character, there will be another contribution to the entropy, viz. ΔS_{el}, associated with the first normal stress difference:

$$-T\Delta S_{el} = \{ \int_V (p_{11}-p_{22})dV\}/V \tag{6}$$

and the total free energy F_t is

$$F_t = -T(\Delta S_v + \Delta S_{el}) = F_v + F_{el} \tag{7}$$

We are now ready to introduce the friction factor, f, which is commonly defined as the ratio of wall shear stress and the kinetic energy per unit volume:

$$f = \tau_w/E_{kin} \tag{8}$$

The decrease in f below a certain value indicates an unstable flow and the onset of turbulence. For viscoelastic liquids, however, f as defined by Eq. 8 is inadequate for use in this context, since the flow characteristics of viscoelastic liquids are quite different from those of purely viscous ones. A more generalized friction factor can be defined as

$$f' = \tau_w/F_t \tag{9}$$

which is applicable to the complete spectrum of liquids, from the purely viscous to the purely elastic.

Let us compare onset of turbulence in water and in a polymer melt, using f'. For water, $F_t = F_v$ and from Eqs. 5, 8 and 9 it may be deduced that $f' = f/g(n)$. Since onset of turbulence in water occurs at $f \sim 0.015$, and g(n) for water is 0.33, than $f' \sim 0.015/0.33 \sim 0.05$.

For the molten polymer, $F_t = F_{el}$ and from Eqs. 6 and 9, $f' \sim p_{12}/(p_{11}-p_{22}) \sim 1/S_R$ where S_R is the recoverable elastic deformation. The first instabilities in a polymer melt flow are observed (Ref. 6) at $S_R \sim 10$, and so $f' \sim 0.1$. It is remarkable that values of f' are so similar for such extremely rheologically different liquids as water and polymer melt.

Consideration of f' and the "stored" energy in a flowing liquid thus illuminate the fact that air, water, mercury and glycerin behave similarly insofar as they are all non-elastic and become turbulent at Re $\sim$ 2300; and that molten polymer flow is turbulent even at Re < 1. (Ref.6)

Some Considerations of the Boundary Layer and the Bursting Rate

The laminar sublayer is of special interest to our approach since it is here that the velocity gradient is greatest, as is also the entropy difference. Suppose the sublayer has a thickness δ_o, and the velocity of the liquid just outside the sublayer is the slip velocity $v_o \equiv (\tau_w/\rho)^{\frac{1}{2}}$. We showed previously (Ref. 1a) that the entropy difference across the sublayer $\Delta S(\delta_o)$ was

$$-T\Delta S(\delta_o) = 0.5\rho v_o^2 \tag{10}$$

which also indicates that $\Delta S(\delta_o)$ is proportional to τ_w. The energy $-T\Delta S(\delta_o)$ of a unit volume of liquid is of entropic origin and is dimensionally equivalent to a force $\mathcal{F}$ acting on it through a distance δ_o:

$$-T\Delta S(\delta_o) \sim \mathcal{F}\delta_o \tag{11}$$

Thus $\mathcal{F}$ may be understood to represent the force necessary to cause a unit volume of liquid, stationary at the wall, to jump across the boundary. From Eqs. 10 and 11 it can be seen that $\mathcal{F} \propto \tau_w/\delta_o$; the experimentally observed effects of increasing the flow velocity are an increase in τ_w and a decrease in δ_o, and so the force $\mathcal{F}$ - a measure of the tendency of the fluid to generate turbulent motion - increases rapidly with increase in velocity.

We can combine these ideas with the general description of turbulence on the microlevel given by Levich (Ref. 7). Energy is dissipated in turbulence by small scale eddies, for which Re $\sim$ 1. If the rate of energy dissipation per unit time per unit volume is ε, then the Kolmogorov-Oboukhov relation

$$v_o \sim (\varepsilon\delta_o/\rho)^{1/3} \qquad (12)$$

is valid. This result was obtained on the basis of dimensional analysis only; it can however also be derived from our approach as follows. If the mini-vortex in the final stage of turbulence is under the influence of force $\mathcal{F}$ and moves with velocity v_o, then the dissipation rate ε is

$$\varepsilon = \mathcal{F}v_o \sim 0.5\rho v_o^3/\delta_o \qquad (13)$$

from Eqs. 10 and 11. Rearrangement of Eq. 13 gives $v_o \sim (2\varepsilon\delta_o/\rho)^{1/3} = 1.26(\varepsilon\delta_o/\rho)^{1/3}$ which is essentially the Kolmogorov-Oboukhov relation, Eq. 12. This shows that the rather novel quantities ΔS and $\mathcal{F}$ are related to experimentally observable flow properties and are consistent with a description of turbulence such as that of Levich.

Consideration of $\mathcal{F}$ also contributes to the understanding of the "bursting rate" (BR) from the wall region, as observed by Kline (Refs. 8 and 9) and Donohue (Ref. 10). We can connect $\mathcal{F}$ and BR if BR is taken to be a measure of the tendency toward turbulence, which is how we interpreted $\mathcal{F}$ earlier. In the language of the present approach, BR is a measure of the tendency for "discharging" large local entropy imbalances $\Delta S(\delta_o)$ across the boundary layer. Therefore there should be a proportionality between $\mathcal{F}$ as defined in Eq. 11 and the BR measured experimentally. This can be done if we use the relation $v_o\delta_o/\nu \sim 1$ previously mentioned. From Eqs. 10 and 11, $\mathcal{F} = 0.5\rho v_o^2/\delta_o$; substituting for δ_o, we arrive at

$$\mathcal{F} = 0.5\rho v_o^3/\nu \qquad (14)$$

Column 5 of Table 2, which uses the data of Donohue (Ref.10) for BR, v_o and ν, shows this proportionality between BR and $\mathcal{F}$.

It is noteworthy that BR/$\mathcal{F}$ varies so little, given the variation in BR and the qualitative change in the nature of the flowing liquid. Both BR and $\mathcal{F}$ are evidently proportional to v_o^3; it is equally evident moreover that $\mathcal{F}$ depends also on ν (Eq. 14). This suggests that BR has a similar dependence upon ν. The BR data of Donohue are represented in Fig. 5, which shows the polymer solution curve below the water curve; this is what is expected if BR also has a ν^{-1} dependence. The same point is made in comparing columns 5 and 6 of Table 2. We are led therefore to suggest that BR is smaller in the drag-reducing system not (as Donohue et al. suggest) because of its drag-reducing properties but simply because the kinematic viscosity is greater. If this is so, than a reduction of BR would be observed in a more viscous Newtonian liquid, e.g. ethylene glycol.

We can profitably comment also on the streak spacing in turbulence, also investigated by Donohue. The non-dimensional streak spacing $\lambda^+ = \lambda v_o/\nu$ was found to be 100 for water and for polymer solutions at the onset of drag reduction. Taking $\delta_o v_o/\nu \sim 1$, this means that $\lambda/\delta_o \sim 100$. We had shown previously (Ref. 1a) that, under these conditions, $\delta_o/d \sim$ const., where d is the diameter of the polymer molecule in solution. In this way the size of the polymer molecule determines the streak spacing at the first appearance of drag reduction. This possibility, not noticed in the experimental studies, suggests a further link between micro- and macro-level properties of drag-reducing systems.

Comparison with Other Theories of Drag Reduction

The present theory (Ref. 1a) regards macromolecules in dilute solution as particles of mass m in Brownian motion, whose mean velocity u_{rms} is given by

$$0.5mu_{rms}^2 = 1.5kT \quad (15)$$

Such a translational motion of a particle is phenomenologically equivalent to diffusion, whether or not there is a concentration gradient.

At first sight this has little to do with the drag reduction theories of Virk (Ref. 11) and Astarita (Ref. 12). Virk deals with wall shear stress and the size of the macromolecule, Astarita with its relaxation time. All three, however, can be interrelated. On the one hand, the maximum relaxation time τ_{max} is given by (Ref. 13a)

$$\tau_{max} = f_o N^2 a^2 / 6\pi^2 kT \quad (16)$$

and the diffusion coefficient D by (Ref. 13b)

$$D = kT/f_o N \quad (17)$$

where a and N are characteristic of a given polymer of a given molecular weight, f_o being characteristic of the solvent. Combining Eqs. 16 and 17 we have

$$\tau_{max} = Na^2/6\pi^2 D \quad (18)$$

which connects the diffusion coefficient and the relaxation time. On the other hand, Virk showed that

$$v_o d/\nu = \text{const.} \quad (19)$$

at the onset of drag reduction. The Einstein-Stokes treatment of diffusion (Ref. 14) for spherical particles gives

$$D = kT/3\pi\eta d \quad (20)$$

where η is the viscosity of the medium. In other words, $D \propto d^{-1}$ which, when combined with Eq. 19 gives $v_o \propto \nu D$. It should be noted that the derivation of Eq. 20 uses the equation of Brownian motion (Eq. 15) explicitly.

Thus it is seen that τ_{max}, d, v_o and D are all inter-connected, and that a "time-based" theory (Astarita) does not exclude a "dimension-based" theory (Virk). All these approaches attempt to interpret macroscopic events in terms of microscopic properties.

Acknowledgment

The authors are indebted to the National Research Council of Canada for an operating grant (A9359) which they adknowledge with thanks.

References

(1) a) Hlaváček, B., Sangster, J. and Stanislav, J.F. "A physical model for the origin of turbulence". Chem. Eng. J., 10, pp. 95-98 (1975).
b) Hlaváček, B. and Stanislav, J.F. "Thermodynamical aspects of the origin of turbulence". Proc. VIIth International Congress on Rheology, pp. 296-297. Chalmers University of Technology, Gothenburg, Sweden (23-27 August 1976).

(2) Astarita, G. "An introduction to non-linear continuum thermodynamics". Ch. 2. Societa Editrice di Chimica. (1975).

(3) Brush, S.G. "The kind of motion we call heat". Book 1, pp. 186-187. North-Holland. (1976).

(4) Glansdorff, P. and Prigogine, I. "Thermodynamic theory of structure, stability and fluctuations". Wiley-Interscience. (1971).

(5) Hlaváček, B., Carreau, P.J. and Schreiber, H.P. "A thermodynamic interpretation of instability phenomena in polymer fluids". Delivered at International Conference on Polymer Processing. Massachusetts Institute of Technology, Cambridge (15-18 August 1977).

(6) Tordella, J.P. "Rheology: theory and applications". (Ed. T.R. Eirich) Vol. 5, Ch. 2. Academic Press. (1969).

(7) Levich, V.G. "Physicochemical hydrodynamics". Ch. 1. Prentice-Hall. (1962).

(8) Kim, H.T., Kine, S.J. and Reynolds, W.C. "The production of turbulence near a smooth wall in a turbulent boundary layer". J. Fluid Mech., 50, pp. 133-160 (1971).

(9) Offen, G.R. and Kline, S.J. "A proposed model of the bursting process in turbulent boundary layers". J. Fluid Mech., 70, pp. 209-228 (1975).

(10) Donohue, G.L., Tiederman, W.G. and Reischman, M.M. "Flow visualization of the near-wall region in a drag-reducing channel flow". J. Fluid Mech., 56, pp. 559-575 (1972).

(11) Virk, P.S., Merrill, E.W., Mickley, H.S. and Smith, K.A. "The critical wall shear stress for reduction of turbulent drag in pipe flow". In "Modern developments in the mechanics of continua". Ed. S. Eskinazi, pp. 37-52. Academic Press. (1966).

(12) Astarita, G. "Possible interpretation of the mechanism of drag reduction in viscoelastic liquids". Ind. Eng. Chem. Fund., 4, pp. 354-356. (1965).

(13) Bueche, F. "Physical properties of polymers". a) p. 208, b) p. 66. Interscience. (1962).

(14) Glasstone, S. "Textbook of physical chemistry". pp. 260-261. Van Nostrand. (1946).

Table 1

The ratio $g(n) = F_v/E_{kin}$ of the free energy to the average kinetic energy for liquids obeying the power law $\tau = K\dot{\gamma}^n$.

n		g(n)
2.0		0.46
1.8		0.44
1.6		0.42
1.4		0.40
1.2		0.37
1.0	----Newtonian----	0.33
0.8		0.29
0.6		0.24
0.4		0.17
0.2		0.085
Blasius *		0.02

* $v(r)/v_o = 8.56 \quad (R-r)v_o\rho/\eta \quad ^{1/7}$

Table 2

The proportionalith between $\mathcal{F}$ and the bursting rate BR in water and PEO solutions.

$$\mathcal{F} = 0.5\rho v_o^3/\nu$$
$$\rho = 10^3 \text{kg m}^{-3}$$

polymer concentration, ppm	$v_o \times 10^3$ $m\ s^{-1}$	$\nu \times 10^6$ $m^2\ s^{-1}$	BR $m^{-1}\ s^{-1}$	$BR/\mathcal{F}$ $kg\ m^{-2}\ s^{-2}$		$(BR/v_o^3) \times 10^{-7}$ $m^{-4}\ s^2$	
0	4.3	0.89	7.5	0.17	0.16	9.4	8.3
0	7.3	0.89	28.0	0.13		7.2	
0	11.3	0.89	107.	0.13		7.4	
0	11.3	1.02	134.	0.19		9.3	
139	6.7	1.25	14.2	0.12	0.12	4.7	4.5
139	6.7	1.25	17.3	0.14		5.8	
139	9.5	1.25	33.9	0.10		4.0	
139	9.2	1.43	26.8	0.10		3.4	

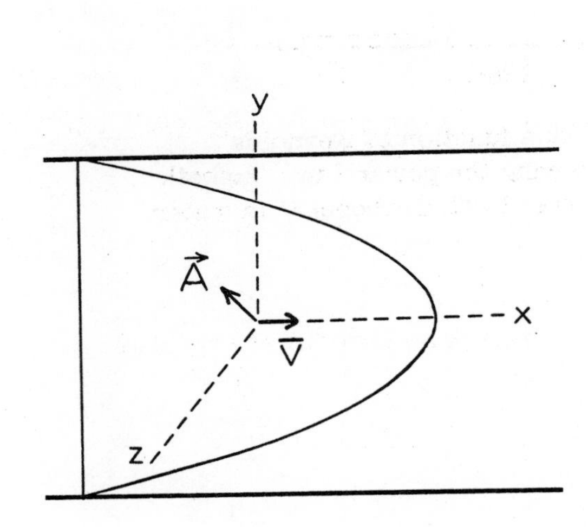

Fig.1 Coordinate system for liquid in pipe flow.

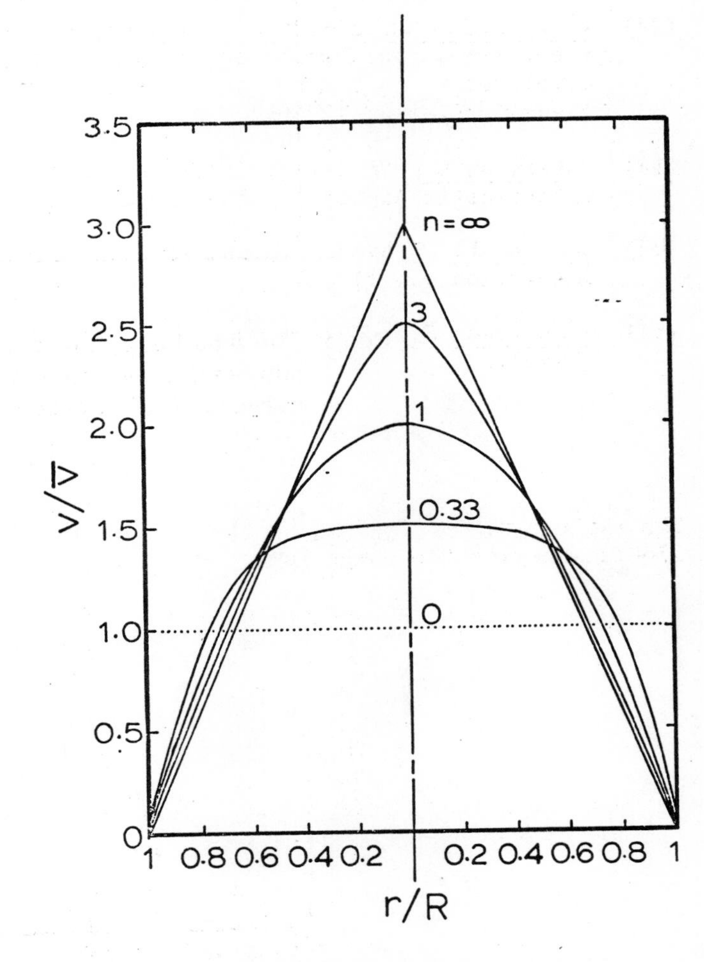

Fig.2 Velocity profile of liquid in pipe flow, as a function of the power-law exponent n.

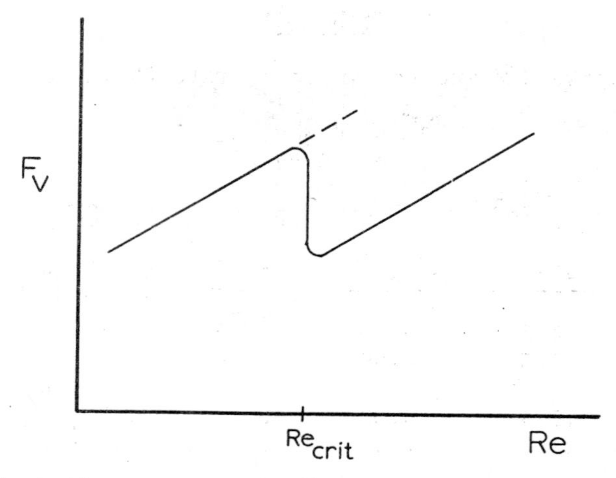

Fig.3 Free energy of a flowing liquid as a function of the Reynolds number.

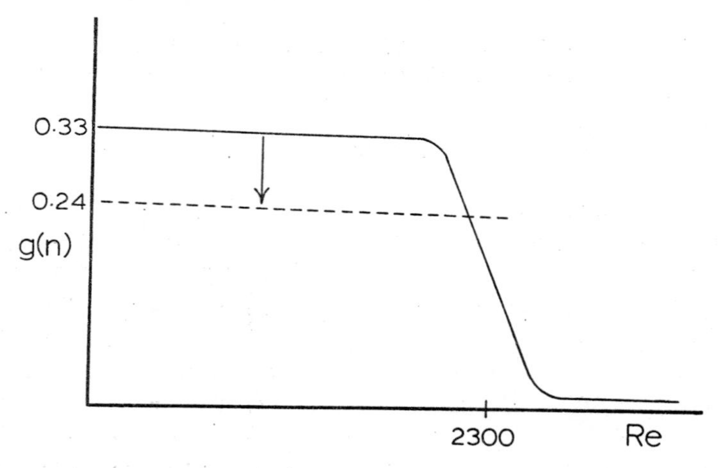

Fig.4 The function g(n) of eq.5 as a function of Reynolds number for two liquids obeying the power law. Smooth curve: water. Dashed curve: 0.3% Carbopol C in water.

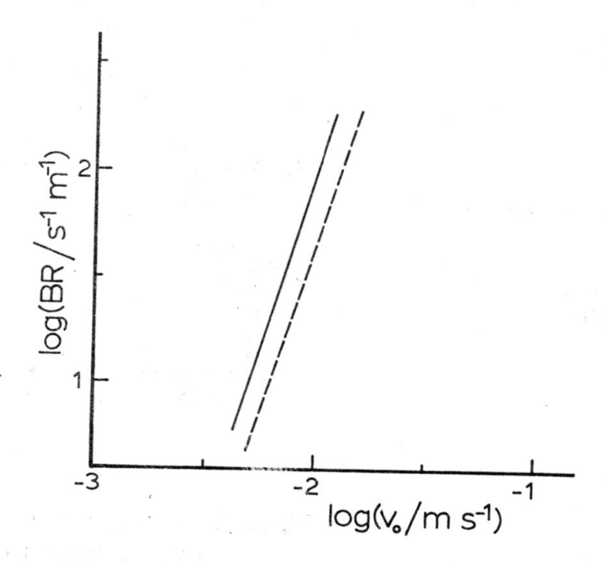

Fig.5 The bursting rate in water (solid line) and polymer solution (dashed line). Data from Ref. 10.

Second International Conference on

Drag Reduction

August 31st - September 2nd, 1977

PAPER E2

DRAG REDUCTION IN WATER BY HEATING

E. Reshotko,

Case Western Reserve University, U.S.A.

Summary

A significant reduction of drag is available to water vehicles with on-board propulsion systems if the reject heat of the propulsion system is discharged through heating the laminar flow portion of the hull.

Calculations for a flat plate based on an e^9 transition criterion show that the prospective drag reduction is strongly dependent on vehicle speed and the thermal efficiency of the power plant and only mildly a function of vehicle length. With a power plant efficiency of 17%, at speeds up to 10 knots, the drag reduction is less than 10%. However, for a vehicle speed of 25 knots there can be a 25% drag reduction and at 50 knots a reduction of about 60% in drag relative to that of the unheated plate. These effects are realized with overheats of the order of 40°C or less so that prospective vehicle skin temperatures pose no problem.

Held at St. John's College, Cambridge, England.
Organised and sponsored by BHRA Fluid Engineering, Cranfield, Bedford, MK43 0AJ.

NOMENCLATURE

A	reference area for drag
C_{D_F}	friction drag coefficient
c	specific heat of water
c_f	skin-friction coefficient
c_h	Stanton number
D	drag force
D_F	friction drag
d	body diameter
L	vehicle length
P_A	power available for heating
P_D	drag power
P_H	power required to heat laminar portion of vehicle hull
Pr	Prandtl number
q	dynamic pressure, $\frac{1}{2}\rho_\infty u_\infty^2$
Re	Reynolds number
T_w	wall temperature
T_∞	free-stream temperature
u_∞	free-stream velocity
w	lateral length (or perimeter) of vehicle surface
x	longitudinal distance along vehicle surface
x_{tr}	longitudinal distance to transition
δ^*	boundary layer displacement thickness
η_{eff}	effectiveness in supplying heat where needed
η_{th}	thermal efficiency of power plant
ρ_∞	free-stream density

INTRODUCTION

It was noted many years ago in experiments at low subsonic speeds (Refs. 1, 2) that the transition location of the flat plate boundary layer in air is advanced as a result of plate heating. Based on this observation it had long been suspected that heating would have the opposite effect in water, namely that it would delay the onset of transition. This is because heating in water reduces the viscosity near the wall resulting in a fuller, more stable velocity profile than the Blasius. Cooling in water (and heating in air) on the other hand tends to give an inflected velocity profile which is less stable than the Blasius profile.

These suspicions remained untested until confirmed by the analysis of Wazzan, Okamura and Smith (Refs. 3,4) and by the more complete parallel-flow stability formulation of Lowell and Reshotko (Ref. 5). The results of references 3-5 for minimum critical Reynolds number are shown in Fig. 1. Both sets of calculations predict significant boundary layer stabilization (increased minimum critical Reynolds number, decreased disturbance amplification rates, etc.) with moderate heating, but display a maximum and subsequent decrease as the wall to free-stream temperature difference is further increased.

The calculated sensitivity of the stability characteristics to small amounts of heating has been experimentally confirmed by Strazisar et al (Refs. 6,7). In Fig. 2 experimentally determined neutral stability curves for nominal uniform wall temperature differences of $T_w-T_\infty = 0, 5, 8°F$ are compared with curves calculated by the method of Ref. 5. The experimental results are curves faired through the measured neutral points which have not been shown for the sake of clarity. Both the calculated curves and experiment readily show that with increase in T_w-T_∞, $(Re_{\delta^*})_{min\ crit}$ increases and the band of frequencies that are amplified becomes narrower. Note that while the theoretical neutral curves according to Lowell's parallel flow calculation nest within each other, this does not happen experimentally until Re_{δ^*} exceeds 860. A wall overheat of 5°F results in an almost 50% increase in $(Re_{\delta^*})_{min\ crit}$ indicating a doubling in length to the minimum critical point. The recent work of Barker and Jennings (Ref. 8) in a constant diameter pipe shows considerable increase of transition Reynolds number in the entrance flow boundary layers with heating.

These results portend exciting possibilities for drag reduction in water using surface heating. These possibilities are explored in the present paper.

DRAG REDUCTION ANALYSIS

For a vehicle with an on-board propulsion system

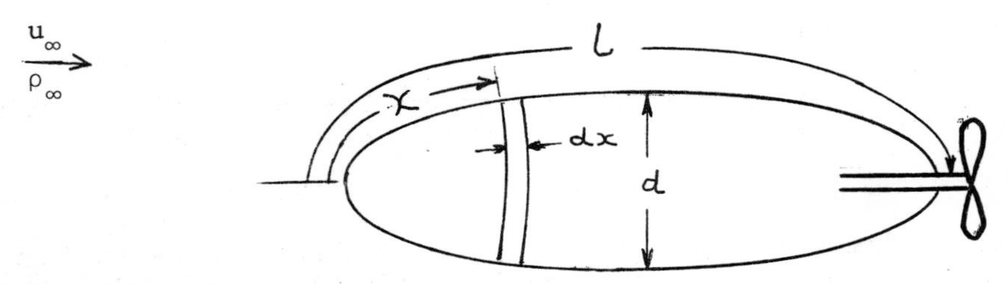

the friction drag is

$$D_F = q \left[\int_o^{x_{tr}} c_{f_\ell} \, wdx + \int_{x_{tr}}^{L} c_{f_t} \, wdx \right] \tag{1}$$

where q is the dynamic pressure, c_{f_ℓ} and c_{f_t} are respectively the laminar and turbulent friction coefficients, wdx is the area element at length x, L is the vehicle length and x_{tr} is the transition location.

The total drag can be written

$$D = D_F \left(\frac{D}{D_F}\right) \tag{2}$$

where $\frac{D}{D_F}$ is the ratio of total to friction drag. For an axisymmetric body this ratio is a function of the fineness ratio of the configuration. Hoerner (Ref. 9) suggests that

$$\frac{D}{D_F} = 1 + 1.5 \left(\frac{d}{L}\right)^{3/2} + \ldots \tag{3}$$

The drag power can then be written

$$P_D = Du_\infty = q\, u_\infty \left(\frac{D}{D_F}\right) [C_{D_F} A] \tag{4}$$

where $[C_{D_F} A]$ is the quantity in brackets in equation (1)

The power available for heating is related to the thermal efficiency of the power plant as follows:

$$P_A = \left(\frac{P_D}{\eta_{th}} - P_D\right)\eta_{eff} = P_D\left(\frac{1}{\eta_{th}} - 1\right)\eta_{eff} \tag{5}$$

where η_{eff} is the effectiveness of transmitting the reject heat to the water in the desired manner.

If one considers heating only the laminar portion of the hull then the power required to accomplish such heating is

$$P_H = \rho_\infty u_\infty\, c\Delta T \int_0^{x_{tr}} c_{h_\ell}\, wdx \tag{6}$$

where c is the specific heat of water, c_{h_ℓ} is the laminar Stanton number for the heated boundary layer at $\Delta T = T_w - T_\infty$.

Applying the available heating power P_A to the laminar portion of the flow ($P_A = P_H$) yields after some simplification

$$\left[1 + \frac{\int_{x_{tr}}^{L} c_{f_t}\, wdx}{\int_0^{x_{tr}} c_{f_\ell}\, wdx}\right] = \frac{\frac{c\Delta T}{u_\infty^2}\left\{\frac{\int_0^{x_{tr}} c_{h_\ell}\, wdx}{\int_0^{x_{tr}} \frac{c_{f_\ell}}{2}\, wdx}\right\}}{\left\{\frac{D}{D_F}\left(\frac{1}{\eta_{th}} - 1\right)\eta_{eff}\right\}} \tag{7}$$

The left side is the ratio of overall friction drag to the laminar friction drag and is configuration dependent. The right side depends on the dimensionless ratio $\frac{c\Delta T}{u_\infty^2}$ and on the bracketed parameter in the denominator related to the amount of reject heat that can be transferred to the boundary layer. The bracketed parameter in the numerator is a Reynolds analogy factor which is configuration dependent. In order to close the calculation, a relation is needed between ΔT and transition Reynolds number $Re_{x_{tr}}$ which is also dependent on configuration.

Example - The Flat Plate

In order to quantitatively evaluate the prospective drag reduction due to heating, it is necessary to choose a particular configuration. The flat plate is chosen because of its great simplicity and because some information on transition with surface heating is available. The results should be representative of what can be obtained for slender shapes having pressure gradients that are not too large.

For a flat plate (w = const)

$$\int_0^{x_{tr}} c_{f_\ell}\, dx = 1.328 \frac{x_{tr}}{\sqrt{Re_{x_{tr}}}}$$

$$\int_{x_{tr}}^{L} c_{f_t}\, dx = 0.074 \left(\frac{L}{Re_L^{1/5}} - \frac{x_{tr}}{Re_{x_{tr}}^{1/5}}\right) \qquad (8)$$

and by Reynolds analogy

$$c_{h_\ell} = \frac{c_{f_\ell}}{2} Pr^{-2/3} \qquad (9)$$

Thus for the case of the flat plate, equation (7) becomes

$$\left[1 + \frac{.074}{1.328}\left(\frac{Re_L^{4/5} - Re_{x_{tr}}^{4/5}}{Re_{x_{tr}}^{1/2}}\right)\right] = \frac{c\Delta T}{u_\infty^2} \frac{Pr^{-2/3}}{\left\{\frac{D}{D_F}\left(\frac{1}{\eta_{th}} - 1\right)\eta_{eff}\right\}} \qquad (10)$$

The left side of equation (10) is the ratio of overall friction drag to laminar friction drag for a flat plate.

The variation of transition Reynolds number $Re_{x_{tr}}$ with overheat ΔT depends on the choice of transition criterion. A criterion that has been shown to give plausible trends is the e^9 criterion of Smith (Ref. 10) and Van Ingen (Ref. 11). For low speed flows, these authors correlated transition Reynolds number over plates, wings and bodies with the amplitude ratio using linear stability theory of the most unstable frequency from its neutral point to the transition point. They found that the transition Reynolds number $Re_{x_{tr}}$ as predicted by assuming an amplification factor of e^9 was seldom in error by more than 20%. Wazzan et al (Ref. 4) have calculated and presented such a curve for heated flat plates in water a portion of which is shown in Fig. 3*. Although not quite shown on the figure, $Re_{x_{tr}}$ reaches a maximum value of about 260×10^6 at an overheat of about 43°C. Overheats up to 40°C are considered herein. With this information, it is now possible to carry out the drag reduction calculations based on equation (10).

Such calculations have been performed for plate speeds up to 24.4 m/sec (80 fps), for plate lengths of 3.05 m (10 ft), 15.24 m (50 ft), 30.48 m (100 ft), 152.4 m (500 ft) and 304.8 m (1000 ft), and for values of $\{\frac{D}{D_F}(\frac{1}{\eta_{th}} - 1)\eta_{eff}\}$ of 2, 5 and 9. Since the product $\frac{D}{D_F}\eta_{eff}$ might be very close to unity, one may view the aforementioned values of the "efficiency factor" $\{\frac{D}{D_F}(\frac{1}{\eta_{th}} - 1)\eta_{eff}\}$ as approximately corresponding to η_{th} = 0.33, 0.17 and 0.10 respectively.

For each value of the efficiency factor results are presented in Figs. 4-6 for $\frac{D}{D_{\Delta T=0}}$, the ratio of the drag with heating to that without using the reject heat for drag reduction purposes, the corresponding laminar fraction of the plate x_{tr}/L, the wall temperature rise of the laminar region, and finally the ratio of the computed drag with heating to that for fully laminar flow over the entire plate.

Generally speaking the drag reduction becomes noticable as speeds exceed 10 m/sec (∿ 20 knots). Although the drag ratio is not a strong function of length, the overheat in the laminar region increases quite significantly with vehicle length.

* The author wishes to thank Dr. Carl Gazley for providing the information for Fig. 3.

Opportunities for drag reduction without large overheat are best for short vehicles. For $\eta_{th} \approx 0.33$ (Fig. 4), even though drag reductions of about 40% are achieved at vehicle speeds of 75 m/sec, only about 30-45% of the length is laminar and the drag is 10 to 30 times the laminar drag.

A lower thermal efficiency leads to greater drag reduction but with larger overheats. For $\eta_{th} \approx 0.17$ (Fig. 5), drag reductions of about 60% are attainable for vehicle speeds of 25 m/sec ($\sim$ 50 knots) while for $\eta_{th} \approx 0.10$ (Fig. 6), the indicated drag reductions at this same speed are 70-75%. Even in this latter situation, with about 9 times the drag power available for heating, full laminarization is not attained. The lowest drag is still three times the laminar drag of the plate.

The indication from the calculations is that full laminarization can be attained in a number of cases (Fig. 7) but only if η_{th} gets below about 0.03. Since the e^9 transition curve (Fig. 3) has a maximum value of $Re_{x_{tr}}$ below 3×10^8, vehicles with length Reynolds numbers above 3×10^8 cannot be completely laminarized.

For a plate of given length at a prescribed speed, the fuel consumption (proportional to D/η_{th}, the slope of a line through the origin in Fig. 7) increases as η_{th} is reduced. But it is far below that of the unheated plate.

Real Configurations

Real vehicle configurations involve additional factors not considered in this flat-plate calculation. Favorable pressure gradient for example can be very effective in delaying transition while regions of adverse gradient are otherwise. Non-uniform longitudinal heating distributions can result in a more optimal use of the available heat. Effects of surface roughness on transition are possibly more pronounced for heated surfaces than for unheated. These factors are presently being studied both experimentally and analytically by a number of investigators for the purpose of obtaining an objective evaluation of the practical capabilities of this relatively simple and readily available means of drag reduction.

CONCLUDING REMARKS

The prospects for drag reduction by heating in water have been examined for flat-plate configurations using an e^9 transition criterion. At speeds of the order of 20 knots or higher, significant drag reduction for vehicles with on-board propulsion systems is possible through using the waste heat from the propulsion system to heat the laminar portions of the hull. At a power plant efficiency of 33%, a 35-40% drag reduction is possible for a vehicle speed of 50 knots. As the efficiency of the power plant decreases, more waste heat is available leading to further reduction in drag. It is not likely however that a completely laminar configuration will be obtained by heating alone since full laminarization by heating is attainable only for thermal efficiencies below about 3%.

The prospects for drag reduction by heating are such as to justify further investigation of the various elements of the phenomenon and the factors affecting its practical application.

REFERENCES

1. Frick, C.W. Jr. and McCullough, G.B.: Tests of a Heated Low Drag Airfoil. NACA ARR, (Dec. 1942).
2. Liepmann, H.W. and Fila, G.H.: Investigations of Effects of Surface Temperature and Single Roughness Elements on Boundary Layer Transition. NACA Rept. 890, (1947).
3. Wazzan, A.R., Okamura, T.T. and Smith, A.M.O.: The Stability of Water Flow over Heated and Cooled Flat-Plates. J. Heat Transfer, Vol. 90, No. 1, pp. 109-14, (1968).
4. Wazzan, A.R., Okamura, T.T. and Smith, A.M.O.: The Stability and Transition of Heated and Cooled Incompressible Laminar Boundary Layers. Proc. 4th Int. Heat Transfer Conf., ed. U. Grigull and E. Hahne, Vol. 2, FC 1.4, Amsterdam: Elsevier, (1970).

5. Lowell, R.L. and Reshotko, E.: Numerical Study of the Stability of a Heated Water Boundary Layer. Report FTAS/TR-73-93, Case Western Reserve University, (Jan. 1974).

6. Strazisar, A.J., Prahl, J.M. and Reshotko, E.: Experimental Study of the Stability of Heated Flat Plate Boundary Layers in Water. Report FTAS/TR-75-113, Case Western Reserve University, (Sept. 1975).

7. Strazisar, A.J. and Reshotko, E.: Stability of Heated Laminar Boundary Layers in Water. AGARD Symposium on Laminar-Turbulent Transition, Paper No. 10, Copenhagen, (May 2-4, 1977).

8. Barker, S.J. and Jennings, C.G.: The Effect of Wall Heating on Transition in Water Boundary Layers. AGARD Symposium on Laminar-Turbulent Transition, Paper No. 19, Copenhagen, (May 2-4, 1977).

9. Hoerner, S.F.: Fluid Dynamic Drag. 2nd edition, (1958).

10. Smith, A.M.O. and Gamberoni, N.: Transition, pressure gradient and stability theory. Report ES 26388, Douglas Aircraft Co., (1956).

11. Van Ingen, J.L.: A suggested semi-empirical method for the calculation of the boundary layer transition region. Report VTH-74, Dept. of Aero. Eng'g. University of Technology, Delft, (1956).

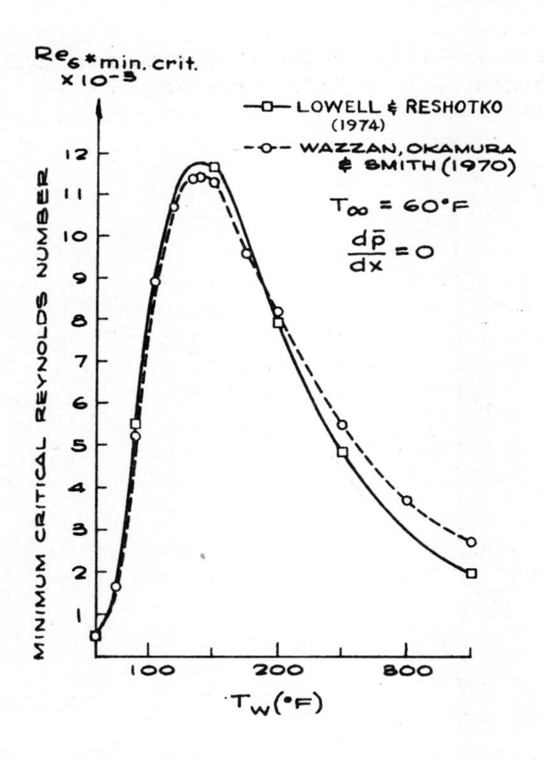

Fig. 1 Effect of wall temperature on minimum critical Reynolds number (from Ref. 5)

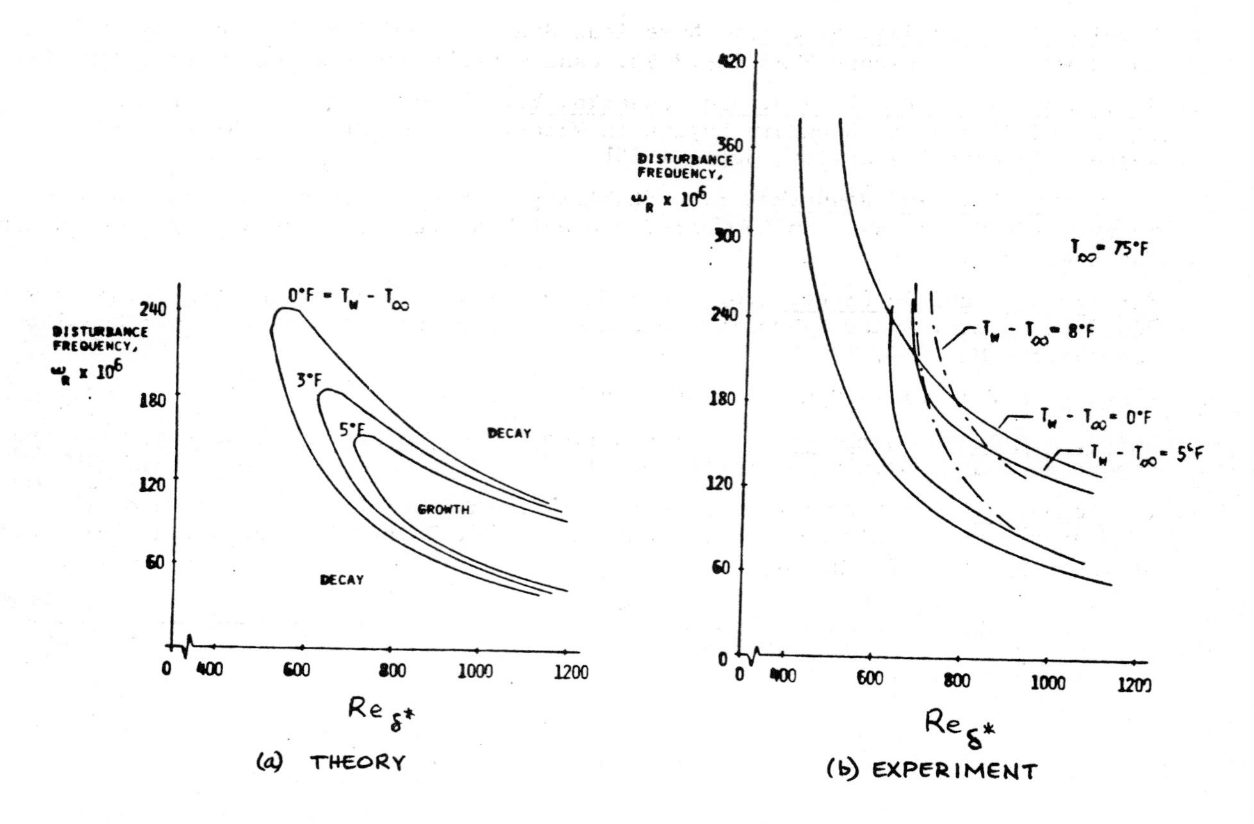

Fig. 2 Neutral stability characteristics for uniform wall temperature, T_∞ = 75°F (from Ref. 7)

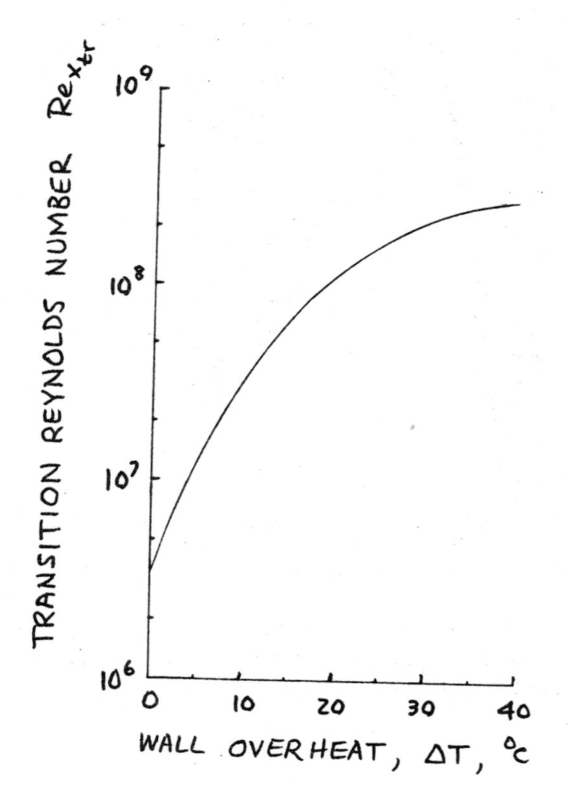

Fig. 3 Variation of transition Reynolds number for a flat plate with uniform wall overheat according to an "e^9" transition criterion, T_∞ = 60°F (data of Ref. 4)

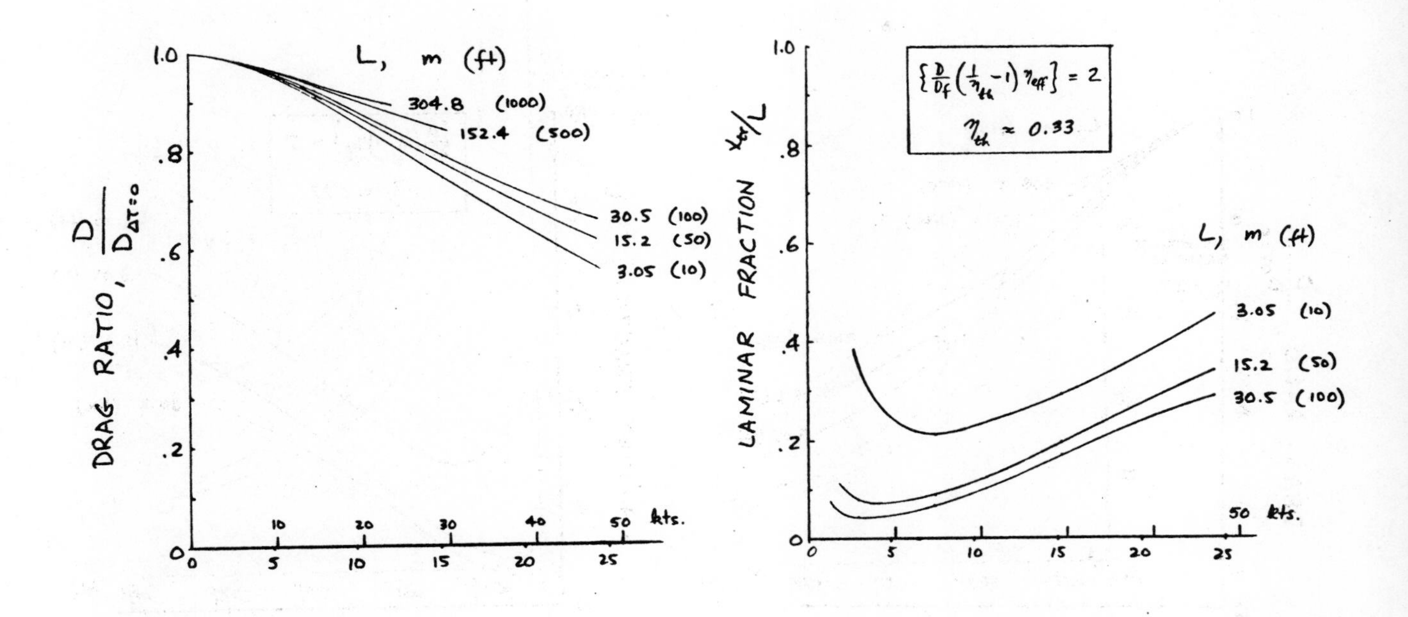

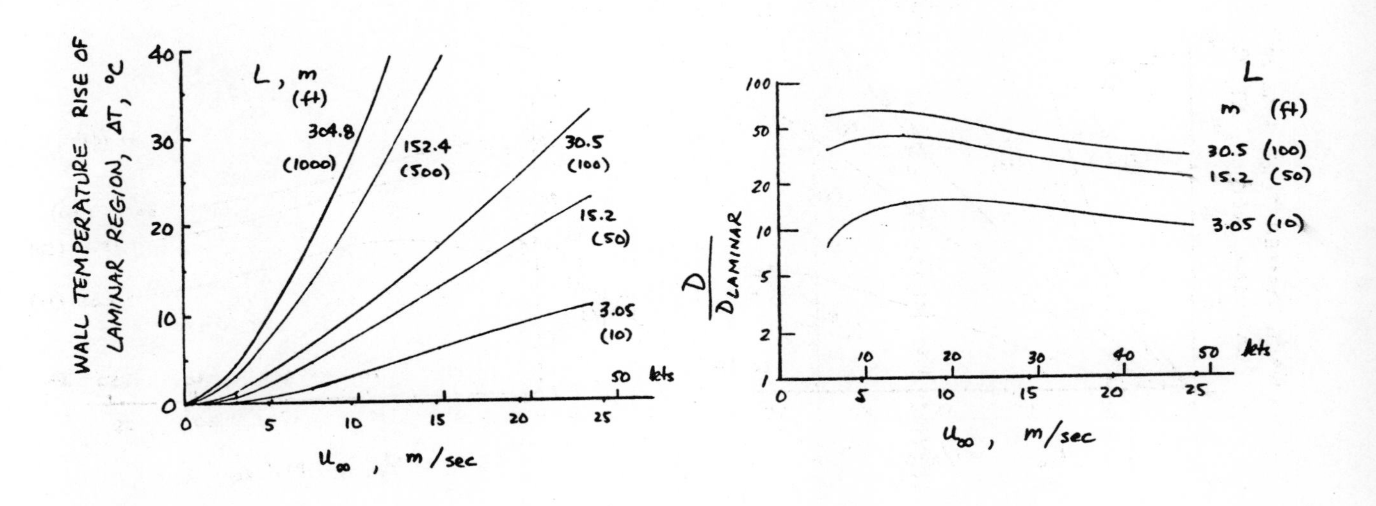

Fig. 4 Drag reduction by use of reject heat of propulsion system for transition delay. $\{\frac{D}{D_F}(\frac{1}{\eta_{th}} - 1)\eta_{eff}\} = 2$, ($\eta_{th} \approx 0.33$).

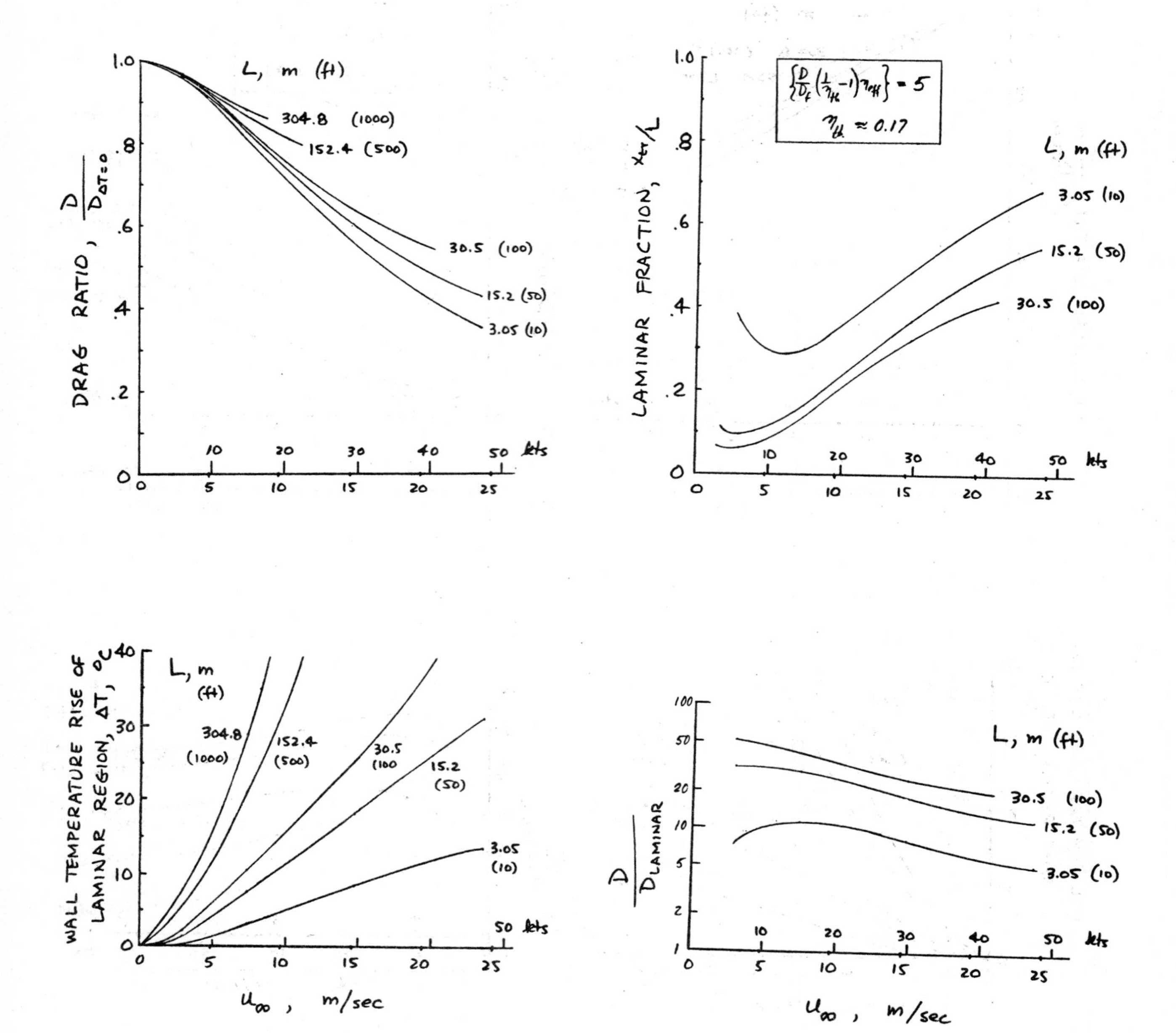

Fig. 5 Drag reduction by use of reject heat of propulsion system for transition delay. $\{\frac{D}{D_F}(\frac{1}{\eta_{th}} - 1)\eta_{eff}\} = 5$, ($\eta_{th} \approx 0.17$).

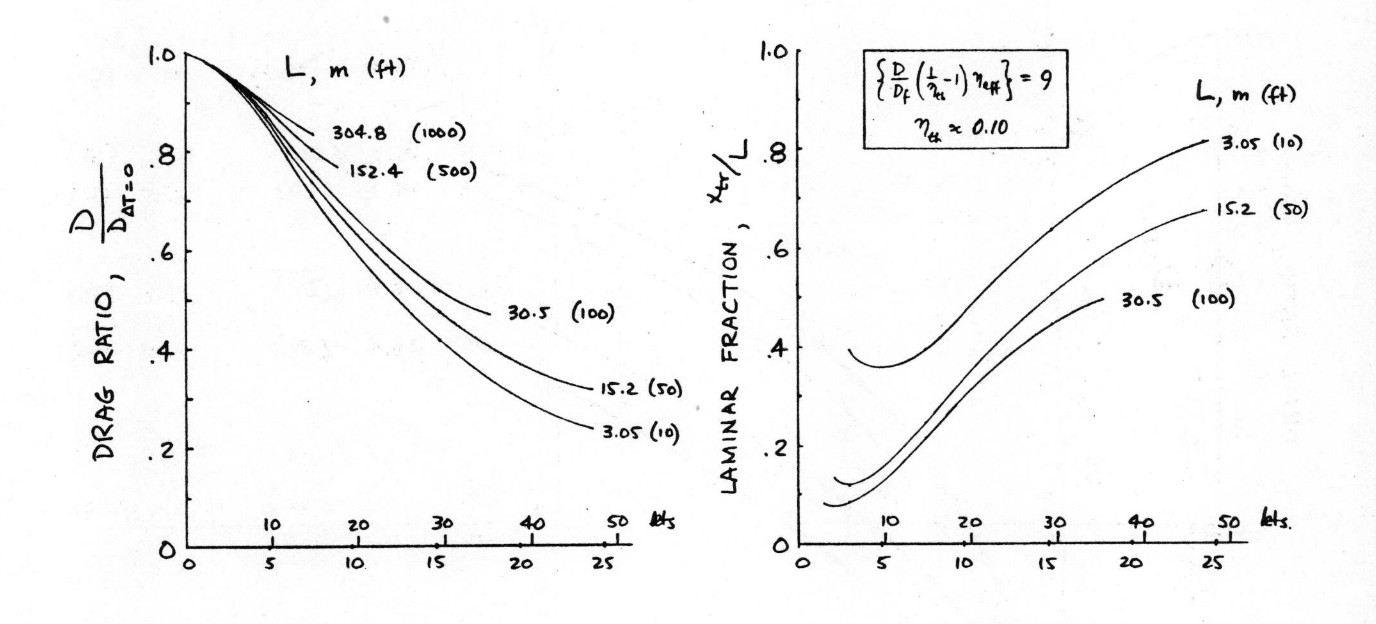

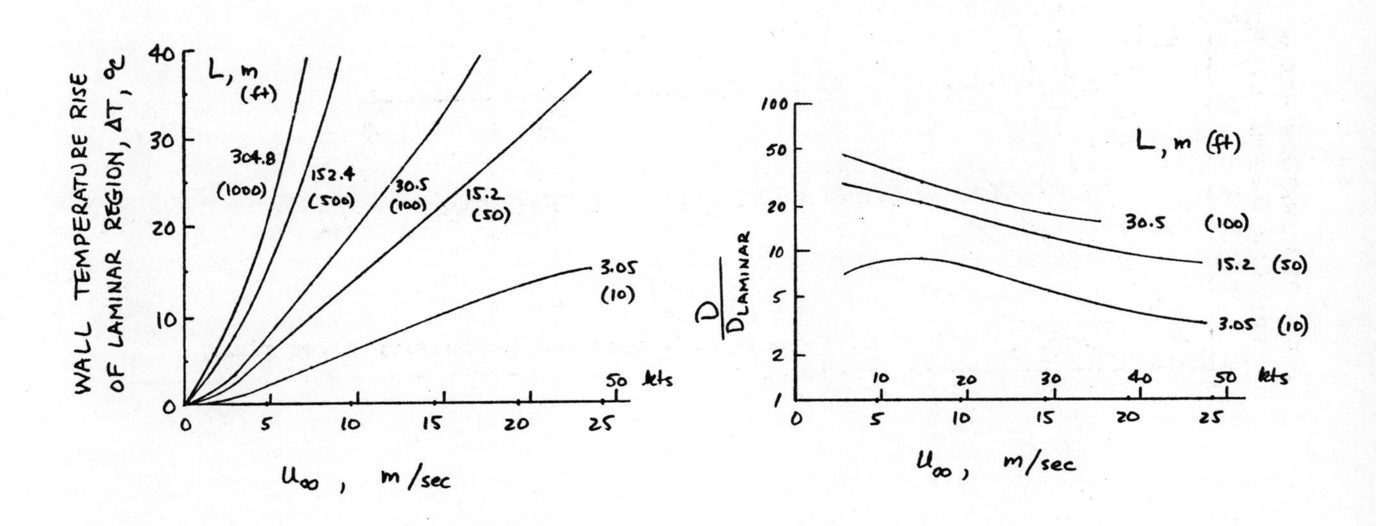

Fig. 6 Drag reduction by use of reject heat of propulsion system for transition delay. $\{\frac{D}{D_F}(\frac{1}{\eta_{th}} - 1)\eta_{eff}\} = 9$, $(\eta_{th} \approx 0.10)$.

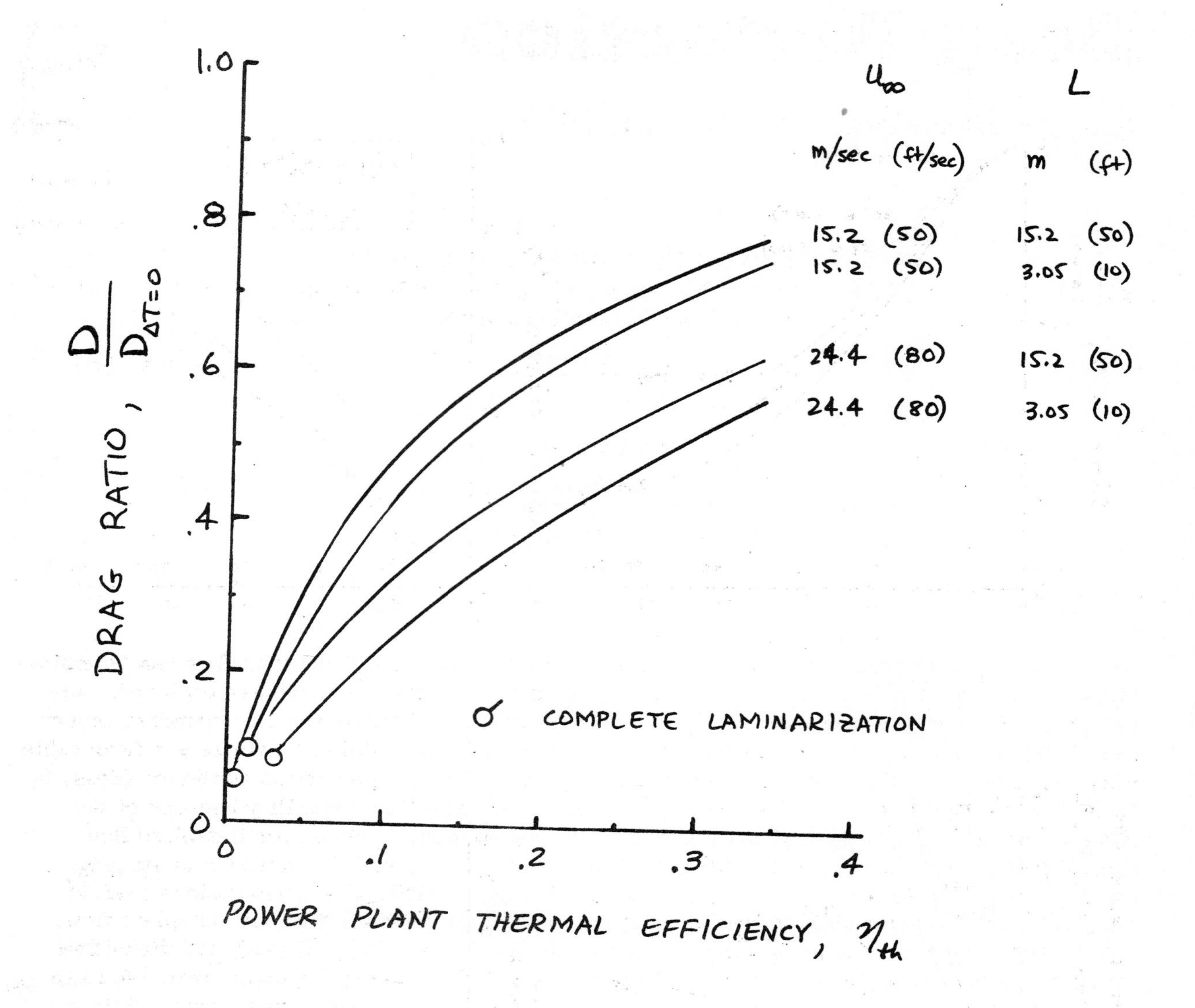

Fig. 7 Effect of thermal efficiency of propulsive power plant on drag reduction.

Second International Conference on

Drag Reduction

August 31st - September 2nd, 1977

PAPER E3

THE COMBINED EFFECTS OF PRESSURE GRADIENT AND HEATING ON THE STABILITY AND TRANSITION OF WATER BOUNDARY LAYERS

A.R. Wazzan,

University of California, U.S.A.,

and

C. Gazley, Jr.,

The Rand Corporation, U.S.A.

Summary

Appreciable drag reduction is possible if extended regions of laminar flow can be maintained. Although a variety of techniques for boundary-layer control have been explored, only recently has the effect of heat transfer on the stability and transition of water boundary layers been investigated. In spite of an early experiment (Ref.1) which did not indicate any favourable effects of heating on the stability of a water boundary in a tube, speculation continued (Refs. 2, 3, 4) that the heating of water boundary layers might increase their stability because of the large variation of viscosity with temperature. An approximate analysis for flat-plate flow (Ref.5) and numerical computations (Refs. 6, 7, 8) have confirmed the increase in stability for stagnation, flat-plate, and separating flows. Spatial amplification computations (Ref.9) also indicate an appreciable effect of heating on boundary layer transition for flat-plate flow.

This paper presents additional computations for the stability and predicted transition characteristics of water boundary-layer 'wedge' flows for Hartree β's ranging from -0.15 to +0.40 and surface temperatures up to 67°C (120°F) above the ambient temperature. Both the minimum critical Reynolds number and the predicted transition Reynolds number of these 'similar' boundary layers increase as the surface temperature is increased above the ambient level.

The interacting effects of pressure gradient and surface heating on stability and predicted transition may be approximately characterized by a boundary-layer shape parameter such as $H = \delta^*/\theta$. The computed distance Reynolds number for neutral stability and predicted transition are given as a function of H. Although this stability/transition map has been formed by computations for similar boundary layers, it can usefully be employed in the analysis of heated bodies with nonsimilar boundary-layer development. Several examples are presented. In order to maintain an extended region of laminar flow, it is apparent that the boundary-layer development should follow a path in which the shape parameter is kept as low as possible over as great a range of Reynolds number as possible.

Held at St. John's College, Cambridge, England.

Organised and sponsored by BHRA Fluid Engineering, Cranfield, Bedford, MK43 0AJ.

NOMENCLATURE

A	amplitude of disturbance
a	amplification
c	disturbance velocity
c_p	specific heat
H	boundary-layer shape parameter = δ^*/θ
k	thermal conductivity
L	characteristic length
n	e exponent, natural logarithm of amplification ratio
Pr	Prandtl number
R	Reynolds number based on boundary-layer thickness
Re	Reynolds number
t	percent freestream turbulance level, also time
T	temperature
u, U	velocity in x direction
v	velocity in y direction
v', u'	velocity fluctuations
x	distance in flow direction
y	distance normal to wall
α_i	spatial amplification rate
β	Hartree parameter
δ	boundary-layer thickness
δ^*	boundary-layer displacement thickness
μ	viscosity
ν	kinematic viscosity
ω	dimensionless frequency
ψ	stream function
ϕ	disturbance velocity amplitude
τ	shear
θ	boundary-layer momentum thickness
Θ	disturbance temperature amplitude

Subscripts

e	condition at edge of boundary layer
w	wall condition
x	conditions at station x
∞	freestream conditions
δ^*	based on boundary-layer displacement thickness
(crit)	neutral stability
(e^9)	condition for amplification ratio = e^9

I. INTRODUCTION

The design of high performance hydrodynamic and aerodynamic bodies requires among other factors a knowledge of the process of transition from laminar to turbulent flow and of the dependence of transition on the pressure distribution, the surface temperature, the surface roughness, freestream turbulence, etc. When surface roughness is not a factor and the freestream turbulence level, t, is very low, say $t = u'/u_\infty \leq 0.2\%$, transition in boundary-layer flows results from the growth of Tollmien-Schlichting waves. The growth of Tollmien-Schlichting waves, in flows where the parallel flow assumptions are reasonably valid, can be determined from linear instability theory. That the theory of small disturbances is capable of predicting the growth (or decay, etc.) of Tollmien-Schlichting waves has been demonstrated as early as 1942 by the well-known measurements of Schubauer and Skramstad (Ref. 10) and most recently by the measurements of Rogler and Reshotko (Ref. 11). It has been demonstrated that under these conditions, the transition Reynolds number, or bounds on the transition Reynolds number, can be predicted using linear instability theory [Liepmann (Ref. 12), Smith and Gamberoni (Ref. 13), Jaffe et al. (Ref. 14).] The premise behind these forecasting techniques is that the flow remains laminar until the disturbance amplitude (the Tollmien-Schlichting wave with the most dangerous frequency) is amplified by a factor e^n where $7.0 \leq n \leq 10$. Recently, Mack* showed that the well-known flat-plate transition results of Dryden can be predicted from linear instability theory using an amplification ratio e^n where

$$n = -8.43 - 2.4 \ln \frac{t}{100}$$

where t is the percent turbulence level; the fit is best in the range $.08 < t < 1.0$. It is to be expected that the total amplification ratio at transition is lower for higher freestream turbulence levels if we assume that transition occurs when u'/u_∞ reaches a fixed level, say 4 percent. Within the framework of linear instability theory, disturbances begin to amplify once the critical Reynolds number $Re_{x(crit)}$ is reached. The process terminates, transition occurs, at Re_x where the initial disturbance amplitude is amplified by a factor e^n; let us denote this Reynolds number by $Re_{x(e^9)}$. Within linear instability theory, $Re_{x(crit)}$ for a body of revolution, e.g., is directly dependent on the local boundary layer which is determined by the local pressure gradient, the body shape, and the surface boundary conditions, e.g., surface heating, cooling, suction, blowing, etc. On the other hand, $Re_{x(e^9)}$, which is the result of integrated spatial amplification rates, is dependent both on the local critical Reynolds number and on the width of the neutral stability curve at each station; that is to say, $Re_{x(e^9)}$ is dependent on the local boundary layer as well as on the history of the boundary-layer development.

If the characteristics of the local boundary layer can be well represented by the shape parameter $H(x) \equiv \delta^*/\theta$, the variation of H with Re_x for a body of revolution, e.g., may well serve to determine the stability of the boundary layer to small disturbances if a relationship does exist between H and $Re_{x(crit)}$ and/or $Re_{x(transition)}$. Relationships between H (or U" and β which are related to H) and Re_x do exist for certain boundary layer flows; for example, two-dimensional wedge flows (Refs. 7, 9, 15, 16). However, relationships between H and $Re_{x(transition)}$ are not yet available in the literature. Such a relationship would be an invaluable tool in the design of high performance hydrodynamic and aerodynamic bodies; the development of such a relationship is the object of this paper. In arriving at this relationship, it is assumed that

$$Re_{x(transition)} \equiv Re_{x(e^9)} \quad .$$

Second, it is also assumed that the stability characteristics of a body of revolution at a given station can be approximated by the stability characteristics of a two-dimensional wedge flow with the same pressure gradient as that of the body of revolution at the station in question. With these two assumptions in mind, we proceed then to formulate (compute) a relationship between $Re_{x(crit)}$ and $H(Re_x)$ and a second relationship between $Re_{x(e^9)}$ and $H(Re_x)$ for adiabatic and heated two-dimensional wedge flows. This task requires knowledge of (1) the boundary layer on adiabatic and heated two-dimensional wedge flows and (2) the stability characteristics (e.g., critical

---

*Private communication.

Reynolds number, spatial amplification rates, etc) of those boundary layers. This information is also reported in this paper.

II. ANALYSIS

DETERMINATION OF THE MEAN FLOW

The mean flow profiles for heated wedge flows in water with $T_e = 19.4^oC$ (67^oF) and $19.4^oC \le T_W \le 86^oC$ ($67^oF \le T_W \le 187^oF$) have been calculated as outlined by Kaups and Smith (Ref. 2). In these calculations all fluid properties are allowed to vary with temperature only (a good assumption for water boundary layers at moderate pressures). In cases where buoyancy effects are not important, viscosity is found to be the most important variable fluid property, Fig. 1. Therefore, in formulating the stability problem, only viscosity variations with temperature have been taken into account. The variation of the following boundary-layer characteristics $(\delta^*/x)\sqrt{Re_x}$, $(\theta/x)\sqrt{Re_x}$, $(c_f/2)\sqrt{Re_x}$, and $Nu/(\sqrt{Re_x}\, Pr^{1/3})$ with $(T_w - T_e)$ are shown in Fig. 2 and the variation of $H = \delta^*/\theta$ with $(T_w - T_e)$ is shown in Fig. 3.

FORMULATION OF THE LINEAR STABILITY PROBLEM

Neglecting temperature fluctuations, assuming viscosity is a function of temperature only, and taking all other fluid properties constant, Wazzan et al. (Ref. 6) found the linearized parallel flow stability problem of water boundary layers with heat transfer can be adequately treated by solving the Orr-Sommerfeld equation modified to include the variation of viscosity with temperature:

$$(U - c)(\phi'' - \alpha^2\phi) - U''\phi + \frac{i}{\alpha R}[\mu(\phi'''' - 2\alpha^2\phi'' + \alpha^4\phi) + 2\mu'(\phi''' - \alpha^2\phi') + \mu''(\phi'' + \alpha^2\phi)] = 0 \qquad (1)$$

All quantities in Eq. (1) are dimensionless where the reference values for velocity, length, and viscosity are the edge velocity U_e, the boundary-layer thickness δ, and the edge viscosity μ_e. R is the Reynolds number based on δ, $R = U_e\delta/\nu_e$. The prime indicates differentiation with respect to y where $y \equiv y^*/\delta$ with y^* the physical distance measured normal to the surface. ϕ is the amplitude of the disturbance which is described by the stream function $\psi = \phi(y)\, e^{i\alpha(x - ct)}$ where α and c are complex quantities with αc, the frequency, taken real. In this case, the amplification is purely spatial; the amplification factor is of the form $e^{-\alpha_i x}$. The proper boundary conditions for this problem are

$$\left.\begin{array}{l}\phi = \phi' = 0\\ T' = 0\end{array}\right\}\text{at } y = 0 \qquad\qquad \left.\begin{array}{l}\phi = \phi' = 0\\ T' = 0\end{array}\right\}\text{at } y \to 1 \qquad (2)$$

The Effect of Temperature Fluctuations on the Stability of Water Boundary Layers with Heat Transfer within the Parallel Flow Assumption

In deriving Eq. (1), temperature and viscosity fluctuations were neglected. We now demonstrate the validity of this assumption. Considering temperature to be the only state variable but allowing for viscosity as well as temperature fluctuations, Eq. (1) within the parallel flow assumption, is replaced by the more general stability equations (Ref. 17):

$$(U - c)\Theta - \overline{T}'\phi = \frac{i}{\alpha}\frac{1}{RPr}(\Theta'' - \alpha^2\Theta) \qquad (3)$$

and $$(U - c)(\phi'' - \alpha^2\phi) - U''\phi = -\frac{i}{\alpha R}\left\{\bar{\mu}(\phi'''' - 2\alpha^2\phi'' + \alpha^4\phi) + 2\bar{\mu}'(\phi''' - \alpha^2\phi') + \bar{\mu}''(\phi'' + \alpha^2\phi) + \left[U'\bar{T}''\frac{d^2\mu}{d\bar{T}^2} + U'(\bar{T}')^2\frac{d^2\bar{\mu}}{d\bar{T}^2} + 2U''\bar{T}\frac{d^2\bar{\mu}}{d\bar{T}^2} + U'''\frac{d\bar{\mu}}{d\bar{T}} + U'\frac{d\bar{\mu}}{d\bar{T}}\alpha^2\right]\theta + 2U''\frac{d\bar{\mu}}{d\bar{T}}\Theta' + U'\frac{d\bar{\mu}}{d\bar{T}}\Theta''\right\} \qquad (4)$$

where Eq. (3) is the energy equation, Eq. (4) is the momentum equation, Pr is the mean Prandtl number, $\bar{T}$ is the mean temperature, $T' = \Theta(y)\,e^{i\alpha(x - ct)}$ is the temperature fluctuation, $\bar{\mu}$ is the mean viscosity, μ' is the viscosity fluctuation with $\mu \equiv \bar{\mu}(T) + (d\bar{\mu}/dT)T' + O(T'^2)$. A crude order of magnitude analysis of the new terms appearing in the right-hand side of the momentum equation [Eq. (4)] vis a vis the right-hand side terms of Eq. (1) is made as follows. At a point in the boundary we assume $\phi \sim \Theta$, $\phi' \sim \Theta'$, $\phi'' \sim \Theta''$, $\alpha \sim 1/\delta$, $y \sim \delta$, $U \sim \bar{T}$, $d/dy \sim 1/\delta$. Now setting for example $a \equiv \bar{\mu}''\phi$, and $b \equiv U'\bar{T}''(d^2\bar{\mu}/d\bar{T}^2)\Theta$, we find

$$(a/b) \simeq \frac{\frac{\bar{\mu}}{\delta^2}\frac{\phi}{\delta^2}}{\frac{U}{\delta}\frac{\bar{T}}{\delta^2}\frac{\bar{\mu}}{\bar{T}^2}} = \frac{1}{\delta} >> 1 \quad . \qquad (5)$$

Repeating the above analysis for other terms of Eq. (4), we find that the ratio of new terms in Eq. (4), arising from allowing for the viscosity and temperature fluctuations, to terms in Eq. (1) is always of order δ and hence can be neglected. These conclusions were recently confirmed by Lowell and Reshotko (Ref. 18). These authors computed the stability of the flat-plate boundary layer in water with heat transfer. All fluid properties were allowed to vary with temperature in computing the mean flow as well as in computing its stability characteristics. These new results differed only slightly from the earlier results of Wazzan et al. (Refs. 6, 9), primarily due to the difference in ambient temperature (see below). Therefore, it appears that for water boundary layers, with moderate rates of heat transfer, it is sufficiently accurate to compute the effect of heat transfer on stability by using the simple stability equation, namely, Eq. (1).

III. RESULTS

Equation (1) was solved for several wedge flows, $\beta = -0.15, -0.10, -0.05, 0, 0.10, 0.20, 0.30,$ and 0.40, at $T_w - T_e = 0, 2.8, 11.1, 16.7, 27.8, 44.4,$ and 66.7^oC (0, 5, 20, 30, 50, 80, and 120°F) with $T_e = 19.4^oC$ (67°F). The critical Reynolds number $Re_{x(crit)}$ and the spatial amplification rates α_i as a function of Re_x were computed. The variation of $Re_{\delta*(crit)}$ with $(T_w - T_e)$ and with H are shown in Figs. 4 and 5. Re_{crit} initially increases with increasing ΔT ($\equiv T_w - T_e$) or with decreasing H. However, as ΔT is continuously increased, Re_{crit} attains a maximum value and decreases as ΔT is further increased or H is further decreased. Since the effect of surface heating on boundary-layer stability is due primarily to the variation of viscosity with temperature, the results are thus dependent not only on the temperature difference $\Delta T = (T_w = T_e)$ but also on the temperature level, T_e. This is evident in Fig. 4 where previous results for $T_e = 15.6^oC$ (60°F) are compared with the present results for $T_e = 19.4^oC$ (67°F). Even this relatively slight change in ambient temperature results in an appreciable difference in the rate of change of viscosity with temperature, and consequently in the predicted stability.

DISCUSSION OF STABILITY RESULTS

Rayleigh theorems for inviscid instability state that for a boundary-layer flow, the necessary and sufficient condition for amplified and neutral inviscid instability is that $U''(y)$ must vanish somewhere in the boundary layer, i.e., the mean velocity profile must be inflected. In addition, some correlation appears to exist between Re_{crit} and the location of the inflection points (Refs. 9, 19). In fact, in adiabatic flows, $T_w = T_e$, Re_{crit} decreases as the location of the inflection point moves away from the wall. That is, in adiabatic flows, Re_{crit} decreases as β becomes more negative. In flows with zero or favorable pressure gradient ($\beta \geq 0$) where the profiles are not inflected, the boundary-layer characteristic of greatest importance to its stability characteristics is $U''(y)$ and to a lesser extent $U''(0)$ (References 7, 15-17). In adiabatic flows it is known that for $\beta \gtrless 0$, where $U''(0) \gtrless 0$, Re_{crit} for $\beta < 0$ is smaller than Re_{crit} for $\beta > 0$. In fact, when a $U'(y)$ distribution is the result of pressure gradient effects only (adiabatic flows), a strong correlation exists between Re_{crit} and β [or $U''(y)$ or simply $U''(0)$] or the shape parameter H (Refs. 7, 15, 16). The shape parameter provides a simple and convenient means of generalizing stability computations; Fig. 6 shows results for isothermal wedge flows (Ref. 20) and for suction (Refs. 21, 22) and it is apparent that the effects of suction are essentially the same as pressure gradient. Previous results for heated wedge flows (Refs. 6, 7-9) are shown in Fig. 7; here also the effects of a boundary-layer modification by heating is qualitatively similar to the effects of pressure gradient except for relatively large temperature differences.

The critical Reynolds number thus exhibits a simple variation with $U''(0)$ or with H; Re_{crit} increases as H decreases (Figs. 4, 5, and 6) or as $U''(0)$ decreases and/or becomes increasingly more negative. This dependence on U'' or H, in fact, is to be expected. An inspection of the Orr-Sommerfeld equation shows the boundary-layer characteristics that directly influence the eigen values, and hence Re_{crit}, are $U''(y)$ and $U(y)$, with $U''(y)$ being the dominant term. Therefore, it may be assumed that the heating of water boundary layers, which produces variations in H similar to those produced through the effect of pressure gradient alone, leads to increased stability and particularly to increasing critical Reynolds number. This assumption, however, is found to be only partially true (Figs. 5 and 7). These figures show that with initial heating H decreases and Re_{crit} increases in agreement with the trend observed in Fig. 5. However, Figs. 5 and 7 show that although H decreases monotonically with increasing surface temperature, $Re_{\delta^*(crit)}$ exhibits a maximum (at least for flows with $\beta \leq 1.0$). The difference in the variation of $Re_{\delta^*(crit)}$ with H, when produced through the effect of pressure gradient alone (adiabatic flows) or through the combined effects of pressure gradient and surface heating, can be qualitatively understood through an examination of the Orr-Sommerfeld equation.

In the adiabatic case, only U'' and U appear in the Orr-Sommerfeld equation and hence the monotonic variation of $Re_{\delta^*(crit)}$ with U'' or H. In the nonadiabatic case, the small disturbance equation is a modified Orr-Sommerfeld equation that contains not only U'' and U but also μ, μ', and μ''. This alters the nature of the problem in two ways. In the adiabatic case the pressure gradient affects $R_{\delta^*(crit)}$ mainly through the mean velocity of the term U'' (which can be represented by some function of H), whereas in the nonadiabatic case heating affects $Re_{\delta^*(crit)}$ not only through the mean velocity or the U'' term (which can still be represented by some function of H) but also through the terms μ, μ', and μ that appear in the modified Orr-Sommerfeld equation. Second, the nature of the eigen function ϕ (and consequently all eigen values and properties depending on the eigen values such as Re_{crit}) is different in the two cases. In the adiabatic case, the differential equation for ϕ (the Orr-Sommerfeld equation) includes only the function ϕ and the even derivatives ϕ'' and ϕ'''', whereas in the nonadiabatic case, the differential equation for ϕ (the modified Orr-Sommerfeld equation) includes not only ϕ and the even derivatives ϕ'' and ϕ'''' but also the odd derivatives ϕ' and ϕ'''. Therefore, in the case of heating, although U'' or H still characterizes the boundary layer, neither present a complete relationship between the mean flow and the eigen function ϕ, and hence $Re_{\delta^*(crit)}$. Physically, this may be interpreted as follows: with heating, initially H decreases rapidly indicating a decrease in momentum loss (Fig. 3) and the stability, e.g., $Re_{\delta^*(crit)}$, increases. However, with still increased heating rates, H continues to decrease but at a much slower rate (Fig. 3). In the meantime, μ, μ', and μ'' continue to vary appreciably with increasing T_w. In fact μ, which has a destabilizing effect (Refs. 6, 8), continues to decrease monotonically with heating, whereas μ' and μ'', and in particular $\mu'(0)$ and $\mu''(0)$, which have a stabilizing effect (Refs. 8), reverse their trend in variation with temperature (change from increasing with T_w to decreasing with T_w) near the temperature where Re_{crit} exhibits a maximum for wedge flows with $\beta < 1.0$. Therefore, for high heating rates the variation of H with T_w becomes negligible and the variation of μ, μ', μ'' with T_w dominate the effect of heating on Re_{crit}. Therefore, at high heating rates it is expected that $Re_{\delta*(crit)}$ will exhibit a maximum with T_w.

In the case of $\beta = 1.0$, the maximum in $Re_{\delta*(crit)}$ with T_w is not observed (Fig. 7) because the unstable zone (region contained within the neutral curve) is rather limited, and in the initial stages of heating when H is fast decreasing with T_w the unstable zone is fast approaching a point. In fact, just at about the temperature when the variation of H with T_w begins to slow down considerably and μ, μ', and μ'' begins to dominate, the unstable region shrinks to zero and the flow becomes totally stable.

In spite of this discussion on the relative importance of H and/or the μ, μ', and μ'' to the stability characteristics of a given boundary layer, Figs. 5, 6, and 7 show that over a large range of H values, a decrease in H results in increased stability, e.g., increasing $Re_{\delta*(crit)}$.

LINEAR STABILITY THEORY OF TWO-DIMENSIONAL DISTURBANCES AND TRANSITION

Of more practical importance than the variation of Re_{crit} with surface temperature and pressure gradient is the effect of these two parameters on transition. Although the transition process is complex and involves nonlinear processes in the final stages of breakdown to turbulence, much insight into the process of transition and its dependence on, e.g., heat transfer and pressure gradient can be gained from a study of the effect of these parameters on linear instability characteristics, such as Re_{crit} and the spatial amplification rates, local and integrated values, particularly in the case of slowly amplifying boundary layers.

Because of the three-dimensionality of turbulence, early workers tended to ignore the role of two-dimensional linear amplification mechanisms, better known as Tollmien-Schlichting mechanisms or TS mechanisms. Schubauer and Skramstad (Ref. 10) and Liepmann (Ref. 12) verified, however, the features of the TS mechanism in flat-plate flow. On the other hand, Emmons (Ref. 23) verified the existence of three-dimensional turbulent spots prior to transition. Criminale and Kovasznay (Ref. 24) and Brooke (Ref. 25) demonstrated, for various oblique TS waves, that localized areas of initially intensified disturbances should develop with strong two-dimensional features in the linear regime.

Reconciliation between early TS amplification and the final three-dimensionality of turbulence was achieved when transition was recognized to begin with linear TS amplification and to terminate with turbulent spots and wedges overcoming the mean laminar flow (Ref. 26). This model was reinforced when the qualitative effects of cooling, heating, suction, pressure gradient, Mach number, etc., on the stability of TS waves (theoretical studies and experimental observations) and on transition (experimental observations) were often found to be parallel.

Freestream Turbulence--Boundary-Layer Interation

Disturbances that may excite or feed TS waves include (Refs. 15, 27):

I. Temperature--density--entropy mode
II. Vorticity--turbulence mode
III. Sound mode

Mode I interacts with the boundary layer because of the growth of the boundary layer in the freestream direction. This mode is therefore not important in instability studies within the framework of the parallel flow assumptions.

Mode II can disturb the boundary layer across stream lines because as it enters the layer it becomes distorted and stretched (Ref. 28). Some measurements by Hall (Ref. 29) and by Klebanoff (Ref. 30) suggest that a boundary layer exhibits a variable receptivity towards freestream vorticity fluctuations. Aside from these observations, the effects of Mode II on TS waves is virtually unknown.

According to Obremski et. al. (Ref. 15), sound of frequency ω_s, Mode III, excites regular, coherent TS waves of the same frequency but different wave length, and that the boundary layer has a non-zero receptivity to acoustic disturbances with frequency in the amplified TS range. Furthermore, when the primary acoustic frequency falls in the TS susceptibility region, ω_{TS}, the onset of transition can be dramatically changed (Ref. 31). Since the susceptible dimensional frequency band ω^*_{TS} scales primarily with U_∞^2/ν, i.e.,

$$\omega_{TS} = \frac{\omega^* \nu}{U_\infty^2} \quad \text{(dimensionless)}$$

a change in freestream velocity U_∞ (or a change in unit Reynolds number U_∞/ν) would change the onset of transition. Schubauer and Skramstad (Ref. 10) and Spangler and Wells (Ref. 32) in subsonic flow, and Kendall (Ref. 33) in supersonic flow, verified the strong influence sound has on transition; the elimination or reduction of sound sources was found to greatly increase the transition Reynolds number. Aside from these observations on the effect of sound on transition, the process through which parts of the freestream sound energies become internalized as growing TS waves is not well understood.

Assessment of Transition

A disturbance growing according to linear instability theory sooner or later reaches a state where (1) the linear theory ceases to be valid, and nonlinear processes commence; (2) the boundary layer becomes locally turbulent--turbulent spots are formed and grow and increase in number; and (3) these spots spread into the neighboring laminar flow until the mean flow becomes fully turbulent. Therefore, satisfactory assessment of the beginning of transition for approximately two-dimensional boundary layers requires at least three elements (Ref. 15): (a) adequate knowledge of the input disturbance and the corresponding boundary-layer receptivity; (b) knowledge of the development of the mean profiles and access to their stability characteristics; (c) information on the length of the nonlinear processes and secondary instability as dependent on pressure gradient, heat transfer, etc. Since the information required in element (a) often is not available in the literature, one usually characterizes a disturbance in terms of the ratio of its amplitudes $(A_{x(2)}/A_{x(1)})$ at two locations, x_1 and x_2.

According to Klebanoff et al. (Ref. 30), stage (1) is reached, for a flat plate, when the rms velocity fluctuation u' in the boundary layer reaches $(u'/U)_{max} \simeq .015$, but that the first appearance of turbulence spots is expected at $(u'/U)_{max} \simeq 0.2$. That is, beyond the onset of nonlinearity an amplification factor of 10 to 15 times ($\simeq e^{2.5}$) is required (Ref. 15). Liepmann (Ref. 12) hypothesized that at the breakdown to turbulence the Reynolds stress $\tau = -\rho\overline{u'v'}$, due to the amplified fluctuations u', becomes comparable in magnitude to the maximum mean laminar shear stress, $\tau_L = \mu\, \partial u/\partial y$ in the boundary layer. The ratio τ/τ_L is given by

$$\tau/\tau_L = \frac{2}{c_{f_L}} \left\{ kb(u'/U_e)^2 [a(x)]^2 \right\}_{max}$$

where

$$b = v'/u'$$

and

$$k = \overline{u'v'}/uv$$

where u and v are the velocities in the x and y direction, respectively. For a given frequency ω^* the amplification a is given by

$$a(x,\omega^*) = \exp\left[-R_L \int_{x_n}^{x} (\alpha_i/R)(U_e/U_\infty)\, dx \right]$$

where R_L = freestream length Reynolds number, $R_L = (U_\infty L/\nu_\infty)$,

α_i = spatial amplification rate,

$x = x^*/L$,
ω = nondimensional frequency $\omega^* \nu / U_e^2$, and
L = characteristic length.

Smith (Ref. 13) reduced Liepmann's criterion for transition to an explicit dependence on the local laminar skin friction coefficient, the disturbance input at the neutral point x_n, and the total amplification ratio $(A_{x(t)}/A_{x(n)})$ where n refers to the neutral point and t to the transition point. Smith studied available transition data for attached boundary layers where the freestream turbulence level was low. Assuming linear theory valid up to the transition point, Smith showed that the ratio of the disturbance amplitude at transition $A_{x(t)}$ to that at the neutral point $A_{x(n)}$ is given by $(A_{x(t)}/A_{x(n)}) \equiv a(x_t, \omega_t^*) \simeq e^9$. Later, more accurate calculations (Ref. 14) showed $(A_{x(t)}/A_{x(n)}) \equiv e^{10}$. In any event, since in the nonlinear zone the amplification to transition is $\simeq e^{2.5}$, we find that for boundary layers with low freestream disturbance levels the linear TS amplification of about $e^{7.5}$ does control to a large extent the major part of the development of the disturbance to the beginning of transition and that element (b), of the transition process, paragraph 1 of section on "Assessment of Transition" appears, at least in this case, to dominate elements (a) and (c).

In spite of the dominant role of element (b), however, the role of element (a) remains extremely important. For example, when Spangler and Wells (Ref. 32) minimized sound disturbances in their measurements of transition in a low-speed boundary-layer channel, their $R_{x(t)}$ exceeded five millions! These results, where a mixture of vorticity and sound disturbances are present, cannot be predicted using any of the presently available forecasting techniques (Ref. 34). Hence a knowledge of the receptivity of the boundary layer to vorticity and sound disturbances is needed for further progress, and much attention need be given element (a). In the absence of such information and in view of the fact that the TS mechanism may describe, in the absence of the effects of surface roughness, vibration, and sound, the substantial growth of disturbances up to the emergence of the final three-dimensional turbulent spots and wedges and the beginning of transition, it is not unreasonable to employ, for the present, linear theory in bracketing the Reynolds number at the beginning of transition for two-dimensional and axisymmetric boundary-layer flows (in axisymmetric flows x is replaced by s, the distance measured along the body surface).

Therefore, in certain boundary-layer flows where the linear mechanism dominates the growth of disturbances to transition, the transition Reynolds number can be bound by $Re_{x(crit)}$ on the lower side and by $Re_{x(e^9)}$ on the upper side. Computations of the values of $Re_{x(e^9)}$ for heated wedge flows are shown in Fig. 8 as a function of temperature difference and the trends are seen to be similar to those for the critical Reynolds number. Computations for both the critical and predicted transition Reynolds numbers are shown as a function of the shape parameter in Fig. 9.

A plot of $Re_{x(crit)}$ and $Re_{x(e^9)}$ vs. H for two-dimensional wedge flows with and/or without heating (as shown in Fig. 9) can perhaps be used as a guideline in bracketing $Re_{x(trans)}$ on a body of revolution, for example. This can be accomplished by computing for the body of revolution H vs. Re_x. If this locus of Re_x vs. H falls between the two curves labeled $Re_{x(crit)}$ and $Re_{x(e^9)}$ in Fig. 9, the flow over the body may be considered to be completely laminar. If the locus of Re_x vs. H crosses the $Re_{x(e^9)}$ curve, the boundary layer is assumed to undergo transition at the Re_x of the intersection point. If the locus lies very close to the $Re_{x(crit)}$ curve, then the body is overheated, whereas if the locus lies very close to the $Re_{x(e^9)}$ curve, then more heating would be preferable (to maintain laminar flow). These remarks, of course, may not hold completely since, as stated earlier, in heated flows H alone does not totally determine the stability, and hence the transition behavior of the body. Further confidence in this suggested analysis can be gained as measurements of $Re_{x(transition)}$ vs. H become available and are used to check the validity of the trends indicated in Fig. 9.

The format of Fig. 9 has been chosen so as to allow easy application to specific problems. The path of boundary-layer development over a given shape on this diagram remains the same, since the shape parameter is only a function of the relative position on a (unheated) body. The path of boundary-layer development then simply moves up or down as the size and velocity of the body are changed.

As a test of these rather speculative comments, the wedge-flow computations have been compared with three cases of the development of nonsimilar boundary layers. The first of these is the development of a boundary layer in a heated tube; for this case, experimental data exist (Ref. 35). The boundary-layer development on the tube wall does not correspond to the similar type of boundary layer computed for the wedge flows; the local value of β increases parabolically from zero near the tube entrance to a positive value at the end of the tube. The path of the boundary-layer development on the Re_x-H plot of Fig. 9 would thus follow, not a vertical line as would be followed by the similar boundary layer on a wedge, but rather a line which is initially near vertical and curves slightly to the left corresponding to the increase in the local value of β (and decrease in the shape parameter H) as the flow proceeds along the pipe. The experimental measurements of Barker and Jennings (Ref. 35) were made in a 6.1m (20 ft) long, 0.1m (4 in.) diameter tube using water at about $11^{\circ}C$ ($52^{\circ}F$). Their measurements correspond to conditions of velocity and wall temperature necessary to maintain a laminar boundary layer along the length of the tube. To attain these conditions, they found that it was necessary to follow carefully a "laminar path" in which velocity and wall temperature were simultaneously increased. Their data, describing this "laminar path," are shown in Fig. 10 in comparison with wedge-flow computations for $Re_{x(e^9)}$ for the mean value of β in the tube (equal to two-thirds the value at the downstream end of the tube) and for an ambient temperature of $11^{\circ}C$ ($52^{\circ}F$). Agreement is remarkable up to a transition Reynolds number of $Re_x = 31 \times 10^6$ corresponding to a wall temperature about $5.5^{\circ}C$ ($10^{\circ}F$) above the ambient temperature. As the wall temperature is increased further, however, the measured values become increasingly less than the predicted ones. Whether this is due to buoyancy-induced secondary flows, to dirt deposits on the tube wall, or to downstream effects has yet to be determined.

The second case is a much more severe test of the wedge-flow computations. Computations of the boundary-layer development on a very blunt body of revolution (Ref. 36)* are shown in Fig. 11 against a background of the wedge-flow computations shown in Fig. 9.** The boundary-layer development for four unit Reynolds numbers over the unheated body are shown; also presented is the case of the body heated $5.5^{\circ}C$ ($10^{\circ}F$) above the ambient temperature. For this body, the departure from similarity is seen to be extreme. The initial path of boundary-layer development is initially almost horizontal on the Re_x-H plane with H increasing very rapidly as the boundary layer develops over the shoulder of the body; H then reaches a maximum and decreases rapdily on a higher horizontal path. The computed value at which an amplification ratio of e^9 occurs is shown on each of these paths (except the lowest unit Reynolds number where the maximum amplification ratio attained is $e^{7.8}$). It will be noted that for this body, the predicted transition point occurs somewhat downstream of the point predicted by a simple application of the wedge-flow correlation. This is presumably because of the effect of the boundary-layer's history--effectively coming from a very stable condition into the region of instability.

The third case, less severe than the second, concerns the boundary-layer development on a relatively slender body of revolution--a 13:1 Reichardt forebody. Figure 12 shows the paths of boundary-layer development on this body at a single unit Reynolds number and several surface temperatures. The boundary-layer development on this body, over most of its length, more closely approximates the similar wedge flows. The computed positions of an amplification ratio of e^9 are shown on these curves and are seen to coincide closely with the wedge-flow computations for $Re_{x(e^9)}$.

---

*This shape was suggested to us by Professor A. J. Acosta of the California Institute of Technology.

**The isothermal wedge-flow lines of Fig. 9 are reproduced in Fig. 10 as dashed lines.

REFERENCES

1. Siegel, R. and Shapiro, A. H: "The effect of heating on boundary-layer transition for liquid flow in a tube." ASME Preprint No. 53-A-178. Paper presented at Annual Meeting, New York (1953).

2. Kaups, K. and Smith, A. M. O.: "The laminar boundary layer in water with variable properties." Paper presented at the Ninth ASME-AIChE National Heat Transfer Conference, Seattle (1967).

3. Poots, G. and Ragget, G. F.: "Theoretical results for variable property, laminar boundary layers in water." Int. J. Heat Mass Trans., 10, pp. 597-610 (1967).

4. Schlichting, H.: "Boundary-layer theory." Sixth Edition. McGraw-Hill Book Co., New York (1968).

5. Hauptmann, E. G.: "The influence of temperature dependent viscosity of laminar boundary-layer stability." Int. J. Heat Mass Trans., 11, pp. 1049-1052 (1968).

6. Wazzan, A. R., Okamura, T. T. and Smith, A. M. O.: "The stability of water flow over heated and cooled flat plates." J. Heat Trans., 90, pp. 109-114 (1968).

7. Wazzan, A. R. and Keltner, G.: "Effect of heat transfer on the temporal growth of Tollmein-Schlichting waves in stagnation water boundary layers." AIChE Symposium Series, 69, 131 (1972).

8. Wazzan, A. R., Keltner, G., Okamura, T. T. and Smith, A. M. O.: "Spatial stability of stagnation water boundary layers with heat transfer." Phys. Fluids, 15, pp. 2114-2118 (1972).

9. Wazzan, A. R., Okamura, T. T. and Smith, A. M. O.: "The stability and transition of heated and cooled incompressible laminar boundary layers." Fourth International Heat Transfer Conference, II, Paper No. FC-14 (1970).

10. Schubauer, G. B. and Skramstad, H. K.: "Laminar-boundary-layer oscillations and transition on a flat plate." NACA Adv. Conf. Rept. (April 1943), later Tech. Rept. No. 909.

11. Rogler, H. L. and Reshotko, E.: AFOSR Scientific Report AFOSR TR 74-0795, Case Institute of Technology (January 1974).

12. Liepmann, H. W.: "Investigations of laminar boundary-layer stability and transition on curved boundaries." NACA Adv. Conf. Rept. No. 3H30 (August 1943), later W-107.

13. Smith, A. M. O. and Gamberoni, H .: "Transition, pressure gradient and stability theory." Report ES26388, Douglas Aircraft Co. (1956).

14. Jaffe, N., Okamura, T. T. and Smith, A. M. O.: "The determination of spatial amplification factors and their application to predicting transition." AIAA Journal, 8, p. 301 (1970).

15. Obremski, H. J., Morkovin, M. V. and Landahl, M.: "A portfolio of stability characteristics of incompressible boundary layers." With contributions from A. R. Wazzan, T. T. Okamura, and A. M. O. Smith, AGARDograph No. 134, NATO, Paris (1969).

16. King, W. S.: "The effects of wall temperature and suction on laminar boundary layer stability." The Rand Corporation, R-1863-ARPA (April 1976).

17. Wazzan, A. R.: "Spatial stability of Tollmien-Schlichting waves." Prog. Aerospace Sci., 16, 2, p. 99 (1975).

18. Lowell, L., Jr. and Reshotko, E.: "Numerical study of the stability of a heated water boundary layer." Report FTAS TR73-93, School of Engineering, Case Institute of Technology (January 1974).

19. Taghavi, H. and Wazzan, A. R.: "Spatial stability of some Falkner-Skan profiles with reversed flow." The Physics of Fluids, 17, 12, p. 2181 (1974).

20. Wazzan, A. R., Okamura, T. T. and Smith, A. M. O.: "Spatial and temporal stability charts for the Falkner-Skan boundary-layer profiles." Douglas Aircraft Co. Report No. DAC-6708 (September 1968).

21. Hughes, T. H. and Reid, W. H.: "The stability of laminar boundary layers at separation." J. Fluid Mech., 23, pp. 737-747 (1965).

22. Tsou, F. K. and Sparrow, E. M.: "Hydrodynamic stability of boundary layers with surface mass transfer." Appl. Sci. Res., 22, pp. 273-286 (1970).

23. Emmons, H. W.: "The laminar-turbulent transition in a boundary layer." Part I, J. Aero. Sci., 18, p. 490 (July 1951).

24. Criminale, W. O., Jr. and Kovasznay, L. S. G.: "The growth of localized disturbances in a laminar boundary layer." J. Fluid Mech., 14, 59 (1962).

25. Brooke, T.: "The development of three-dimensional disturbances in an unstable film of liquid flowing down an inclined plane." J. Fluid Mech., 10, p. 401 (1961).

26. Klebanoff, P. and Tidstrom, K. D.: "Evolution of amplified waves leading to transition in a boundary layer with zero pressure gradient." NASA Tech. Note D-195 (1959).

27. Kovasznay, L. S. G.: "A new look at transition." Aeronautics and Astronautics, Pergamon (1960).

28. Morkovin, M. V.: "Critical evaluation of transition from laminar to turbulent shear layers with emphasis on hypersonically traveling bodies." U.S. AFFDL TR-68-149 (1969).

29. Hall, G. R.: "Interaction of the wake from bluff bodies with an initially laminar boundary layer." AIAA Journal, 5, p. 1386 (1967).

30. Klebanoff, P. S., Tidstrom, K. D. and Sargent, L. M.: "The three-dimensional nature of boundary-layer instability." J. Fluid Mech. 12, 1 (1962).

31. Loehrke, R. I., Morkovin, M. V. and Fejer, A. S.: "New insights on transition in oscillating boundary layers." Fluid Dynamics Symposium, McMaster University, Hamilton, Ontario, Canada (1970).

32. Spangler, J. G. and Wells, C. S., Jr.: "Effect of freestream disturbances on boundary-layer transition." AIAA Journal, 6, p. 543 (1968).

33. Kendall, J. M.: "Supersonic boundary layer stability experiments." Jet Propulsion Laboratory, Space Programs Summary 37-39, p. 147 (May 1966).

34. Granville, P. S.: "Comparison of existing methods for predicting transition from laminar to turbulent flow on bodies of revolution." Naval Ship Research and Development Center, TN 111 (August 1968).

35. Barker, S. J. and Jennings, C. G.: "The effects of wall heating on transition in water boundary layers." Paper presented at AGARD Symposium on Laminar Turbulent Transition, Copenhagen (May 2-4, 1977).

36. Schiebe, S. R.: "Measurements of the cavitation susceptibility of water using standard bodies." Project Report 118, Anthony Falls Hydraulics Laboratory, University of Minnesota (February 1972).

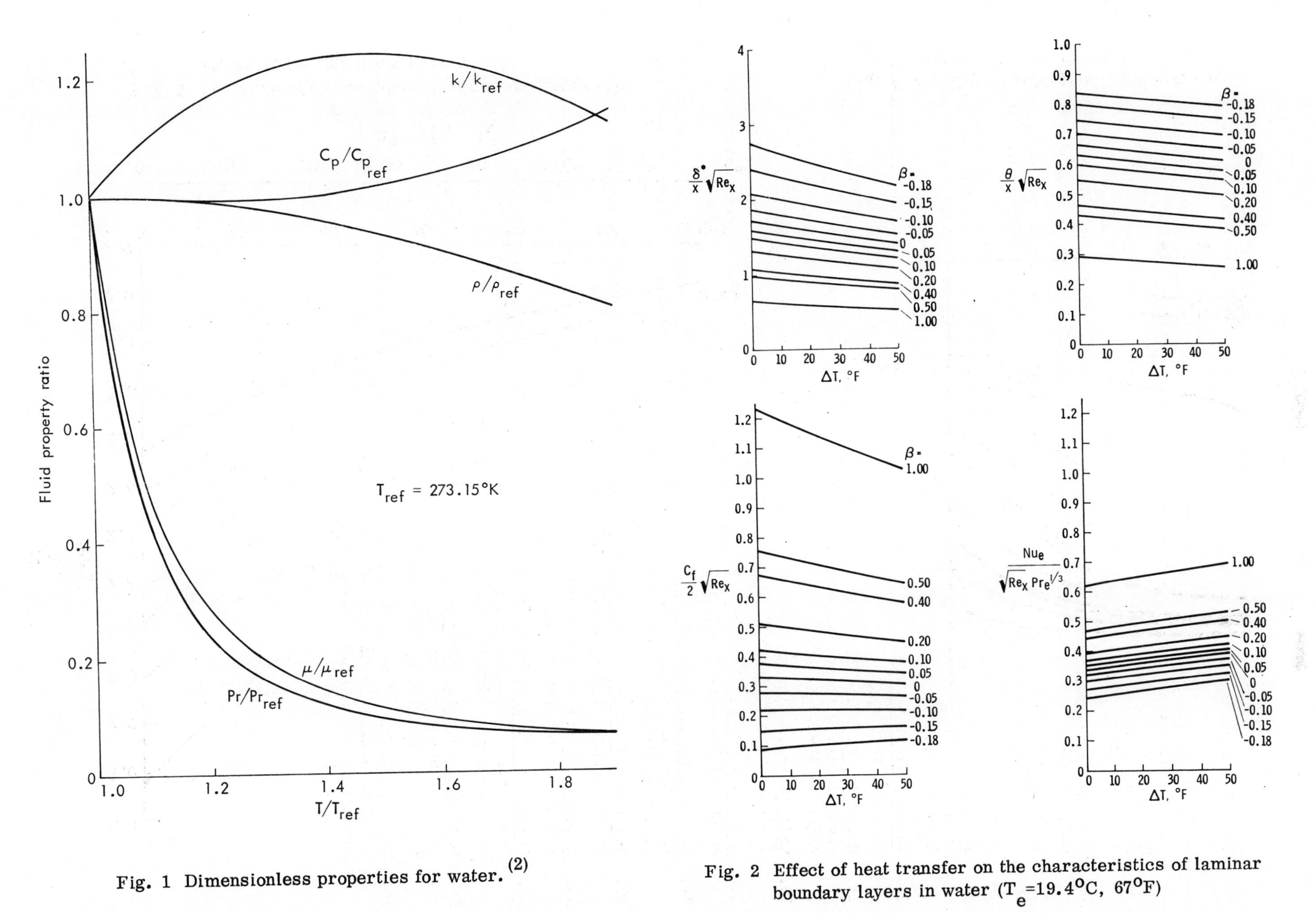

Fig. 1 Dimensionless properties for water.(2)

Fig. 2 Effect of heat transfer on the characteristics of laminar boundary layers in water (T_e=19.4°C, 67°F)

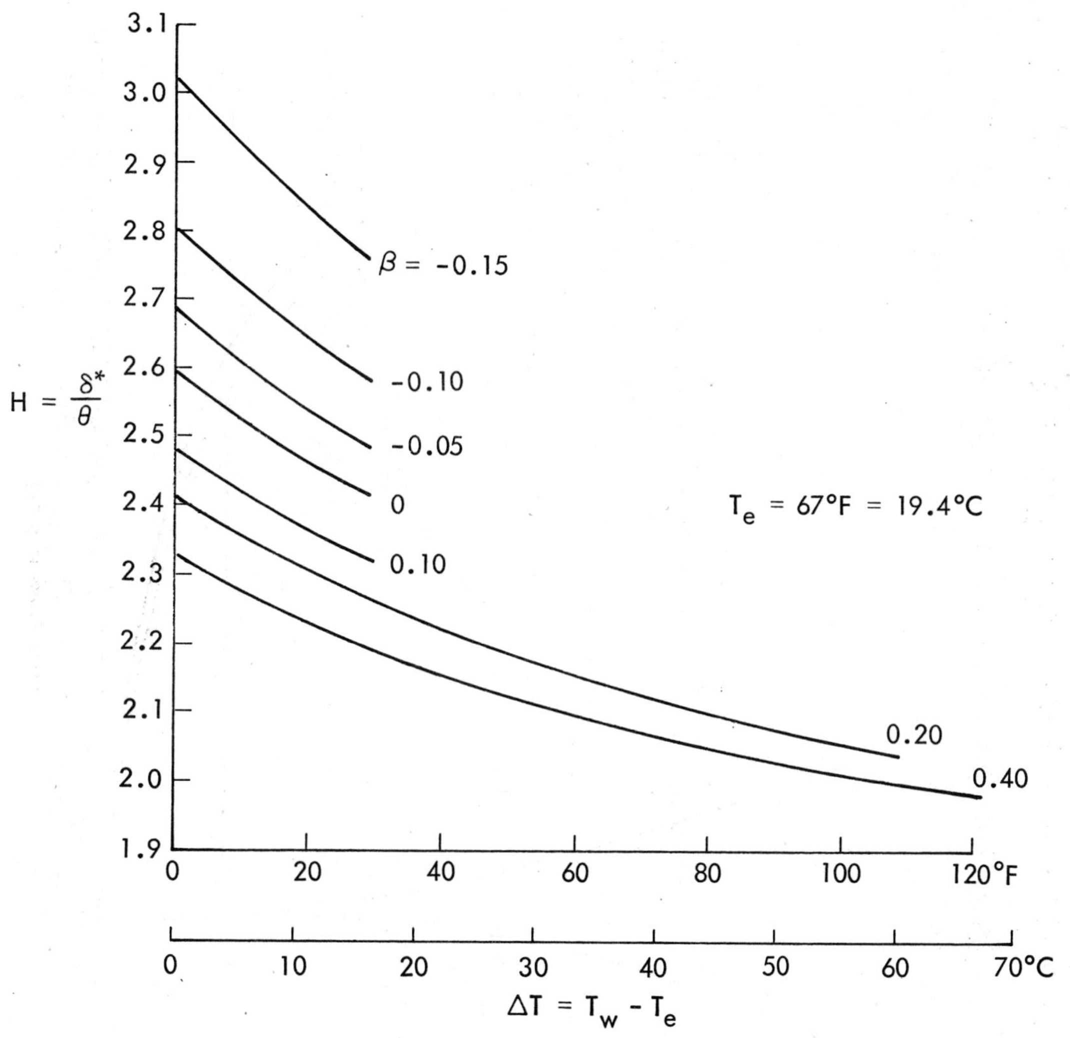

Fig. 3 Variation of the boundary-layer shape parameter for heated wedge flows in water.

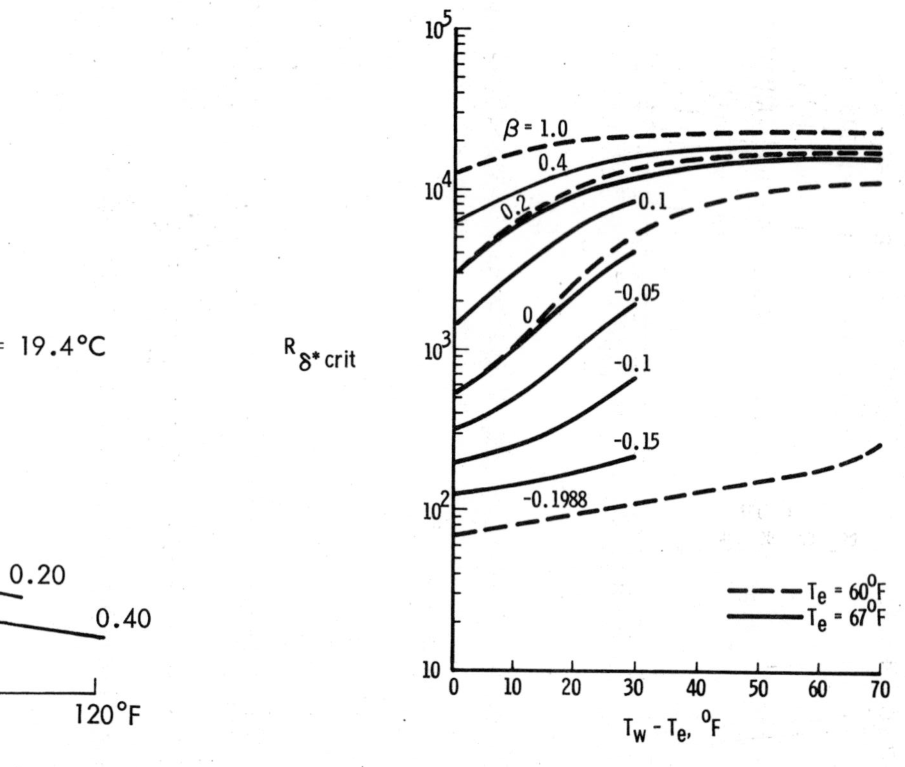

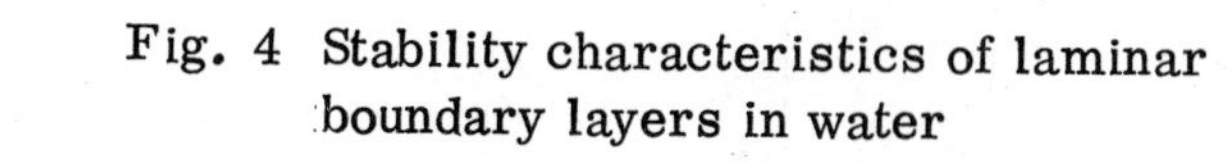
Fig. 4 Stability characteristics of laminar boundary layers in water

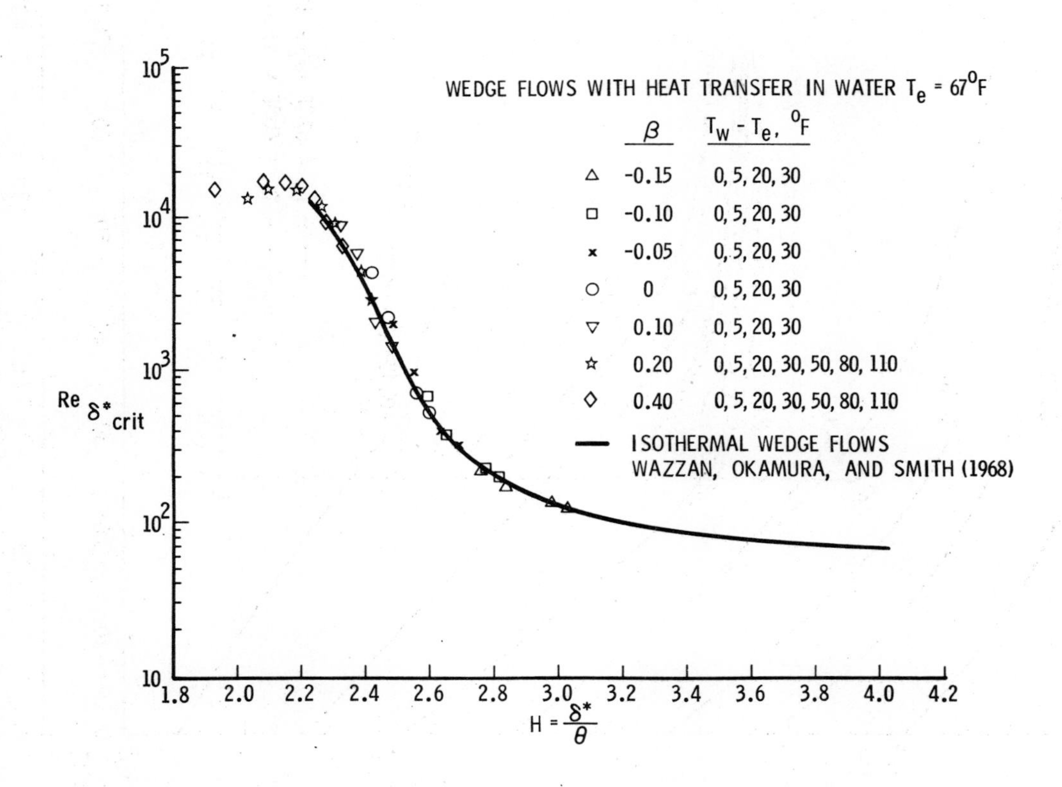

Fig. 5 Critical Reynolds number for heated wedge flows in water

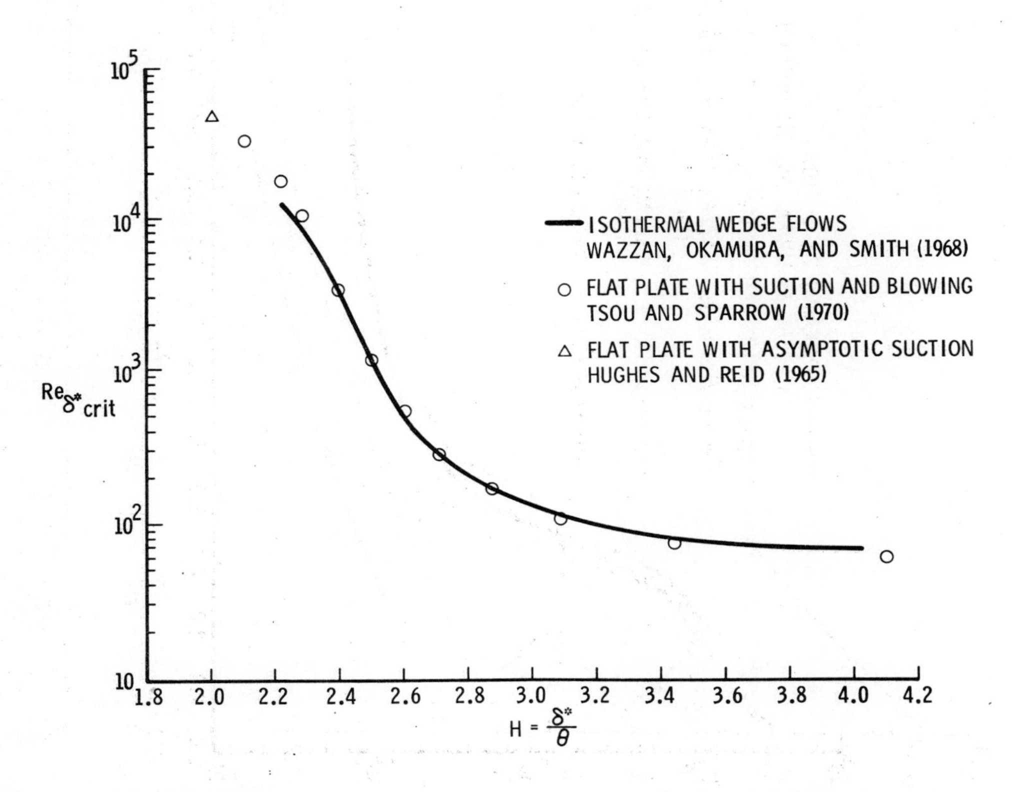

Fig. 6 Critical Reynolds number as a function of the boundary-layer shape parameter, isothermal wedge flows and flat plate with suction and blowing

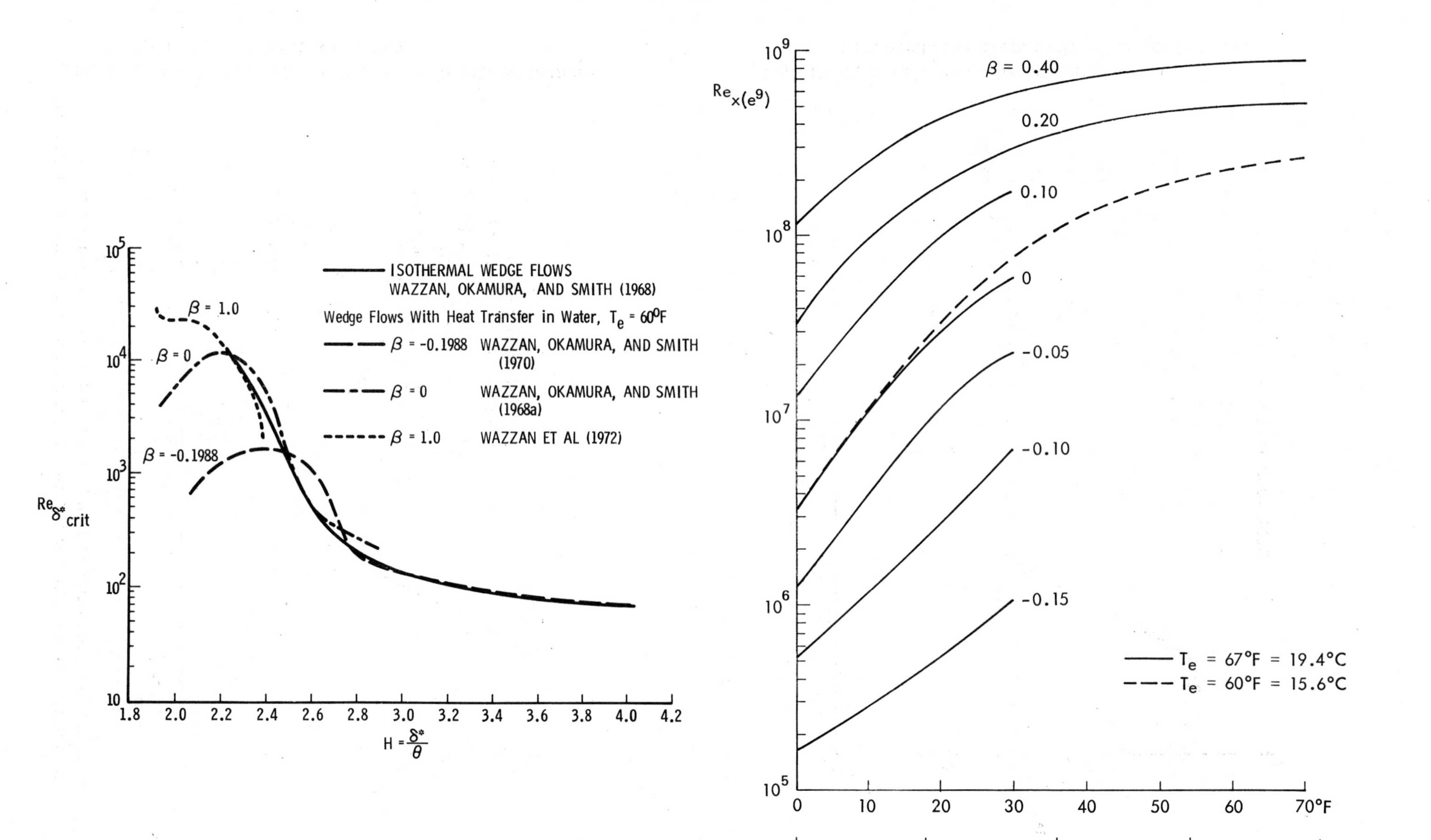

Fig. 7 Previous results for heated wedge flows in water

Fig. 8 Variation of $Re_{x(e^9)}$ forheated wedge flows in water

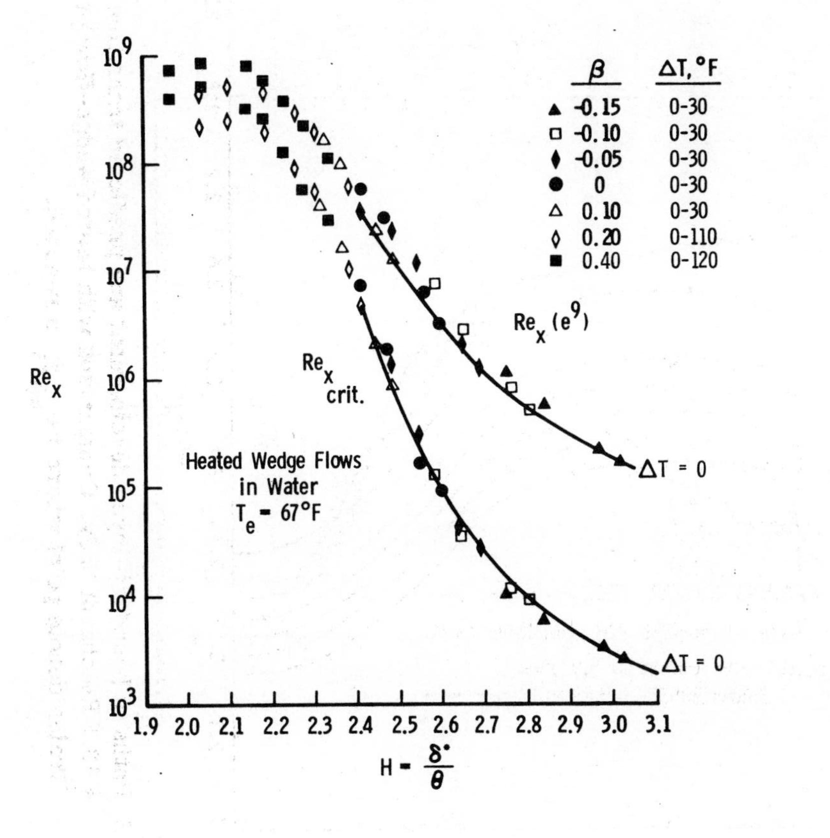

Fig. 9 Critical and predicted transition Reynolds numbers for heated wedge flows in water

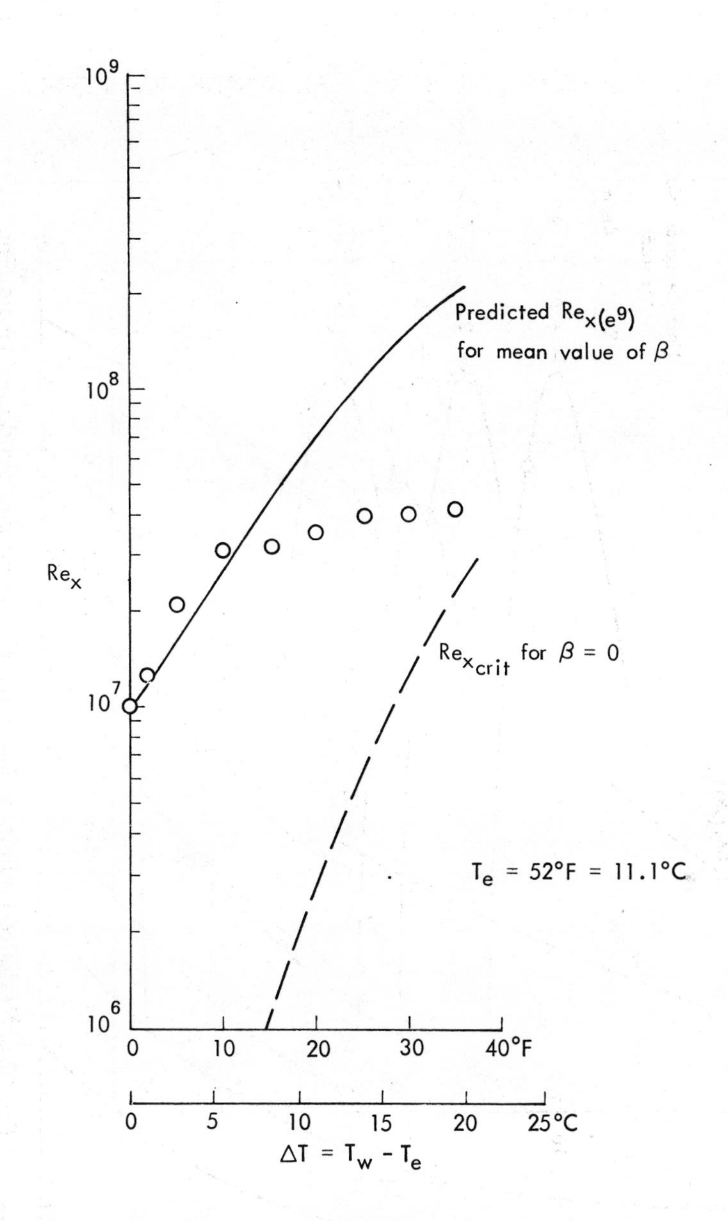

Fig. 10 Comparison of tube experiments (35) with prediction from wedge-flow computations

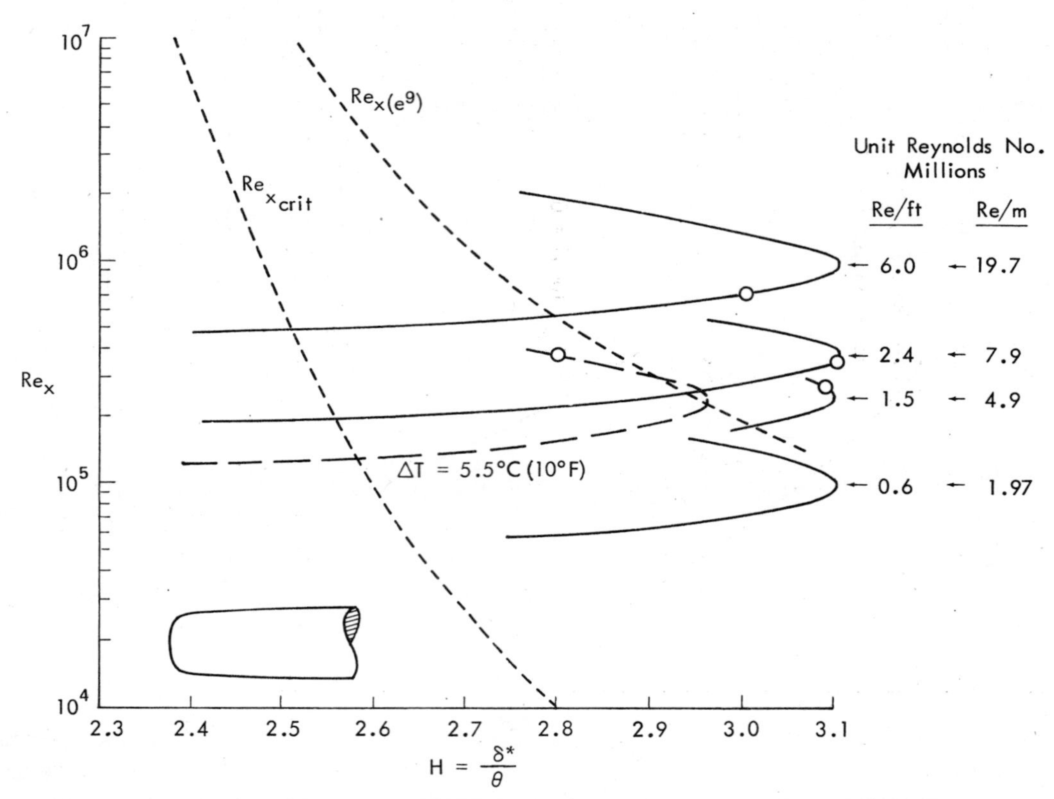

Fig. 11 Paths of boundary-layer development and predicted transition for a very blunt body.[36] Comparison with heated wedge-flow computations. Circles denote points where $Re_{x(e^9)}$ is reached

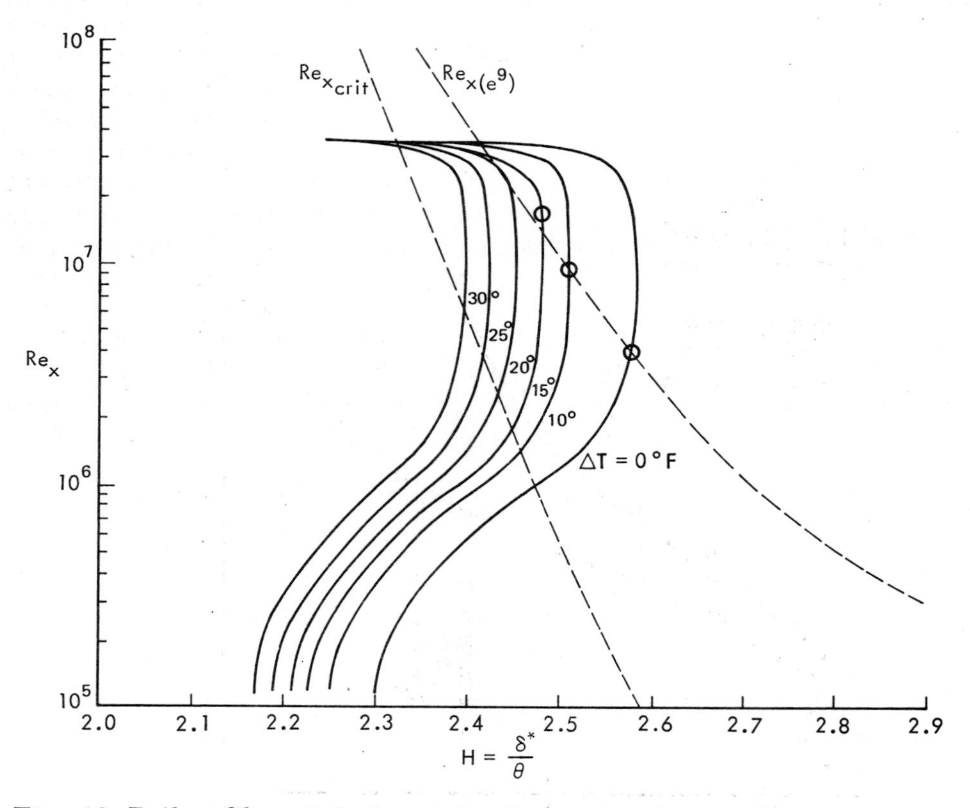

Fig. 12 Paths of boundary-layer development and predicted transition for a 13:1 Reichardt body. Comparison with heated wedge-flow predictions. Circles denote point where $Re_{x(e^9)}$ is reached.

Second International Conference on

Drag Reduction

August 31st - September 2nd, 1977

PAPER E4

THE EFFECTS OF UNSTEADY POTENTIAL FLOW ON HEATED LAMINAR BOUNDARY LAYERS IN WATER: FLOW PROPERTIES AND STABILITY

W.S. King, J. Aroesty, L.S. Yao and W. Matyskiela,

The Rand Corporation, U.S.A.

Summary

The laminar boundary layer over heated bodies undergoing acceleration or deceleration in water has been investigated to determine the combined influence of heating, pressure gradient and unsteadiness on boundary-layer stability. For simplicity, the boundary-layer flow over accelerating wedges or cones has been analyzed for values of the acceleration parameter, $\varepsilon_1 = (x/U_1^2)\,(dU_1/dt)$, which are small but not negligible. It is shown that boundary-layer stability analysis, in this range of ε_1, requires solution of the unsteady boundary-layer equations, but that stability characteristics may be determined using the classic Orr-Sommerfeld approach.

Velocity profiles have been calculated by a perturbation expansion in ε_1 about the quasi-steady flow, and the stability of the resulting profiles has been calculated by an accurate and convenient modification of the Dunn-Lin theory, specialized to constant density, variable jpr property flow. The results confirm that acceleration is stabilizing, while deceleration is destabilizing, and show that the additional stabilizing effect of acceleration is diminished by both positive velocity gradient and heating. A value of ε_1 of only 0.1 produces a four-fold increase in the critical Reynolds number for a flat plate and two-fold increase for a highly favourable pressure gradient. Heating reduces these changes by as much as a factor of two. When positive velocity gradient and heating are combined, they are stabilizing and each enforces the other; however, when they are combined with unsteadiness, they oppose the additional stability enhancement of acceleration. This is more striking when the wall compatibility condition, which indicates that unsteadiness, heating, and positive velocity gradient act in concert to produce a trend towards stability, is considered. The probable explanation for this variance is discussed.

Held at St. John's College, Cambridge, England.
Organised and sponsored by BHRA Fluid Engineering, Cranfield, Bedford, MK43 0AJ.

I. INTRODUCTION

Unsteady laminar boundary layers are found in a wide variety of practical circumstances, but the quantitative evaluation of the unsteady effects is not yet widely considered. The effects of unsteadiness are usually assessed by employing the quasi-steady approximation. However, for a slightly more vigorous unsteadiness, the calculation procedure has been to consider small departures from the quasi-steady state. This theory is well developed and is discussed in Refs. 1 through 3 for arbitrary magnitudes of slowly-varying unsteadiness. The accuracy of this theory is presently estimated from order of magnitude estimates, or from the calculation of further terms in the solution sequence. However, a recent theoretical study of fluctuating free-stream perturbations to flat-plate flow via matched asymptotic methods suggests strongly that the quasi-steady approach is uniformly valid when the frequency or flow length is small (Ref. 4). Nevertheless, it appears that an assessment of the accuracy of the technique must await ultimate comparison with results from the newly fledged numerical schemes for unsteady flow. Until this is done, the quasi-steady approximation sequence used here must be considered the practical method of choice for realistic mild unsteadiness (Ref. 5).

This paper discusses the effects of unsteadiness on laminar boundary-layer flows and their stability as governed by the Orr-Sommerfeld equation.* It presents the required conditions for the validity of the quasi-steady approximation to be valid for both the boundary-layer flow and its hydrodynamic stability. It is well known that a critical parameter for boundary-layer flow is the ratio of diffusion time to flow time; the criterion for nearly quasi-steady flow is that this parameter should be small. It is also shown that the criteria for validity of the Orr-Sommerfeld equation are based on modifications of this ratio, and are not nearly as stringent. Examples of heated wedge flows in water are presented and discussed that show the profound effect of even slowly-varying unsteadiness on both laminar boundary-layer flow and its stability.

II. ANALYSIS

When the flow external to the boundary layer varies with time, variations in boundary-layer momentum are produced, diffused to the wall, and dissipated. It is through this mechanism that the boundary-layer flow accommodates variation in free-stream velocity. If the boundary layer accommodates instantaneously, the flow may be considered quasi-steady. If, on the other hand, the boundary layer accommodates slowly, the flow is unsteady and the rate of accommodation is important. A measure of how rapidly the boundary layer accommodates is the ratio of diffusion time to flow time. In the following discussion this is the critical parameter, and all comparisons of unsteadiness in boundary-layer flow to unsteadiness in boundary-layer stability are based on its magnitude.

The analysis will first focus on boundary-layer flow in that the quasi-steady flow parameters will be derived and numerical examples will be presented, and next on boundary-layer stability in that a criterion for nearly quasi-steady flow will be developed and numerical examples of the effect of unsteadiness on critical Reynolds numbers will be shown. The specific flows that are used as examples are two-dimensional Falkner-Skan flows with wall heating, variable viscosity, and Prandtl number corresponding to water.

A. BOUNDARY-LAYER FLOW

The unsteady equations for a two-dimensional laminar boundary-layer flow at an incompressible fluid with heat transfer and variable viscosity are:

$$\frac{\partial u}{\partial t} + u\frac{\partial u}{\partial x} + v\frac{\partial u}{\partial y} = -\frac{\partial p}{\rho\,\partial x} + \frac{\partial}{\partial y}\left[\mu(T)\frac{\partial u}{\partial y}\right] \qquad (1)$$

$$\frac{\partial T}{\partial t} + u\frac{\partial T}{\partial x} + v\frac{\partial T}{\partial y} = \frac{\nu_e}{Pr}\frac{\partial^2 T}{\partial y^2} \qquad (2)$$

where Pr is based on freestream properties.

---

*C. Von Kerczek of NSRDC discussed the effect on transition in a presentation at the Rand Workshop on Low Speed Boundary Layer Transition, Santa Monica, CA, Sept. 1977.

To assess the effect of unsteadiness on boundary-layer flow, the relative order of magnitude of the first term on the left-hand side of the momentum equation and the viscous term on the right side must be determined. The order of magnitude of the viscous term is

$$\nu \frac{\partial^2 u}{\partial y^2} = O\left(\frac{\nu U_1}{\delta^2}\right)$$

where U_1 is the external velocity with an arbitrary time dependence, and δ is the momentum layer thickness.

The order of magnitude of the unsteadiness is estimated by the variation in time of the freestream velocity,

$$\frac{\partial u}{\partial t} = O\left(\frac{\partial U_1}{\partial t}\right) .$$

Then, the ratio of magnitude of the two terms is

$$\frac{\delta^2}{\nu} \frac{\partial U_1}{\partial t U_1} = \frac{\text{unsteady effect}}{\text{viscous effect}}$$

A more physical interpretation consistent with that described in the Introduction is obtained by noting that

$$\frac{\partial U_1}{\partial t U_1} \sim \frac{1}{t_f}$$

where t_f is a characteristic flow time, i.e., the time it takes the external flow to change from one state to another. Further, ν/δ is a diffusion velocity and δ^2/ν is the time it takes the momentum to diffuse to the wall. Therefore, the ratio of the magnitude of the unsteady effect to the viscous effect is related to the ratio of characteristic times,

$$\frac{\delta^2}{\nu} \frac{\partial U_1}{\partial t U_1} \sim \frac{t_d}{t_f} .$$

If the boundary-layer thickness is related to the local Reynolds number $[\delta \sim x(Re)^{-\frac{1}{2}}]$, the quasi-steady parameter of Refs. 1 through 3 is obtained:

$$\frac{\delta^2}{\nu} \frac{\partial U_1}{\partial t U_1} \sim \frac{x}{U_1^2} \frac{\partial U_1}{\partial t} = \varepsilon_1 . \tag{3}$$

When this parameter is small, or $t_d/t_f << 1$, the flow in the boundary layer is quasi-steady. A similar parameter is discussed in Ref. 4. For flows that are nearly quasi-steady, a systematic series expansion procedure can be developed. A detailed discussion of this procedure as it applies to general unsteady flow is given in Ref. 2. However, a special version of nearly quasi-steady flow and this technique are illustrated in the example to be discussed in the next section.

1. Unsteady Boundary-Layer Flow over Heated Bodies

The theory developed in Refs. 2 and 3 will be extended to examine the unsteady effect on a flow over heated bodies in water with constant freestream and wall temperatures. Although the general theory will be developed, this discussion is concerned with nearly quasi-steady flows where the time variation is limited to slowly-varying functions and wedge flows. The attractive attribute of this theory is that for nearly quasi-steady flows the unsteady component of the freestream velocity can have arbitrary magnitude.

Following Ref. 3, a boundary-layer scaling will be employed that allows the quasi-steady component to be determined as the zero-order solution. The scaling to transform Eqs. (1) and (2) is:

$$\eta = \sqrt{\frac{U_1}{2\nu x}}\, y$$

$$\psi = \sqrt{2\nu x U_1}\, f(\eta,x,t) \tag{4}$$

$$\theta = \frac{T - T_e}{T_w - T_e}$$

where U_1, the inviscid unsteady velocity along the surface is given by $U_1 = U_e(x)\, A(t)$.

Note that this form of the time-dependent inviscid surface velocity corresponds to unsteady irrotational flow, in the absence of displacement effects. $U_e(x)$ corresponds to the surface velocity when a body translates with unit velocity into a fluid at rest and A(t) is the actual time-dependent velocity of translation. Thus, $U_e(x)$ is to be determined from a conventional potential flow solution corresponding to unit body velocity. The transformed equations are:

$$(Nf_{\eta\eta})_\eta + \Big(1 + M(x)\Big) ff_{\eta\eta} + 2M(x)(1 - f_\eta^2)$$

$$= 2x(f_\eta f_{\eta x} - f_{\eta\eta} f_x) + \frac{2U_{1t}x}{U_1^2}\left(f_\eta + \frac{\eta}{2} f_{\eta\eta} - 1\right) \tag{5}$$

$$+ \frac{2x}{U_1} f_{\eta t}$$

$$\frac{\theta_{\eta\eta}}{Pr} + \Big(1 + M(x)\Big) f\theta_\eta = 2x(f_\eta \theta_x - \theta_\eta f_x)$$

$$+ \frac{2U_{1t}x}{U_1^2}\left(\frac{\eta}{2}\theta_\eta\right) + \frac{2x}{U_1}\theta_t \tag{6}$$

where $M(x) = U_{1x}x/U_1$ and N is the ratio of boundary layer to freestream viscosity. N can be approximated by the empirical expression for water used in Ref. 6.

$$N = \mu/\mu_e = (a + br + cr^2 + dr^3 + er^4)^{-1}\frac{\mu_{ref}}{\mu_e} \tag{7}$$

where $r = T/491.69^{\circ}R$ and $\mu_{ref}/\mu_e = 4.339\ \text{lb/hrft}/\mu_e$. The appropriate boundary conditions are:

$$\begin{array}{lll} t \geq 0 & & \\ \eta \to \infty & f_\eta \to 1 & \theta \to 0 \\ \eta = 0 & f = f_\eta = 0 & \theta = 1 \end{array} \tag{8}$$

The general velocity and temperature distributions are functions of three independent variables, i.e.,

$$f = f(\eta,x,t) \tag{9}$$

However, for special cases where f is a slowly-varying function of time, one will find it convenient to express the velocity distribution as

$$f = f(\eta,x;\varepsilon_n) \tag{10}$$

where now f depends on a sequence of general unsteady parameters

$$\varepsilon_n = \left(\frac{x}{U_1}\right)^n \frac{\partial^n U_1}{U_1 \partial t^n} \qquad n = 1,2,3 \ldots \tag{11}$$

The first parameter, ε_1, is an obvious expansion parameter (the ratio of diffusion time to flow adjustment times) that is explicitly displayed in the equations. For $n > 1$, the parameters are ratios of various diffusion times to various flow times and a result of substituting the expansion of f into the governing equations and requiring a self-consistent set of differential equations.

When a body has constant acceleration, the parameters ε_n for $n > 1$ are precisely zero. For flows that have nearly constant acceleration, some consideration must be provided for ε_n when $n > 1$. It is assumed here that $(\varepsilon_1)^n > \varepsilon_n$ for $n > 1$. This allows the velocity and temperature distributions to be expressed to first order as

$$\begin{aligned} f(\eta,x,t) &= f_0(\eta,x) + \varepsilon_1 f_1(\eta,x) \\ \theta(\eta,x,t) &= \theta_0(\eta,x) + \varepsilon_1 \theta_1(\eta,x) \end{aligned} \tag{12}$$

After substituting Eq. (12) into Eqs. (5) and (6), one obtains the set of differential equations representing the zero and first-order approximations to a nearly quasi-steady flow.*

<u>Zero Order</u>

$$(N_0 f_{0\eta\eta})_\eta + \Big(1 + M(x)\Big) f_0 f_{0\eta\eta} + 2M(x)(1 - f_{0\eta}^{\,2}) = 2x(f_{0\eta} f_{0\eta x} - f_{0\eta\eta} f_{0x}) \tag{13}$$

$$\frac{\theta_{0\eta\eta}}{Pr} + \Big(1 + M(x)\Big) f_0 \theta_{0\eta} = 2x(f_{0\eta}\theta_{0x} - \theta_{0\eta} f_{0x}) \tag{14}$$

<u>First order</u>

$$\begin{aligned} (N_0 f_{1\eta\eta})_\eta &+ \Big(1 + M(x)\Big) f_0 f_{1\eta\eta} - 2\Big(1 + M(x)\Big) f_{0\eta} f_{1\eta} + \Big(3 - M(x)\Big) f_{0\eta\eta} f_1 \\ &= 2x\Big(f_{0\eta} f_{1\eta x} + f_{1\eta} f_{0\eta x} - f_{0\eta\eta} f_{1x} - f_{1\eta\eta} f_{0x}\Big) \\ &+ 2(f_{0\eta} + \frac{\eta}{2} f_{0\eta\eta} - 1) - (N_1 f_{0\eta\eta})_\eta \end{aligned} \tag{15}$$

$$\begin{aligned} \frac{\theta_{1\eta\eta}}{Pr} &+ \Big(1 + M(x)\Big) f_0 \theta_{1\eta} - 2\Big(1 - M(x)\Big) f_{0\eta} \theta_1 \\ &= - f_1 \theta_{0\eta}\Big(3 - M(x)\Big) + 2x\Big(f_{0\eta}\theta_{1x} + f_{1\eta}\theta_{0x} - \theta_{0\eta} f_{1x} - \theta_{1\eta} f_{0x}\Big) + \eta\theta_{0\eta} \end{aligned} \tag{16}$$

The boundary conditions are:

---

*Note that the viscosity ratio, N, is also expanded in the form $N = N_0 + \varepsilon_1 N_1$.

$$\eta = 0: \quad f_0 = f_{0\eta} = f_1 = f_{1\eta} = 0;\ \theta_0 = 1;\ \theta_1 = 0$$
$$\eta \to \infty: \quad f_{0\eta} \to 1;\ f_{1\eta} \to 0;\ \theta_0 \to 0;\ \theta_1 \to 0 \tag{17}$$

In a later section a special version of these equations will be specialized to wedge flows where $U_e \sim x^m$.

B. BOUNDARY-LAYER STABILITY

At present there is no theory to predict the hydrodynamic stability of an unsteady flow when unsteady effects are large. For cases where the time-dependent component of the freestream velocity varies slowly, one could develop a theory involving multiple time scales, but this approach has not yet been carried out for the spatial stability problem. In this section, characteristic times for the Orr-Sommerfeld equation will be derived and used to establish a heuristic criterion to define the region where unsteadiness modifies the stability analysis.

Using this criterion, it will be shown that the stability of an unsteady boundary layer can be studied by using an approach that employs unsteady boundary-layer flow theory and quasi-steady stability analysis. This approximation, together with an approximate solution of the Orr-Sommerfeld equation, will be employed to discuss the stability of the flows presented in the previous section.

Referring to Ref. 7 for a heuristic discussion of the Orr-Sommerfeld equation for a time-dependent mean flow, one will find that the constant property two-dimensional version of the equation is

$$\frac{\delta}{U_1 i\alpha}\frac{\partial}{\partial t}(\phi'' - \alpha^2\phi) + (u - c)(\phi'' - \alpha^2\phi) - u''\phi = \left(\frac{1}{i\alpha R}\right)[\phi'''' - 2\alpha^2\phi'' + \alpha^4\phi] \tag{18}$$

where the superscript prime denotes a derivative with respect to (y/δ). All velocities are scaled with U_1; R, the Reynolds number, is based on boundary-layer thickness; and α is the wave number.

The approach that will be used in this section will be to investigate the Orr-Sommerfeld equation in three regimes: the outer inviscid region, the inner viscous region near the wall and the critical layer. These are the important regions for a uniformly valid solution (Ref. 8), when matched asymptotic expansion methods are applied.

1. The Inviscid Stability Equations

In the limit of large αR, the equation reduces to

$$\frac{\delta}{U_1 i\alpha}\frac{\partial}{\partial t}(\phi'' - \alpha^2\phi) + (u - c)(\phi'' - \alpha^2\phi) - u''\phi \cong 0 \tag{19}$$

To determine the relative effect of an unsteady freestream velocity, the magnitude of the first two terms must be compared. The magnitude of the ratio of the unsteady term to a convection term is at most

$$\frac{\dfrac{\delta}{U_1 i\alpha}\dfrac{\partial}{\partial t}\phi''}{(u - c)\ \phi''} = 0\left(\frac{\delta}{\alpha}\ \frac{\partial U_1}{\partial t U_1^2}\right) \tag{20}$$

where the magnitude of the convection term is approximated by

$$(u - c)(\phi'' - \alpha^2\phi) = 0(1) \quad . \tag{21}$$

For the purpose of comparison with the boundary-layer quasi-steady parameter, it is convenient to reduce Eq. (20) to a multiple of the ratio of diffusion time to flow time. The new parameter is

$$\frac{\delta}{\alpha}\frac{\partial U_1}{\partial t U_1^2} = \frac{\delta^2}{\alpha R \nu}\frac{\partial U_1}{\partial t U_1} = \frac{t_d}{t_f}\cdot\frac{1}{\alpha R} \tag{22}$$

and for quasi-steady flow in the inviscid stability equation this parameter must be small. Note that this restriction is less stringent than the boundary-layer condition since $\alpha R \sim 10^3$ for a flat-plate boundary layer.

2. The Wall Viscous Region

Following Ref. 8, a scaling appropriate for the viscous region near the wall is

$$\tilde{\eta} = y(R\alpha)^{\frac{1}{2}};\quad \frac{F(\tilde{\eta},t)}{(R\alpha)} = \phi(y,t) \quad . \tag{23}$$

Employing the new dependent and independent variables to evaluate the unsteady term and the first viscous term which is the dominant viscous term of Eq. (18), it can be shown that since

$$\frac{\delta}{U_1\alpha}\frac{\partial}{\partial t}(\phi'' - \alpha^2\phi) = \frac{\delta}{U_1\alpha}\frac{\partial}{\partial t}\left(F_{\tilde{\eta}\tilde{\eta}} - \frac{\alpha^2 F}{R\alpha}\right)$$

and (24)

$$\frac{\phi''''}{\alpha R} = F_{\tilde{\eta}\tilde{\eta}\tilde{\eta}\tilde{\eta}} \quad ,$$

it can be concluded that the reduced viscous term is of unit order and the relative magnitude of the unsteady term is at most

$$\frac{\delta}{U_1^2\alpha}\frac{\partial U_1}{\partial t} = \frac{t_d}{t_f}\cdot\frac{1}{\alpha R} \tag{25}$$

Thus the unsteady contribution for mild unsteadiness is the same order of magnitude in the wall viscous as it is in the outer inviscid region.

3. The Region Around the Critical Layer

The next region to investigate is the region near the critical layer. The most convenient way to analyze this region is to use the following change in dependent and independent variables:

$$\eta = (y - y_c)(\alpha R)^{1/3};\quad \phi(y,t) = G\frac{(\eta,t)}{(\alpha R)^{1/3}} \quad . \tag{26}$$

These definitions scale the unsteady term and the first viscous term. The results are:

$$\frac{\delta}{U_1\alpha}\frac{\partial}{\partial t}(\phi'' - \alpha^2\phi) = \frac{\delta}{U_1\alpha}(\alpha R)^{1/3}\frac{\partial}{\partial t}\left(G_{\eta\eta} - \frac{\alpha^2 G}{(\alpha R)^{2/3}}\right);\quad \frac{\phi''''}{\alpha R} = G_{\eta\eta\eta\eta} \tag{27}$$

The resulting order of magnitude of the ratio of these two terms is

$$\frac{\delta}{\alpha}\frac{(\alpha R)^{1/3}}{U_1^2}\frac{dU_1}{dt} \cong (t_d/t_f)(\alpha R)^{-2/3} \quad . \tag{28}$$

This parameter must be small for the quasi-steady approximation to be accurate in the critical layer. From this sequence of estimates it is seen that when t_d/t_f is small, then the unsteady laminar velocity profile can be obtained as a correction to the quasi-steady flow (nearly quasi-steady) and the stability of the resulting profiles may be studied using classical Orr-Sommerfeld analysis.

4. Simplified Stability Analysis of Unsteady Flow over Heated Surfaces

Ultimately, a complete analysis of the stability characteristics of specific unsteady flows, via numerical solution of the Orr-Sommerfeld equation would be useful. For this preliminary study, stability analysis is restricted to the estimation of minimum critical Reynolds number. Lin's original simplified analysis (Ref. 9), and Dunn's subsequent modification for compressible flow (Ref. 10) have been widely used in simplified stability analyses. Recently, it was shown that a modifed version of this theory and formulae is appropriate for water boundary layers with variable viscosity (Ref. 11). In addition, certain numerical constants in the original Dunn-Lin formulae have been revised on the basis of accurate numerical results which were not yet available to Dunn and Lin. While the original Dunn-Lin theory is now known to be inappropriate for compressible flow, its treatment of property variations is satisfactory. For completeness, the modified Dunn-Lin relations of Ref. 11 are:

$$v(c)\left(1 - 2\lambda(c)\right) = .58 \tag{29a}$$

where

$$v(c) = -\frac{\pi u''_c u'_w c}{u'^3_c} \tag{29b}$$

$$\lambda = \frac{u'_w}{c}\left[\frac{\nu_w/\nu_e}{c}\right]^{\frac{1}{2}}\left(\frac{3}{2}\int_0^{\eta_c}\sqrt{\frac{c-u}{\nu/\nu_e}}\,d\eta\right) - 1 \approx .4\left\{\left(1 - \frac{u'_c}{u'_w}\right) + .5\left(1 - \frac{\nu(c)}{\nu_w}\right)\right\} \tag{30}$$

$$R_{e_{crit}} \approx 28\,\frac{u'_w}{c^4}\left(\frac{\nu_w}{\nu_e}\right). \tag{31}$$

The equations, together with the boundary-layer velocity profiles will be used to calculate $Re_{crit} = U_1\delta_1/\nu_e$ evaluated at the minimum point of neutral stability.

III. UNSTEADY BOUNDARY-LAYER FLOW OVER HEATED WEDGE FLOWS

The approach developed and discussed in Section II has been used to investigate the stability of heated wedge flows under conditions of positive or negative acceleration. This section describes the application of the general method and numerical results to this class of flows. In this way, the combined effects of heating, pressure gradient, and unsteadiness can be studied.

A. UNSTEADY BOUNDARY-LAYER FLOW

When the flow is steady and the external velocity is represented by $U_e \sim x^m$ (wedge flow), the boundary-layer flow is self-similar. The boundary-layer equations become ordinary differential equations, and the variable $M(x)$ in Eq. (5) is replaced by the constant parameter β, that is equal to $2m/(1 + m)$. For nearly quasi-steady flow and $U_e \sim x^m$, the departure from a self-similar flow is governed by the departure from a quasi-steady flow, and this is controlled by the magnitude of the parameter ε_1. Thus, for the special case of wedge flows, Eq. (12) reduces to

$$\begin{aligned} f(\eta,x,t) &= \left(\bar{f}_0(\bar{\eta}) + \varepsilon_1 f_1(\bar{\eta})\right)\Big/\left(\sqrt{1+m}\right) \\ \theta(\eta,x,t) &= \theta_0(\bar{\eta}) + \varepsilon_1\theta_1(\bar{\eta}) \end{aligned} \tag{32}$$

where $\bar{\eta} = \eta\sqrt{1 + m}$. The new scaling by the constant $\sqrt{1 + m}$ introduced into Eq. (32) is to facilitate comparison to related works, i.e., Refs. 2 and 12. If Eq. (32) is substituted into Eqs. (13) through (16), the following sets of ordinary differential equations are obtained:

Zero Order:

$$(N_0\bar{f}_0'')' + \bar{f}_0\bar{f}_0'' + \beta[1 - (\bar{f}_0')^2] = 0$$

$$(N_0\bar{f}_1'')' + \bar{f}_0\bar{f}_1'' - 2\bar{f}_0'\bar{f}_1' + (3 - 2\beta)\bar{f}_0''\bar{f}_1 = (2 - \beta)\left[\bar{f}_0' + \eta\frac{\bar{f}_0''}{2} - 1\right] + (\alpha N_0^2\theta_1\bar{f}_0'')' \tag{33}$$

First Order:

$$\frac{\theta_0''}{Pr} + \bar{f}_0\theta_0' = 0$$

$$\frac{\theta_1''}{Pr} + \bar{f}_0\theta_1' - 2(1 - \beta)\bar{f}_0'\theta_1 = -(3 - 2\beta)\theta_0'\bar{f}_1 + (2 - \beta)\frac{\eta\theta_0'}{2} \tag{34}$$

The boundary conditions are the same as those shown in Eq. (17). The first-order equations for $\bar{f}_1$ and θ_1 have homogeneous boundary conditions and would have a trivial solution if the right-hand sides of Eqs. (33) and (34) were zero. From this it is implied that the effects of unsteadiness decrease as the pressure gradient becomes more favorable because the magnitude of the inhomogeneous term decreases. The impact of heating is not so clear. Although it is plausible, it is not definite that heating diminishes the effects of acceleration because it tends to drive the inhomogeneous terms to zero. The effects were more striking when one considers the compatibility relation at the wall, which indicates that the effects of unsteadiness are enhanced by both favorable pressure gradients and wall heating. (This can be confirmed by evaluating Eqs. (33) at $\bar{\eta} = 0$.) However, the numerical results demonstrate that this is a localized condition. The compatibility condition is a bound on the dissipation momentum at the wall; however, at any distance from the wall this bound is opposed by the convection of momentum. These are the two effects that influence the magnitude of $\bar{f}_1$, and the results demonstrate this trend.

Equations (33) and (34) are solved numerically, and the resulting parameters are given in Table 1 for a heated water boundary layer with an ambient temperature of 520°R. Moreover, the skin friction coefficient, C_f, the Nusselt number, Nu, the displacement thickness, δ_1, and the momentum thickness, δ_1, are written as

$$\frac{\delta_2}{x}\sqrt{\frac{Re_x(1 + m)}{2}} = \int_0^\infty (\bar{f}' - \bar{f}'^2)\, d\bar{\eta} = \delta_{2,0} + \varepsilon\delta_{2,1} \tag{35}$$

$$C_f\frac{\mu_e}{\mu_w}\sqrt{\frac{Re_x}{2(1 + m)}} = \bar{f}_0''(0) + \varepsilon_1\bar{f}_1''(0) \tag{36}$$

$$Nu = \sqrt{\frac{Re_x(1 + m)}{2}}\ \theta_0'(0) + \varepsilon_1\theta_1'(0) \tag{37}$$

$$\frac{\delta_1}{x}\sqrt{\frac{Re_x(1 + m)}{2}} = \int_0^\infty (1 - \bar{f}')d\bar{\eta} = \delta_{1,0} + \varepsilon\delta_{1,1} \tag{38}$$

B. RESULTS

In view of the previous discussion and that of Ref. 2, the results presented in Table 1 are not surprising. In fact, the results for $\beta = 0$ and $T_w - T_e = 0$ are identical to those given in Ref. 2. The new results are those for $\beta \neq 0$ and $T_w - T_e \neq 0$. In general, we can conclude that accelerating flows demonstrate increased skin friction coefficient and decreased Nusselt number. These conclusions are consistent with those presented in Ref. 2. Further, acceleration results in a decreased displacement thickness and a small increase in momentum thickness. The physical explanation for this is that acceleration causes an increase in the convection of momentum, and the wall shear must increase to dissipate this excess momentum.

Similarly for a nearly quasi-steady accelerating flow, the unsteady contribution to the heat transfer is affected principally by the thermal inertia, which resists heat transfer and leads to a decrease in Nusselt number. Similar findings have been reported in Ref. 13. There is a diminution of these trends as the pressure gradient becomes more favorable. Naturally, the contrary is true for decelerating flows.

A word of caution regarding flows with adverse pressure gradient: Note the relatively large magnitude of $\overline{f_1}''(0)$ in comparison to that of a flat plate. This implies that the unsteadiness correction is larger, and the region of validity of the two-term expansion may be more limited for adverse gradients.

The stability results were obtained by applying Eqs. (29) through (31) to determine the critical Reynolds number (based on displacement thickness, Re_{crit}). The effect of flow unsteadiness in the range $-.05 < \varepsilon < .1$ was considered for a variety of pressure gradients and surface overheat. These results are summarized in Table 2 and are illustrated in Figs. 1 through 6.

Using Re_{crit} as an indication of stability, it can be seen that for $\beta = 0$, and $T_w - T_e = 0$, the effects of unsteadiness are most significant. This parameter is reduced by 35 percent when ε is $-.05$, increased by 200 percent when ε is $+.05$, and increases by almost 600 percent when ε doubles from .05 to 1.

From Figs. 1 through 5, we observe that for those situations in which pressure gradient and surface overheat is small, even mild unsteadiness ($|\varepsilon| \leq .05$) can have powerful effects on the value of the critical Reynolds number and the location of the point of neutral stability. However, larger values of favorable pressure gradient and surface overheat which are already characterized by increased stability, exhibit less sensitivity to unsteadiness. From the viewpoint of the boundary-layer wall compatibility condition alone, this result is surprising. However, further reflection and reconsideration of the known effects of combined heating and pressure gradient on stability (Ref. 14) suggest that the situation may not be very different here. For combined heating and pressure gradient, it is known that the relative effect of additional surface overheat on stability decreases with increasing favorable pressure gradient or surface overheat. In fact, it has been shown that large values of overheat can decrease the stability of flat-plate and adverse gradient flows.

Note that the shape of the curves changes when ΔT increases. As suggested above, the larger the value of ΔT, the less sensitive Re_{crit} is to further heating. At $\beta = .5$, for example, Re_{crit} exhibits only weak dependence on further surface overheat and unsteadiness when $\Delta T > 30^\circ$.

Figure 5 shows an adverse pressure gradient case ($\beta = -.05$) in which the effects of unsteadiness are again significant. This arises because unsteadiness dominates the wall region, and stability is primarily controlled by details of the velocity profile between the critical layer and wall.

Figure 6 shows a collection of the results for the various pressure gradients, wall heating, and degree of unsteadiness, and a function of the shape factor $H = \delta_1/\delta_2$. A fair degree of correlation is demonstrated, indicating that H is to a rough approximation, a universal parameter that can be used to estimate the combined effects of these several parameters on stability. The line on this figure is the result of exact numerical computations (Ref. 15) for steady isothermal wedge flows. Similar computations for steady heated wedge flows in water (Ref. 14) have already demonstrated that H is a useful correlation parameter for the combined effects of pressure gradient and heating under conditions of moderate heating.

IV. CONCLUSION

Under conditions which are likely to occur in practice, the present nearly quasi-steady approach to the calculation of unsteady laminar boundary layer velocity profiles should be highly accurate. Our analysis of the stability of unsteady flow is not yet as precise as it will be after deeper study involving mutliple time scales. However, the heuristic arguments presented here suggest that there is an important practical flow regime where classical Orr-Sommerfeld analysis can still be used to define the stability characteristics of nearly quasi-steady velocity profiles. Our analysis and computation, performed on this basis, infer that even mild unsteadiness can have a powerful impact on flow stability. This impact is particularly strong at realistic levels of pressure gradient and heating in water.

REFERENCES

1. Rott, N.: "Theory of time-dependent laminar flows." in "Theory of Laminar Flows," F. K. Moore (ed.). Princeton University Press, pp. 395-438 (1964).
2. Moore, F. K.: "Unsteady laminar boundary layer flow." NACA Technical Note 2171 (1951).
3. King, W. S.: "Low frequency, large-amplitude fluctuations of the laminar boundary layer." AIAA Journal, 4, 6 (1966).
4. Riley, N.: "Unsteady laminar boundary layers." SIAM Review, 17, 2, pp. 274-296 (1975).
5. Ackerberg, R. C. and Phillips, J. H.: "The unsteady boundary layer on a semi-infinite flat plate due to small fluctuations in the magnitude of the free-stream velocity." J. Fluid Mech. 51, Part I, pp. 137-158 (1972).
6. Kaups, K. and Smith, A. M. O.: "The laminar boundary layer in water with variable properties." Douglas Aircraft Company Paper 3780, presented at Ninth ASME-AIChE Material Heat Transfer Conference, Seattle, Washington, (6-9 August 1967).
7. Shen, S. F.: "Some considerations on the laminar stability of time-dependent basic flows." Aerosp. Sci., 28, pp. 397-404, 417 (1961).
8. Betchov, R. and Criminale, W. O., Jr.: "On analytic solutions of the Orr-Sommerfeld equation," in "Stability of Parallel Flows," Academic Press, New York (1967).
9. Lin, C. C.: "The theory of hydrodynamic stability." Cambridge Univ. Press., Cambridge (1955).
10. Dunn, D. W. and Lin, C. C.: "On the stability of the laminar boundary layer in a compressible fluid." J. Aero. Sci., 22, pp. 455-477 (1955).
11. Aroesty, J. and King, W. S.: "Accurate and fast estimate of the stability of heated water boundary layers using a modification of Lin's method." Paper presented at Low-Speed Boundary-Layer Transition Wrokshop: II, Sant Monica, California, (September 13-14, 1976).
12. Moore, F. K.: "The unsteady laminar boundary layer of a wedge and a related three-dimensional problem." Proceedings Heat Transfer and Fluid Mechanics Institute, pp. 99-118 (1957).
13. Lighthill, M. J.: "The response of laminar skin friction and heat transfer to fluctuations in the stream velocity." Proc. Roy. Soc., A224, pp. 1-23 (1954).
14. Wazzan, A. R. and Gazley, C., Jr.: "The combined effects of pressure gradient and heating on boundary-layer stability and transition." Paper presented at Low-Speed Boundary-Layer Transition Workshop: II, Santa Monica, California, (September 13-14, 1976).
15. Wazzan, A. R., Okamura, T. T., and Smith, A. M. O.: "Spatial and temporal stability charts for the Falkner-Skan boundary-layer profiles." Douglas Aircraft Company Report No. DAC-67086 (September 1968).

Table 1.
Boundary-Layer Characteristics.

β	Tw - Te	$\overline{f}_0''(0)$	$\theta_o'(0)$	$\overline{f}_1''(0)$	$\theta_1'(0)$	$\delta_{1,0}$	$-\delta_{1,1}$	$\delta_{2,0}$	$-\delta_{2,2}$
0	0	0.46947	0	1.1999	0	1.218	.7273	.46917	.00597
	30	0.65699	-1.0161	1.4275	.31512	1.070	.6232	.4459	.0338
	40	0.72072	-1.0394	1.4919	.31156	1.028	.59127	.4378	.0406
	60	0.84796	-1.0826	1.6063	.30943	0.9525	.53284	.4216	.0513
0.1	0	0.58694	0	0.97274	0	1.080	.44490	.4354	.00304
	30	0.8062	-1.072	1.1533	.45502	0.9491	.39775	.4111	.0273
	40	0.8797	-1.0948	1.2046	.44761	.9115	.381542	.4029	.0331
	60	1.0252	-1.1371	1.2959	.43780	.8441	.35034	.3866	.0423
0.2	0	0.6866	0	0.8208	0	.9838	.30339	.4082	.00138
	30	0.932558	-1.11379	0.96722	.52867	.8632	.28058	.3836	.0230
	40	1.0143	-1.1365	1.0087	.5200	.8287	.2714	.3753	.0281
	60	1.1752	-1.178	1.0825	.507496	.7668	.2527	.3591	.0362
0.5	0	.92776	0	.57146	0	.8048	.1277	.3502	.00068
	30	1.2366	-1.2002	.62156	.62266	.7022	.12988	.3255	.0150
	40	1.3378	-1.2222	.64356	.61370	.6732	.12828	.3174	.0199
	60	1.53511	-1.2630	.68228	.59963	.6215	.1234	.3016	.0261
-0.05	0	.400123	0	1.3729	0	1.313	1.0028	.4903	.0077

Table 2.
Critical Reynolds number and shape factor as a function of understeadiness parameter ε, surface overhead, T_w-T_e, and pressure gradient parameter β

		Critical Reynolds Number, $(U_1\delta_1/\nu_e)_{crit}$				Shape Factor, $H = \delta_1/\delta_2$			
β	Tw - Te	ε = 0	ε = 0.05	ε = -0.05	ε = 0.1	ε = 0	ε = 0.05	ε = -0.05	ε = 0.1
0	0	514	1040	332	2940	2.59	2.51	2.67	2.44
	30	4770	10240	2050	18100	2.40	2.34	.246	2.28
	40	7270	13680	3450	21800	2.35	2.29	2.41	2.23
	60	11100	17900	6190	25700	2.26	2.21	2.31	2.16
0.1	0	1380	2750	794	5490	2.48	2.43	2.53	2.38
	30	9460	14490	5690	20300	2.31	2.27	2.36	2.23
	40	12270	17520	7980	23200	2.26	2.22	2.30	2.19
	60	15570	20470	11170	25710	2.18	2.15	2.22	2.12
0.2	0	2880	4800	1750	7610	2.41	2.37	2.45	2.34
	30	13100	17200	9540	21500	2.25	2.22	2.28	2.19
	40	15700	19700	12100	23800	2.21	2.18	2.24	2.15
	60	18300	21700	14800	25200	2.14	2.11	2.16	2.09
0.5	0	7860	9670	6300	11700	2.30	2.28	2.32	2.25
	30	18900	20800	16900	22700	2.16	2.14	2.17	2.13
	40	20600	22300	18800	24000	2.12	2.11	2.14	2.09
	60	21400	22800	19900	24100	2.06	2.05	2.07	2.04
-0.05	0	311	555	216	1681	2.68	2.58	2.78	2.48

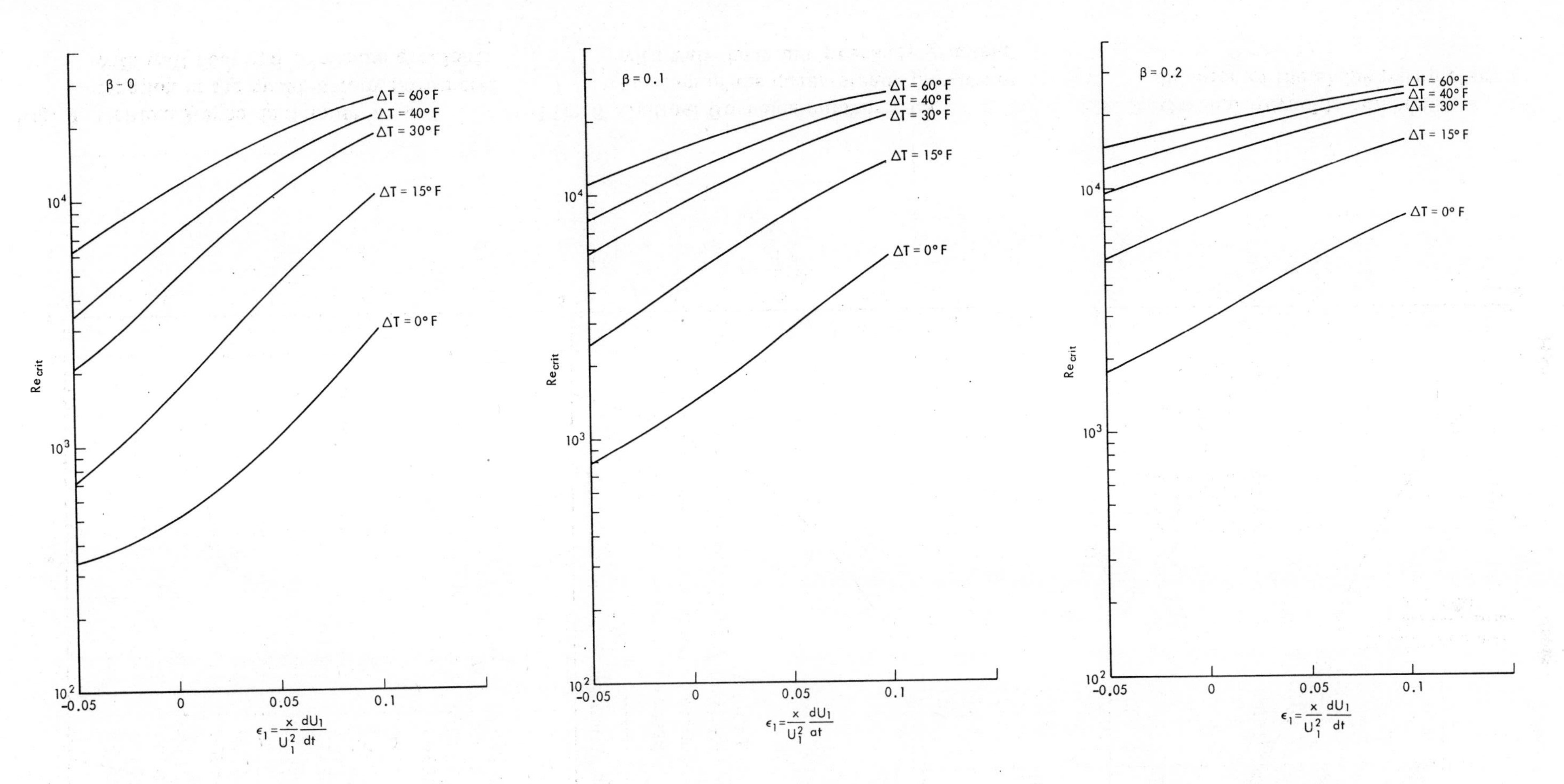

Fig. 1 Critical Reynolds number as a function of the quasi-steady parameter with wall heating and zero pressure gradient

Fig. 2 Critical Reynolds number as a function of the quasi-steady parameter with heat and pressure gradient

Fig. 3 Critical Reynolds number as a function of the quasi-steady parameter with wall heat and pressure gradient

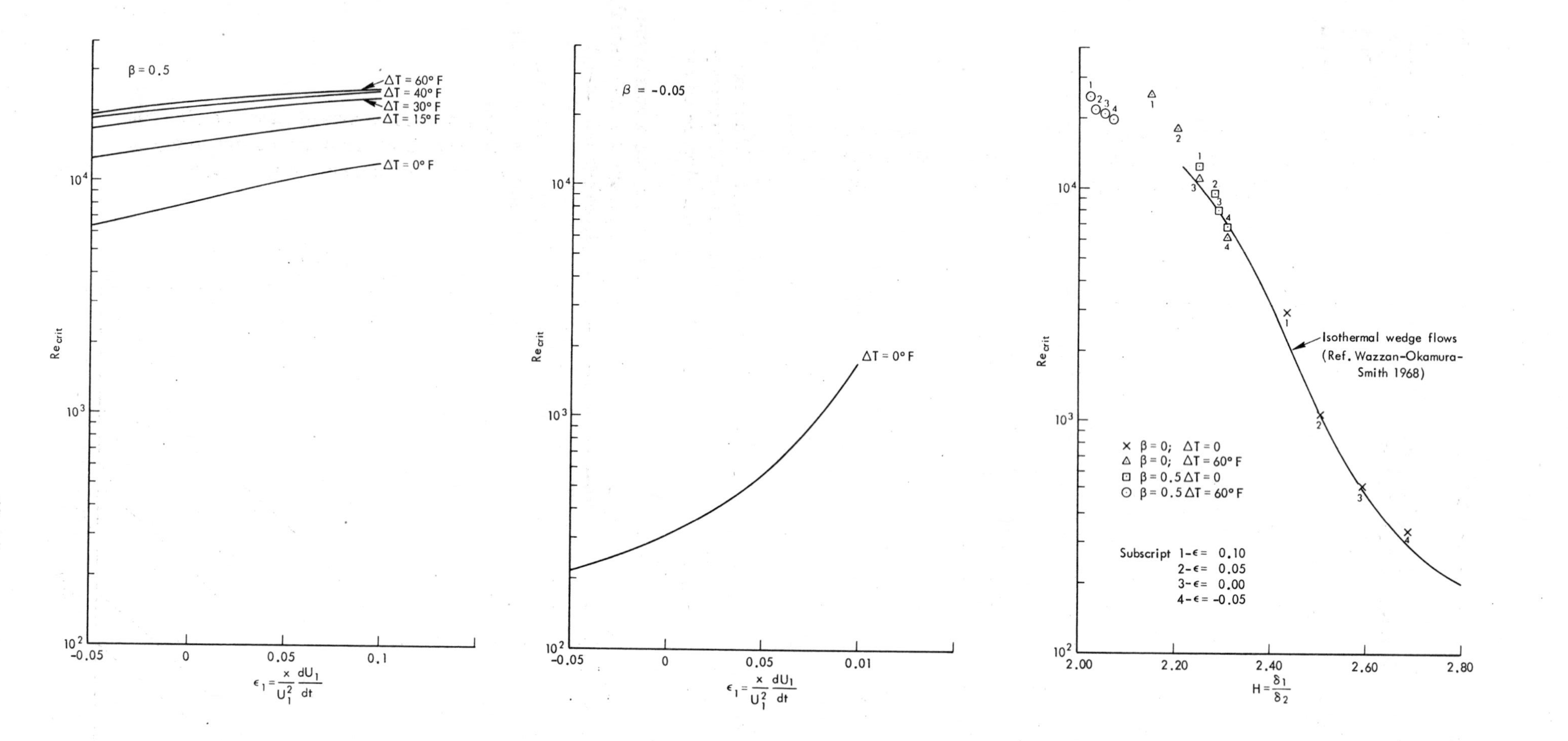

Fig. 4 Critical Reynolds number as a function of the quasi-steady parameter with wall heat and pressure gradient

Fig. 5 Critical Reynolds number as a function of the quasi-steady parameter with wall heat and pressure gradient

Fig. 6 Critical Reynolds number as a function of the shape parameter

Second International Conference on

Drag Reduction

August 31st - September 2nd, 1977

PAPER E5

AN ELECTROSTATICALLY DRIVEN SURFACE FOR FLEXIBLE WALL DRAG REDUCTION STUDIES

L.M. Weinstein,

NASA Langley Research Center, U.S.A.,

and

R. Balasubramanian,

Old Dominion University Research Foundation, U.S.A.

Summary

This paper describes the theoretical analysis, experimental development and surface motion data from an electrostatically driven flexible wall capable of producing high frequency, large amplitude, short wavelength standing wave motions. The driven wall was developed for use in studying flexible wall-turbulent boundary layer interactions.

The theoretical analysis considers the coupled electrochemical system using a continuum model. For the system under consideration the electrical field is uncoupled from the mechanical system leading to a simplified analysis. The structural problem is a non-linear (moderately) large amplitude problems and details of the analysis are presented.

Surface motion measurements, made using a modified schlieren type optical system, are discussed and details of the surface motion under different excitations are presented. Fair agreement is obtained between experimental measurements and theoretical calculations.

Held at St. John's College, Cambridge, England.

Organised and sponsored by BHRA Fluid Engineering, Cranfield, Bedford, MK43 0AJ.

Nomenclature

a	waveheight
b	width of electrostatic wall
c	wavespeed
h	thickness of membrane surface
ℓ	length of bay of membrane surface
m	mode number
t	time
w	deflection of membrane surface
E	Young's modulus of membrane surface
H_1, H	Height of air cavity; Distance from electrode to surface
L	maximum dimension of electrostatic wall
V	applied voltage
$a+$	waveheight in law of wall coordinated (non-dimensional height)
c_o	speed of light
$\hat{n}$	normal vector to deflected surface
$\tilde{u},\tilde{v}$	inplane deflections of membrane surface along x-y axis
$\underline{w}$	nondimensional deflection
w_1	perturbation deflection field
w_x, w_y	slopes of the deflected surface
$\tilde{E}$	electric field vector
$\tilde{E}_o$	primary field vector
$\tilde{E}_1$	perturbation field vector
E_x, E_y, E_z	electric field components in x, y, z axis
N_x, N_y	mid plane force in x, y axis
N_{xy}	mid plane shear
p_o^e	pressure load on the membrane surface
$\overline{p}_o^e$	non-dimensional pressure $(= p_o^e \frac{\ell}{Eh})$
${}^m p_{gen}^e$	generalized nondimensional pressure due to the mth mode
u_∞	free stream velocity
γ	nondimensional thickness $(= \frac{h}{\ell})$
θ	local angle, rad.
λ	wavelength
$\lambda+$	non dimensional wavelength in law of wall coordinates
ν	Poisson's ratio of membrane material
ρ	density of membrane material
τ	non-dimensional time $(= t/\ell\sqrt{E/\rho})$
υ	volume of air cavity
ω	angular frequency of applied a.c. voltage
Ω	non-dimensional frequency $(\frac{\omega t}{\tau})$

Introduction

Over the past decade, a large number of theoretical and experimental studies on the applicability of compliant or flexible walls for turbulent boundary layer drag reduction have been reported in the literature. Many of the experiments suffer from lack of repeatability. Also, in most of these experiments relatively little attention has been paid to the motion of the flexible wall (which probably is responsible for any favorable effects). Recently, Bushnell, et al. (1) have proposed a mechanism of favorable compliant wall interaction based on consideration of interruption of the "burst" cycle due to the modulations imposed by the moving wall. For favorable interruption of the burst cycle, the wall motion required (as suggested by the Bushnell model) has an amplitude a^+ of the order of 1-5 in law of the wall coordinates, a wavelength λ^+ of the order of 100 for a nondimensional wavespeed (C/u_∞) of 0.4. Experimental and theoretical studies conducted at Langley (2) indicate that at least for low speed, low dynamic pressure air experiments, "passive" walls are incapable of producing the required motion. A recent summary paper (3), suggests that low speed air experiments be conducted with controlled or active wall surface surface motion to assess the nature of possible turbulent boundary layer modifications due to the wall motion. The present paper discusses the design, construction and testing of such an active wall capable of producing two dimensional wall motions in the range suggested by the Bushnell model.

Previous studies using active walls have confined attention only to a limited range of wall motion parameters. Kendall (4) reported measurements on a wavy wall which had a wavespeed range between $-0.5u_\infty$ to $+0.5u_\infty$; the amplitude of the waves in law of the wall coordinates was of the order $a^+ \sim 600$, wavelength $\lambda^+ \sim 18000$. (corresponding to λ = 100mm, a = 3.17mm, u_∞ = 4.5 m/s). Mattout (5) reported experiments in water. The surface motion characteristics of his driven wall were $\lambda^+ \approx 17000$, $a^+ \simeq 430$ (i.e. λ = 41.2mm - 500mm; a = 1mm - 2mm, u_∞ = 1.5 m/s - 7.5 m/s.). Saeger and Reynolds (6) made measurements in a two dimensional fully turbulent channel flow again with wall parameters in the range of the other experiments. Merkulov and Yu Savehenko (7) also reported active wall experiments with the same range of surface motion parameters.

All of the above active wall experiments used mechanical drivers. (usually, camshafts for driving the surface at discrete points.) These drive systems are inadequate for producing the high frequency motion believed necessary and hence, an alternate system capable of producing these high frequencies is required. The present study describes an electrostatic wall designed to operate in a frequency range of 200 Hz - 10 KHz with two dimensional standing waves of wavelengths 1.8mm. A thin aluminized mylar membrane (thicknesses 6.3µm and 2.5µm) is the surface and a series of transverse electrodes etched on a P.C. board are the exciters. The surface is supported at discrete lines by transverse ribs. A detailed analysis of the electrostatic driven wall is presented herein along with the experimental surface motion data.

Analysis of the electrostatic wall system

The schematic of the electrostatic system is shown in Fig. 1a. The output from the transformer T is biased at the centre-tap and connected to terminals A and B as indicated. The conducting membrane surface of width 'b', thickness 'h', and isotropic properties (Young's modulus = E, density = ρ , Poisson's ratio = ν.) is structurally supported at separations of length 'ℓ' where $\ell \ll b$; the surface is electrically grounded. The equally spaced electrodes, at a distance H from the membrane surface, are sequentially connected to terminals A and B. Each periodic bay of the conducting surface has a sealed air cavity of volume v ($H_1 \times \ell \times b$) beneath it.

The electrostatic wall configuration has been designed to achieve an operating frequency range of 200 Hz - 10 KHz. The largest dimension of the models examined is 0.4m. Under these design conditions the non-dimensional speed ($\omega L/c_o$) is $\ll$ 1; therefore, the field between the wall and the electrodes can be considered as quasistatic.

For the electrostatic wall the electric field equations are only weakly coupled to structural motions. The assumption is made that the electric field, E, and the deflection, w, can be uncoupled through the following relations

$$\tilde{E} = \tilde{E}_o + \tilde{E}_1 (w, \tilde{E}_o)$$

$$w = w(\tilde{E}_o) + w_1(w(\tilde{E}_o), \tilde{E}) \qquad \text{(i)}$$

$$\text{where } \tilde{E}_1 << \tilde{E}_o \text{ and } w_1 << w(\tilde{E}_o)$$

The analysis of the primary field is carried out in appendix A. For voltage distribution V_A and V_B (Fig. 1b) it is shown that the electric field distribution at the membrane surface is given by,

$$E_x = E_y = 0$$

$$E_z = \frac{H}{\pi\varepsilon} f(x) \qquad \text{(ii)}$$

where $f(x) = Q_1 g_1(x) + Q_2 g_2(x)$

and $$g_1(x) = \frac{1}{(\ell/4 - x)^2 + H^2} + \frac{1}{(3\ell/4 + x)^2 + H^2} + \frac{1}{(5\ell/4 - x)^2 + H^2} + \frac{1}{(9\ell/4 - x)^2 + H^2} + \frac{1}{(7\ell/4 + x)^2 + H^2}$$

$$g_2(x) = \frac{1}{(x + \ell/4)^2 + H^2} + \frac{1}{(3\ell/4 - x)^2 + H^2} + \frac{1}{(5\ell/4 + x^2) + H^2} + \frac{1}{(7\ell/4 - x)^2 + H^2} + \frac{1}{(9\ell/4 + x)^2 + H^2} \qquad \text{(iii)}$$

and

$$Q_1 = (c_1 + 2c_2) V_A + (c_1 - 2c_2) V_B$$

$$Q_2 = (c_1 + 2c_2) V_A + (c_1 - 2c_2) V_B.$$

where $\qquad$ (iv)

$$c_1 = \frac{\pi\varepsilon}{\cosh^{-1} 8H^1/\ell}, \quad c_2 = \frac{2\pi\varepsilon}{\cosh^{-1} (2)} \quad \text{and} \quad H^1 = H + \ell/8$$

The force density causing primary deformation of the membrane can be estimated as

$$p_o^e = \frac{\varepsilon_o \varepsilon_r E_z^2}{2} \qquad \text{(v)}$$

where $\varepsilon_o \varepsilon_r$ is the dielectric constant of the intervening medium between the electrodes and the wall.

The undamped structural motion of the simply supported periodic bay is governed by the dynamic equation of motion,

$$\rho h \frac{\partial^2 w}{\partial t^2} + \frac{Eh^3}{12(1-\nu^2)} \left(\frac{\partial^2}{\partial x^2} + \frac{\partial^2}{\partial y^2}\right)^2 w - \left(N_x \frac{\partial^2 w}{\partial x^2} + N_y \frac{\partial^2 w}{\partial y^2} + N_{xy} \frac{\partial^2 w}{\partial x \; \partial y}\right) = p_o^e \qquad \text{(vi)}$$

where N_x, N_y, N_{xy} are the midplane forces. The membrane is assumed to have zero initial tension; the presence of the air cavity under the membrane and the specific nature of the primary field $\widetilde{E}_o$, dictates a choice of the deflection shape for the membrane as

$$w_o = \sum_m w_m = \sum_m f_m(t) \sin \frac{2\pi\, mx}{\ell} \sin \frac{\pi y}{b} \qquad \text{(vii)}$$

In Appendix B, the midplane forces are obtained as,

$$N_x = \frac{Eh}{8(1-\nu^2)} \sum_m n_m^2\, f_m^2(t)\, [(1+\nu\beta_m^2) - (1-\nu^2) \cos \frac{2\pi y}{b}]$$

$$N_y = \frac{Eh}{8(1-\nu^2)} \sum_m n_m^2\, f_m^2(t)\, [(\nu + \beta_m^2) - (1-\nu^2)\, \beta_m^2 \cos \frac{4\pi mx}{\ell}] \qquad \text{(viii)}$$

$$N_{xy} = 0$$

where

$$\beta_m = \frac{\ell}{2mb}, \qquad n_m = \frac{2\pi m}{\ell}. \qquad \text{(ix)}$$

The equation (vii) is non-dimensionalised with

$$\tau = \frac{t}{\ell}\sqrt{\frac{E}{\rho}}, \quad \Omega = \frac{\omega\tau}{t}$$

$$a_m = \frac{f_m(t)}{\ell}, \quad \gamma = \frac{h}{\ell}, \quad \bar{p}_o^e = \frac{p_o^e\, \ell}{Eh} \qquad \text{(x)}$$

The modal analysis of (vii) yields the following system of equations.

$$\frac{d^2 a_m}{d\tau^2} + a_m \,[G_m \sum_n n^2 a_n^2 R_n + K_m] = m_{p_{gen}}$$

$$m = 1,2\ldots\infty \qquad \text{(xi)}$$

where $m_{p_{gen}} = \frac{4}{\ell b} \int_0^\ell \int_0^b \bar{p}_e \sin \frac{\pi y}{b}\, dy \sin \frac{2\pi mx}{\ell}\, dx,$

$$G_m = \frac{2\pi^4 m^2}{(1-\nu^2)},$$

$$R_n = 1 + 2\nu\, \beta_n^2 + \beta_n^4 + \frac{1-\nu^2}{2} - \beta_n^4\, \delta_{mn} \frac{(1-\nu^2)}{2},$$

$$\delta_{mn} = 1, \; m = n$$

$$\delta_{mn} = 0, \; m \neq n \qquad \text{(xii)}$$

and $$K_m = 16\, \gamma \frac{\pi^2 m^4 \,{}^2 (1+\beta_m^2)}{12(1-\nu^2)}$$

The set of ordinary differential equations (xi) can be solved for 'a' using a numerical procedure (variable order Adam-Bashforth scheme). With the knowledge of a the deflection w_o is fully known.

The perturbation to the electric field $\tilde{E}_1$ due to known w is shown to be of the form

$$\begin{aligned} \tilde{E}_x &\approx w_x \tilde{E}_{oz} \\ \tilde{E}_y &\approx w_y \tilde{E}_{oz} \\ \tilde{E}_z &\simeq (w_x^2 + w_y^2)^{1/2} \tilde{E}_{oz} \end{aligned} \qquad \text{(xiii)}$$

in appendix C. When w_x and w_y are less than 0.1 the modified field does not cause any change in the estimation of 'w', and hence is safely ignored in subsequent analysis.

Approximate technique for design of the electrostatic wall.

An investigation of the equation system (xi) indicates that

$${}^1p_{gen} < {}^2p_{gen}$$

and

$$G_1 \, a_1 R_1 + k_1 << G_2 \sum_{n=1}^{2} (n^2 \, a_n^2 R_n) + k_2$$

and for $\ell/b << 1$, $\beta_m^2 << 1$. (xiv)

Thus for preliminary design purposes it is sufficient to take a one term approximation for w. Thus

$$w = w_1 \sin\frac{2\pi x}{\ell} \sin\frac{\pi y}{b} \qquad \text{(xv)}$$

The equation for w is now given as,

$$\frac{d^2a}{dt^2} + a\,[G_{11}a^2 + k_1] = {}^1p_{gen} \qquad \text{(xvi;a)}$$

where

$$k_1 = \frac{16\,\gamma^2\pi^4}{12(1-\nu^2)},$$

$$G_{11} = 2\pi^4 \left(\frac{1}{(1-\nu^2)} + \frac{1}{2}\right) \qquad \text{(xvi;b)}$$

For a voltage form at the terminals

$$\begin{aligned} V_A &= V_o + V_1 \sin\omega t \\ V_B &= V_o - V_1 \sin\omega t \end{aligned} \qquad \text{(xvii)}$$

the generalised force is of the form

$${}^1p_{gen} = p_1 \sin \Omega\tau \qquad \text{(xviii)}$$

where p_1 is a complicated function of H and ℓ and is given as,

$$p_1 = \frac{6400 * \varepsilon * v_1 v_0 (H/\ell)**2 * c_1 * c_2}{9.81 * \pi * Eh\ell} \text{ (FX)} \qquad \text{(xixa,b)}$$

and

$$\text{(FX)} = 10^{-5} \int_0^1 g_1(\tfrac{x}{\ell})\ g_2(\tfrac{x}{\ell})\ \sin\frac{2\pi x}{\ell}\ d(x/\ell)$$

ε is given in pf/m, v is given in kv. Using a simple program we have evaluated Fx for two design

(i) $H/\ell = 0.070$ with $Fx = 0.175410$
(ii) $H/\ell = 0.168$ with $Fx = 0.0106204$

Finally, equation (xvia) is modified using the following transformations

$$\theta = \Omega\tau\ , \quad u = \frac{a\sqrt{G_{11}}}{\Omega}$$

$$\beta_1 = k_1/\Omega^2, \quad p_2 = p_1\sqrt{G_{11}}/\Omega^3 \qquad \text{(xx)}$$

leading to the following equation

$$\frac{d^2u}{d\theta^2} + \beta_1\ u + u^3 = p_2\ \cos\theta \qquad \text{(xxi)}$$

The stability of periodic solutions for a given family of β_1, p_2 can be studied by investigation of the nature of the Hill's equation associated with (xxi). In the case of the walls we have designed we find that the periodic solutions of (xxi) are stable solutions for the domain of interest.

Preliminary designs

Case 1. The seperation distance 'H' between the electrodes and the membrane was H = 0.127 mm

Case 2. For this case the seperation distance H was H = 0.305 mm.

For both cases the membrane used was thin mylar (a)h = 6.35μm, (b)h = 2.54μm whose dimensions and properties were

$E = 3.59 \times 10^7$ kgf/m^2

$\rho = 138$ kg/m^3

$\nu = 0.3$

$\ell = 1.814$mm

$b = 203.2$ mm

The harmonic solution of (xxi) for both cases leads to a simplified cubic equation, i.e.,

$$(\beta_1 - 1)\quad u + 3/4\ u^3 = p_2 \qquad \text{(xxii)}$$

Model Design

1. Geometry selected

The support distance or the wavelength was selected as $\ell = 1.814$mm, as a compromise to the scale suggested by the Bushnell criterion. A threaded rod with 14 threads per inch was used to wind the threadings in the fabrication of the model. The seperation between the electrodes and the membrane was either 0.127mm or 0.305mm and this was dictated by the diameter of the nylon threads used as membrane supports. The overall model sizes were chosen so that drag measurements can be made in a small, low speed windtunnel facility at a later date. The largest model had a dimension of 0.4m long x 0.2m wide.

2. Wall materials

In order to obtain a reasonable range of amplitudes, the following materials given in order of increasing E, were chosen as likely candidates for wall material:

1) Rubber (Neoprene)
2) low modulus plastics (i.e., alathon, polyethelyne)
3) Mylar.

Even though the modulus of mylar was fairly higher than desired, it was included in the testing for reasons that it was available in thicknesses of 6.35μm and 2.54μm, and is easily aluminised, with a good optical quality.

The theoretical amplitudes are shown in fig. 2 as a function of material thickness, for a 0.300mm spacer configuration, with the applied voltage difference at the maximum operating level.

3. Voltage Limitations

The electrostatic wall system breaks down when arcing occurs in the air gap between the wall and the electrodes. The r.m.s. breakdown voltage gradient in air as given by Paschen's law is 3.1KV/m. The voltage distribution giving the maximum force density on the membrane is $V_0 = V_1$. Thus the operation voltage for the set up is

$$V_0 = 3.1\ H\ KV,$$
$$V_1 = 3.1\ H\ KV, \text{ where H is in mm.}$$

If a gas such as SF_6 were used in place of air the breakdown voltage can be raised to $V_1 = V_0 \simeq 6H$ KV.

SF_6 was used in some experiments but the quality of the membrane surface available to us was not perfect (contained pinholes) and the SF_6 diffused out in a fairly short time.

Fabrication Technique

The array of electrodes was etched on a printed circuit board, as shown in Fig. 3. The p-c boards were coated with either epoxy (which had to be degassed to avoid air bubbles and hence potential arcing situations between electrodes) or lacquer to avoid breakdown between the electrodes. Vent holes were drilled in the p-c board between each pair of electrodes. The p-c board was then placed on a holder with threaded rods on two sides, and the nylon string were wound such that the thread went between each second electrode, and the drilled holes were midway between the nylon threads. The nylon threads were bonded by spraying a coat of lacquer over the board and an epoxy border was constructed at the edge of the model. The surface of the p-c board was lightly coated with epoxy (or in some cases with thick grease of thickness ≈ 25.4μm) and the membrane, held in a frame, was lowered to the board. The holes in the p-c board allowed the trapped air in each chamber to escape until the surface was straight between the supports. The vent holes were then covered with tape to trap the remaining volume of air, and the membrane was taped at the edges.

Experimental procedures

1. Optical set up

The optical system used to examine the models is shown in Fig. 4. This is similar in principle to a schlieren set up, with a graded filter giving a reasonable dynamic range. The system was calibrated by using a rotating mirror. The calibration curves are shown in Fig. 5. The optical system has a dynamic range of 0.078 rad. (±0.039 rad.) and is reasonably linear (repeat scans give slightly different curves, hence part of the nonlinearity of 5b is due to signal noise.) The r.m.s. noise corresponds to ≈ 0.001 rad, limiting readings to this level. The set up was adjusted so that for zero tilt (level surface) the reflected light is in the centre of the filter, and the average d-c level on the model is adjusted to give 0.018 rad./volt on the oscilloscope. The surface quality of the models were not perfect and hence the uncertainty of data is in the range $\pm 25\%$.

The entire optical system was placed in a motor driven box, on a track, to obtain the variation of angle with axial position. If the low frequency part of the filter response were plotted with position, the average surface quality is obtained. Some typical surface profiles are shown in Fig. 6. Fig. 6a shows about 3 structural wavelengths; the membrane had partially collapsed due to leaks in the model and temperature changes. Fig. 6b shows surface quality of early designs; even though they are apparently not very smooth, the conversion of angles to height indicate generally less than 3μm RMS variations.

The optical system could also show the high frequency motion (up to 10KHz) at different fixed points, or by moving the system slowly along the model the envelope of motion with position could be measured. Fig. 7 shows typical photographs; 7a is at a fixed point, while 7b is the envelope of several waves. Comparison of the waveform obtained at one point support the theoretical assumption that

$$w \approx w_1 \sin \omega t \sin \frac{\pi y}{b} \sin \frac{2\pi x}{\ell}$$

The envelope photograph shows a generally similar waveform.

Experimental Data

The optical system was used to determine the following surface motion characteristics for several models:

(1) uniformity of amplitude
(2) quality of the surface
(3) local maxima of surface angles.

Test Results and Discussions

The theoretical response curves for the two design cases are shown in Fig. 8, 9 and 10. Fig. 8 shows the deflection amplitudes for Case 1 (H = 0.127mm), with the applied voltage variation $V_1 = V_0$ = 30V to 350V. The breakdown p.d. for this configuration is about 350 volts. Curve A is the response curve for a 6.25μms (1/4 mil) thick mylar and curve B is the response curve for 2.5μms thick (1/10 mil) mylar. The applied voltage has an a.c. frequency of 300 Hz. Fig. 9 shows the curves for a 0.305mm gap configuration (Case II of design) at 300Hz a.c. voltage. The voltage is varied from 85V - 900V. The breakdown p.d. for this configuration is at about 900 volts. In Fig. 10 the frequency response on the variation of amplitude with frequency for case 1 with a given potential difference applied to the wall, is shown for the 2.5μms thick mylar. This structure has a natural frequency at about 1400 Hz. The curves for V = 85 volts and 250V show that the resonant frequency is shifted away from the natural frequency as the applied load level increases indicating a hard spring type behavior. The 6.25μms thick mylar has a natural frequency of 3500 Hz and hence the variations in amplitude with frequency are negligible at the range of tests that were conducted. Fig. 11 shows the variation of amplitude of surface motion at a typical wave of the surface, with applied voltage. The a.c. component of the applied voltage has a frequency of 300 Hz. The line shows the theoretically computed values of amplitude. At low voltages there is a slight disagreement between the theoretical and experimental values. This can be ascribed to problems of resolution and the overall accuracy of the measurements. Fig. 12 shows the variation of amplitude with frequency of the same configuration for various applied voltages. The rolloff shown in this figure is obviously in disagreement with theory. Hence an effort was made to check the accuracy of measurements. The frequency response of the amplifier measuring the output of the optical system was found to roll off past 400 Hz which was the cause of the data roll off. Stray cable capacitance and the high resistance of the phototube were the cause of such a large time constant for the system. For later data the time constant was made small enough so that accurate measurements up to 10KHz could be made.

For the 0.127mm spacer design two sets of measurements were made (i) using 6.25μm thick mylar as the membrane and (ii) using 2.54μm thick mylar as the membrane. From the point of view of obtaining reasonably uniform waveforms and of fairly large amplitudes this design was thought to be more appropriate (smaller air gaps ensures better satisfaction of the incompressibility of air pocket.) The variation of amplitude with frequency is shown in figures 13 and 14 for the 2.5μms, 6.25μms thick mylar membranes at various voltages. Fig. 13 shows that for the 2.5μms thick mylar there is a slight gain in amplitude with frequency at the applied voltages of 60, 100 and 150 volts, and the trend continues well up to 6KHz. The theoretical curve shown in Fig. 10, however, predicts that the response curve should have traversed through the resonant frequency within this range of frequencies. The obvious disagreement between the two is attributable to the fact that the simply supported end conditions are not fully met in the model. It is very likely that the ends are more nearly fixed than simply supported. The response curve shown in Fig. 14 indicates larger scatter in the data than any we have discussed above. The amplitudes are in some cases about 1/5 of that predicted by the theory. Here the disagreement is mainly attributable to residual pretensions in the structure, limiting the response. In many of the models that we tested we did find that some models agreed better with the theoretical predictions than the others and we have qualitatively identified the ones which performed poorly as the ones with large unknown pretensions in the structure. However the theory does give a useful upper limit for the zero pretension surface amplitudes.

The variation with applied voltages for the 2.5μm and 6.25μm thick mylar surfaces are shown in Figs. 15 and 16. The data in Fig. 16 show large deviations from that predicted by the theory. The curves in Fig. 15 are in excellent agreement with theory.

Although the preceeding discussion was restricted to standing waves with wavelength 1.8mm, it is not certain that these type waves, or this wavelength will give the favorable interaction with turbulent flow hoped for. It may be that true traveling waves are needed rather than standing waves, or that yet smaller length waves are needed. The same type of electrostatic driver system can be used to make traveling waves (supported every few wavelengths to hold the fixed volume) if four drivers, with 90° phase shifts, are used. Also, the wavelength can be made smaller, the size being limited only by fabrication techniques. The model developed herein is the only active or driven wall approach, known to the present authors, capable of short wavelength, moderate amplitude high frequency motion.

Conclusions

An electrostatically driven wall, capable of producing moderately large amplitudes at short wavelengths and at large frequencies, has been built and tested. The theoretical analysis shows excellent agreement with the laboratory experiments. The electrostatic wall can now be used as an active wall for drag reduction experiments. Many of the fabrication difficulties have been solved in the later models, and experiments show promising wall motion performance.

List of References

1. R. L. Ash, D. M. Bushnell, L. M. Weinstein and R. Balasubramanian; "Compliant Wall surface motion and its effect on the structure of a turbulent boundary layer". Fourth Biennial symposium on turbulence in liquids, University of Missouri, Rolla (1975).

2. R. Balasubramanian: "Analytical and numerical investigation of structural response of complaint wall materials". Final Report Supplement I, Old Dominion University, Norfolk, (April 1977).

3. D. M. Bushnell, J. N. Hefner and R. L. Ash: "Effect of compliant wall motion on turbulent boundary layers", I.U.T.A.M., Symposium on Structure of Turbulence and Drag reduction, Wash. D.C., June 1976.

4. J. M. Kendall: "The turbulent boundary layer over a wall with progressive surface waves", J. Fluid Mechanics, 41, 2, pp.259-281, (1970).

5. Mattout R. B. Cottenceau: "Etude Experimentale D'Une Paroi Souple Activee En tunnel Hydrodynamique Mesures Globales", Tn. No. 71-C1-09, Society Bertin & Cie. (1971).

6. Saeger, J. C. and W. C. Reynolds: "Perturbation Pressures over Travelling Sinusoidal Waves with Fully Developed Turbulent Stress Flow". Technical Report FM-9, Dept. of Mech. Eng, Stanford Univ. (1971).

7. V. I. Merkulov, Yu, I. Savchenko. "Experimental Investigation of Fluid Flow Along a Travelling Wave." Hydradynamic Problems of Bionics, Bionica, Keiv, 4, p.3-120, (1970).

8. V. V. Bolotin: "The Dynamic Stability of Elastic Systems", Ed. Julius J. Brandstatter, Holden-Day Inc. (1964).

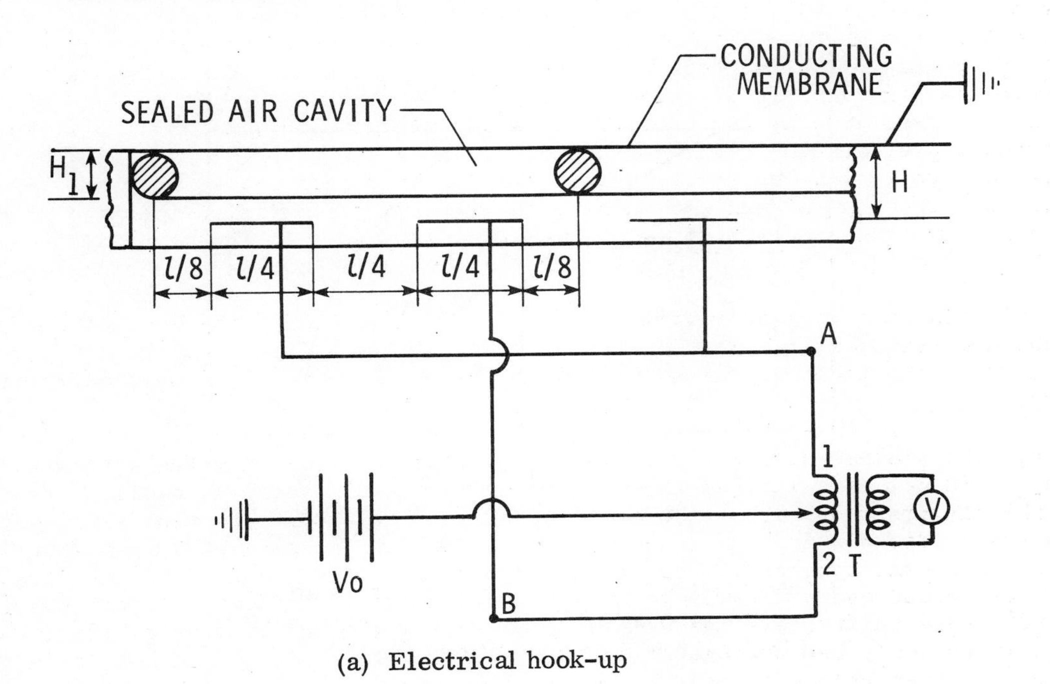

(a) Electrical hook-up

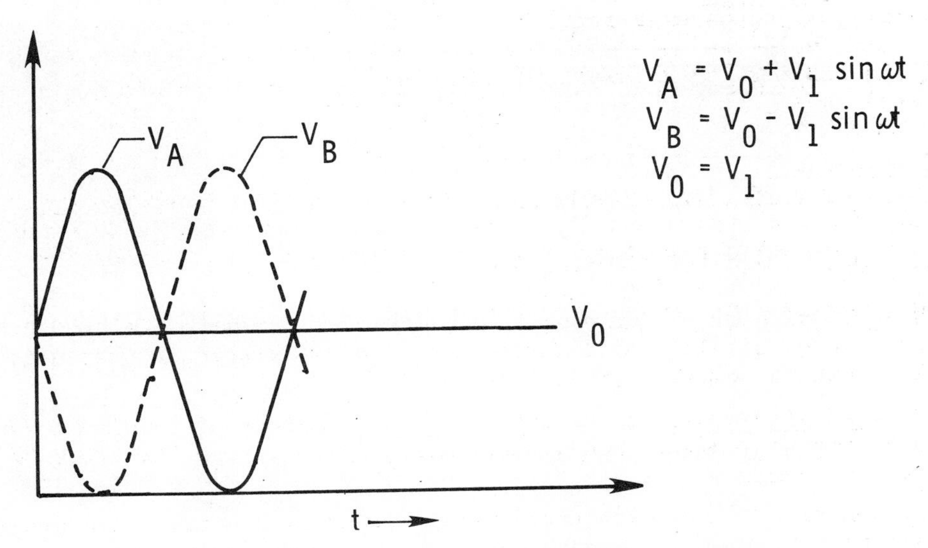

(b) Voltage distribution in the terminals A and B

Fig. 1. The electrical arrangement of the electrostatic wall system.

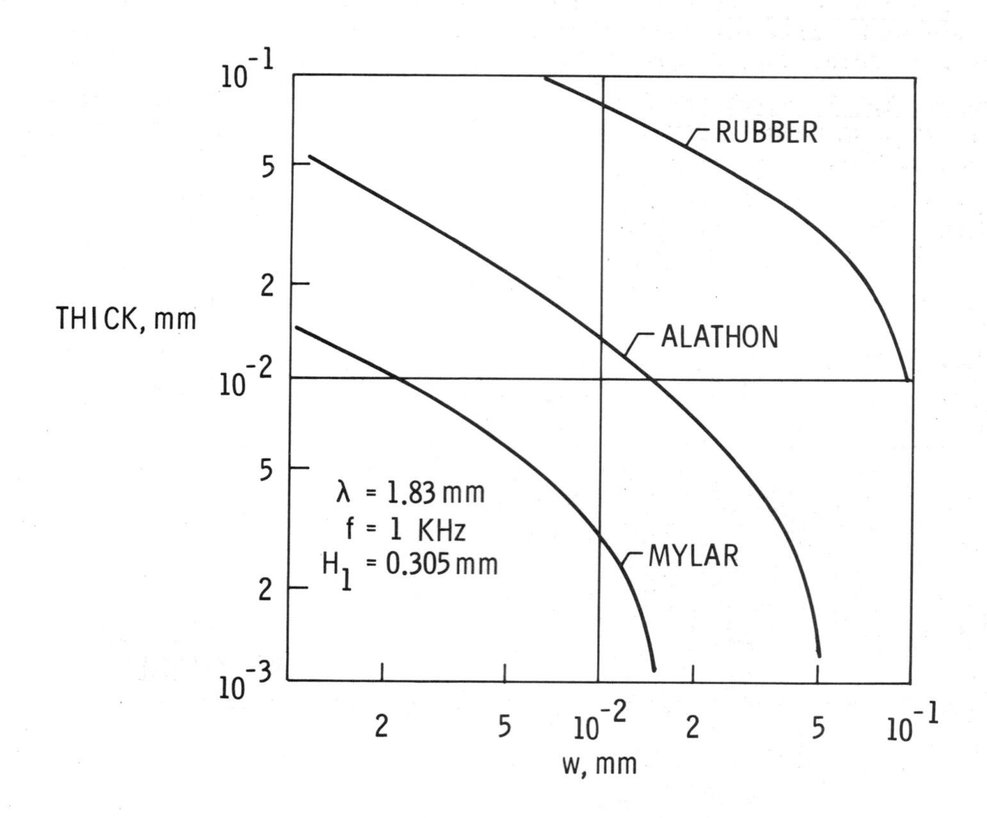

Fig. 2. Theoretical estimate of maximum amplitude possible for different thicknesses at the maximum design voltage in air.

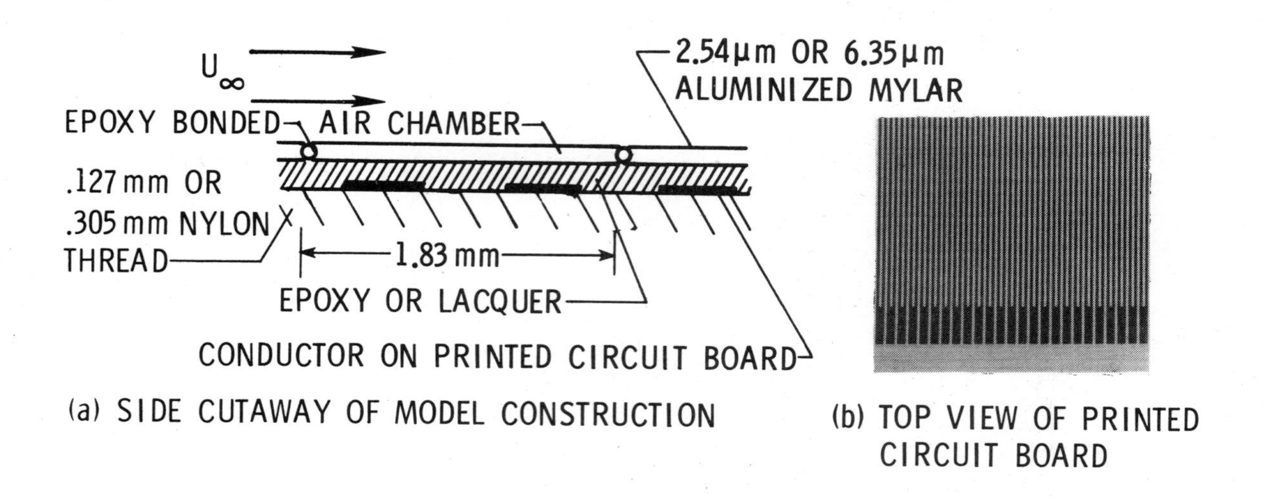

Fig. 3. Model construction details.

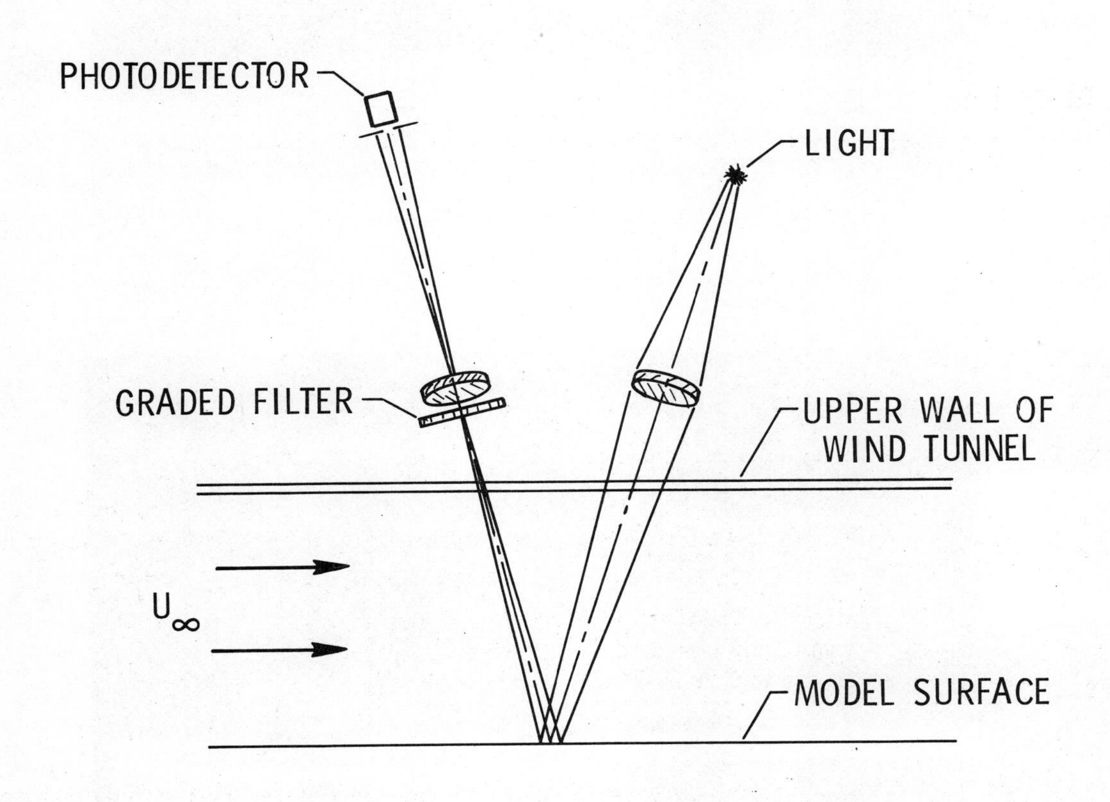

Fig. 4. Optical system to examine surface motion.

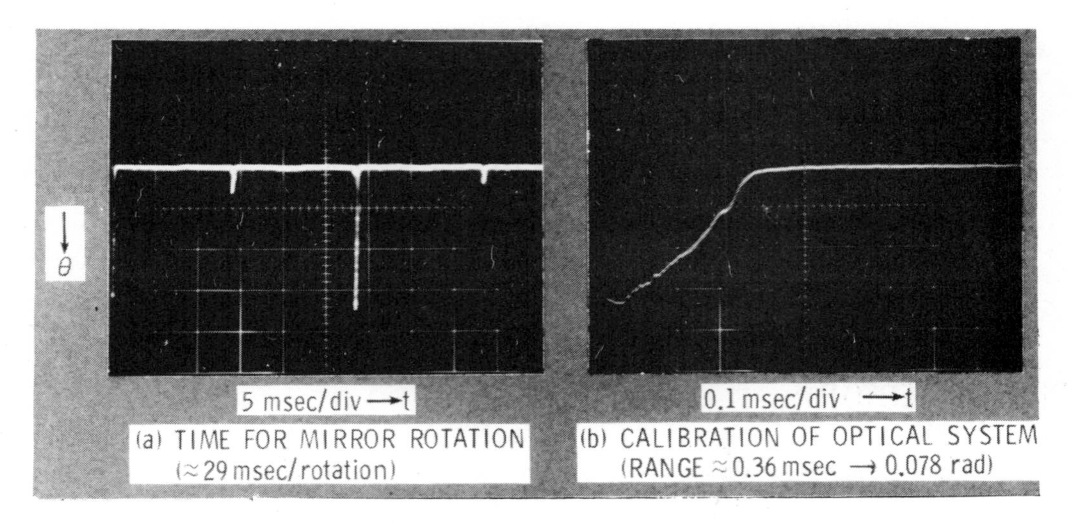

Fig. 5. Response of optical system.

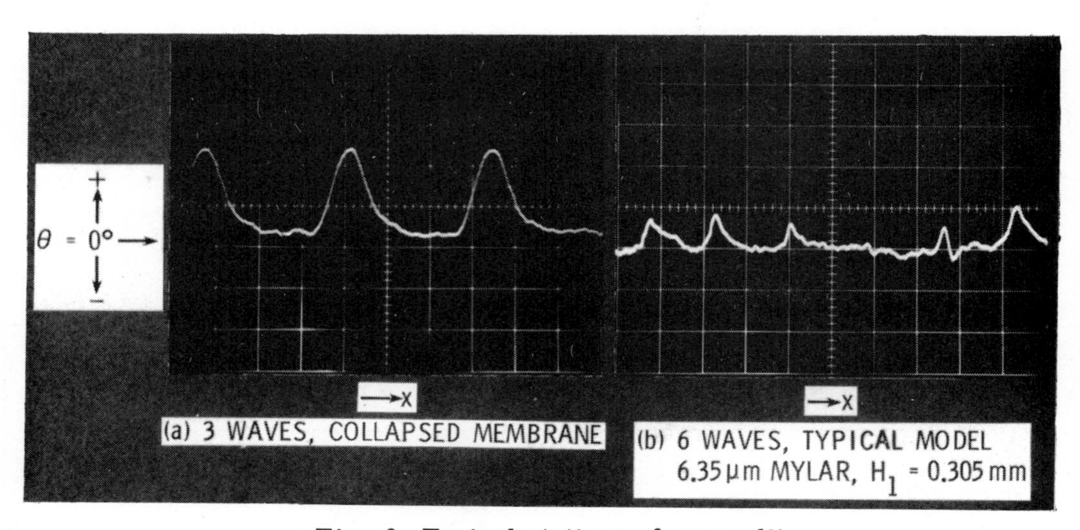

Fig. 6. Typical static surface quality.

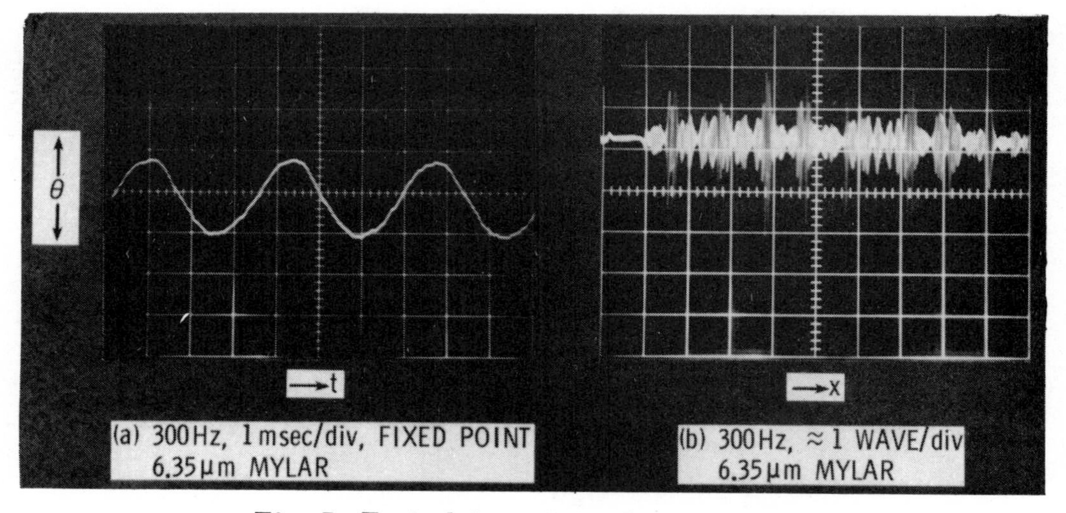

Fig. 7. Typical dynamic surface response.

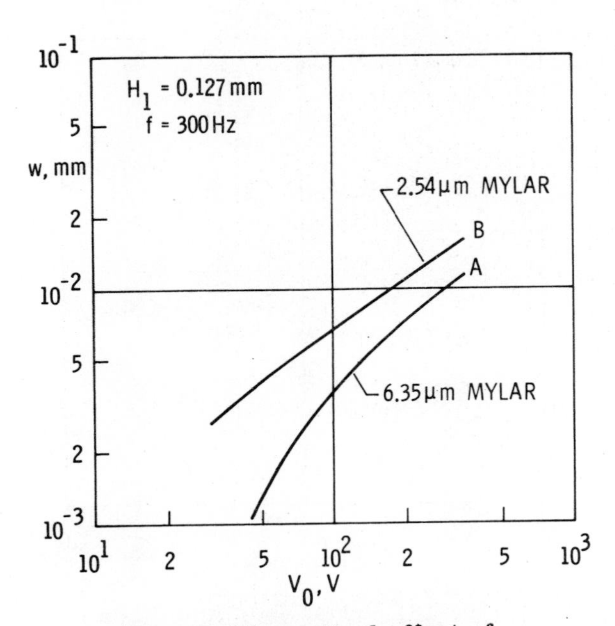

Fig. 8. Theoretical effect of voltage on amplitude.

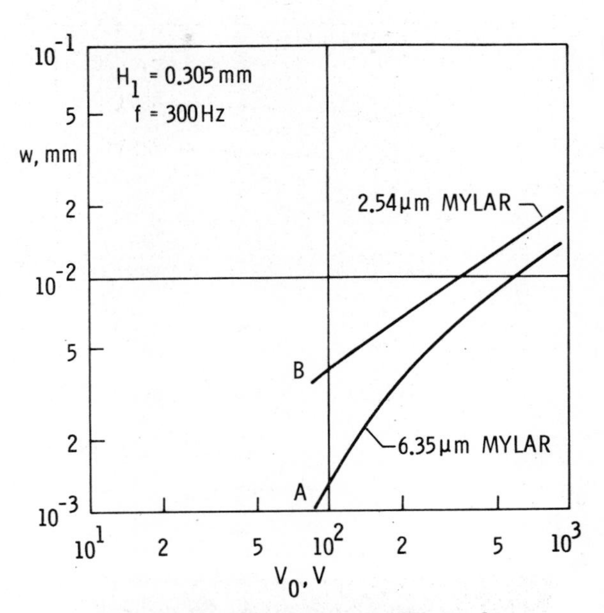

Fig. 9. Theoretical effect of voltage on amplitude.

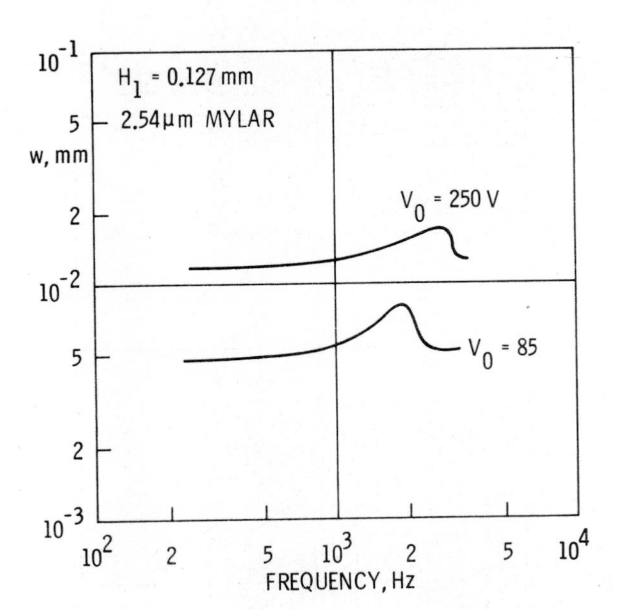

Fig. 10. Theoretical variation of amplitude with frequency.

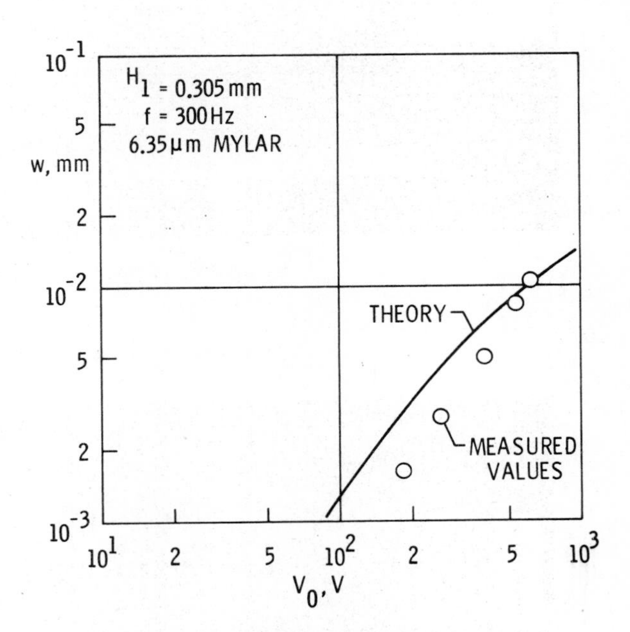

Fig. 11. Variation of amplitude with voltage.

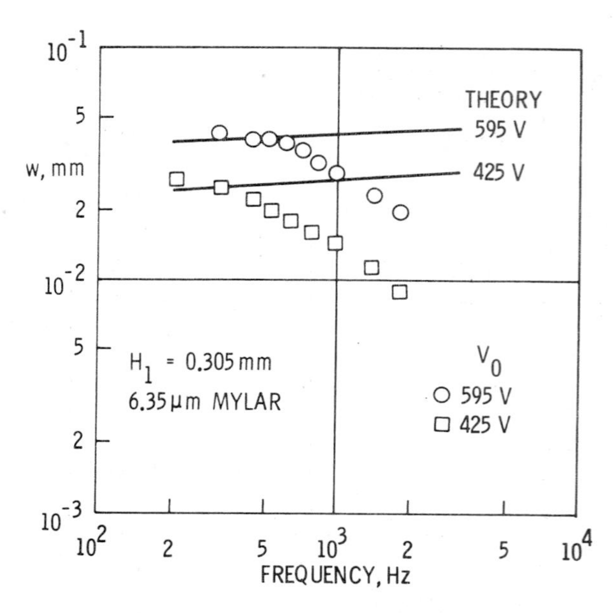

Fig. 12. Variation of amplitude with frequency.

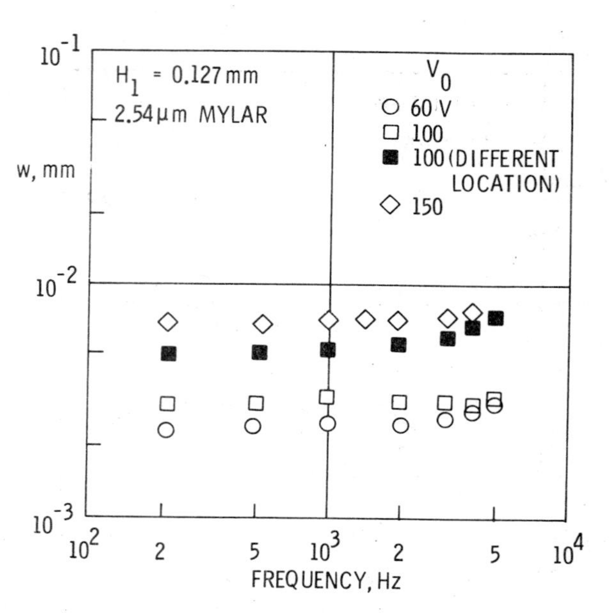

Fig. 13. Variation of amplitude with frequency.

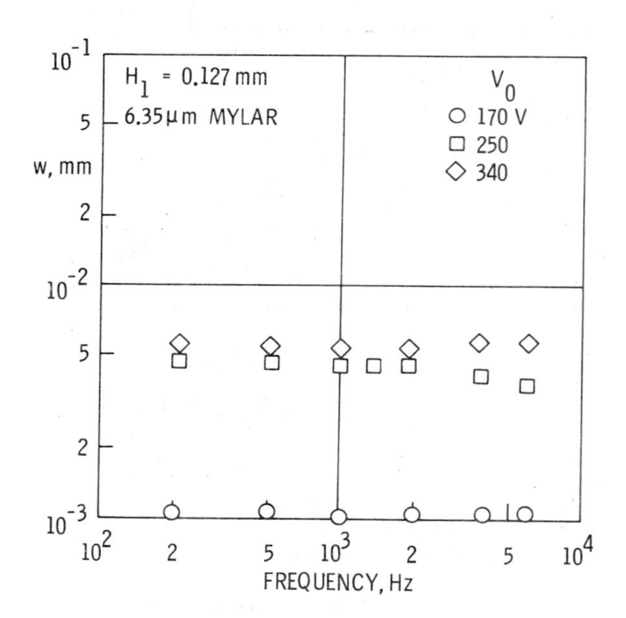

Fig. 14. Variation of amplitude with frequency.

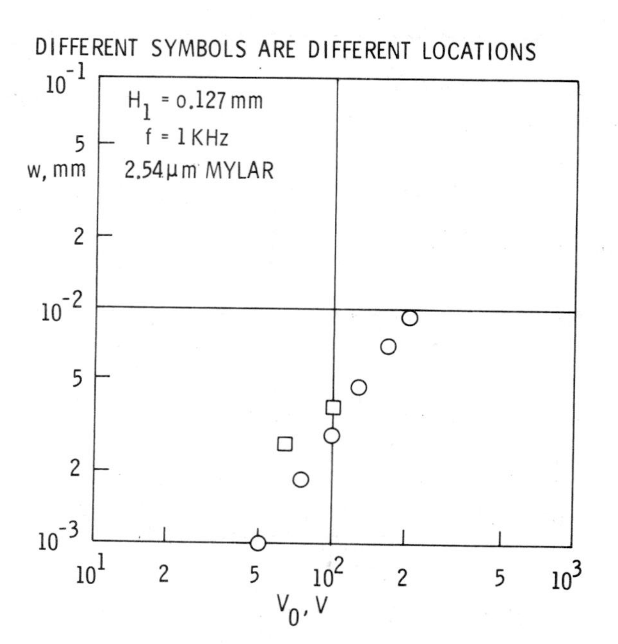

Fig. 15. Variation of amplitude with voltage.

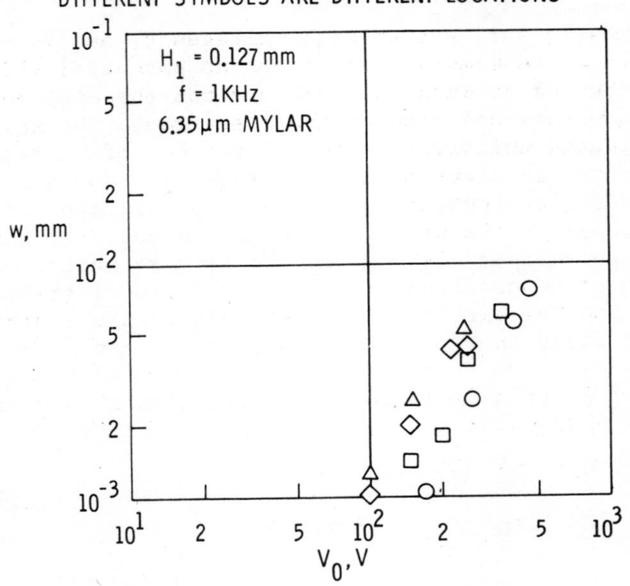

Fig. 16. Variation of amplitude with voltage.

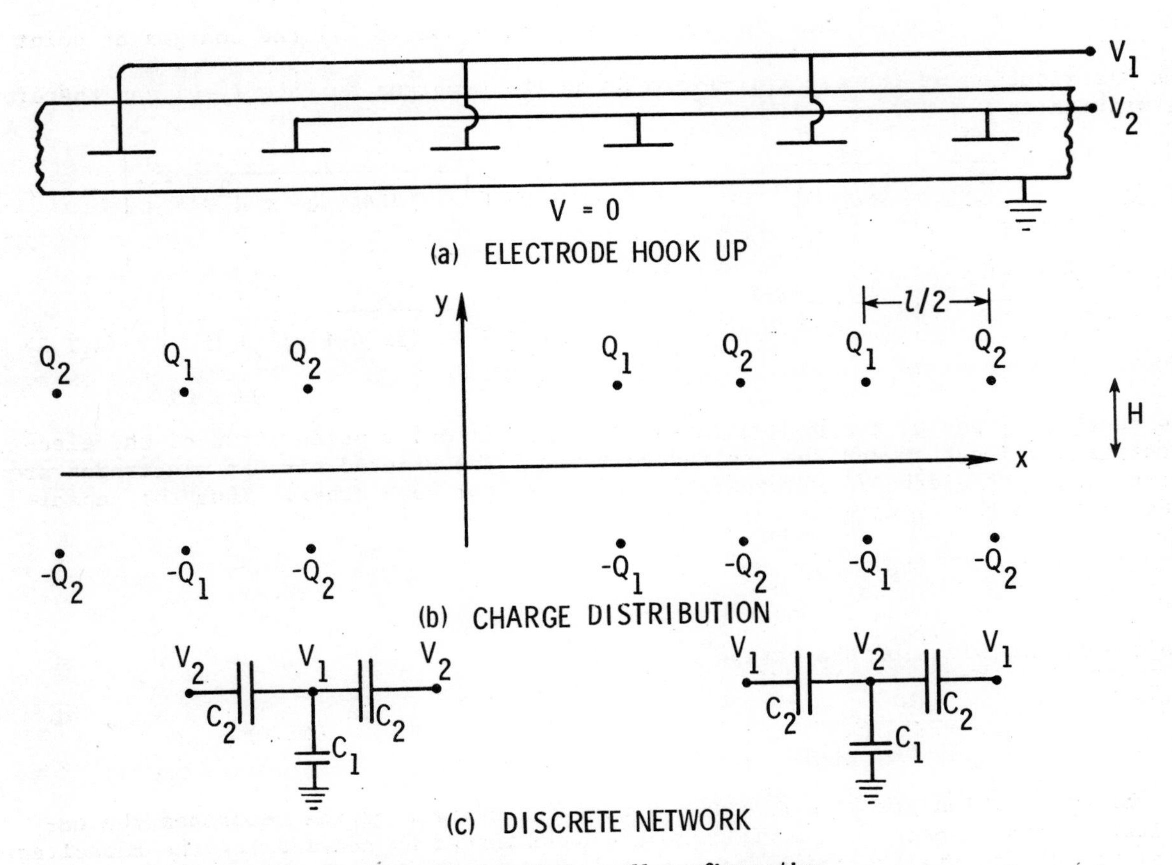

Fig. 17. The electrostatic wall configuration.

Appendix A

Consider the electrostatic wall system with voltages V_1 and V_2 on the terminals A and B as shown in Fig. 17a. The membrane is at ground potential (Fig. 17a). Let the electric field distribution between the membrane and the electrodes cause a charge distribution in the membrane and also on the electrodes. We shall assume that charges Q_1 are distributed uniformly on the electrodes of voltage V_1 and charges Q_2 are uniformly distributed on electrodes of voltage V_2. The distribution of the electric field in the region bounded by the membrane of zero potential and the electrodes can be studied by the method of image charges. In Fig. 17b the charges Q_1 and Q_2 and the image charges $-Q_1$ and $-Q_2$ are kept at height +H and -H respectively from the membrane and at seperations of $\ell/2$. The field distribution in the region between the membrane and the electrodes is periodic in the x direction and hence we evaluate the electric field in one periodic bay.

We shall consider the electric field in the y direction at a point p(x,y) due to the charges and the image charges.

$$E_y^1\Big|_p = \frac{Q_1(H-y)}{2\pi\varepsilon\,[(x-\ell/4)^2 + (H-y)^2]}, \qquad \text{(A.1.1)}$$

due to Q_1 at $(+H,\ell/4)$

and

$$E_y^2\Big|_p = \frac{-Q_1\,(H+y)}{2\pi\varepsilon\,[(x-\ell/4)^2 + (H+y)^2]}$$

due to $-Q_1$ at $(-H,\ell/4)$

The total field at p is thus the sum of the fields due to all the charges at point p.

The distribution of the electric field Ey on the membrane surface (y=0) can therefore be written as

$$E_y = \frac{-Q_1}{\pi\varepsilon}\left[\frac{1}{(\ell/4-x)^2 + H^2} + \frac{1}{(5\ell/4-x)^2 + H^2} + \frac{1}{(3/4\ell+x)^2 + H^2} + \cdots\right]$$

$$\frac{-Q_2}{\pi\varepsilon}\left[\frac{1}{(\ell/4+x)^2 + H^2} + \frac{1}{(3\ell/4-x)^2 + H^2} + \frac{1}{(5\ell/4 + x)^2 + H^2}\right] \qquad \text{(A.1.2)}$$

$$0 \le x \le \ell$$

The charges Q_1 and Q_2 can be evaluated from geometrical consideration of the electrostatic wall system and the applied voltages. The printed circuit electrodes are short strips of width $\ell/4$ at seperations of $\ell/2$ from each other. Thus the capacitance between the electrodes is

$$C_2 = \frac{2\pi\varepsilon}{\cosh^{-1}(\ell/2d)} \qquad \text{(A1.3)}$$

where d is the width of the strips or

$$C_2 = \frac{2\pi\varepsilon}{\cosh^{-1}(2)} \qquad \text{(A1.4)}$$

For the wires which are at a distance H from the surface of the membranes the capacitance with respect to the membrane can be obtained by considering the capacitance of a wire of the diameter equal to the strip width and whose centre is at a distance $H1 = (H + \ell/8)$ from the membrane surface or

$$C_1 = \frac{\pi\varepsilon}{\cosh^{-1}(H1/\ell/8)} = \frac{\pi\varepsilon}{\cosh^{-1}(8H1/\ell)} \qquad \text{(A1.5)}$$

Finally, to evaluate the charges Q_1 and Q_2 consider a discrete network shown in Fig. 17c. From Fig. 17c,

$$(V_1-V_2)\,2C_2 + C_1V_1 = Q_1 \tag{A1.6}$$

$$2C_2(V_2-V_1) + C_1V_2 = Q_2 \tag{A1.7}$$

Appendix B

For a simply supported plate of length ℓ and width b, the dynamic equation of motion (ref. 8) is given as

$$\rho h \frac{\partial^2 w}{\partial t^2} + \frac{Eh^3}{12(1-\nu^2)}\left(\frac{\partial^2}{\partial x^2}+\frac{\partial^2}{\partial y^2}\right)^2 w - N_x\frac{\partial^2 w}{\partial x^2}$$

$$-N_y\frac{\partial^2 w}{\partial y^2} - N_{xy}\frac{\partial^2 w}{\partial x\partial y} = P \tag{B.1}$$

During the deflection of the plate the middle plane of the plate suffers in-plane strains ε_{xx}, ε_{xy}, ε_{yy}. The midplane forces are,

$$\begin{aligned} N_x &= \frac{Eh}{(1-\nu^2)}(\varepsilon_{xx}+\nu\varepsilon_{yy})\\ N_y &= \frac{Eh}{(1-\nu^2)}(\varepsilon_{yy}+\nu\varepsilon_{xx})\\ N_{xy} &= \frac{Eh}{2(1+\nu)}\varepsilon_{xy}\end{aligned} \tag{B.2}$$

The midplane strains are related to the components of deformations, i.e.

$$\begin{aligned}\varepsilon_{xx} &= \frac{\partial\tilde u}{\partial x}+\frac{1}{2}\left(\frac{\partial w}{\partial x}\right)^2\\ \varepsilon_{yy} &= \frac{\partial\tilde v}{\partial y}+\frac{1}{2}\left(\frac{\partial w}{\partial y}\right)^2\\ \varepsilon_{xy} &= \frac{\partial\tilde u}{\partial y}+\frac{\partial\tilde v}{\partial x}+\frac{\partial w}{\partial x}\frac{\partial w}{\partial y}\end{aligned} \tag{B.3}$$

where w is the lateral displacement and $\tilde u$ and $\tilde v$ are the inplane displacements.

It is customary to neglect longitudinal inertia effects while considering the inplane motions, since the inplane frequencies are generally larger than the out of plane natural frequencies. The dynamic equations of inplane motion are thus,

$$\begin{aligned}\frac{\partial N_x}{\partial x}+\frac{\partial N_{xy}}{\partial y} &= 0\\ \frac{\partial N_y}{\partial y}+\frac{\partial N_{xy}}{\partial x} &= 0\end{aligned} \tag{B.4}$$

Substituting for N_x and N_y from (B.2) and (B.3), one obtains,

$$\begin{aligned}\frac{\partial^2\tilde u}{\partial x^2}+\left(\frac{1-\nu}{2}\right)\frac{\partial^2\tilde u}{\partial y^2}+\frac{1+\nu}{2}\frac{\partial^2\tilde v}{\partial x\partial y}+F_x(x,y,t) &= 0\\ \frac{\partial^2\tilde v}{\partial y^2}+\left(\frac{1-\nu}{2}\right)\frac{\partial^2\tilde v}{\partial x^2}+\frac{1+\nu}{2}\frac{\partial^2\tilde u}{\partial x\partial y}+F_y(x,y,t) &= 0\end{aligned} \tag{B.5}$$

where,

$$F_x(x,y,t) = \frac{\partial w}{\partial x}\left[\frac{\partial^2 w}{\partial x^2} + \frac{1-\nu}{2}\frac{\partial^2 w}{\partial y^2}\right] + \left(\frac{1+\nu}{2}\right)\frac{\partial w}{\partial y}\frac{\partial^2 w}{\partial x \partial y}$$

$$F_y(x,y,t) = \frac{\partial w}{\partial y}\left[\frac{\partial^2 w}{\partial y^2} + \frac{1-\nu}{2}\frac{\partial^2 w}{\partial x^2}\right] + \frac{1+\nu}{2}\frac{\partial w}{\partial x}\frac{\partial^2 w}{\partial x \partial y} \qquad \text{(B.6)}$$

The air cavity behind the plate dictates the conservation of volume

$$\iint w \; dxdy = 0 \qquad \text{(B.7)}$$

The nature of the excitation on the plate is periodic with period 2π in spanlength 'ℓ' and hence the choice of an appropriate deflection mode shape is

$$w = \Sigma w_m = \sum_m f_m(t) \sin\frac{2m\pi x}{\ell} \sin\frac{\pi y}{b} \qquad \text{(B.8)}$$

with

$$\beta_m = \frac{\ell}{2mb},$$

$$\eta_m = \frac{2\pi m}{\ell} \qquad \text{(B.9)}$$

Equation B.6 becomes,

$$F_x = \sum_m f_m^2(t) \frac{\eta_m^3}{4}\left\{\sin\frac{4\pi mx}{\ell}\left[(\nu\beta_m^2 - 1) + \cos\frac{2\pi y}{b}(1+\beta_m^2)\right]\right\}$$

$$F_y = \sum_m f_m^2(t) \frac{\eta_m^3\beta_m}{4}\left[\sin\frac{2\pi y}{b}\left\{(\nu-\beta_m^2) + \cos\frac{4m\pi x}{\ell}(1+\beta_m^2)\right\}\right] \qquad \text{(B.10)}$$

We seek solutions $\tilde{u}$, $\tilde{v}$ for (B.5), such that

$$\tilde{u}(x,y,t) = \sum_m A_x^m \sin\frac{4m\pi x}{\ell} + \sum_m C_x^m \sin\frac{4m\pi x}{\ell}\cos\frac{2\pi y}{b} + \tilde{u}_o(x,y,t)$$

$$\tilde{v}(x,y,t) = \sum_m A_y^m \sin\frac{2\pi y}{b} + \sum_m C_y^m \sin\frac{2\pi y}{b}\cos\frac{4m\pi x}{\ell} + \tilde{v}_o(x,y,t) \qquad \text{(B.11)}$$

where $\tilde{u}_o(x,y,t)$, $\tilde{v}_o(x,y,t)$ are chosen that they are a particular solution of

$$\frac{\partial^2 \tilde{u}_o}{\partial x^2} + \frac{1-\nu}{2}\frac{\partial^2 \tilde{u}_o}{\partial y^2} + \frac{1+\nu}{2}\frac{\partial^2 \tilde{v}_o}{\partial x \partial y} = 0$$

$$\frac{\partial^2 \tilde{v}_o}{\partial y^2} + \frac{1-\nu}{2}\frac{\partial^2 \tilde{v}_o}{\partial x^2} + \frac{1+\nu}{2}\frac{\partial^2 \tilde{u}_o}{\partial x \partial y} = 0 \qquad \text{(B.12)}$$

with the restraints

$$\tilde{u}_o(o,y,t) = \tilde{u}_o(\ell,y,t) = 0$$

$$\tilde{v}_o(x,o,t) = \tilde{v}_o(x,b,t) = 0$$

A particular solution for (B.12) is

$$\tilde{u}_o \equiv \tilde{v}_o \equiv 0 \qquad \text{(B.13)}$$

Substituting in (B.5), the expressions for the various derivatives of $\tilde{u}$ and $\tilde{v}$ using (B.11) yields,

$$A_x^m = -\frac{\eta_m}{16} f_m^2(t)\ (1-\nu\beta_m^2)$$

$$A_y^m = \frac{\eta_m}{16\beta_m} f_m^2(t)\ (\nu-\beta_m^2)$$

$$C_x^m = \frac{\eta_m}{16} f_m^2(t) \tag{B.14}$$

$$C_y^m = \frac{\eta_m\beta_m}{16} f_m^2(t)$$

Thus

$$N_x = \left(\frac{Eh}{1-\nu^2}\right) \sum_m \eta_m^2 \frac{f_m^2(t)}{8} \left[(1+\nu\ \beta_m^2) - (1-\nu^2)\cos\frac{2\pi y}{b}\right]$$

$$N_y = \left(\frac{Eh}{1-\nu^2}\right) \sum_m \eta_m^2 \frac{f_m^2(t)}{8} \left[(\nu+\beta_m^2) - (1-\nu^2)\ \beta_m^2 \cos\frac{4\pi mx}{\ell}\right]$$

$$N_{xy} \equiv 0 \tag{B.15}$$

Appendix C

Due to the primary field $\tilde{E}_o$ the membrane wall deforms from its original position at $z = 0$, to $z = w(x,y,t)$ i.e. the equation of the membrane surface is,

$$f(x,y,z,t) = z - w(x,y,t) = 0 \tag{C.1}$$

or

$$\nabla f = i_z - \frac{\partial w}{\partial x} i_x - \frac{\partial w}{\partial y} i_y \tag{C.2}$$

The normal vector at the surface is

$$\tilde{n} = \frac{i_z - w_x i_x - w_y i_y}{\sqrt{1 + w_x^2 + w_y^2}} \tag{C.3}$$

Since the membrane and the electrodes are conductors they cannot support any internal fields. Thus

$$\tilde{E} \times \tilde{n} = 0 \tag{C.4}$$

With assumption that the total field consists of the primary field E_{oz} and the perturbation fields e_x, e_y, e_z

$$\tilde{E} = e_x i_x + e_y i_y + (\tilde{E}_{oz} + e_z) i_z \tag{C.5}$$

Thus at the membrane surface

$$z = 0, \quad e_y w_x = e_x w_y$$

$$w_x(e_z + E_{oz}) + e_x = 0 \tag{C.6}$$

and at $z = H$,

$$e_x = e_y = 0 \tag{C.7}$$

The total field in the domain should satisfy

$$\nabla\cdot E = \nabla \times E = 0 \tag{C.8}$$

An approximate solution satisfying the above is given as,

$$e_x \approx \Sigma\, w_x^m\, \tilde{E}_{oz} \quad \sinh\left(\frac{2\pi m(H-z)}{\ell}\right) / \sinh \frac{2m\pi H}{\ell}$$

$$e_y \approx \Sigma\, w_y^m\, \tilde{E}_{oy} \quad \sinh\left(\frac{2\pi m(H-z)}{\ell}\right) / \sinh \frac{2m\pi H}{\ell}$$

$$e_z \approx -\Sigma\, w_x^m\, \tilde{E}_{oz} \quad \frac{\cosh \frac{2\pi m(H-z)}{\ell}}{\sinh \frac{2m\pi H}{\ell}} \qquad \text{(C.9)}$$

Thus the field distribution at $z = 0$ is given as

$$e_x \approx w_x E_{oz}$$

$$e_y \approx w_y E_{oz}$$

$$e_z \approx -\Sigma\, w_x^m\, E_{oz} \coth \frac{2m\pi H}{\ell} \qquad \text{(C.10)}$$

When w_x and w_y are small the total field differs very little from the primary field. As an example for a deflection $w \approx 0.025$mm, the total field is 95% of the primary field for the configurations tested in the laboratory.

Second International Conference on

Drag Reduction

August 31st - September 2nd, 1977

PAPER F1

THE POSSIBLE RELEVANCE OF PERSISTENT EXTENSIONAL FLOW ON THE INTERPRETATION OF DRAG REDUCTION PHENOMENA

A. Keller,

University of Bristol, U.K.,

and

M.R. Mackley,

University of Sussex, U.K.

Summary

We review recent experiments carried out by us that illustrate the basic conditions necessary to extend polymer molecules in simple centro-symmetric flow fields. We show that high chain extension, observed experimentally as flow birefringence, can only be achieved in Persistently Extensional flows. It is shown that localized molecular extension in these flows occurs in the form of threads or sheets of oriented molecules, which is an essential consequence of the nature of the flows studied.

By examining 'two dimensional' persistently extensional flows, significant modification to the flow patterns were observed with the presence of polymer. These changes could be explained in terms of localized enhancement of extensional viscosity. We speculate on the relevance of our general findings concerning chain extension on drag reducing effects in general and those observed by use in a Couette apparatus when Taylor vortices are operating.

Held at St. John's College, Cambridge, England.

Organised and sponsored by BHRA Fluid Engineering, Cranfield, Bedford, MK43 0AJ.

Introduction

Our original interest in polymer chain extensions comes from attempts to understand the so called shish kebab crystal structure of polyethylene (see general Ref.1-3). These studies lead us on to the basic question of what types of flows and under what conditions can high polymer chain extension be achieved in flowing systems. We believe our findings on this subject may be of relevance to the understanding of the Toms effect (Ref.4), and the objective of this paper is to review the experiments we have conducted over the last seven years and hope that others more knowledgeable in the complexity of turbulent flow can make the relevant connections towards understanding drag reduction behaviour in truly turbulent flows.

It was suggested concurrently with the recognition of shish-kebab structures that the predominantly extended chain backbone was produced by the chain extension of polymer molecules whilst being subjected to some form of flow induced orientation (e.g. Ref.3). The definitive hydrodynamic situation became evident when Pennings, van der Mark and Booij (Ref.5) studied the growth of the shish kebab morphology under well defined flow conditions. He used a conventional Couette apparatus and discovered that normal simple shearing flow did not modify the crystal growth of a polyethylene/xylene solution from that of typical growth in a quiescent solution; namely, lamella single crystals of a postulated chain folder character (see general Ref.3). If the rotation rate of the inner cylinder was increased such that a secondary flow known as 'Taylor vortices' (Ref.6) was superimposed on the simple shearing flow, Pennings found that there was a dramatic change in the crystal growth and fibrous crystals of a shish kebab morphology were grown. Pennings et al. (Ref.5) explained this remarkable result by saying that when simple shearing flow alone existed the polymer chains were essentially undeformed and the flow had little or no effect on crystal growth. When the Taylor vortices were operating the hydrodynamic situation was very different. In this case the region between adjacent vortices shown in figure 1 from (Ref.7) contain regions of extensional flow. In these regions chains have the ability to be appreciably extended which causes the intrinsic change in the crystal morphology. The realization that extensional flows have greater power to extend polymer molecules than simple shearing flows had previously been postulated theoretically by Ziabicki (Ref.8) and later by Peterlin (Ref.9).

Rotation free flows

Taylor vortices, whilst being more amenable to study than normal turbulent flow, are still relatively complex in that they are a superposition of a circumferential toroidal motion on the usual simple shearing flow. Following a suggestion by F.C.Frank, F.R.S., we developed a flow system that was capable of producing an extensional flow without simple shearing components and where high velocity gradients could readily be achieved. Up to this date, with the notable exception of G.I.Taylor's work in 1934 (Ref.10), little attention had been paid to the experimental generation of pure extensional flows. The system developed at Bristol consisted simply of two mutually opposed jets immersed in a 'sea' of polymer solution, as shown schematically in figure 2 (Refs. 11 and 12). Flow could be directed such that the flows either impinged onto the symmetry plane of the geometry or alternatively the 'reverse' situation would produce inflow into both jets. The former case produced essentially axial compression and the latter case axial extension. A very important feature of the flows was that they were both centrosymmetric having a stagnation point at the centre of symmetry.

In terms of chain extension two important realizations were made from the double jet work and later work conducted using a 'Taylor four roll mill' apparatus (Ref.13).

Firstly, it was made apparent from the work of Ziabicki (Ref.8) and Peterlin (Ref.9) that in order to stretch dilute solution polymer chains significantly a pure extensional velocity gradient $S = \partial V_x/\partial x$ of magnitude $S\tau > 1$ was required, where τ is equal to a characteristic relaxation time of the polymer molecule and is an increasing function of molecular weight. Chain extension can be observed in both the double jet and four roll mill apparatus, by observing flow birefringence. It was found that flow birefringence corresponding to high chain extension was only detected for conditions where the product $S\tau$ was greater than unity, in qualitative accordance with theory. Typically τ ranges from 10^{-3} - 1 sec, consequently velocity gradients of 10^3 - 1 sec were required to obtain high chain extensions. In the double jet experiments for polyethylene/xylene solutions (Ref.11) it was concluded from the observed magnitude of the flow birefringence that nearly complete chain alignment of the polymer chains could be achieved for a velocity gradient of about 10^3 sec^{-1}. This contrasts very strongly with usual flow birefringence studies of flexible polymers in solution whilst subject to a simple shearing flow where very small chain extensions are observed for velocity gradients up to the practical limit of about 10^6 sec^{-1}.

The second important physical point to emerge from these studies was initially unexpected, but on reflection intuitively obvious. In both the double jet and four roll mill experiment highly localized birefringence was observed (Refs. 11 and 13). This again contrasts with normal flow birefringence studies where, for example in the Couette apparatus, a low intensity uniform general flow birefringence exists over the whole region between the inner and outer walls of the Couette apparatus.

The position of the observed localized flow birefringence is summarized by figure 3. In each flow situation the birefringence is localized in a region on and near the 'outgoing' symmetry plane or axis of the flow, producing sheets or threads of oriented molecules. The reason for the localization is most readily understood by referring to the situation in the four roll mill introduced first into polymer studies in Ref.13 and developed further in Ref.14 which also includes an attempt to define the degree of localization in terms of τ.

With reference to figure 4, consider the deformation of a fluid element entering a four roll mill on a line $y = y_o$.

The velocity gradient of the pure shearing flow is given by

$$S = \frac{\partial V_x}{\partial x} = - \frac{\partial V_y}{\partial y}$$

Assume the fluid element is of length ℓ_o in the direction of x, at $y = y_o$ and time $t = 0$. The length ℓ of the element after time t is governed by

$$\int_{\ell_o}^{\ell} \frac{\partial x}{x} = St \quad \text{and} \quad \int_{y_o}^{y} \frac{\partial y}{y} = -St$$

The strain of the fluid element ℓ/ℓ_o at time t is given by

$$\ell/\ell_o = y/y_o$$

Thus the strain of the fluid element is inversely proportional to the distance from the 'outgoing' symmetry plane. In order to stretch polymer chains the fluid element that contains the chain must be extended by an amount at least equal to the strain necessary to stretch the chain from the random coil to that of the extended chain. Thus typical polymers require a strain $\partial\ell/\ell_o$ of about 10^2 - 10^3. These high strains will only be achieved very close to the outgoing symmetry plane of the flow. Similar argument apply to all centrosymmetric flows. We therefore reach the conclusion that high chain extension in centrosymmetric flows will by necessity be localized. This condition can be expressed by the statement that $St \gg 1$, where t is the time that the molecule is subjected to the pure extensional flow.

Flows with rotation

The three flows that we have considered are examples of 'pure' extensional flows. We now consider the effect of an additional rotational component to the flow. This situation is very clearly illustrated by observing the flow between two co-rotating rollers, see figure 5 and Ref.15. Here, unlike the previous cases, the asymptotes of the centrosymmetric flow between the rollers do not intersect at right angles. In fact the degree of obliquity of the asymptotes gives an accurate measure of the degree of rotation of theflow.

Because the flow is two dimensional and essentially incompressible we can say that the flow in the region of the stagnation point can be represented by a stream function

$$\phi = Ax^2 + By^2,$$

where lines of constant ϕ are stream lines and the velocity at any point is given by

$$(V_x, V_y, V_z) = (\frac{-\partial\phi}{\partial y}, \frac{\partial\phi}{\partial x}, 0).$$

There are three parameters that we define and consider for the above flow.

Principal strain rate S defined as maxima of $\partial V_r/\partial r$,

$$S = A - B$$

Rotation ω defined as $\omega = \frac{1}{2}$ curl V,

$$\omega = A + B = -S \cos 2\alpha,$$

where α is the asymptotic angle shown in figure 5.

Persistent strain rate σ

$$\sigma^2 = S^2 - \omega^2$$

The persistent strain rate is the strain rate that a fluid element would 'persistently' be subject to. In the case of the oblique col flow it is the strain rate along the outgoing asymptote given by

$$\sigma = S \sin 2\alpha.$$

When a polymer chain enters the flow between the rollers the molecule will quickly align itself parallel to the outgoing asymptote and it will then be subjected to a persistent strain rate given by σ. The condition for high chain extension can now be written by $\sigma\tau > 1$ and $\sigma t >> 1$. This means that the persistent strain rate, not the principal strain rate S, is the important chain stretching parameter.

Two cases are of particular interest.

$\omega = 0$ $\quad \sigma = S$ $\quad$ pure shearing flow,

$\omega = S$ $\quad \sigma = 0$ $\quad$ simple shearing flow.

We see that for simple shearing flow the persistent strain rate is zero; this is why high chain extension of flexible chains in simple shearing flow is not possible.

We shall now consider the effect that polymers have in modifying the oblique col flow, figure 6 (Ref.17). When a Newtonian fluid (glycerol) is used the asymptotic angle between incoming and outgoing asymptotes of the flow do not change with increasing roller speed. However if we use a 1% polyethylene oxide (WSR 301)/water solution we note that as the velocity gradient increases the asymptotic angle changes. From this result we can deduce that ω/S increases. Flow birefringence studies show that when the polymer is present, as expected by the argument given earlier, we see localized chain extension on the outgoing symmetry sheet, figure 7 (Ref.15). To explain our findings we now appeal to the theories of Peterlin (Ref.9), which indicate that in extensional flows chain extension will produce considerable enhancement of extensional viscosity. From our flow birefringence observations we know this will occur only in the localized regions of high chain extension. The enhanced viscosity in this region could simply be thought of as decreasing the value of S causing a resultant increase in ω/S.

The above analysis suggests that localized enhanced extensional viscosity can modify the flow. A more dramatic example of the way polymers can modify flow is given in a recent paper by Berry and Mackley (Ref.16). The flow under examination is that produced by six rollers arranged symmetrically and with roller rotation direction shown in figure 8. By altering roller settings many different flow patterns can be explored in a systematic way. In the context of this paper the important point to note is the comparison of the behaviour of a Newtonian fluid and that of a 1% polyethylene oxide solution for different roller speeds. The relevant flow patterns and corresponding flow birefringence are shown in figure 9. We see that for the same roller settings the topology of some flows are completely different when polymer is present; some flow patterns remain unchanged. In this experiment the polymer has not merely 'perturbed' the flow as in the two roll mill. In some situations the flow has been completely changed by the presence of the polymer. The key to understanding the reason for the remarkable behaviour of polymers in the six roll mill is obtained by studying the associated flow birefringence. Again we observe localized flow birefringence in the region of the outgoing asymptotes of the flow. This orientation will produce localized enhanced extensional viscosities in the region of the asymptotes of the outflows. The polymer has had the effect of 'breaking the symmetry' between the inflows and the outflows. Using this property we are quantitatively able to explain how some of the flows are dramatically modified and others unchanged by the presence of polymer (Ref.16). Essentially the same point is being brought out by experiments with two counter-rotating rollers currently being analysed at Bristol where the 'barrier' effect of the localized birefringent sheet constituted by highly aligned chains is recognisable even by inspection (C.Farrel,unpublished).

Drag Reduction

Our own published attempts to understand polymer drag reduction are concerned with monitoring the drag behaviour of a low concentration polymer solution in a Couette apparatus (Ref.7). The experimental system consists simply of a rotating inner cylinder and a freely suspended outer cylinder from which torque measurements can be made. By comparing the behaviour of water and a 50 ppm polyethylene oxide (WSR 301) solution it was found that at rotation rates above $\omega \approx 4\ s^{-1}$ Taylor vortices were observed to form in the manner shown in figure 1 . With increasing rotation rate a progressive decrease in drag was observed when the polymer was introduced,fig.10. This effect is very similar to the Toms effect, although in this situation we do not have full turbulence.

The analogies with this experiment and the fibrous growth of polyethylene crystals, already anticipated by Peterlin (18), is quite striking. Both require Taylor vortices to be present and both are more effective with higher molecular weight material. It is very tempting to associate the growth of fibrous crystals and drag reduction to one of the same mechanisms; namely chain extension by the persistent extensional flow produced between adjacent vortices.

Observations of the profile of the Taylor vortices generated between the inner and outer cylinder did not show any detectable difference when polymer was present, although some evidence does suggest that the vortices may have increased in speed. At present we are unable to understand the mechanism which causes the drag reduction when Taylor vortices are present: certainly the vortex motion is an essential part of the flow; we also see from these experiments that drag reduction is not caused by a delay in the onset of the instability. Frank (Ref.17) has already speculated on the significance of localized persistent extensional flows in terms of general turbulent drag reduction. We can imagine turbulence as a complex time varying array of vortices. At any given time there will be stagnation points in the flow between adjacent co-rotating vortices; in these positions sheets or threads of oriented molecules will be present; it may be this complex network of extended polymer chains which in some way modifies the energy cascade through the turbulence spectrum to reduce drag by up to 50%.

Turbulent drag reduction is not restricted to polymers, anisotropic particles (Ref.19) and asbestos fibres (Ref.20), for example, have been found capable of producing the effect. In this respect there is a very attractive feature about centrosymmetric persistent extensional flows. Particles, polymer chains or fibres when subjected to a centrosymmetric persistent extensional flow will be 'focused' by the flow towards the outgoing asymptote of the flow. Consequently a barrier of particles will be formed in the region of the exit asymptote of the flow, in a similar way to the accumulation of flow birefringence observed for the polymer case. This effect is very clearly illustrated when observing the formation of weather fronts (Ref.21). A diffuse weather front will only form a sharp boundary if it is subjected to a centrosymmetric persistent extensional flow; other vortex dominated flows will lead to the further diffusion of the front. This very general property of 'flow focusing' may be of significant importance in understanding turbulent drag reduction, the barrier sheets formed by these flows may be a controlling factor.

References

1. Keller, A. and Willmouth, F.M.,"Some macroscopic properties of stirring induced crystals of polyethylene", J.Macromol.Sci.(Phys.), B,6, pp.493-537 (1972).
2. Pennings, A.J., van der Mark, J.M.A.A. and Kiel, A.M., "Hydrodynamically induced crystallization of polymers from solution III", Kolloid Z.u.Z.Polymere, 237, pp.336-358 (1970).
3. Keller, A., "Polymer Crystals", Rep.Prog.Phys.,XXXI, 2, pp.623-704 (1968).
4. Toms,B.A., "Some observations on the flow of linear polymer solutions through straight tubes at large Reynolds number","Proc.First Intern.Cong. on Rheology", North-Holland Publishing Company (1948).
5. Pennings, A.J., van der Mark, J.M.A.A. and Booij, H.C., "Hydrodynamically induced crystallization of polymers for solution II", Kolloid Z.u.Z.Polymere, 236, pp.99-111 (1970).
6. Taylor, G.I., "Stability of a viscous liquid contained between two rotating cylinders", Phil.Trans.Roy.Soc.CCXXIII, pp.289-343 (1923).
7. Keller, A., Kiss, G. and Mackley, M.R., "Polymer drag reduction in Taylor vortices", Nature, 257, 5524, pp.304-305 (1975).
8. Ziabicki, A., "Studies on orientation phenomena by fiber formation of polymer melts", J.Appl.Polymer Sci., 11, 4, pp.24-31 (1959).
9. Peterlin, A., "Hydrodynamics of macmolecules in a velocity field with longitudinal gradient", J.Polymer Sci., 134, pp.287-290 (1966).
10. Taylor, G.I., "The formation of emulsions in definable fields of flow",Proc.Roy. Soc.Lond., A146, pp.501-523 (1934).
11. Mackley, M.R. and Keller, A., "Flow induced polymer chain extension and its relation to fibrous crystallization", Phil.Trans.Roy.Soc.(Lond.), 278, 1276, pp.29-66 (1975).
12. Frank, F.C., Keller, A. and Mackley, M.R.,"Polymer chain extension produced by impinging jets and its effect on polyethylene solution", Polymer, 12, pp.467-473 (1971).
13. Crowley, D.G., Frank, F.C., Mackley, M.R. and Stephenson, R.G., "Localized flow birefringence of polyethylene oxide solutions in a four roll mill", J.Polymer Sci. (Physics), 14, pp.1111-1119 (1976).
14. Pope, D.P. and Keller, A., "Alignment of macromolecules in solution by elongational flow. A study of the effect of pure shear in a four roll mill", Colloid and Polymer Sci.,in the press.
15. Frank, F.C. and Mackley, M.R., "Localized flow birefringence of polyethylene oxide solutions in a two roll mill", J.Polymer Sci.(Physics), 14, pp.1121-1131 (1976).
16. Berry, M.V. and Mackley, M.R., "The six roll mill: unfolding an unstable persistently extensional flow", Phil.Trans.Roy.Soc.(Lond.), in press.
17. Frank, F.C., "Persistently extensional flow and the flow alignment of long chain molecules", Invited lecture at the 50th anniversary of the Canadian Pulp and Paper Institute, Montreal (1975).
18. Peterlin, A.,"Molecular model of drag reduction by polymer solutes", Nature, 222, 373 (1969).
19. Radin, I., Zarkin, J.L. and Patterson, G.K.,"Drag reduction in solid fluid systems", J.American Inst.Chem.Eng., 21, 2, pp.358-371 (1975).

20. Peyser, P., "The drag reduction of chrysotile asbestos dispersions", J.Appl. Polymer Sci., 17, 2, pp.421-431 (1973).
21. Petterson, S., "Introduction to Meteorology", published by McGraw-Hill, pp.210-235 (1958).

Table I

Flow number	1	2	3	4	5	6	7	8	9	10
Ω_I	2.5	2.5	2.0	2.0	2.0	2.0	2.0	2.0	2.0	2.5
Ω_{II}	2.5	4.0	4.0	3.5	2.75	2.0	1.0	0.5	0	0.5
Ω_{III}	2.5	2.5	4.0	4.0	4.0	4.0	4.0	4.0	4.0	2.5

Data for flows numbered 1 to 10 on figure 9
Ω_I, Ω_{II}, Ω_{III} are the experimental roller speeds in radians/sec for each series (8a, 8b & 8c) corresponding to each flow number (each value of Ω has an estimated error of ±6%).

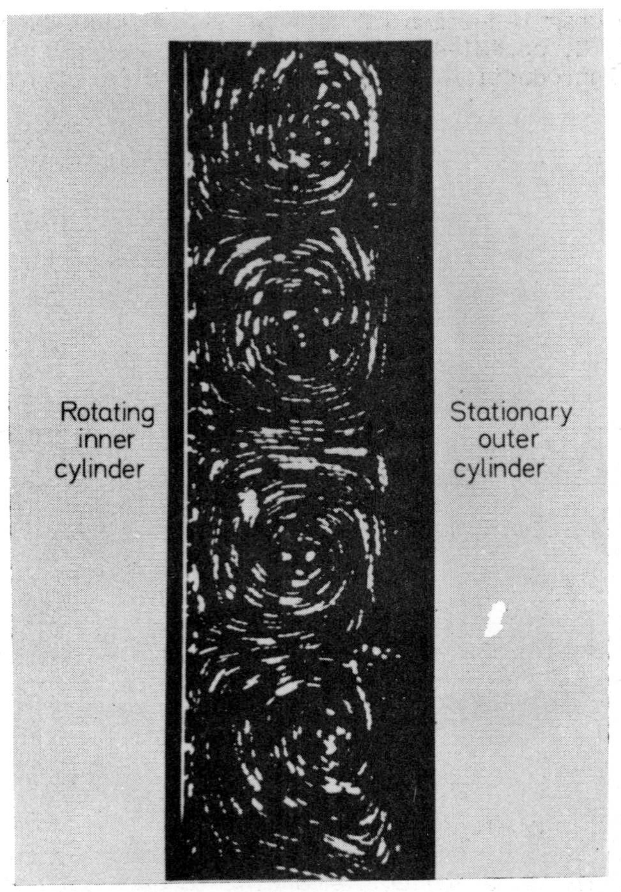

Figure 1 Photographic profile view of Taylor vortex pattern generated in a Couette apparatus; inner cylinder diameter 4 cm and gap width 1.0 cm (from Ref.7).

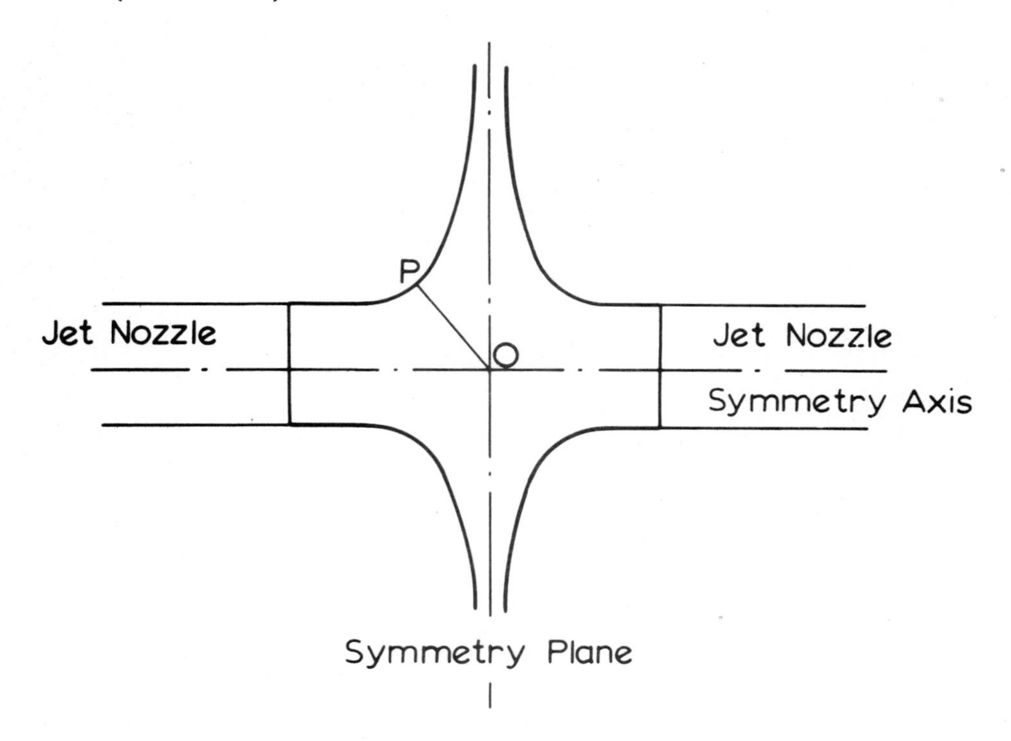

Figure 2 Diagrammatic representation of double jets (from Ref.11).

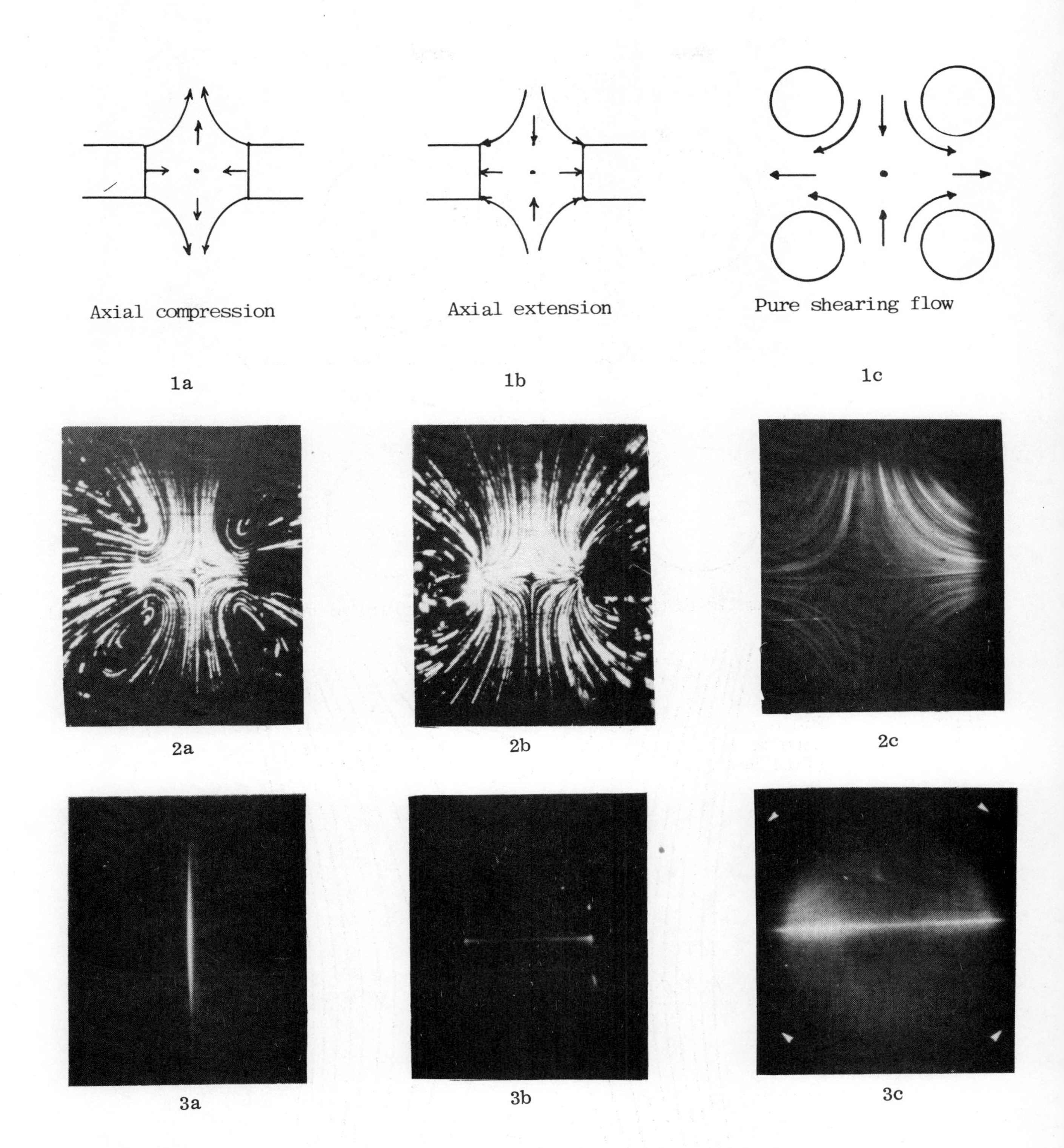

Figure 3 Localized birefringence observed for a) axial compression, b) axial extension, and c) pure shearing flow. For each case figure shows i) schematic diagram of flow, ii) observed streamlines, and iii) associated flow birefringence. Streamline and flow birefringence photographs obtained from Refs. 11 & 13.

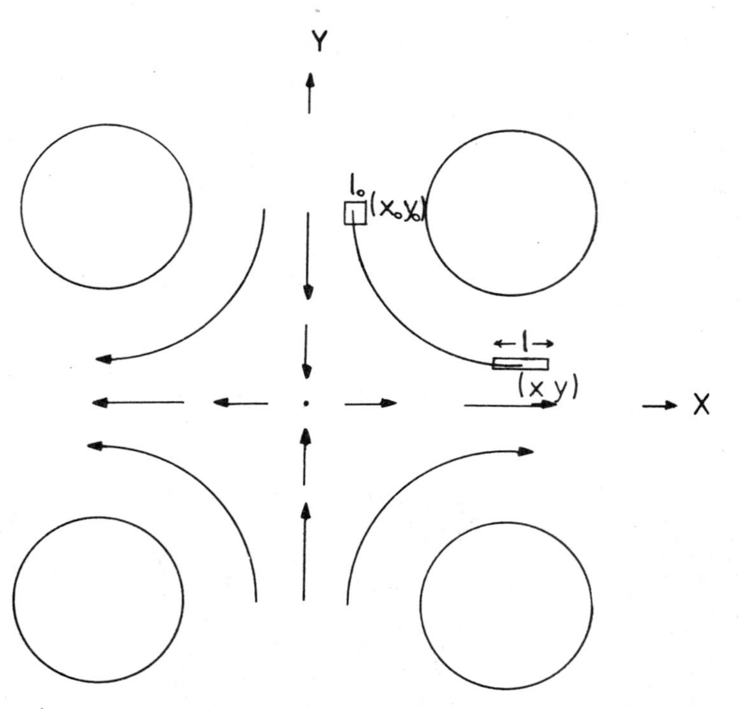

Figure 4 Schematic diagram of deformation of fluid elements in a four roll mill.

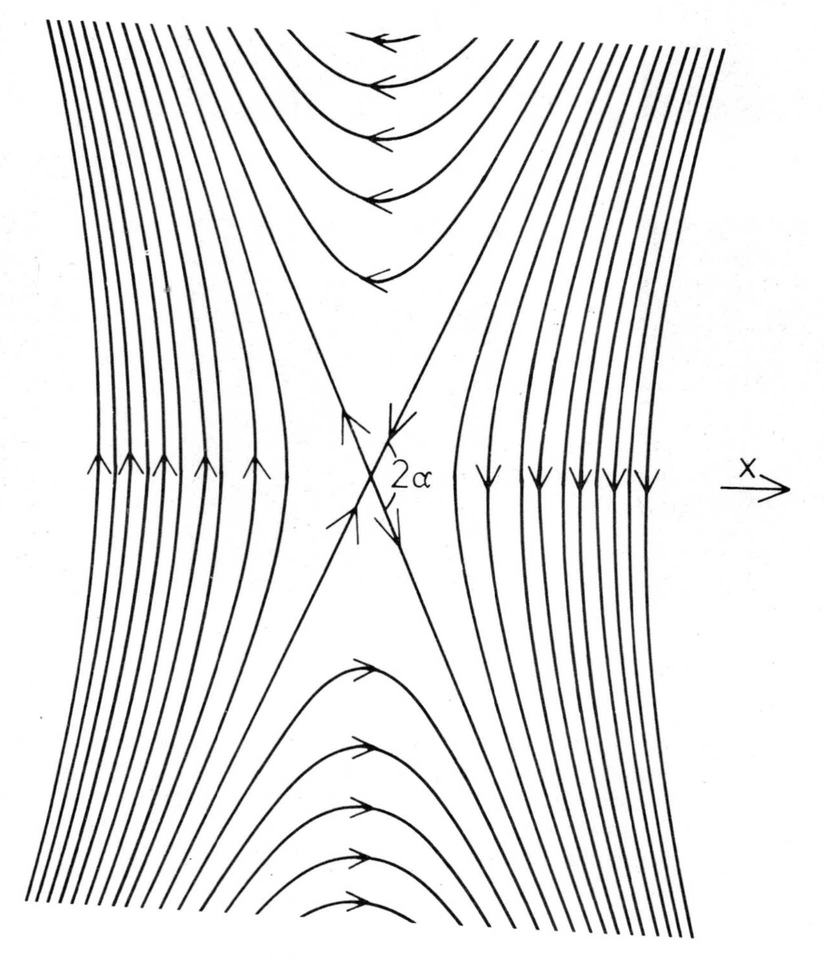

Figure 5 Schematic diagram of idealized flow between two co-rotating rollers. From Ref.1 .

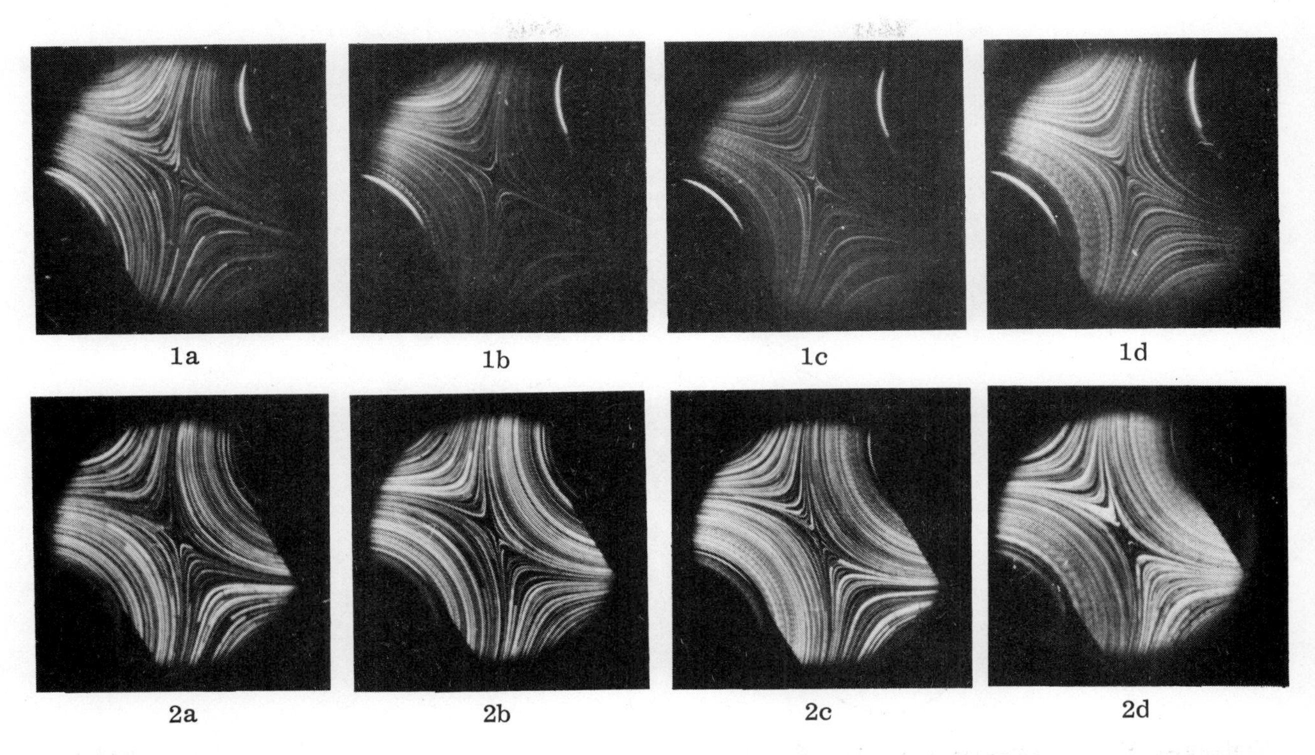

Figure 6 Streamline photographs of i) water/glycerol and ii) polyethylene oxide solution between co-rotating rollers. Roller rotation rate (rad.s^{-1}) for each sequence, a = 1.5, b = 6.3, c = 12.6, d = 18.8. From Ref.15.

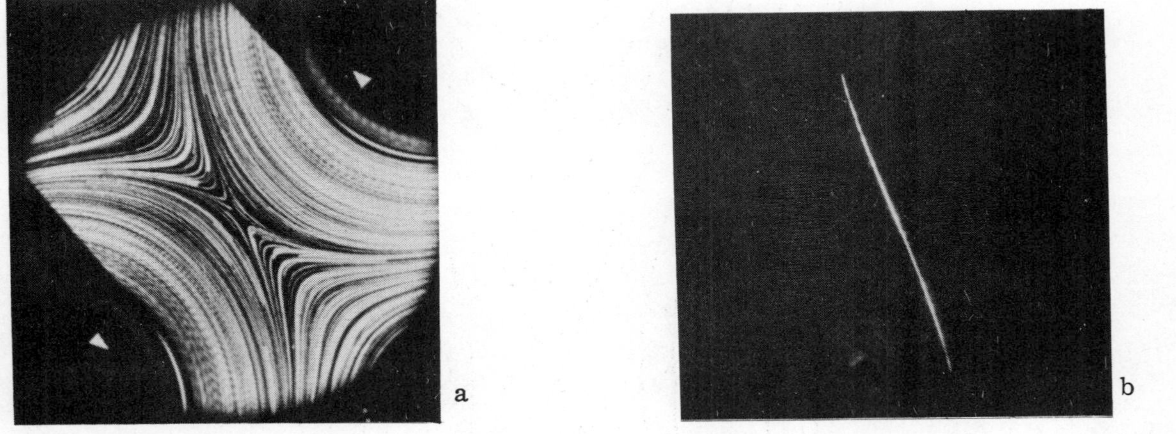

Figure 7 Localized flow birefringence (Fig.7a) and corresponding streamline photograph (7b) observed in co-rotating two roll mill for polyethylene oxide solution. From Ref.15.

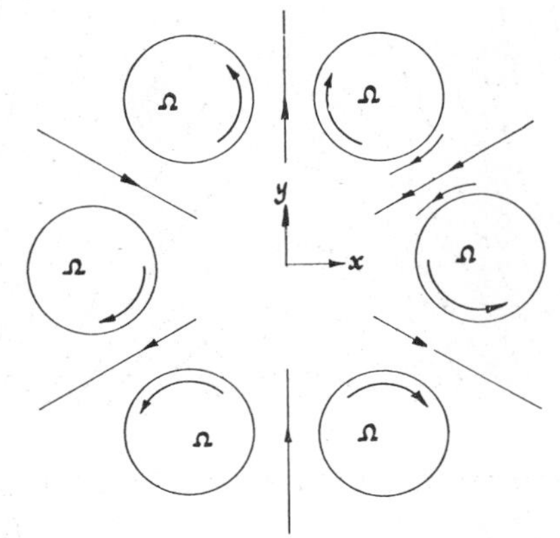

Figure 8 Schematic diagram of the "six roll mill assembly". From Ref.16.

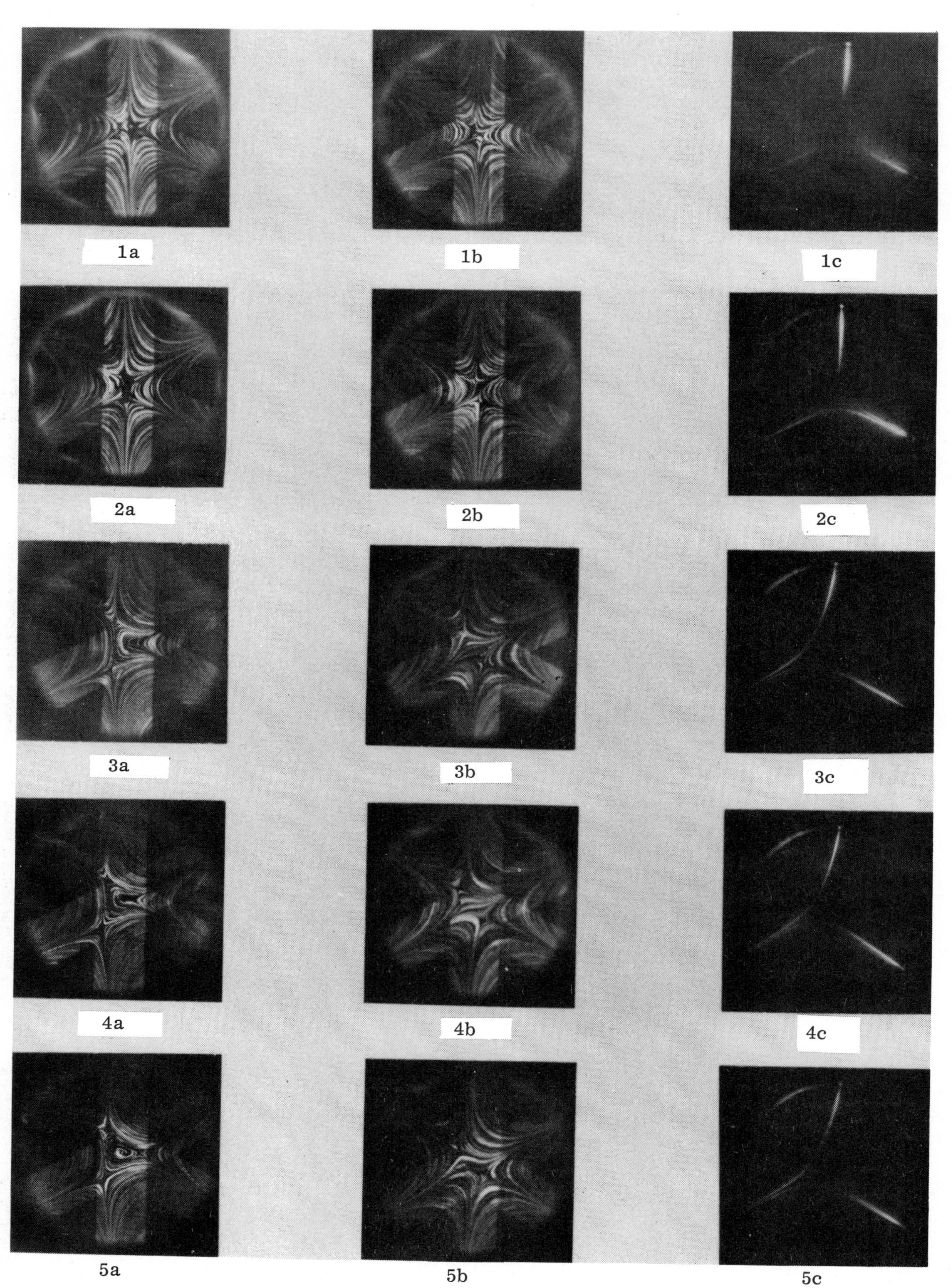

Figure 9 Streamline patterns of a) glycerol and b) 2% polyethylene oxide solution in the six roll mill corresponding to different roller speeds given in Table I. c) Corresponding flow birefringence observations of the polyethylene oxide solution. From Ref.16.

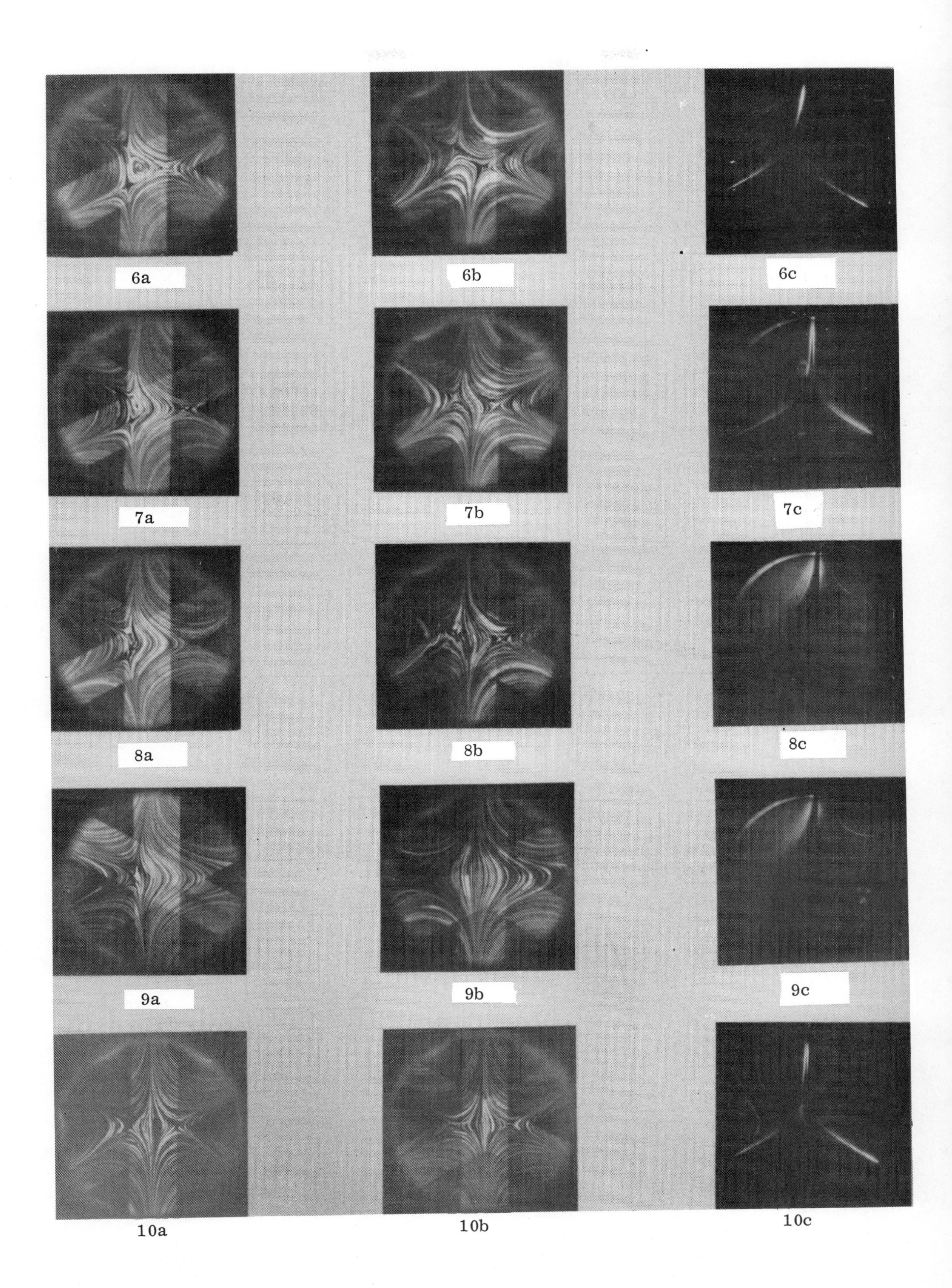
6a
6b
6c
7a
7b
7c
8a
8b
8c
9a
9b
9c
10a
10b
10c

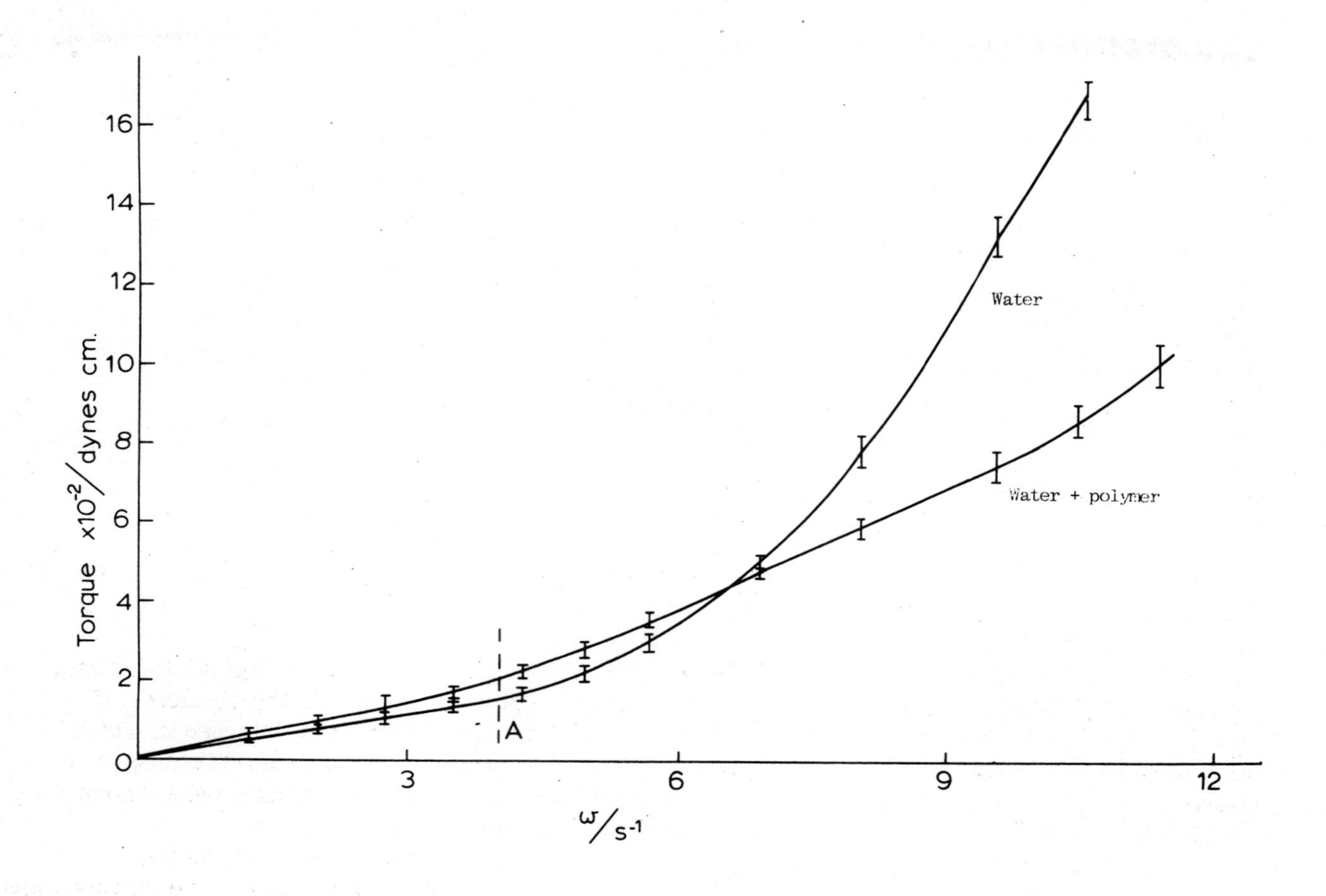

Figure 10 Graph of outer cylinder torque as a function of angular velocity of inner cylinder for Couette apparatus. From Ref.7.

Second International Conference on

Drag Reduction

August 31st - September 2nd, 1977

PAPER F2

DRAG OF CIRCULAR CYLINDERS IN DILUTE POLYMER SOLUTIONS

E. Kit and M. Poreh,

Technion - Israel Institute of Technology, Israel.

Summary

The drag of small circular cylinders in flows of dilute drag reducing polymer solutions, has been investigated experimentally. By changing the velocity of the fluid, the diameter of the cylinders and the concentrations of the polymer, different regimes were detected in which the drag of the cylinders in the solution varied over a wide range, showing from 50% drag reduction to 600% drag increase, relative to the Newtonian drag. Simultaneous measurements in a pipe flow with the same solutions always showed drag reduction.

Examination of the data suggests that two dimensionless parameters, a Reynolds number and a 'polymer number' PN determine the drag coefficient. Four regions are distinguished in the drag coefficient versus Reynolds number curve. A qualitative explanation of the mechanism which determines the different regions is discussed.

Held at St. John's College, Cambridge, England.
Organised and sponsored by BHRA Fluid Engineering, Cranfield, Bedford, MK43 0AJ.

Nomenclature

C	- concentration of polymer additives
C_D	- drag coefficient
D_h	- hydraulic diameter of the rectangular pipe
d	- the wire diameter
f	- friction coefficient in a square pipe
k_N	- characteristic constant for Newtonian case
k_P	- characteristic constant for polymeric regime
L	- the length of the wire
n	- the number of wires
PN	- polymer number $PN = \tau \nu/d^2$
Re	- Reynolds number $Re = Vd/\nu$
S	- the area of the cross-section
U	- the average velocity in the cross-section
Ws	- Weissenberg number $Ws = \tau_1 V/d$
$Ws_{critical}$	- critical Weissenberg number

Greek symbols

Δp	- the pressure drop along the test section
τ	- the relaxation time of the solution
τ_1	- the relaxation time of the polymer
ν	- the kinematic viscosity
ρ	- the density of the solution

INTRODUCTION

The Toms effect, discovered in 1948, has remained an object of interest to scientists. In spite of a large number of studies its mechanism is still imperfectly understood. (Ref. 7). For this reason, attempts have been recently made to study laminar flows with polymer additives involving both shear and logitudinal rate of strain. Interest in these flows was aroused by the marked effects of the longitudinal strain on the viscosity. These effects lend themselves to theoretical analysis (subject to certain restrictions) by molecular theories (Ref.9) as well as by macroscopic rheological approaches such as Maxwell's convection model (Ref.2). However the degree of their correlation with the Tom effect is still an open question. Typical examples are flows through diaphragms (Refs.1,8), flows through porous media (Refs.5,12), and flows past cylinders of small diameter (Refs.3,4,6).

The present work deals with the flow of dilute polymer solution past small circular cylinders. The experimental data on the flow of dilute polymer solution past small cylinders appear to be contradictory. On one hand, direct drag measurements (Refs.3,4) show only an increase of drag in the presence of polymers and a very mild dependence of the drag coefficient on the Reynolds number (See Fig.1). On the other hand, measurements of the Strouhal number (vortex shedding frequency) (Ref.6), show substantial reduction of this parameter in the presence of polymers. This latter observation suggests a reduced drag coefficient, as the drag in this region is approximately proportional to the vortex shedding frequency (Ref.11).

To provide a better understanding of the phenomena, it was decided to conduct measurements of the drag of cylinders over a wide range of Reynolds numbers in which both reduced and increased values of drag might be recorded, as well as to obtain simultaneous measurements of the skin friction reduction in a pipe flow with the same solutions. The experimental results and their analysis are presented in this work.

DESCRIPTION OF THE EXPERIMENTS

The experimental system is described schematically in Fig.2. During the experiments the inlet valves were fully opened to eliminate degradation upstream the test section.

The test section was a rectangular pipe (20 x 9.5 mm) containing a set of taut vertical cylindrical wires (nylon and stainless steel) anchored at the top and bottom. The nylon wires (0.25 and 0.45 mm diameter) were arranged in 12 rows spaced 10 mm apart, each row consisting of 4 wires spaced 2 mm apart. The stainless steel wires (1.3 mm) were arranged in 6 rows spaced 20 mm. Each row consisted of 2 wires spaced 4 mm. apart.

The test solutions were prepared from a master stock solution (2% polyox coagulant) which had been diluted to 1000 ppm before it was poured into the head tank together with the incoming water which created mixing in the tank. The quality of the solution was controlled by measuring the drag reduction of samples from the tank in a calibrated pipe of a small diameter under a constant head.

The pressure drop a long the various sections of the pipes was measured using manometers. The pressure drop along sections containing the wires was about 10 times larger than the corresponding pressure drop along the same section without the wires. This increase in the pressure drop, Δp, is primarily due to the drag acting on the wires. One can thus define a drag coefficient for a single wire using the following equation:

$$C_D = \frac{2\Delta p S}{n \rho V^2 d L}$$

where Δp is the pressure drop along the test section, S is the area of the cross-section, n is the number of wires, d is the wire diameter, U is the average velocity in the cross-section , L is the length of the wire , and ρ is the density of the solution.

Obviously, this drag coefficient is not equal to the drag coefficient of a single wire in a uniform unbounded stream. The difference between the two is due to the increased shear on the walls, the non-uniform velocity in the cross section, and primarily the effect of blockage. This last effect was estimated to increase the drag coefficient by the order of (30-50%). Indeed the measured values of C_D for Newtonian flows were larger by that order of magnitude from the corresponding values for single cylinders reported in the literature. The effect of the polymers, as will be shown later, was to increase the drag coefficients by as much as 600% above the Newtonian value. It may thus be concluded that the measured effect of the polymers on drag of the cylinders is similar to their effect on the drag of a single cylinder in an unbounded flow.

THE EXPERIMENTAL RESULTS

The measured values of the drag coefficient C_D, as defined above, for three sizes of wires, d = 0,25, 0,45 and 1,30 mm are plotted in Figs.3,4 and 5 versus the Reynolds number $Re = Ud/\nu$ where d is the diameter of the wire, V the mean velocity in the pipe and ν for the tested concentration were practically equal to that of water. The drag coefficient for each wire was measured for several concentration up to C = 50 ppm.

The experimental results show a clear dependence of C_D on Reynolds number, wire diameter and concentration. The largest effect, at each concentration, was found in the smallest diameter wire. At small Reynolds numbers a very large drag coefficient was measured almost 7 times larger than that of water. On the other hand the data show at larger Reynolds numbers a significant drag reduction. To the best of our knowledge this set of measurements is the first in which both a drag increase and a drag reduction are obtained at different Reynolds numbers with the same wire and the same solutions.

Unlike the data of James and Gupta (Ref.9), see Fig.1. ,these data suggest that the function C_D (Re) is practically a monotonically decreasing function, except at very high Reynolds numbers. It is noted that the maximum slope of this function, in the log log plot is -1, namely $C_D = k_p Re^{-1}$, where k_p is constant. This slope prevails in some experiments over half a decade, see Fig. 3. When one compares our results with those of James and Gupta it becomes evident that the two sets of measurements were made at different regimes. There is some overlapping, as in the present data a region of $C_D \simeq$ constant is also observed. It is also clear that the two sets of measurements complement, rather than contradict, each other.

In Fig. 3 (b) we have also described the measured values of the skin friction coefficient f (calculated from Darcy - Weisbach Equation) for the rectangular pipe. The Reynolds number Re' in this figure is based on the hydraulic diameter of the pipe. (The figure was plotted so that the simultaneous measurements of the pair C_D and f corresponding to the same average velocity in the pipe would be at the same point of the common velocity scale.)

It is clear from this figure that the same solution which gives both a drag increase and a drag decrease in case of a cylinder, gives only a friction reduction in the pipe.

Now, it was observed that the drag increase depends to a large extend on the preparation and aging of the solution. Figure 6 shows four tests with 20 ppm solutions prepared from the same solution of 2%, which were diluted and tested at different times. Obviously, the phenomenon of drag increase is very sensitive to the aging. The difference in performance of these solutions as skin friction reducers in the pipe and as drag reducers in the region of high Reynolds numbers, on the other hand, was hardly recognized.

DISCUSSION

The experimental results clearly demonstrate a diameter effect which is caused by the polymers. Denoting by τ the characteristic time of the solution (at large strain rates), it follows that

$$C_D = F(Re, PN)$$

where

$$PN = \tau\nu/d^2$$

The dimensionless parameter PN, was initially proposed by Poreh and Paz (Ref.10) and later used by James and Acosta (Ref.3). Note that this number is constant for a given polymer-solution-wire system, that is independent of the velocity.

Although the above general function cannot yet be fully determined, neither analytically nor experimentally, its approximate shape, as well as its form in some regions can be deduced from the available data and general considerations.

Figure 7 shows what the authors believe to be the approximate variation of the drag coefficient C_D(Re, PN) for three values of PN. The broken line, PN_2, corresponds to a small value of PN for which the effect of the polymers is small. The solid line PN_1 ($> PN_2$) demonstrates a large effect of the polymers and suggests that four regions should be recognized.

Region I

When subjected to a straining motion the macromolecules in the solution will elongate and consequently the resistance of the fluid to the straining motion will be increased. This change is expressed in the constitutive equations of the fluid by an increase in the value of the longitudinal viscosity. The elongation of the macromolecules takes place only after the Weissenberg number, $W_S = \tau_1 U/d$, where τ_1 is the relaxation time of the polymer molecule, reaches a critical value.

For a given PN, the limit Re $\to$ O corresponds to $W_S \to 0$. There is therefore a small Reynolds number region, for any PN, in which the effect of the polymers is not recognized and C_D in the polymer solution is equal to C_D in Newtonian fluids:

$$C_D = k_N Re^{-1}$$

Region II

As the Weissenberg number increases above the critical value, the longitudinal viscosity increases very rapidly, and reaches a new high value, which is practically constant for $W_S > 2W_{S\,critical}$. According to Peterlin this value can be several order of magnitudes higher than its original value ν. (Ref.9)

If such a change would occur simultaneously in the entire flow around the cylinder the fluid will behave at the region $Ws > 2Ws_{critical}$ like a viscous fluid with a higher effective viscosity and the drag coefficient would vary as $k_p Re^{-1}$ where $k_p >> k_N$.

However, the transition to this Re^{-1} region in case of a flow around a cylinder is slow as the rate of strain in the field is not uniform. Moreover, the polymer solution is usually heterogeneous. Therefore there is a transition region which can extend over a range of one to two decades of the Reynolds number. In this region two opposite effects determine the change of C_D. On one hand the effective viscosity increases significally as the local Weissenberg number becomes larger than the critical value in a larger portion of the flow field. On the other hand, C_D decreases, as it usually does in the viscous regime, with increasing Reynolds numbers. Apparently the two effects are more or less balanced as both the data of James and Gupta (Ref. 4) and of the authors show that C_D is approximately constant in this region (See Fig. 1).

James and Acosta (Ref. 3) and James and Gupta (Ref. 4), who were not aware of the behaviour of C_D versus Re at higher Reynolds numbers proposed that in this region the viscous effects are negligible and thus C_D is a function of the polymer number PN only, which is a constant for each experiment. The existance of a third region in which $C_D \div Re^{-1}$ does not support their proposition.

It should be noted that the beginning of the second region depends on τ_1, the characteristic time of the molecule, which for very dilute solutions is independent of the concentration. Indeed James and Gupta's data for 5 ppm and 10 ppm (See Fig. 1) show that the effect of the polymers in these two cases starts at the same point but the curves deviate later.

Region III

If the effective viscosity reaches the maximum value, one obtains again, at higher Reynolds numbers based on the Newtonian viscosity, a viscous behaviour $C_D = k_p Re^{-1}$ where $k_p >> k_N$. This can happen only in flows around small cylinders.

Note that line C_D versus Re at the right side of this region crosses the Newtonian line and one obtains there a drag reduction.

Region IV

In Newtonian fluids when the Reynolds number becomes of the order of 100 the flow around the cylinder is still laminar but the drag is essentially a form drag which is independent of the Reynolds number, and vortices are shed in this regime into the wake.

If the Weissenberg number of this unsteady flow is large enough, and this is the case if there is an earlier drag increase, the presence of the macromolecules will effect this unsteady flow, apparently inhibit the shedding of the vortices and in this way decrease the drag.

Since the drag itself is proportional to the frequency of the vortices a drag reduction implies a reduction in that frequency too. The measurements of Kalashnikov and Kudin (Ref.6) show such a frequency reduction and thus complement our observations of a drag reduction in this region.

CONCLUSIONS

We have seen that macromolecules which cause a drag reduction in pipe flow can cause both an increase and a decrease in the drag of small cylinders.

The drag increase occurs when the inertial forces in the flow are small whereas the reduction occurs when the drag is essentially a form drag. In these regions a part of the flow is unsteady and the Weissenberg number based on the time scale of this unsteady motion is larger than the critical Weissenberg number of the polymer.

The relation between this phenomena and drag reduction in pipe flows is not obvious. It may be speculated, however, that there is a similarity between the attenuating effect of the macromolecules on the formation of the vortices behind the wire and formation of bursts in a turbulent boundary layer.

ACKNOWLEDGEMENTS

This work was supported by a grant from the United States-Israel Binational Science Foundation (BSF), Jerusalem, Israel.

REFERENCES

1. Bilgen, E. "On the orifice flow of dilute polymer solutions". Journal de Mecanique, 12, 3, pp. 375-391 (1973).

2. Everage, A.E. and Gordon, R.J. "On the stretching of dilute polymer solutions". AIChE Journal, 17,5, pp. 1257-1259 (September,1971).

3. James, D.F. and Acosta, A.J. "The laminar flow of dilute polymer solutions around circular cylinders". J. Fluid Mech., 42,2, pp. -69-288 (1970).

4. James, D.F. and Gupta O.P. "Drag on circular cylinders in dilute polymer solutions". Chemical Engineering Progress Symposium Series. Drag reduction, 67, 111, pp. 62-73 (1971).

5. James, D.F. and McLaren, D.R. "The flow of dilute polymer solution through porous media". Dept. Mech. Engineering Univ. of Toronto, Annual Rep. (1970).

6. Kalashnikow,V.N. and Kudin,A.M. "Karman vortices in flows of solutions of friction-drag-reducing polymers". Disa Information, 10, pp. 3-6 (October 1970).

7. Lumley,J.L. "Drag reduction in turbulent flow by polymer additives". J. Polymer Sci., Macromolecular Reviews,7, pp. 263-290 (1973).

8. Metzner,A.B. and Metzner,A.P. "Stress levels in rapid extensional flows of polymeric fluids". Rheologica Acta,9, pp. 174-181 (1970).

9. Peterlin, A. "Hydrodynamics of linear macromolecules". Pure and Appl. Chem. 12, pp. 563-586 (1966).

10. Poreh,M. and Paz, U. "Turbulent heat transfer to dilute polymer solutions". J. Heat Mass Transfer. 11, pp. 805-818 (1968).

11. Schlichting, H. "Boundary Theory", McGraw-Hill, N.Y. (1972)

12. Vossoughi, S. and Seyer, F.A. "Pressure drop for flow of polymer solution in a model porous medium". The Canadian Journal of Chemical Engineering. 52, pp. 666-669, (October, 1974).

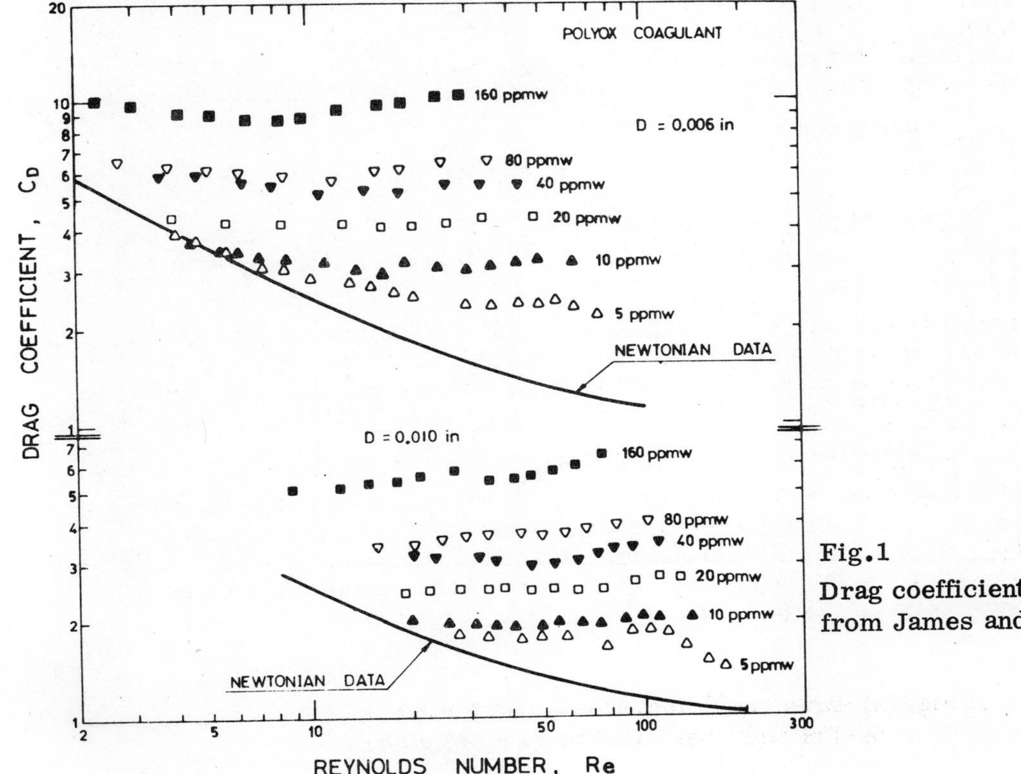

Fig.1
Drag coefficient measurements from James and Gupta (Ref.4)

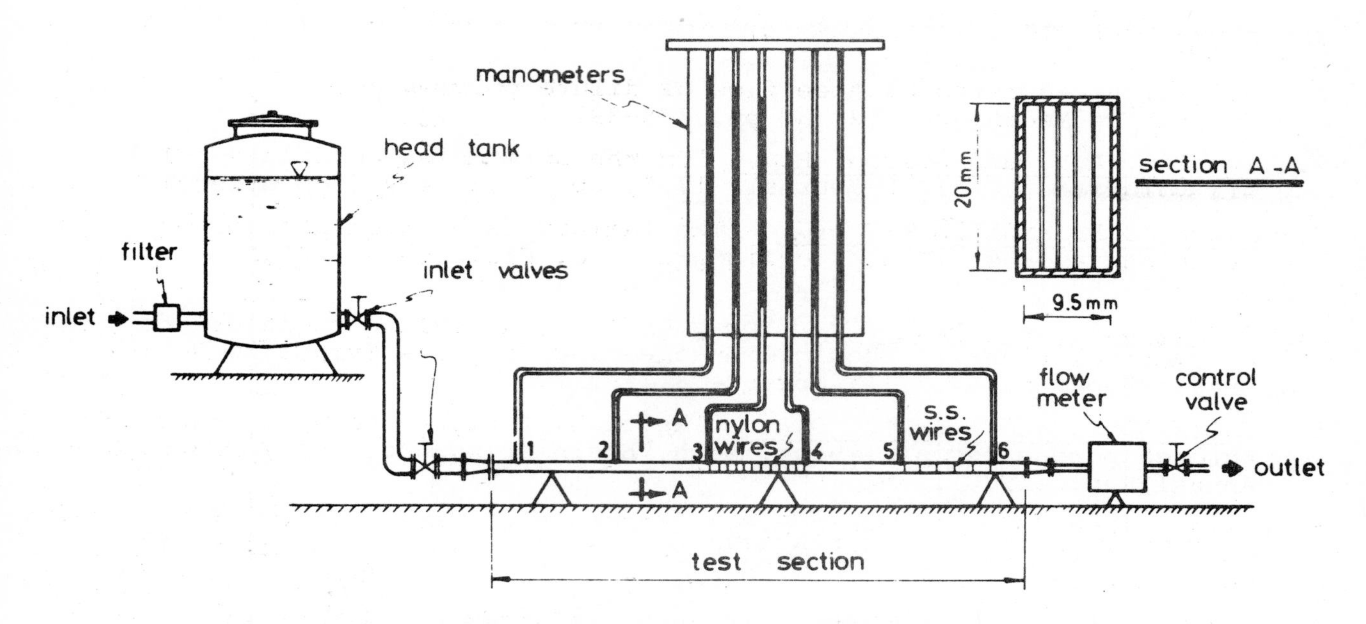

Fig.2 The experimental system

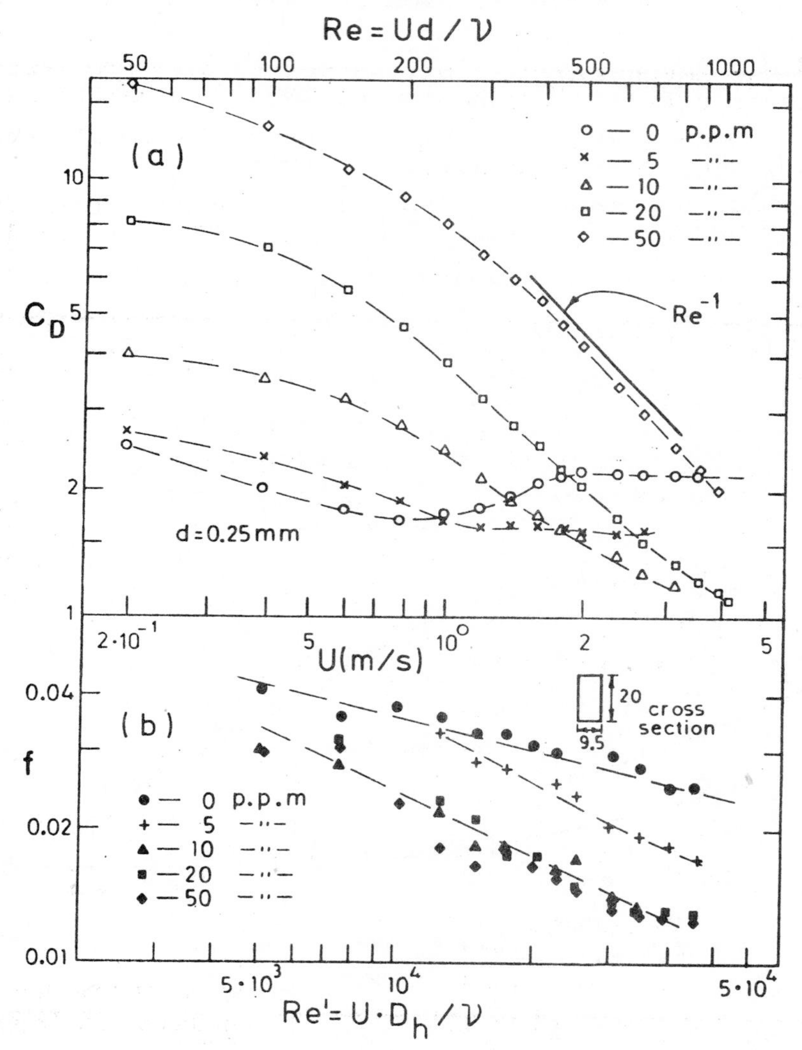

Fig.3a) Drag coefficient of a 0.25mm wire

b) Friction coefficient in a square pipe

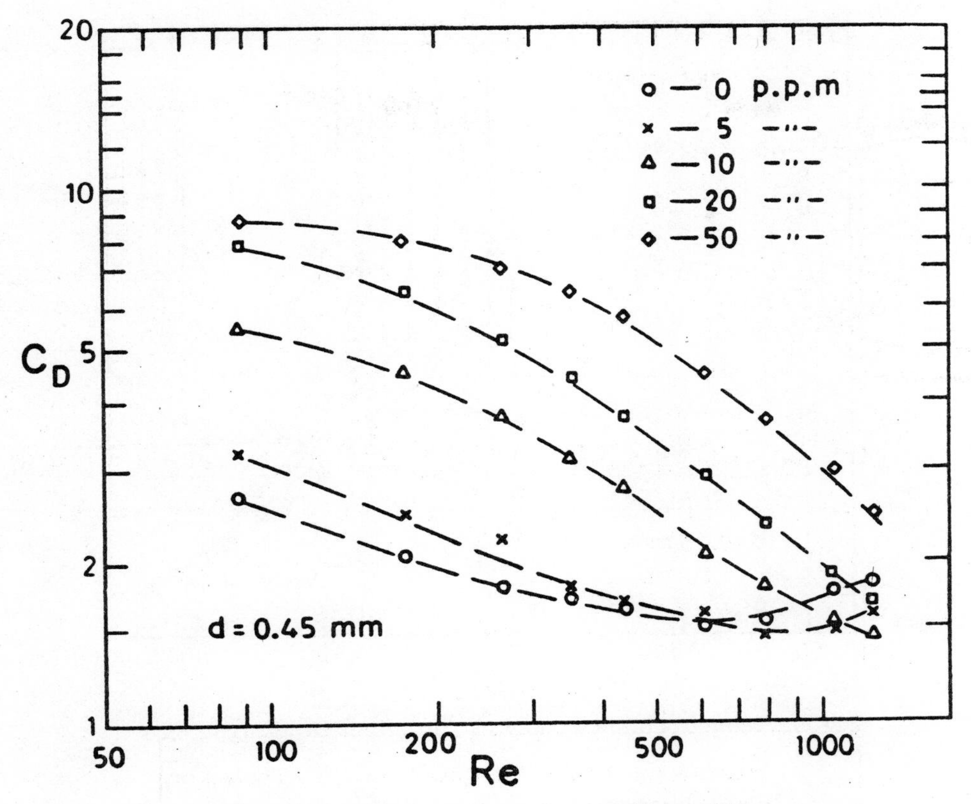

Fig.4 Drag coefficient of a 0.45 mm wire

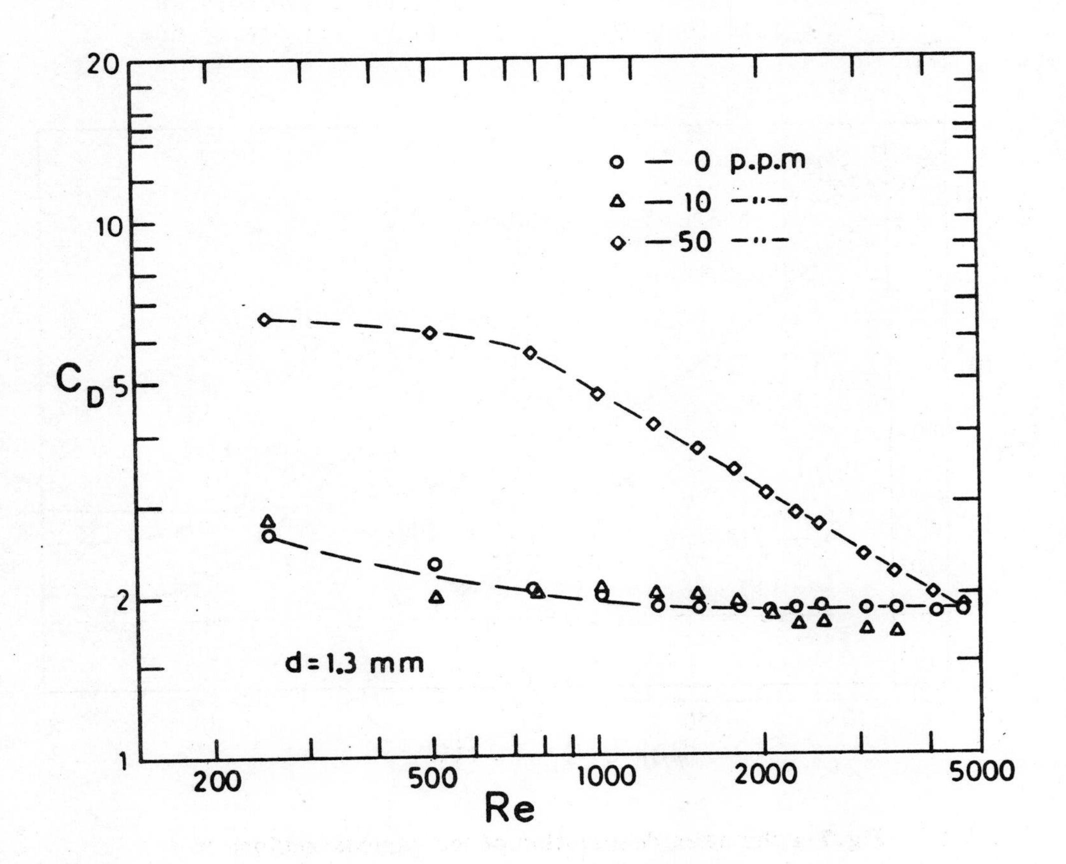

Fig.5 Drag coefficient of a 1.3 mm wire

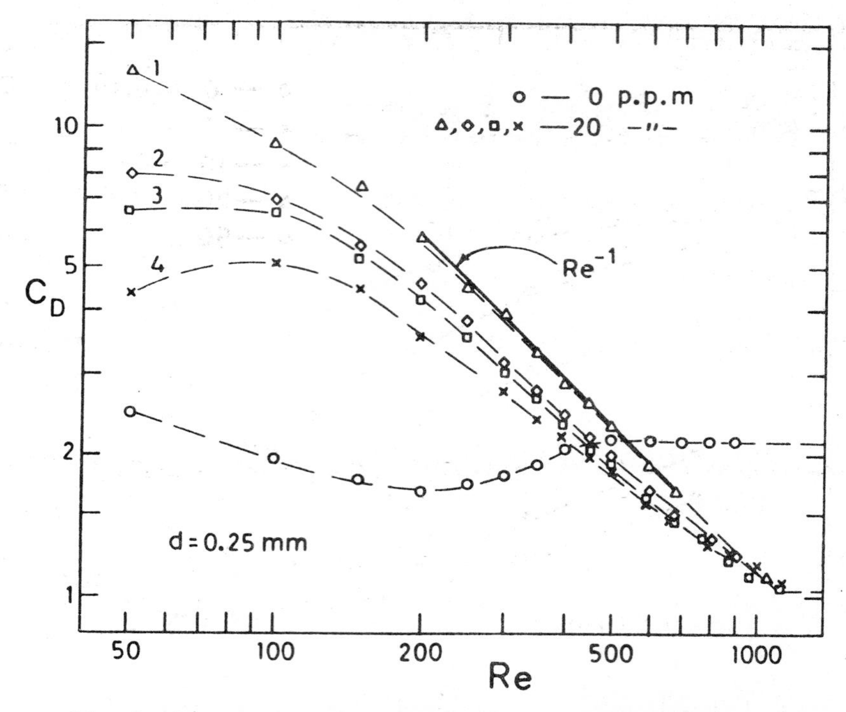

Fig.6 Measurements showing the effect of aging

	Master solution prepared on	Diluted on	Tested on
1)	1975-10-1-08.00	10-1-12.00	10-1-12.30
2)	1975-10-1-08.00	10-1-12.00	10-2-12.30
3)	1975-10-1-08.00	10-15-12.00	10-15-12.30
4)	1975-10-1-08.00	10-15-12.00	10-16-12.30

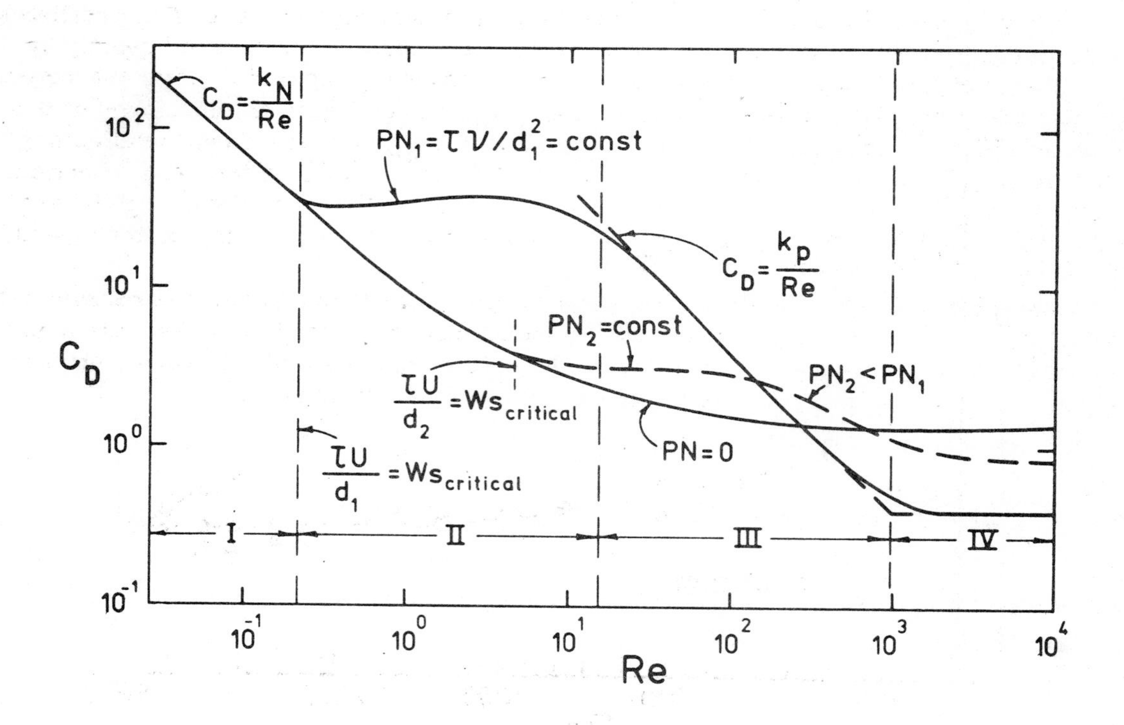

Fig.7 Schematic description of the various regions in the drag coefficient curve

Second International Conference on

Drag Reduction

August 31st - September 2nd, 1977

PAPER F3

DRAG REDUCTION OF AN OSCILLATING FLAT PLATE INJECTING DILUTE POLYMER SOLUTIONS

H. Miyamoto and H. Akiyama,

Keio University, Japan,

and

Y. Tasaka,

Sumitomo Denko Co. Ltd., Japan.

Summary

The drag reduction of a flat plate oscillating with large amplitude was studied, as one representative problem of unsteady flow. The characteristic mutations of velocity profile in boundary layer were investigated, in order to research the influences of injected dilute polymer solutions on the drag-reducing effect.

The drag reduction was obtained by measuring the response amplitude of an oscillating plate submerged in water. Then, it was found that the skin friction was reduced greatly by injecting Poly-Ethylene-Oxide solution from the porous surface of the plate. This characteristic was shown in the correlation between the friction coefficient and the Reynolds number of the oscillating plate or the Strouhal number of it. And discussion was made about the effects of the molecular weight, the injection rate and the concentration of solution, on the drag reduction. In comparison with the former research results of steady flow, the characteristic difference was found such that the maximum point of drag-reducing curve did not appear in this unsteady experiment.

To know the causes of these phenomena, it is necessary to investigate the behaviour of polymer solutions at the surface of the plate. By measuring the profile of the velocity amplitude using a conical hot film probe and photographing the region near the wall, it became obvious that the penetration depth was extended by injecting the polymer solutions.

Held at St. John's College, Cambridge, England.
Organised and sponsored by BHRA Fluid Engineering, Cranfield, Bedford, MK43 0AJ.

NOMENCLATURE

A	response amplitude of model
C	friction constant
C_f	mean friction coefficient (per 1 cycle)
L	model length
Q	injection flow rate
Re	Reynolds number ($= \frac{U_m L}{\nu}$)
S	total surface area of model plate
St	Strouhal number ($= \frac{\omega \delta}{U_m}$)
u	fluid velocity
U_m	mean velocity (per 1 cycle)
U_0	velocity amplitude of the wall ($= A \omega$)
a	amplitude of forced vibration
f	frequency
k	spring constant
m	mass of model plate (in water)
$2\bar{n}$	equivalent damping coefficient
t	time
x	distance parallel to the direction of motion
x_0	displacement of foundation
y	distance perpendicular to the wall
δ	penetration depth of water ($= \sqrt{2\nu / \omega}$)
ν	kinetic viscosity of water
ρ	density of water
ϕ	phase lag ($= y\sqrt{\omega / 2\nu}$)
ω	angular velocity
ω_0	natural angular velocity
$\bar{\omega}_0$	equivalent natural angular velocity

1. INTRODUCTION

A large number of the literatures (Ref.1, 2) which dealt with Toms effect, have been published for several decades, and the greater part of them have taken up steady flow problems without distinction between internal flow and external flow. But actually, if we pay our attention to hypotheses on the drag reduction mechanism, there is no single theory of it which fits all the experimental evidences generally. This fact must await further and many-sided investigations for its proof.

There are many workers, who have studied the drag reduction of external flow especially as a problem of a flat plate which injects polymer solutions into the boundary layer near its leading edge, and some of them not only obtained a large drag reduction but also considered the polymer economy of its injection. For example, Tagori (Ref.3) studied overall ejection of polymer through the porous wall of a flat plate. And Kowalski (Ref.4) and Wu (Ref.5) tested to make the injection rate of polymer pulsate and succeeded in decreasing the polymer consumption under the same drag reduction.

Separately from these remarkable achievements, it is noticeable that there is not enough of the necessary basic data on unsteady problem of this phenomenom. This paper deals with the unsteadiness in boundary layer flows near the flat plate. Concretely, the drag reduction of an oscillating flat plate injecting dilute polymer solutions is studied as one representative problem of unsteady flow. According to many former researches, there are several hypotheses on the drag reduction mechanism; stretching of molecules, suppression of small eddies, interaction of additives, visco-elastic effect and so on. These hypotheses have been considered on case of the disturbance vorticities being suppressed near the wall by the polymer solutions, but this method of experiment is different in that the wall of the oscillating plate has its own large kinetic energy of the vortices. In addition, the effect of visco-elasticity should be more remarkable than in the state of rest.

On the other hand, in the practical use, we should often encounter pulsating or unsteady flow and problems of oscillating object in some fluid, so this experiment must be in a sense suggestive of the characteristic of them. In this paper, with attention paid to these paticular aspects, the mechanism and characters of the drag reduction are discussed dynamically and visually.

2. EXPERIMENTAL APPARATUS

The whole experimental apparatus is shown in Fig.1. The shaker of this apparatus is a direct-drive mechanical vibration system. The torque is delivered by a constant-speed motor (750w), in conjunction with a belt-driven speed changer. This vibration exciter consists of a rotating eccentric cam and a connecting link. The displacement is adjusted by exchanging the intermediate link and the operating frequencies can be provided by the belt-driven speed changer and reciprocation of a V-belt pulley. The amplitude of oscillation is picked up by differential transformers, and recorded on an electro-magnetic oscillograph.

The tank, which is filled with fresh water to a volume of about 0.35m^3, has a water supply pipe and a drain pipe, because it is necessary to change the water intermittently in several minutes of experiment before being contaminated with polymer solutions. Polymer solutions are supplied by the solution container with its head energy and adjusted for the flow rate by a sluice valve.

Figure 2 shows the structure of an oscillating plate, which is made of transparent plastic panels, 300mm in width and 380mm in length, with 210 holes of 0.5mm in diameter bored in one line, totally 17 lines being on each side at every 20mm. Both edges of the model plate are sharpened to minimize the pressure drag component. The polymer solutions are introduced in the inner space of the plate through five entrance tubes, and injected into the boundary layer through the numerous holes.

In this experiment, three kinds of Poly-Ethylene-Oxide (PEO18,15,8) are used as the polymer additives, which have been manufactured by Seitetsu Kagaku Kogyo Co.Ltd: PEO18 which has a weight-average molecular weight of about 4,000,000 ~ 5,000,000; PEO15 which has about 3,000,000 ~ 4,000,000 and PE08 which has about 1,000,000 ~ 1,700,000.

Oscillating tests are performed in lower neighbourhood of the resonance frequency in order to obtain a large amplitude, and practically, the experimental frequencies are set at 2.6Hz to 3.5Hz with an interval of o.16Hz, by adjusting the oscillation of foundation considering the plate spring and model weight. Two plate springs which have different values of spring constant (k) are used; k_1=9.17N/cm, k_2=8.53N/cm. The amplitude of foundation is about 1.0cm, and the response amplitudes of model plate are from 2.0cm to 5.2cm. This experiment is run with four injection flow rates; 15, 30, 45, 60cm^3/s and various polymer concentrations: 0, 50, 100, 200, 300, 500, 700, 1000, 1500, 2000wppm.

Thus, a fully reversing oscillation is obtained as a single degree-of-freedom system, and the flow patterns around the oscillating plate, especially in the neighbourhood of the wall, are observed and also photographed through a window which is set at the side of the tank. And the drag reduction is derived from the mutations of friction coefficient which is calculated as described in the following chapter.

3. DERIVATION OF MEAN FRICTION COEFFICIENT

When the flat plate, which has skin friction, is oscillated in a water region as shown in Fig.2, it may be regarded as a single degree-of-freedom system with viscous damping, excited in forced vibration by motion of foundation.

Now, if the mass of a model plate, which is submerged in water, is put as m and the displacement of foundation as $x_0 = a\cos\omega t$, where ω is angular velocity, the differential equation of motion is as follows

$$m\ddot{x} + C\dot{x} + k(x - x_0) = 0 \tag{1}$$

Here, C is the friction constant and k is the spring constant. When the amplitude of the vibratory displacement and the velocity grow larger, the friction resistance is more correctly expressed by a term proportional to the square of the velocity. Further, the resistance is always opposed to the motion, therefore, the non-linear equation of motion should be written

$$m\ddot{x} + C|\dot{x}|\dot{x} + k(x - x_0) = 0 \tag{2}$$

By substituting sinusoidal displacement of foundation (x_0) into equation (2)

$$m\ddot{x} + C|\dot{x}|\dot{x} + kx = ka\cos\omega t \tag{3}$$

In order to solve the equation (3), the method of Kryloff and Bogoliuboff (Ref.6) is used. Assume the equivalent damping coefficient and the natural angular velocity to be $2\bar{n}$ and $\bar{\omega}_0$ respectively, the equation (3) can be rewritten in the form

$$m\ddot{x} + 2m\bar{n}(A)\dot{x} + m\bar{\omega}_0^2(A)x = ka\cos\omega t \tag{4}$$

Here

$$\bar{n}(A) = \frac{4C\omega A}{3\pi}$$

$$\bar{\omega}_0^2(A) = \omega_0^2$$

Consequently, under a steady-state, the amplitude of the reformed equation (4) is

$$A = \frac{ka}{\sqrt{\left[\bar{\omega}_0^2(A) - \omega^2\right]^2 + \left[2\bar{n}(A)\omega\right]^2}} \tag{5}$$

Rearrange the equation (5) by substituting the values of $\bar{n}(A)$, $\bar{\omega}_0$ into it. Then the friction constant C is

$$C = \frac{3\pi\sqrt{k^2a^2 - A^2(k - m\omega^2)^2}}{8A^2\omega^2} \tag{6}$$

Now, the value of C can be calculated by substituting the experimental data: amplitude of model plate A , spring constant k , angular frequency ω , amplitude of the forced vibration a , and the mass of model plate m . After all, the mean friction coefficient C_f is defined as follows:

$$C_f = \frac{C}{\frac{1}{2}\rho S} \tag{7}$$

where ρ is the density of water and S is the total surface area of model plate.

4. EXPERIMENTAL RESULTS AND DISCUSSIONS

Under every decided conditions of angular frequency (ω), polymer concentration (wppm) and injection rate (Q) of polymer solutions, the friction coefficient (C_f) is obtained by recording the amplitude of exciting force and the response amplitude. In this case, water injections at each decided conditions are the standard for comparison with experimental values of solution injection. Based on these investigated data, the character of drag reduction in state of oscillation is discussed in relation to Reynolds number, Strouhal number and polymer concentration. Simultaneously, the velocity profiles of oscillating layer are investigated by using a conical hot film probe, and the flow patterns near the wall are indicated by photographic observation.

4-1. The relation between Reynolds number and drag reduction

Typical examples of injecting effectiveness of PEO18, according as the concentration and the injection flow rate are changed, are presented in Fig.3-1 ~ 3-4 of C_f versus Reynolds number. Here the friction coefficient C_f is derived from equation (7), and the Reynolds number Re is defined as $A\omega L/\nu$, where L is the length of model plate and ν is the kinetic viscosity of water. In these figures, the thick solid curve means the above mentioned standard value of water (0wppm), injected at the same flow rate of polymer solutions.

All friction coefficients of polymer injection are less than that of water injection especially at rather high concentration, but the concentration dependence of C_f seems to be saturated at high concentration (refer to Fig.3-3 & 3-4); then at 1000 wppm, this monotonous decrease of C_f-Re curve begins to be broken. This fact is discussed later in the relation of C_f versus polymer concentration. As it is shown in Fig.3-2, the dependence of C_f on the injection flow rate appears remarkably at 100 ~ 300wppm, so it is found that if a largely dilute or highly concentrated polymer solution is injected, the injection flow rate will not be so important a factor to the drag reduction. Furthermore, these parameters, i.e., polymer concentration and the injection flow rate do not depend greatly on the drag reducing effect in a range of high Reynolds number, which means a comparatively quick oscillation with large amplitude, and so a large drag reduction and effective injection for it are considered to be attained in the low Reynolds number region ($Re < 1.0 \sim 1.5 \times 10^5$).

Besides, with the injection flow rate kept constant (Q =30cm^3/ s), three different polymers which have various molecular weights (PEO8 < PEO15 < PEO18) are injected at several changed concentrations, and the effect of molecular weight is demonstrated in Fig.4-1 ~ 4-3. They show that the range of friction reducing is extended when the injected solution has a larger molecular weight, and the effect of polymer concentration increases simultaneously.

4-2. The relation between Strouhal number and friction coefficient

Figures 5-1 ~ 5-3 show the drag reducing curve as a change of Strouhal number at various concentrations of PEO solutions which are injected at a rate of 30cm^3/s . Strouhal number is defined as $\omega\,\delta/U_m$, supposed to be a parameter which expresses the character of this oscillating flow. Here δ is the depth of penetration of water ($=\sqrt{2\nu/\omega}$), and U_m is a mean velocity per one cycle oscillation. These figures are also expressed by the change of polymer concentration and molecular weight. For the least molecular weight, the effect of PEO8 concentrations is almost negligible (refer to Fig.5-3), but for the largest molecular weight PEO18, the experimental data change remarkably over a wide range of C_f spreading as Strouhal number increases, and they are considered to depend on the frequency f (cycle per second) and the concentrations of PEO18.

From these points of view, the effect of frequency and polymer concentration is shown in Fig.5-1; the drag reduction has a tendency to increase as the polymer concentration increases, but in a certain range of very high concentrations (1000wppm in this experiment), this drag reducing curve begins to have a maximum peak and descend after it. This fact means that there occurs a virtual saturation of drag reduction.

In Fig.5-1, the dotted line means an iso-frequency line, and the effect of the concentration on the friction coefficient becomes smaller in a high range of frequencies. According to these experimental data, PEO15 which has an intermediate molecular weight between PEO18 and PEO8, shows intermediate values between them.

The effectiveness of injection flow rate is presented in Fig.6, in which three polymer solutions are used having a concentration of 300wppm. It is found that PEO8 can hardly reduce the friction, its data almost agreeing with the water injection and the effect of injection flow rate is the least. So it is suggested that a larger molecular weight of linear macro-molecule polymer solutions brings more drag reduction as far as these experimental data are concerned.

4-3. Effect of polymer concentration on the drag reduction

The concentration of polymer solutions has been a very important parameter in the preceding sections, and here the characteristic curves of C_f versus polymer concentrations are obtained as Fig.7-1 ~ 7-3. Three kinds of polymer solutions are compared with each other at different Reynolds numbers (angular frequencies) under the same injection flow rate of $30cm^3/s$. In each figure, drag reducing effects are increasing in the same order of the three increasing molecular weights.

It is very characteristic that every curve except PEO8 has a steep descent below 200 ~ 300wppm, and so the effective drag reduction is considered to be able to be obtained by injection of rather dilute solutions of about 100 ~ 300wppm. Although these phenomena appear with an increase of Reynolds number, the range of friction reductions becomes less wide at higher Reynolds number under various polymer concentrations.

By the way, it is known that in the steady experimental results of former researches, the C_f-concentration curve has a maximum peak point and decreases after it with an increase of polymer concentration. Nevertheless, no friction increase does appear in this experiment, and it is expected that such a point would not appear or appear under higher polymer concentration than 1000wppm.

4-4. Mean velocity distributions in oscillating direction near the wall

When a flat plate makes a linear harmonic oscillation parallel to itself, a certain distance of δ perpendicular to the wall is considered the depth of penetration and it is regarded as a kind of thickness of boundary layer. In this layer which is carried by the wall, the velocity profile should have the form of damped oscillation. And in this experiment, a distance from the wall is non-dimentionalized by the theoretical depth of penetration of water ($\delta = \sqrt{2\nu/\omega}$).

In this paper, this velocity profile is picked up by a conical hot film probe attached to a fine adjustment carriage and the distance from the wall to the conical nose is measured by its vernier which can be read to 1/10mm with accuracy. The output of hot film is recorded on an electro-magnetic oscillograph by a linearizer and the motion of the wall is picked up by a differential transformer and recorded simultaneously. To obtain an actual size of the velocity amplitude, a charateristic of the conical hot film probe in different solutions is corrected using a calibration curve which is obtained by oscillating the probe in still water and PEO solution.

In view of the whole flow around the oscillating plate, there exists a secondary flow which forms as a twin vortex on each side of the plate. But considering the facts that the velocity amplitudes are measured very near the wall and that the velocity of the secondary flow is further less than it of fluid carried by the wall oscillation, these influences should be negligible in this experimental run.

By way of example of the final results, three cases of velocity amplitudes are compared in Fig.8; the flat plate is oscillating in water with no injection, water injection, and the PEO18 solution-injection through the porous surface of the plate.

A dotted line means a theoretical result of Stokes's 2nd problem (Ref.7). In this experiment, the values of non-injection should be considered as a same case as the Stokes's problem. And the agreement is quite well in Fig.8. But when some fluid is injected through the wall, the condition might have been changed, and so the effect of polymer solution should be discussed in comparison of water and PEO solution injection. Attending these, the remarkable difference is shown as that when PEO solution is injected, the velocity amplitude increases and a flow layer in which a fluid is carried by the wall oscillation, becomes thicker than that of water injection. This fact also means a velocity gradient which is supposed to be an important factor of shear stress near the wall, is reduced by the polymer injection.

The outputs of the hot film probe and the transformer of the wall oscillation are recorded in a same time scale so that a phase lag can be measured, and it is shown in Fig.9. In a region of so-called flow layer, the phase lag of non-injection shows a similar influence with the numerical value of Stokes's solution. And the values of PEO-injection agrees approximately with those of water injection, and they are smaller than those of non-injection. This fact means in effect, that the wall oscillation is propagated more greatly by the injection than the non-injection to the water near the wall, and when PEO solution is injected, the velocity amplitude is larger than the other cases at the same distance from the wall. It is suggested that the polymer molecules have a tendency to adhere to the oscillating plate and thicken the flow layer, so they should decrease the velocity gradient on the wall and reduce the drag of the plate consequently.

4-5. Photographical observation around the model plate

To observe the flow pattern near the oscillating wall, photographic visualization are adopted, and this experiment is done as follows:

(1). Observing the streak line of flow per 1 cycle movement, polystyrene particles which are 0.1 ~ 0.2mm in diameter and are coated with a fluorescent paint, are used as a tracer. Four examples are shown in Fig.10; here Fig.10-1 and Fig.10-2 are showing the case when the flat plate is submerged in a pure water and a PEO18 (200wppm) solution, and Fig.10-3 and Fig.10-4 are showing the flow near the plate oscillating in a water comparing the water injection and the PEO18 solution injection through a porous surface of the plate.

(2). The depth of flow layer is visualized by using aluminum powder, which is carried away by secondary flow from the leading edges to the center of the plate. Thus the difference of the flow layer thickness are shown as in Fig.11-1 and Fig.11-2. On the other hand in Fig.11-3 and Fig.11-4, the aluminum powder is added previously to the injected water and the PEO18 solution and injected into the region near the wall.

(3). To get a general idea of the whole flow around the plate, a mixed solution of polymer and black carbon ink are injected into the boundary layer. These are presented in Fig.12. The upper two small photographs are taken near the dead point of oscillation, and they indicate the difference of injection and of diffusion in shape between PEO solution and water. The lower two photographs are shot from a different angle and the carbon ink solution are injected through a few of the supply tubes controllably.

By flashing the light which passes through a narrow slit over the upper surface, the flow patterns are photographed from a side window of tank using a close telephotographic lens, excepting the method of (3).

In Fig.10-2, PEO solution's own characteristic are shown that it has a tendency to suppress a disturbance from the oscillation and it does not propagate the disturbance to surroundings. This tendency is also shown in Fig.10-4, because comparing Fig.10-3 and Fig.10-4, it is noticeable that the surroundings of plate is more static in Fig.10-4 than Fig.10-3. On the other hand the flow conditions of Fig.10-1 and Fig.10-3 are rather similar, so considering the velocity profile of Fig.8, it is thought the influence of water injection is not so much as the remarkable mutation observed in polymer injection. And another considerable characteristic of PEO solution in Fig.10-2 is to arrange the flow near the wall in direction of the oscillation and particles are adhering to the wall and oscillating with it. In this point, the photograph of Fig.12-2 supports specifically this character which is supposed to be concerned with the difference of diffusion of injected fluids clearly shown in Fig.12.

In Fig.10-3 and Fig.10-4, although the disturbance is propagated much under the water injection, the thickness of flow layer is thinner than it of PEO injection. It soon become obvious in Fig.11; the aluminum powder is floated more from the wall in Fig.11-2 than in Fig.11-1. This fact means the flow layer region is extended when PEO solution is injected, because the distance between the powder traces and the plate wall indicates the thickness of flow layer. And in Fig.11-1 and Fig.11-2, the powder thickness means the condition of boundary layer of pre-mentioned secondary flow which is concerned with the disturbance of surroundings, so this Fig.11-2 supports the above mentioned phenomenom; the suppression of polymer solution to the disturbance. And Fig.11-3 and Fig.11-4 are presenting an oppisite situation; a white gradational part shows the flow layer induced by the oscillating plate with different frequency. Thus Fig.11 concludes clearly that the polymer injection makes the flow layer thickened.

5. CONCLUSIONS

Friction coefficient of a porous flat plate, injecting dilute PEO solutions and oscillating with a large amplitude in water, is obtained experimentally. In addition, visual observation and measurement of mean velocity amplitude profile near the plate are made. Thus the data presented in this and the experience gained in the operation of equipment described lead to the following conclusions:

(1). On the whole, a large quantity of drag reduction is obtained as compared with external steady flows past a flat plate. The drag reduction effect has a tendency to increase as the molecular weight and the injection flow rate increase, but there is a certain limit to this effect under too large flow rate or too high concentration.

(2). When largely dilute polymer solutions are injected, the friction coefficient shows a rapid rate of decrease, which become slower and saturated under higher concentration.

(3). At high Strouhal number or low Reynolds number oscillation, the drag reduction is sensitive to the changes in polymer concentration and injection flow rate.

(4). By injecting polymer solutions, the oscillating flow layer can be thickened and the disturbances can be suppressed near the oscillating flat plate.

6. ACKNOWLEDGEMENTS

The authors wish to extend great appreciation to Professor T.Ando of KEIO University for his helpful suggestions. Their thanks are also due to Mr.A.Uozumi and M.Umeda for their earnest assistances in the experimental works.

REFERENCES

(1). White, A. and Hemmings, J.A.G: "Drag reduction by additives(Review and Bibliography)". B.H.R.A. pp.7-169. (1976).

(2). Hoyt, J.W: "The effect of additives on fluid friction". Trans.A.S.M.E. 94, Series D,2 pp.258-282. (June, 1972).

(3). Tagori, T. and Ashidate, I: "Some experiments on friction reduction on flat plate by polymer solutions". Proc.12th Int.Towing Tank Conf. pp.132-135. (Sept. 1969).

(4). Kowalski, T:"Turbulence suppression and viscous drag reduction by non-newtonian additives". Trans.Royal Inst.Naval Architects, 110, 2 pp.207-219. (April, 1968).

(5). Wu, J: "Surface containing of polymer solutions and pulsative ejection". Nature phys. Sci., 231, 24 pp.150-152. (June 14th, 1971).

(6). Harris, C.M. and Crede, C.E: "Shock and vibration handbook". vol.1,4, pp.15-17. McGraw-Hill. (1961).

(7). Schlichting, H: "Boundary layer theory". pp.85-86. McGraw-Hill. (1968).

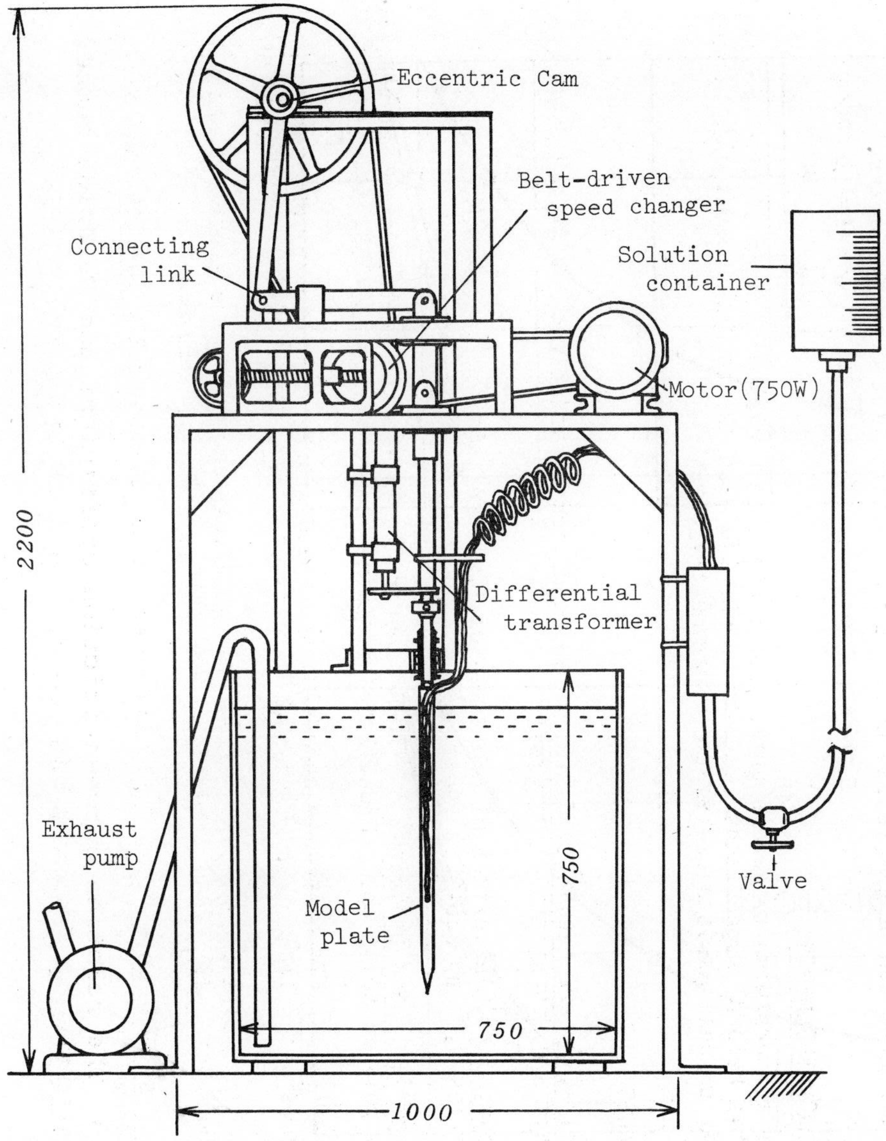

Fig.1 Experimental apparatus

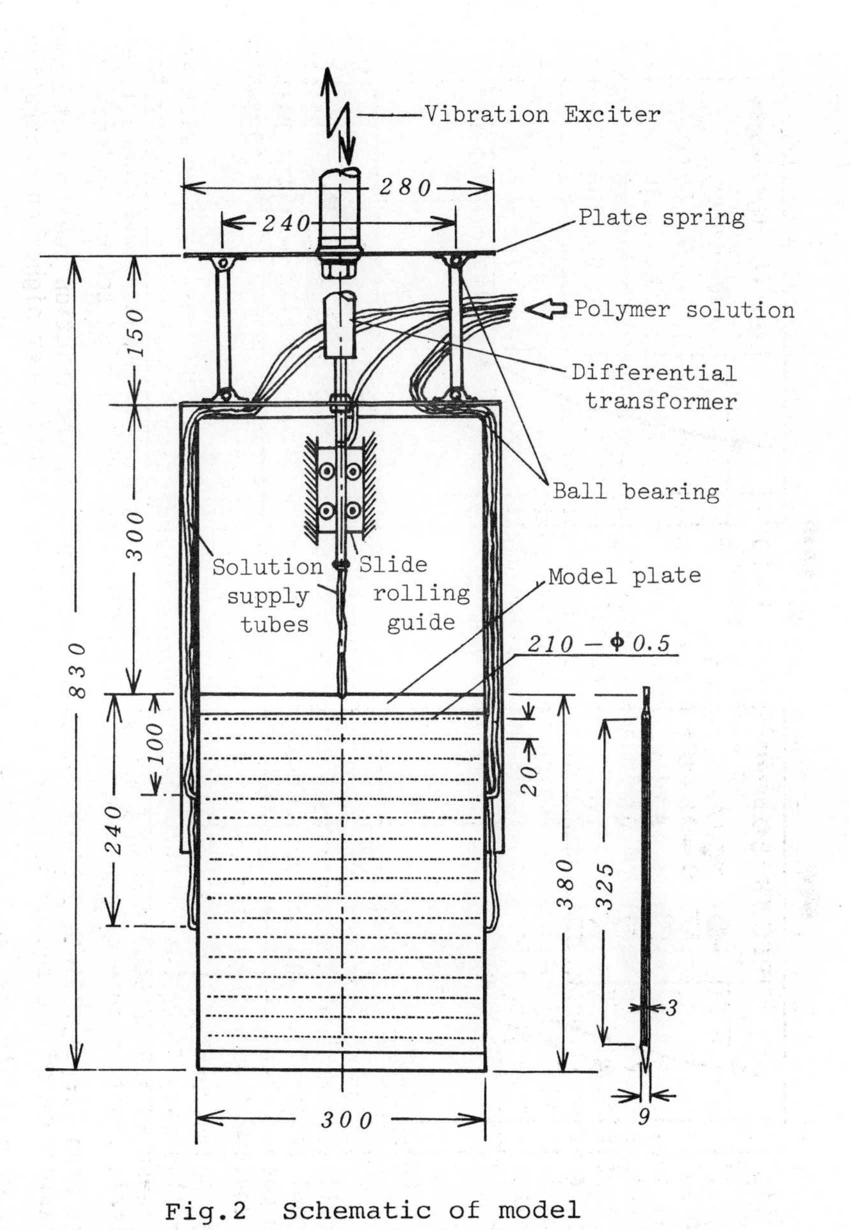

Fig.2 Schematic of model

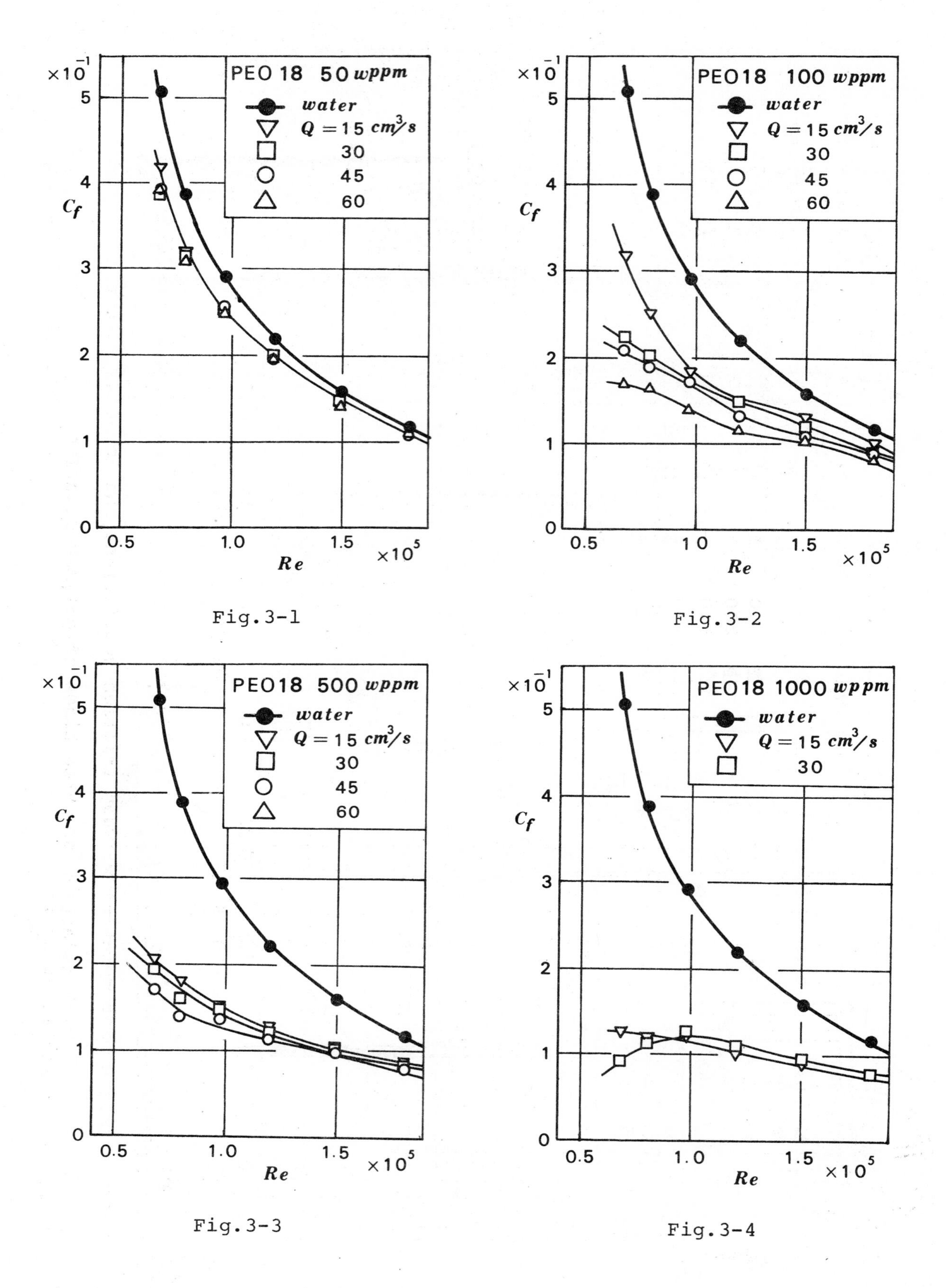

Fig.3 The relation between friction coefficient and Reynolds number

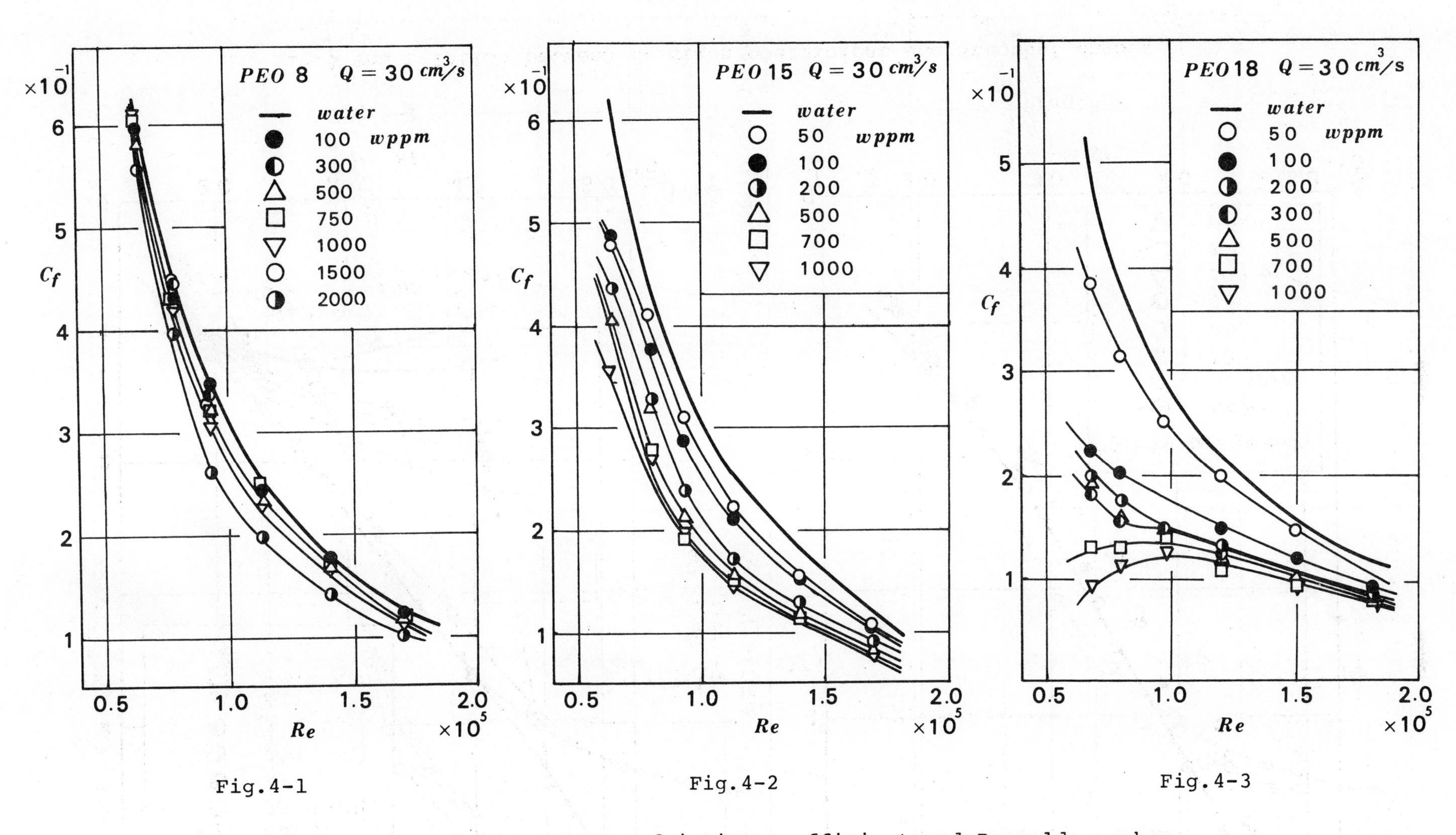

Fig.4-1

Fig.4-2

Fig.4-3

Fig.4 The relation between friction coefficient and Reynolds number with injection of three kinds of polymer solution

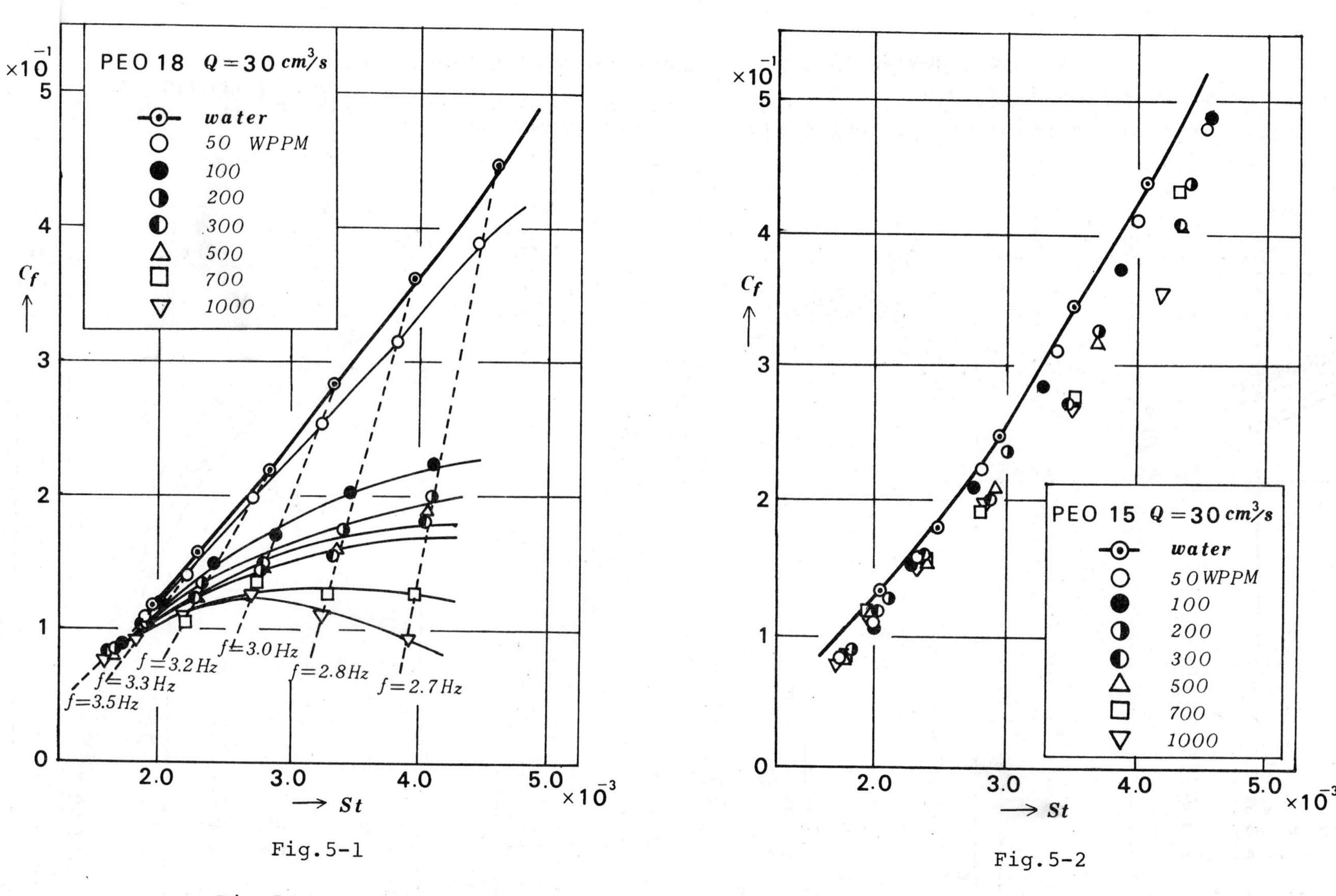

Fig.5-1

Fig.5-2

Fig.5 The relation between friction coefficient and Strouhal number

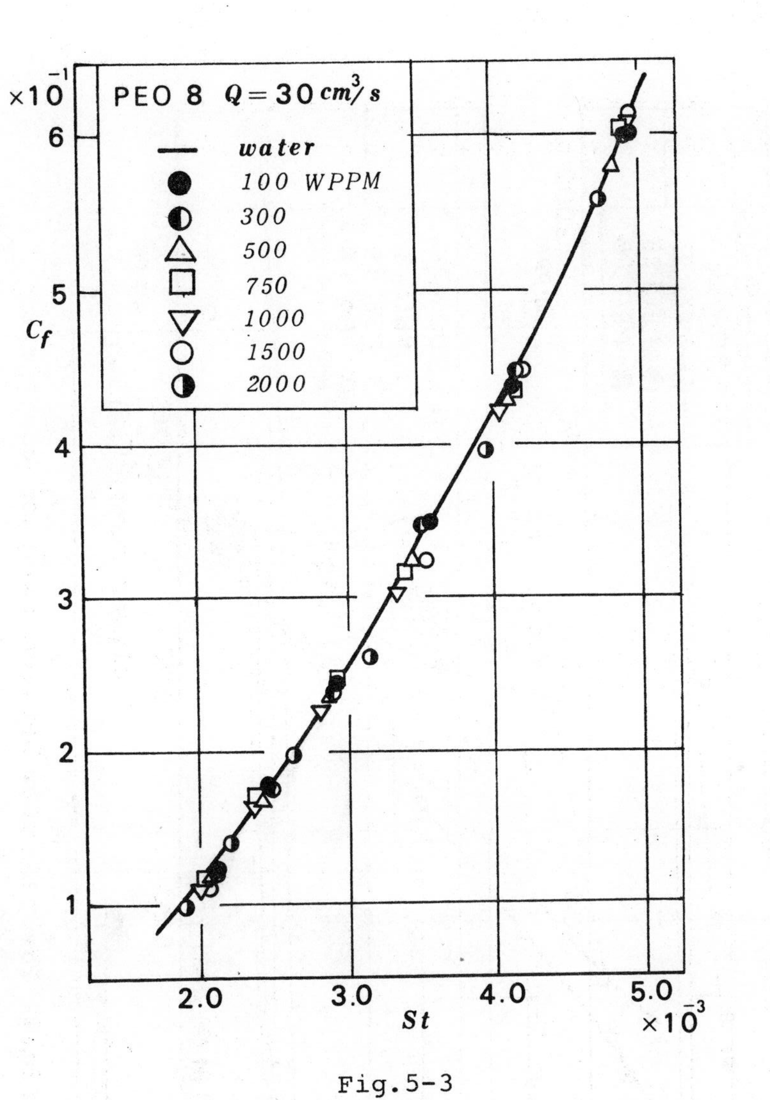

Fig.5-3

Fig.5 The relation between friction coefficient and Strouhal number

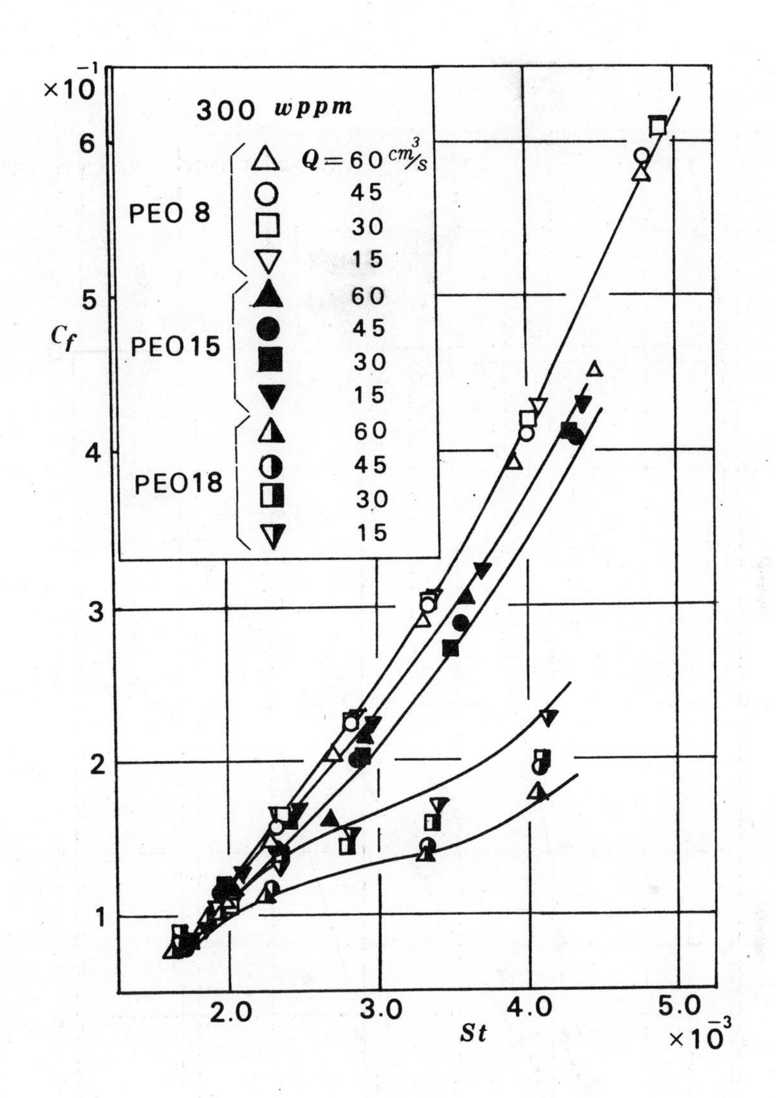

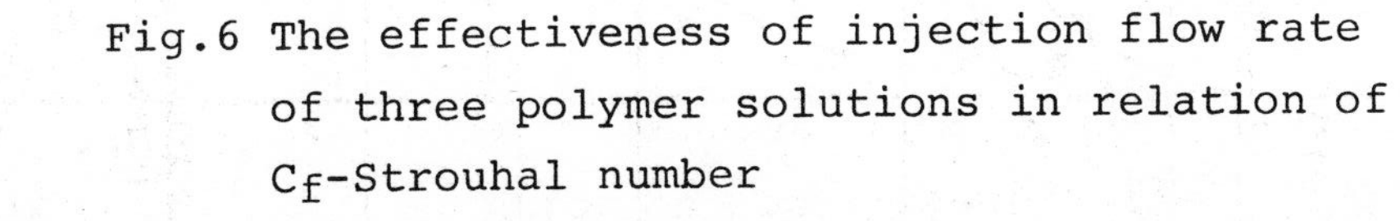

Fig.6 The effectiveness of injection flow rate of three polymer solutions in relation of C_f-Strouhal number

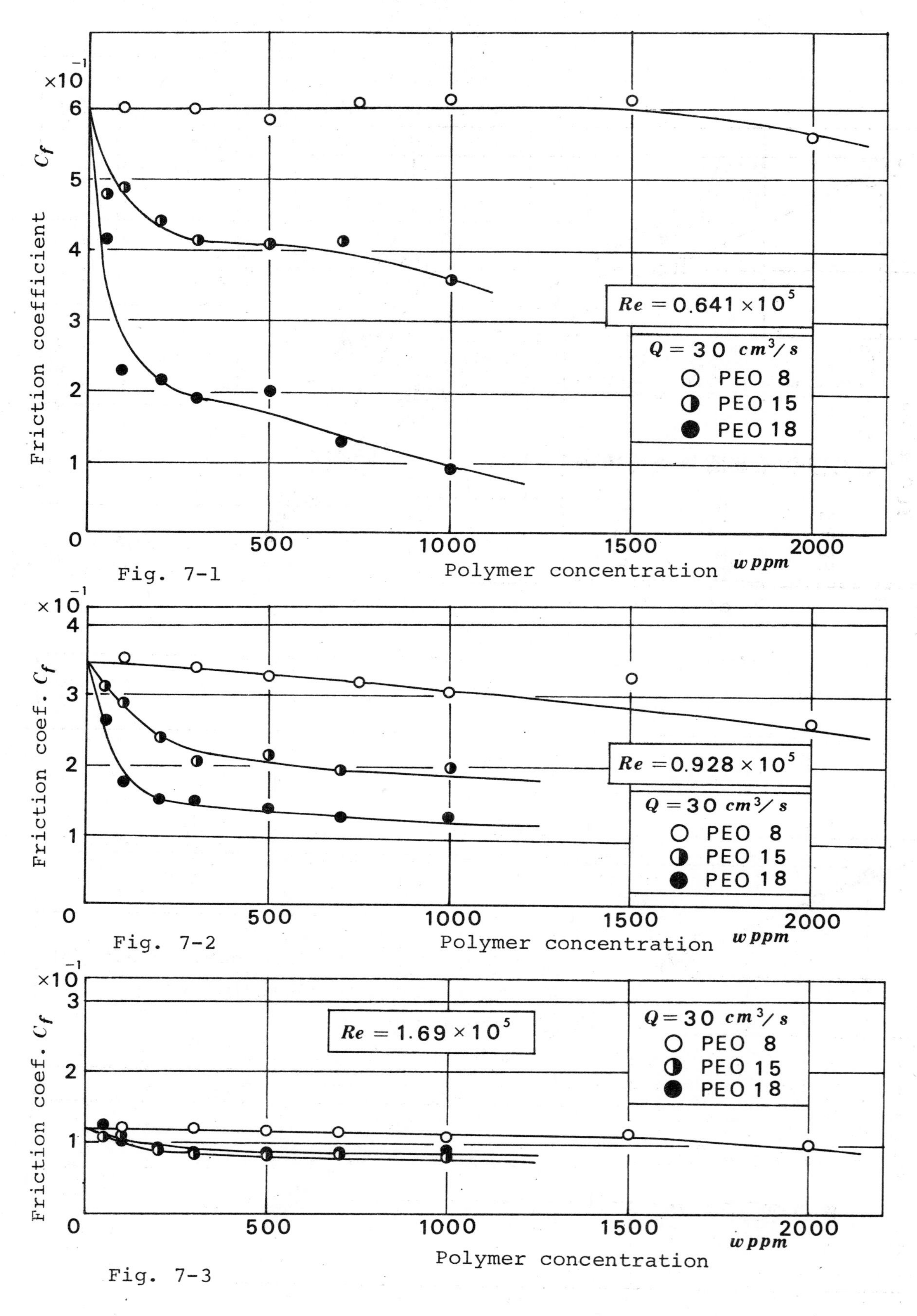

Fig.7 The relation between friction coefficient and polymer concentration

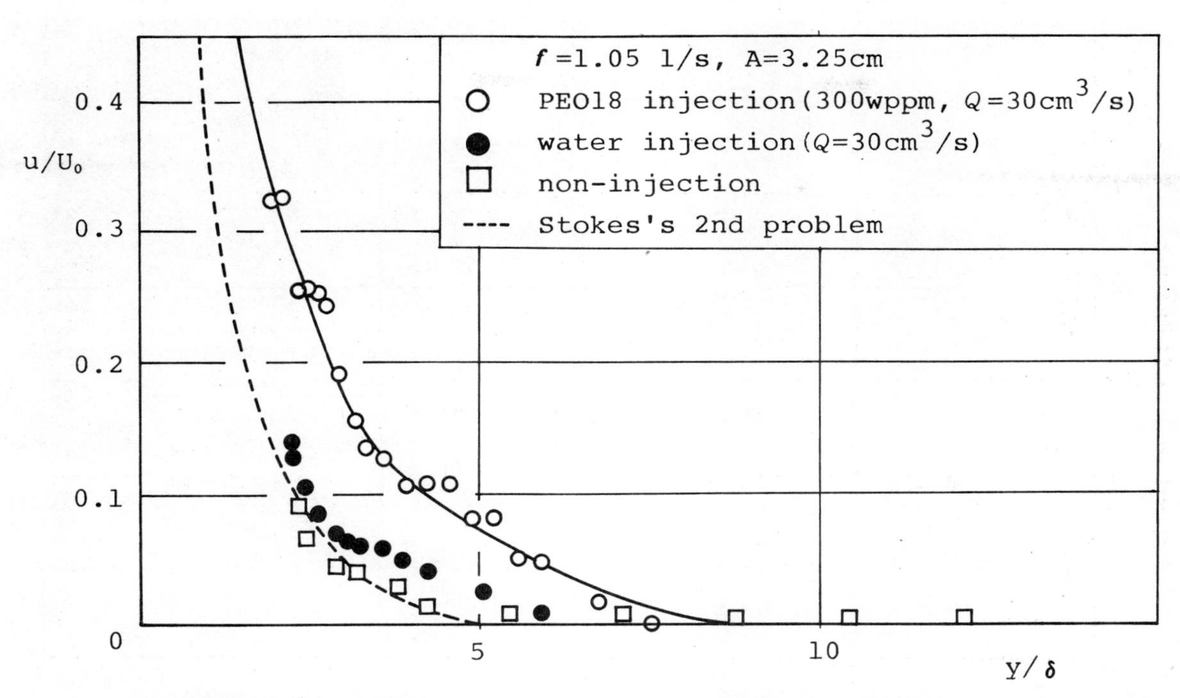

Fig.8 The profiles of velocity amplitude in the neighbourhood of an oscillating wall

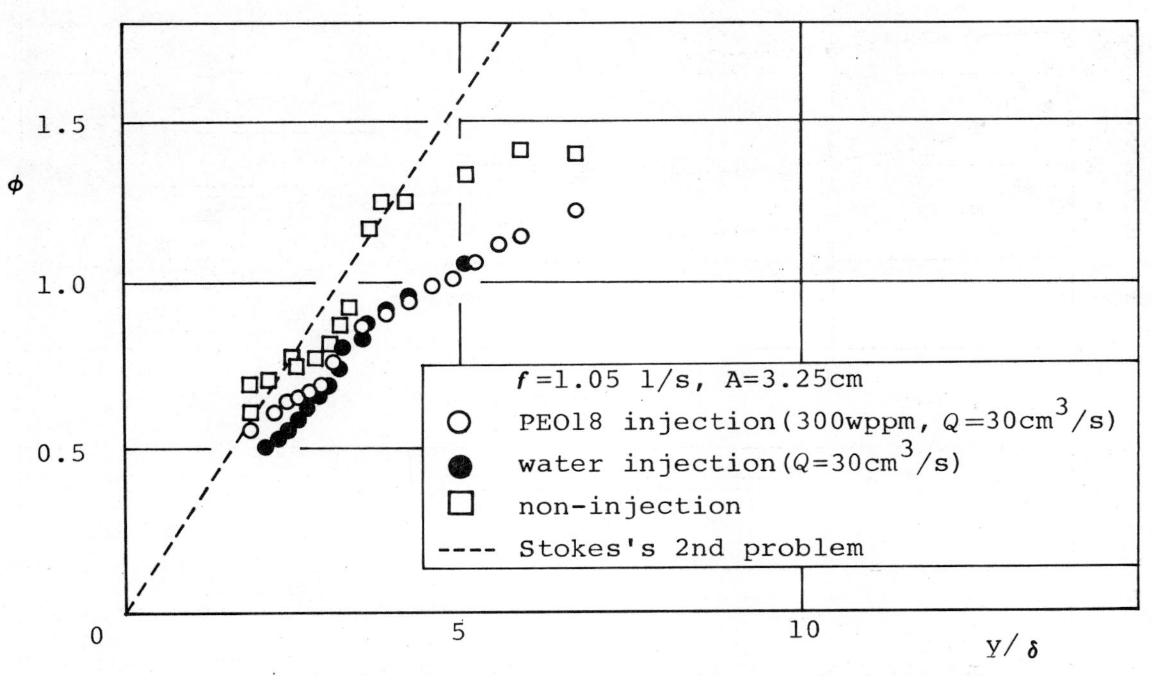

Fig.9 The phase difference in relation of a distance from an oscillating wall

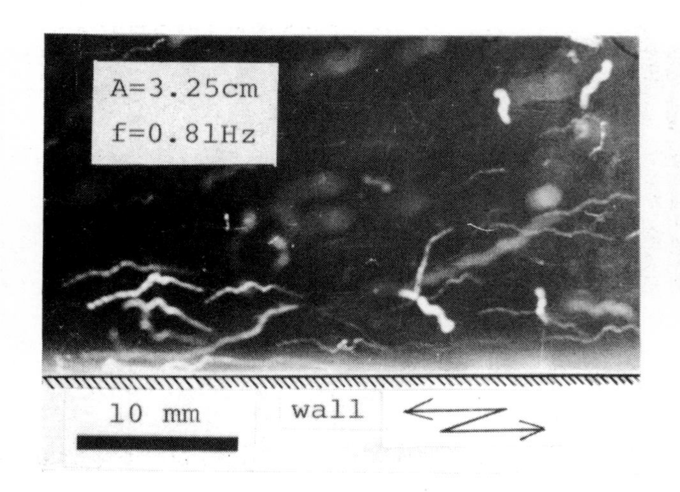

Fig.10-1 Still water

non-injection

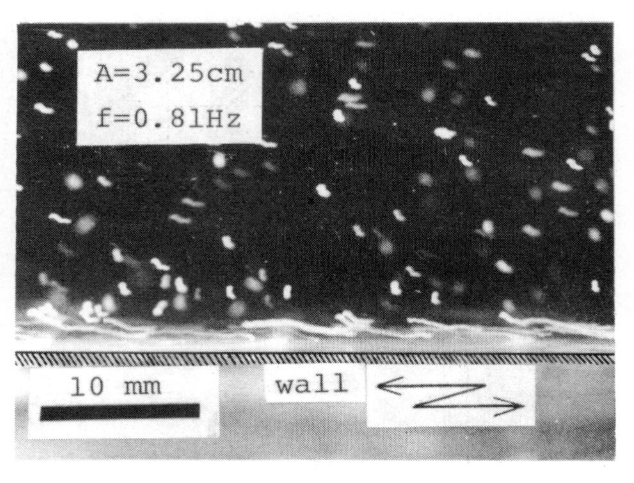

Fig.10-2 Still PEO 200wppm solution

non-injection

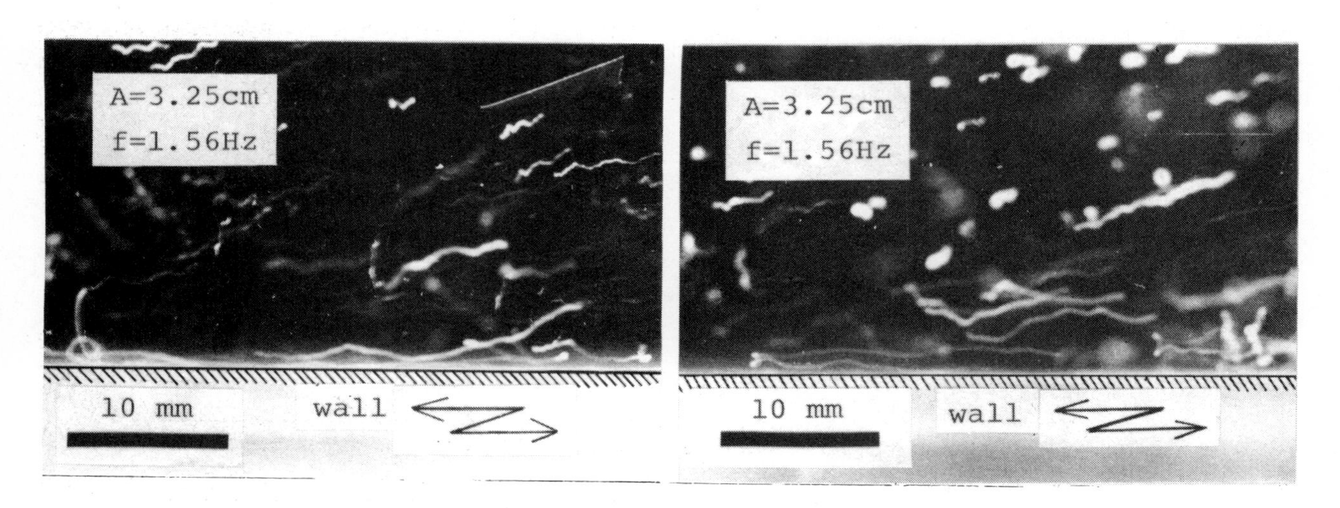

Fig.10-3 Water injection

$Q=20\ cm^3/s$

Fig.10-4 PEO 200wppm injection

$Q=20\ cm^3/s$

Fig.10 The streak line of flow per 1 cycle movement

(Polystyrene particles used as the tracer)

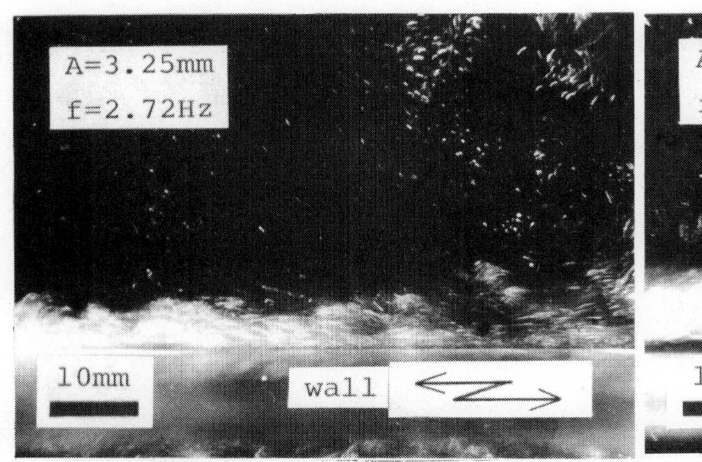

Fig.11-1 Water injection

Q=20 cm^3/s

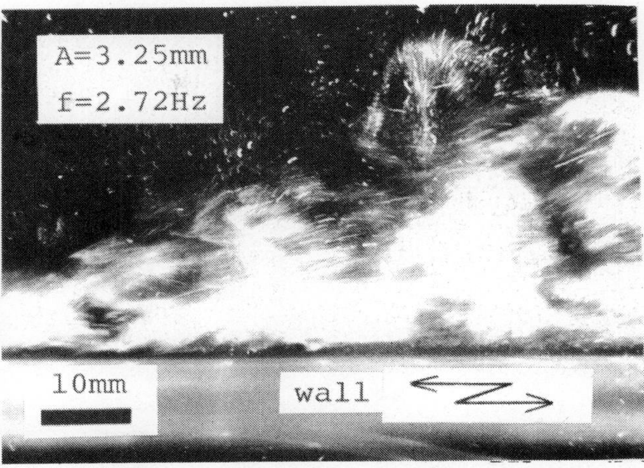

Fig.11-2 PEO 18 300wppm injection

Q=20 cm^3/s

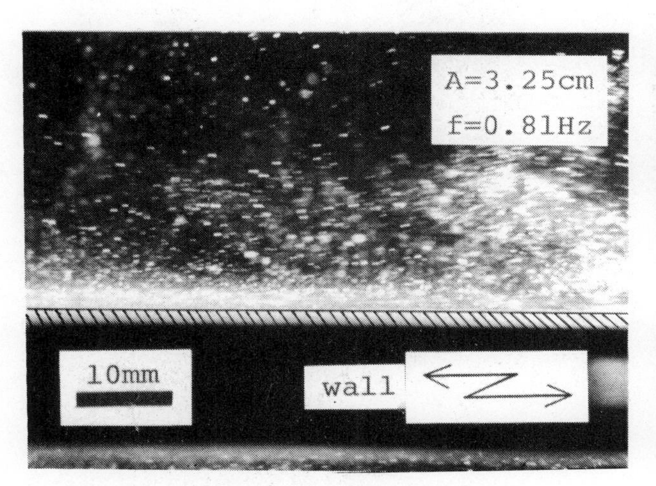

Fig.11-3 Water injection

(containing aluminum powder)

Q=20 cm^3/s

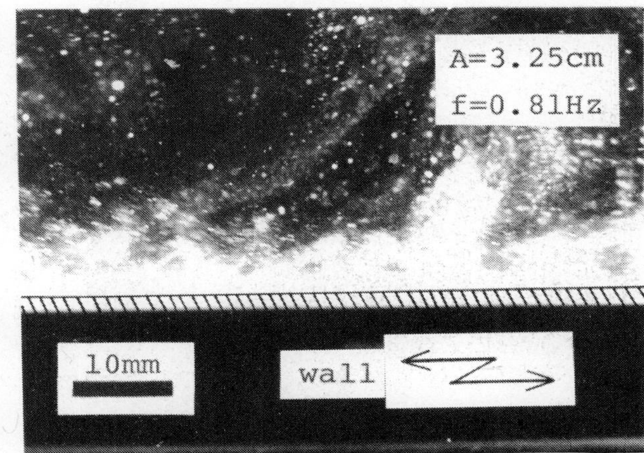

Fig.11-4 PEO 18 300wppm injection

(containing aluminum powder)

Q=20 cm^3/s

Fig.11 The thickness of oscillating flow layer

(Visualization using aluminum powder)

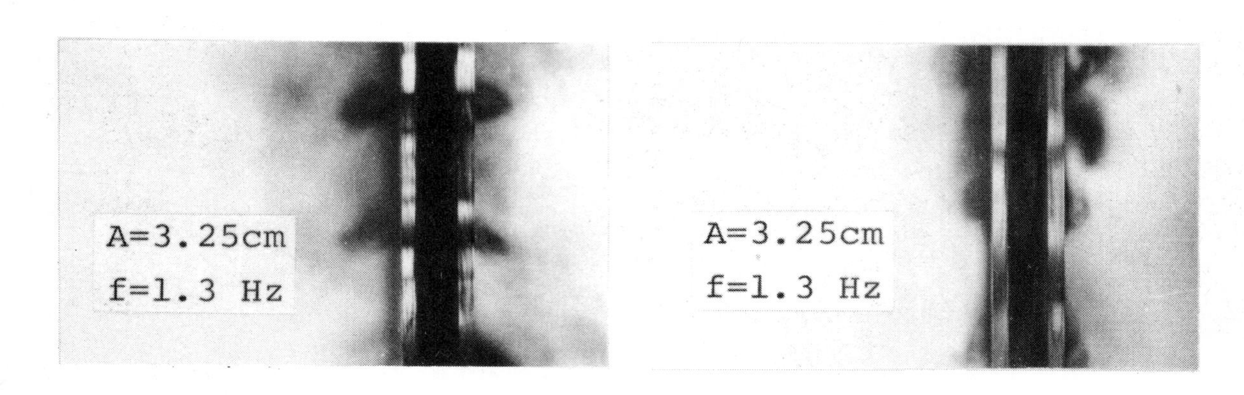

Fig.12-1 Water injection

Fig.12-2 PEO 18 300wppm injection

Fig.12 Looks of the injection pattern
(Visualization using black carbon ink)

Second International Conference on

Drag Reduction

August 31st - September 2nd, 1977

PAPER F4

EFFECT OF CAPILLARY TUBE DIAMETER IN LAMINAR FLOWS OF POLYMER SOLUTIONS

A. Ouibrahim

Laboratoire d'Aerodynamique, France

Summary

Experimental results obtained in different flow situations, indicate that polymer solutions, such as Polyox, behave like viscoelastic fluids, characterised by a time scale parameter. There are some grounds to believe that the geometric dimensions of the flow affect the polymer solution behaviour. The question can be raised of whether a scale parameter exists, such as a characteristic length.

In order to determine the significance of the length parameter, experiments have been conducted by flowing polymer solutions inside small diameter (down to 26 microns) capillary tubes at low and moderate Reynolds number. The relative apparent viscosity of the aqueous polymer solution (ratio between the polymer solution and solvent viscosity) was selected as a convenient non-dimensional parameter. Depending on the polymer concentration, the data indicate that the relative apparent viscosity is a function of the tube diameter. For large concentrations, above a few hundred parts per million, the relative apparent viscosity decreases sharply at first for increasing diameters and reaches a minimum for a particular tube diameter, Dm. Further increase of the diameter leads to a slight increase of the relative apparent viscosity up to an asymptotic value for large enough diameters.

The pressure drop between the tube inlet and outlet is a linear function of the shear rate up to a critical value at which the flow becomes unsteady.

The value of the characteristic length, constructed by using the polymer viscosity and the above mentioned critical shear rate, is in agreement with the capillary tube diameter, Dm, previously defined. The empirical relationship also shows that much smaller capillary tube diameters will be required to observe the relative apparent viscosity decay in the case of low polymer concentrations.

The dependency of the apparent viscosity with the tube diameter may be ascribed to the fibrous structure of these polymer solutions observed by microphotography in all tested concentrations.

This behaviour of the relative apparent viscosity may be described either by considering these polymer solutions as suspensions of non rigid particles, by analogy to blood rheology, or, as a continuum by mean of the micropolar fluid theory.

Held at St. John's College, Cambridge, England.
Organised and sponsored by BHRA Fluid Engineering, Cranfield, Bedford, MK43 0AJ.

INTRODUCTION

Since Toms discovery (ref. 1) of drag reduction effects, solutions of high molecular weight polymers have received (ref. 2) much attention by the scientific community. Applied investigations have been concerned with the type and concentrations of the most effective polymer additives, while theoretical research has been directed towards the determination of the constitutive equations or rheological models capable of describing the polymer solution behaviour in various simple flow conditions. As a step in this direction it is of importance to assess the nature of the parameters characterising the solution properties. Numerous investigations have been carried out which show the existence of a characteristic fluid time parameter (ref. 3 to 7). They have been mainly concerned with extensional flows, such as those occuring at the stagnation point of Pitot tube probes (ref. 3, 4, 5) at the entrance of sharp orifices (ref. 5, 6, 7) and at the exit of small capillary tubes (ref. 8, 9) ; as well as with steady and non-steady shearing flows between parallel plates (ref. 10) and cone-plate arrangements. In Pitot tube and orifice flows, in particular, there exists a characteristic flow time above which the fluid behaves as purely viscous. This characteristic flow time seems to be related to the characteristic time of the fluid.

Since there is some grounds to believe that, at least for some of the most effective drag-reducing additives, the geometric dimensions of the flow influence the polymer solution behaviour (ref. 4, 5, 11), a convenient length parameter associated with the structure of the dissolved polymer may also exist. The present paper is primarily concerned with this aspect of the characterisation problem. Its originality resides on the fact that laminar steady flows of poly(ethylene oxide) solution have been investigated (ref. 12) in very small, down to 26 microns diameter, capillary tubes.

Flow characteristics (pressure versus flow rate) of dilute Polyox WSR 301 solutions were measured in a once-through apparatus. Analysis of the flow curves lead to the following main conclusions :

1) For low and moderate shear rates the fluid behave like a purely Newtonian fluid.

2) For high shear rates the fluid behaves like power law fluids.

3) The transition from the Newtonian to non-Newtonian behaviour is characterised by a critical value of the shear rate.

4) The critical shear rate is associated with the initiation of an unsteady or pulsatile flow regime.

5) The relative apparent polymer viscosity computed in the Newtonian region (1) shows a dependency on the capillary tube diameter :

a) for very low diameters and up to a critical value the apparent viscosity decreases sharply,

b) for diameters larger than the critical the apparent viscosity increases slightly and reaches an asymptotic value consistent with that of usual capillary tube viscometers.

6) The critical shear rate depends also on the tube diameter.

Concurrently with the flow characterization, research was carried out to visualise the structure of the dissolved polymer. The flow field at the entrance of the tubes under varying conditions of flow was observed and photographed using aluminium suspended particles.

As a result of this investigation it is concluded that solutions of Polyox can be characterized by a length parameter whose value can be computed by means of the polymer solution relative viscosity and the critical shear rate. Such a characteristic length is analogous to those which determine the dependency of the apparent viscosity with flow dimensions in the case of suspensions of non-rigid particles (blood) and micropolar fluids.

EXPERIMENTAL SET UP

Experiments were conducted in a once-through system (ref. 12) consisting of a reservoir discharging into a feed chamber to which the capillary tubes were attached. The ratio of the area of the feed chamber and the tubes, the contraction ratio, was between 0.3 10^2 and 0.4 10^4. The tubes discharged freely to atmospheric pressure. Flow rates were determined by weighing the volume of fluid collected during a given time. Due to the small dimensions of the tubes, no pressure taps were used to measure the pressure drop. Instead, the driving pressure at the feed chamber was continously recorded. Temperature in the feed chamber was also monitored. The diameters of the tubes were measured using a travelling microscope and checked by computing their values from flow measurements made with demineralized and filtered water in the Poiseuille regime. Before each test with a given polymer solution and tube a water run was performed for verification purposes. With this experimental set up an accuracy of 2 to 3% of the relative apparent viscosity was obtained for the polymer solutions.

The polymer used in the tests was poly(ethylene oxide), Polyox, grade WSR 301. Four concentrations, 100, 300, 600 and 2 000 ppm were investigated. All polymer solutions but 2 000 ppm were prepared by diluting in demineralized and filtered water a 1 000 ppm master solution mixed 24 hours in advance.

DATA REDUCTION

The general expression for the pressure drop, ΔP, in a capillary tube of diameter, D, and length, L, with a mean velocity, U, and fluid viscosity, μ, is written as,

$$\Delta P = 32\,\mu\,\frac{L}{D}\,\frac{U}{D}\,\left(1 + \frac{\alpha_1}{L/D}\right) + \alpha_0 \qquad (1)$$

where α_0 and α_1 are the correction coefficients of Hagenbach and Couette respectively (ref. 12). Owing to the values of the length to diameter ratios of the tubes and of Reynolds numbers of the tests, Table 1, both correction terms are neglected. Expression (1) reduces to the well known Poiseuille law :

$$\Delta P = 32\,\mu\,\frac{L}{D}\,\frac{U}{D} = \frac{32}{Re}\,\frac{L}{D}\,\rho U^2 \qquad (2)$$

where $Re = \frac{\rho U D}{\mu}$, and $U = \frac{4\,q}{\rho \Pi\, D^2}$, with q being the mass flow rate.

Expression (2) was used to compute the values of the apparent viscosity by,

$$\mu p = \frac{1}{32}\,\frac{D}{L}\,\frac{D}{U}\,\Delta P_p \qquad (3)$$

where the subscript refers to the polymer solution. The apparent viscosity relative to water, subscript w, for a given tube and shear rate, is thus

$$\mu_{ar}\left(\frac{U}{D}\right) = \frac{\Delta P_p}{\Delta P_w} \qquad (4)$$

since the specific mass ρ is the same for water and the polymer solutions (ref. 12).

POLYOX WSR 301 SOLUTIONS

1 - FLOW CURVES

The pressure drop versus shear rate curves present some interesting features which will be discussed next. As shown in Fig. 1 two different flow regions can be defined depending, on whether the shear rate is below or above a critical value $(U/D)_c$. For values below the critical the driving pressure varies linearly with shear rate and the flow is steady. Extrapolation of the linear region towards zero shear rate leads to a finite value of the pressure drop (threshold). In fact, some test results obtained in this region have shown a local non linear dependency of the pressure, as depicted schematically in Fig. 1. For values above the critical, the pressure drop varies with shear rate as for a power law fluid and the flow is unsteady. This unsteadiness can be detected by direct observation of the oscillatory motion of jets and drops exiting the capillary tube. In the intermediate region a sharp or smooth transition between both flow regions occur depending on whether tests are performed by increasing or decreasing the driving pressure, characterizing a hysteresis cycle. The flow discontinuity has been observed for the four tested concentrations. A sharp transition is displayed also in the case of flow of polymer solutions through sharp edged orifices (ref. 5, 6) and is considered (ref. 5) as a characteristic time of the polymer concentration.

Main results for some selected capillary tubes and the different polymers concentrations are summarized in Table 1. The table shows the critical value, $(\frac{U}{D})_c$, of the shear rate at the sharp transition (column 5), the relative apparent viscosity, μ_{ar}, computed at the critical shear rate (column 6), μ_{ar} computed from the pressure versus shear rate slope thus neglecting the non linear initial region (column 7), the maximum shear rate reached during the tests (column 8) and the value of μ_{ar} for the terminal shear rates (column 9).

2 - DATA ANALYSIS

2.1 Relative Apparent viscosity versus shear rate

The dependency of the relative apparent viscosity, μ_{ar}, on the shear rate is obtained from the flow curves by using relations (3) and (4). It can be seen, Fig. 2a to Fig. 2c, that :

i) μ_{ar} decreases with U/D increasing and reaches a minimum for $(U/D)_c$
ii) For values larger than $(U/D)_c$, in the unsteady flow region, μ_{ar} increases
iii) The dependency of $(U/D)_c$ on the tube diameter is shown.
iv) The effect on tube diameter on μ_{ar} for the steady flow region is shown by the portion of the curves situated left of $(U/D)_c$.

2.2 Relative apparent viscosity versus tube diameter

The relative apparent viscosity, μ_{ar}, computed in the steady region at given shear rate is given in Table 2 and represented in Fig. 3. For comparison purposes, Table 1 gives the value by using the pressure versus shear rate slope. The important feature in Fig. 3 is the significant dependency of μ_{ar} on the tube diameter for the 600 and 2 000 ppm solutions. For values below a critical tube diameter Dm, μ_{ar} increases sharply, reaching values several times larger than those usually obtained with capillary tube viscometers. For tube diameters above the critical, μ_{ar} slightly increases before reaching a constant value.

It is interesting to note that the increase of μ_{ar} for low diameter tubes has not been evidenced in the case of 100 and 300 ppm. This apparent abnormal behaviour, as compared with more concentrated solution, will be discussed later in detail.

2.3 Relative apparent viscosity versus concentration

Increasing the polymer concentration increases μ_{ar} over all the range of tube diameters. However, the shape of the curves in Fig. 3, though similar, do not permit collapsing them in a simple curve where μ_{ar} will be a function of concentration. This is due to the fact that the minimum of μ_{ar} occurs for different diameters depending on the concentration. This question will be further considered during the discussion of the results.

2.4 Critical shear rate versus tube diameter

The critical shear rate, $(U/D)_c$, decreases with increasing tube diameter (table 1) and polymer concentration. However, it should be pointed out that the value of $(U/D)_c$ seems to be very much influenced by the L/D ratio of the capillary tubes. Fig. 2 show that the values of L/D affect the flow behaviour in the unsteady region, as a result of the decreasing relative importance of pressure head due to flow unstabilities with respect to total head. This particular question deserves more attention and is presently being investigated.

3 - VISUALISATIONS

3.1 Flow visualisation

The flow in the upstream feed chamber was visualised using aluminium particles. A typical example of the flow streamlines is shown in Fig. 4 for the capillary tubes, number 1, 4 and 6 (Table 1). For very low velocity the flow streamlines are in agreement with those expected for a viscous flow at low Reynolds number. A toroidal vortex will eventually develop and grow with increasing velocities. These patterns are in agreement with those obtained for orifice flow of concentrated and diluted polymer solutions respectively (ref. 5, 19).

For velocities larger than the critical, $(U/D)_c$, the unsteadiness of the flow is characterized, in the upstream chamber, by a pulsatile flow whose period decreases with increasing velocity.

3.2 Fibrous structure visualisation

The dependency of the relative apparent viscosity with tube diameter suggests that the polymer solutions behave like a two phase fluid. In order to gain some understanding on this particular question, drops of the polymer solution were vacuum dried from the liquid state (evaporation) or frozen solid state (sublimation). The dried drops, observed with a microscope, showed an intricate fibrous structure in which the size and the compactness of the fibers depend on the polymer concentration. Visualisations were carried out for concentrations ranging between 10 and 5 000 ppm (ref. 12). Fig. 5 shows the observed structures for 10, 50 and 600 ppm. Concentrations larger than 2 000 ppm are characterised by a very tight cellular structure.

It is interesting to note that solutions of other drag reduction agents such as Polyacrylamide (PAM AP30) did show a similar cellular structure, while solutions of salt and suspensions of carbon particles did not.

DISCUSSION OF THE RESULTS

The analysis of the dependency of the relative apparent viscosity on the capillary tube diameter suggests the existance of a characteristic length parameter associated with the polymer solution which interacts with the characteristic length of the flow (tube diameter). An attempt has been made to try to identify the polymer length parameter by using available data. The only combination of the apparent viscosity and the critical shear rate (ref. 12), giving a length parameter, is :

$$d_c\ (m) = \left[\frac{\mu_p}{\rho} \cdot \frac{1}{(U/D)_c}\right]^{1/2} \qquad (5)$$

with $$\mu_p = \mu_{ar} \cdot \mu_w \qquad (6)$$

This expression, checked against experimental data, Table 3, gives a unique value of the length scale for each polymer concentration, provided the tube diameters for which the computation is carried out are close enough to the critical one, Dm. For 600 and 2000 ppm the values of this parameter d_c are close to the experimental diameters Dm.

The values of d_c, 23 and 30 µm respectively, obtained for 100 and 300 ppm concentrations clearly show that the tube diameters used in the tests with these two solutions were larger than those required to show an increase of the relative apparent viscosity.

The value of d_c (Table 3) is a linear function of the concentration C for $100 \leq C \leq 2000$ ppm (Fig.6^c) and can be represented by the following relation :

$$d_c = k_1 C + k_o \qquad (7)$$

where $k_o = 20$ µm, $k_1 = 35.10^{-3}$ µm/ppm, with an accuracy of 5 to 7 %.

RHEOLOGICAL APPROACH

As a first attempt to describe the behaviour of the relative apparent viscosity of polymer solutions, the rheological characteristics of suspensions of non-rigid particles, such as blood cells in plasma, and micropolar fluids have been considered.

In the first case, it has been shown (Ref.13,14) that, depending on the relative dimensions of the cells and the capillary tubes, the relative apparent viscosity of the suspensions :

a) decreases for tube diameters below a certain value. It behaves like a fluid with two phases, one of which is characterized by an elasticity modulus.

b) increases within an appropriate range of diameters as in a continuum but non uniform fluid ; such an increase corresponds to the "Fahraeus-Lindqvist" effect (Ref. 13, 14).

c) reaches an asymptotic value for large enough diameters.

Moreover, both polymer solutions and blood display relative apparent viscosities which decrease with increasing shear rate before reaching the newtonian behaviour region. In the case of blood, such a dependency results from elongation and disruption of red cell rolls.

Thus, suspensions of non-rigid particles appear to satisfactory describe, at least partially, the key features of polymer flow shown in this paper.

Micropolar fluids, on the other side, though characterised by a scale length parameter (Ref.15 to 18) do not allow for a such clear analogy. The general behaviour of μ_{ar} can be grossly described by micropolar fluid theory provided two different boundary conditions (Ref.12) are considered, on both side of the characteristic parameter d_c, for the microstructure rotation. This introduces a discontinuity of μ_{ar} for a tube diameter equal to d_c. Other aspects of the behaviour of the relative apparent viscosity of Polyox solutions cannot be justified on the basis of micropolar fluid theory.

CONCLUSIONS

It has been shown that the behaviour of polymer (Polyox WSR 301) solutions in small, down to 26 μm, diameter capillary tubes is influenced by the flow length scale (diameter). The relative apparent viscosity of the solutions decreases with increasing tube diameter and reaches a minimum for a particular value of the tube diameter which depends on the polymer concentration.

This unusual behaviour of the apparent viscosity raises the question of whether or not a characteristic length parameter associated with the polymer solution structure can be defined. An attempt to define such a parameter, on the basis of known measured flow properties, has been made.

Other properties of the viscous flow of these polymer solution have also been investigated in this paper. In particular, it is shown that the non-newtonian behaviour at large shear rate is associated with flow instabilities both in the upstream feed chamber and in the discharge section.

Investigations are currently underway to further substantiate some of the above results.

ACKNOWLEDGMENTS

The author wishes to express his gratitude to the scientific and technical staff of the Laboratoire d'Aérodynamique for their continuous support in the course of this work. The Direction de Recherches et Moyens d'Essais (DRME), France, is deeply acknowledged for its financial support.

REFERENCES

1. Toms,B.A. : "Some Observations on the flow of linear polymer solutions throught straight pipes at large Reynolds numbers". Proc. 1st Int. Cong.of Rheology, Vol.2, p.135, North Holland Amsterdam (1948).

2. Hoyt,J.W. :"The effect of Additives on fluid friction". Trans. of the ASME, June (1972).

3. Fruman,D.H., Loiseau,G. and Ville,B. : "Relaxation time measurements of dilute polymer solutions by Pitot tubes" (Mesure du temps de relaxation des solutions diluées de polymères au moyen des tubes de Pitot). Laboratoire d'Aérodynamique Orsay (1971).(in French).

4. Loiseau,G. : "Behaviour of Pitot tubes probes in viscoelastic fluids". (Comportement des tubes de Pitot dans les fluides viscoélastiques). Thèse 3ème cycle, Université de Paris. February (1970). (In French).

5. Piau,J-M. : "Polymers and Lubrication". (Polymères et Lubrification). Colloques Int. of C.N.R.S., n°233, Brest 20-23 May (1974).(In French).

6. Gilles,W.B.: "Orifice flows of Polyethylene oxide solutions". Nature,vol.224 November 8, (1969)

7. Bilgen,E. : "On the orifice flow of dilute polymer solutions". Journal de Mécanique vol. 12, n° 3, September (1973)

8. Bilgen,E. : "Effect of Dilute polymer solutions on discharge coefficients of fluid meters". Fluid Engineering, Heat Transfer, and Lubrication Conference, Detroit, Mich., n° 70-FE-38, May 24-27 (1970).

9. James, D.F. : "A method for measuring normal stresses in dilute polymer solutions". Trans. Soc. Rheology, 19 : 1, 67-80 (1975).

10. Middleman,J. and Gavis,J. : "Expansion and Contraction of capillary jets of viscoelastic liquids". The Physics of Fluids, vol.4, n°8, August (1961).

11. Granville,S. : "Hydrodynamics aspects of drag reduction with additives". Naval Ship Research and Development Center, September (1972)

12. Ouibrahim,A. : "Laminar flow of polymer solutions in the capillary tubes". (Ecoulement laminaire de solutions de polymère dans les tubes capillaires). Thèse 3ème cycle, Université de Paris, June (1976).(In French).

13. Haynes,R.H. : "The Rheology of Blood". Trans.Soc.of Rheology, V,p.85 (1961)

14. Sutera,S.P. : "Biomecanics Course".(Cours de Biomécanique). Université de Paris VI (1972-73).(In French).

15. Pennington,C.J. and Cowin,S.C. : "The effective viscosity of Polar Fluids". Trans.Soc.of Rheology, 14 : 2,p. 219 (1970).

16. Kline,K.A., Allen,S.J. and De Silva,C.N. : "A continuum approach to Blood Flow". Biorheology, Vol.5,pp.111-118, (1968).

17. Hartmann,C. :"Effects of second order in parallel flows of a Micropolar fluid". (Effets du second ordre dans les écoulements par droites parallèles d'un fluide micropolaire). Journal de Mécanique, vol.12,n°1, p.98 (1973).

18. Kline,K.A. : "Predictions from Polar Fluid theory which are independent of spin boundary condition". Trans.Soc.of Rheology, 19 : 1,p. 139 (1975).

19. Oliver,D.R. and Bragg,R. : "Flow patterns in viscoelastic liquids upstream of orifices". The Canadian Journal of Chemical Engineering, vol.51, June (1973).

Table 1 : Values of geometric characteristics of capillary tubes and significant flow parameters for water and Polyox WSR 301 solutions.

1	2	3	4		POLYOX WSR 301 SOLUTIONS																			
					5				6				7				8				9			
N°	D (mm)	L/D	⩽ Re ⩽ Water		$(U/D)_C$ (s^{-1}) Critical shear rate				$\mu_{ar}[(U/D)_C]$				μ_{ar} (slope)				U/D max.				$\mu_{ar}[(U/D\ max)]$			
					100 ppm	300 ppm	600 ppm	2000 ppm	100 ppm	300 ppm	600 ppm	2000 ppm	100 ppm	300 ppm	600 ppm	2000 ppm	100 ppm	300 ppm	600 ppm	2000 ppm	100 ppm	300 ppm	600 ppm	2000 ppm
1	1,006	1161	69	795			467	206			2,16	5,74			2,12	-			1026	313			2,22	6,79
2	0,759	482	91	1348			565	386			2,09	6,03			2,03	-			1230	597			2,40	7,80
3	0,497	702	25	503			900				1,99				1,98				1957				2,10	
4	0,497	591	54	538				600				5,03				4,52				857				5,43
5	0,305	688	19	284		1180	1278			1,42	2,01	4,50		1,42	1,92	4,14		2106	2008	900		1,59	2,13	-
6	0,246	711	25	181	1711		1506		1,18		1,93	4,08	1,18		1,88	3,76	3389		1997	991	1,20		2,08	-
7	0,151	695	19	101		1730				1,44				1,42				2664				1,56		
8	0,100	581	5	50		2113	1000			1,42	1,90	4,17		1,42	1,90	4,03		3371	2584	1178		1,49	1,91	-
9	0,0857	291	6	53		2120	997			1,43	1,92	4,87		1,42	1,89	4,46		5331	4289	1896		1,90	2,23	-
10	0,0646	210	2	62		2923	1041			1,48	1,94	12,53		1,48	1,87	-		8142	5404	1074		1,77	2,59	-
11	0,0540	185	5	52		2976	1035			1,42	2,02			1,40	1,93			8916	5323			1,89	2,62	
12	0,0412	111	5	50		1706	1683			1,61	3,36			1,61	3,1			11556	5182			2,37	4,36	
13	0,0311	149	3	18			1996				3,74				3,70				4116				4,30	
14	0,0264	170	3	12	2768				1,48				-				10759				1,55			

Table 2 : Relative apparent viscosity values, μ_{ar}, at given shear rate in steady flow region.

N°	100 ppm	300 ppm	600 ppm	2000 ppm	
	U/D(s^{-1})	U/D	U/D	U/D	
	1500	947	947	208	312
1			2,15*	5,75	-
2			2,08*	5,64	5,84
3			1,98		
4				5,64	5,26
5		1,43	2,00	5,64	5,20
6	1,17		1,94	5,21	4,81
7		1,45	1,92		
8		1,44	1,91	5,03	5,81
9		1,43	1,92	5,40	5,07
10		1,44	1,94	8,84	9,1
11		1,49	2,01		
12		1,65	3,20		
13			3,80		
14	1,11				

* : Value of μ_{ar} for U/D = 380s^{-1} before unsteadiness.

Table 3 : Length Parameter, d_c, for each concentration, following the relation (5)

	N°	$(U/D)_C$ (s^{-1})	$\mu_{ar}[(U/D)_C]$	d_c (µm)
100 ppm	14	2768	1,48	23
300 ppm	12	1706	1,61	30
600 ppm	13	1996	3,74	43
	12	1683	3,36	44
	11	1035	2,02	44
	10	1041	1,94	43
	9	1000	1,92	43
	8	1000	1,92	43
2000 ppm	4	600	5,03	91

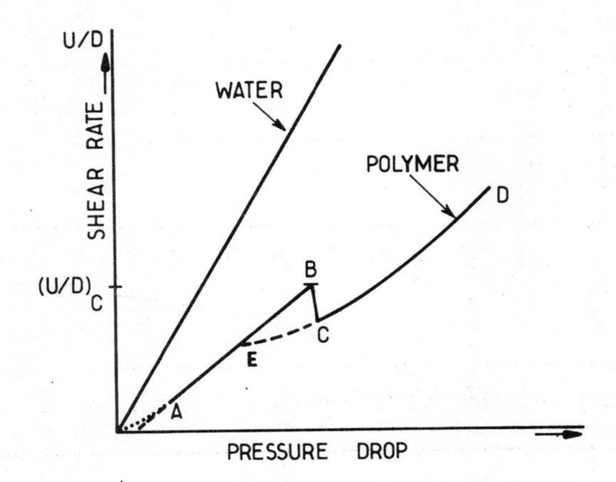

Fig.1 Typical flow curve for Polyox WSR 301 solutions in a capillary tube. Hysteresis cycle shown by the path: EBC increasing pressure; CE decreasing pressure.

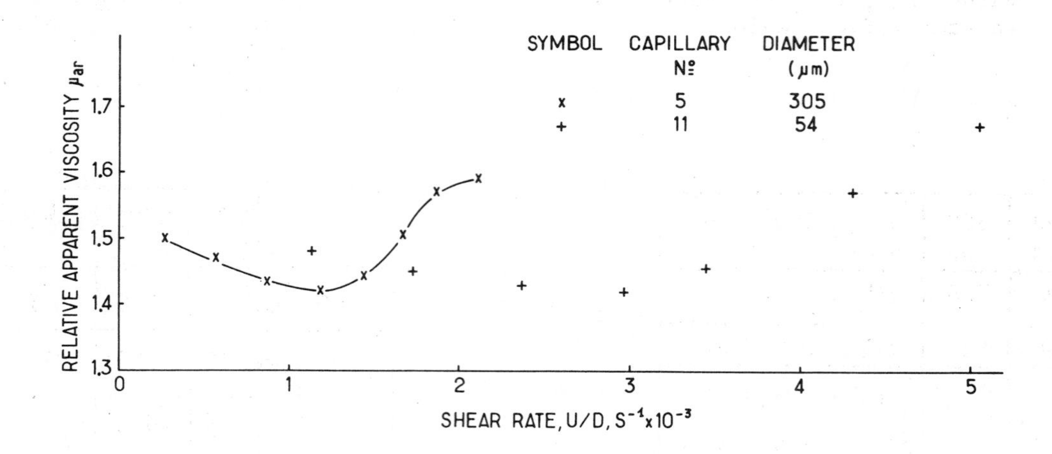

Fig.2 a) **Relative apparent viscosity versus shear rate for 300 ppm**

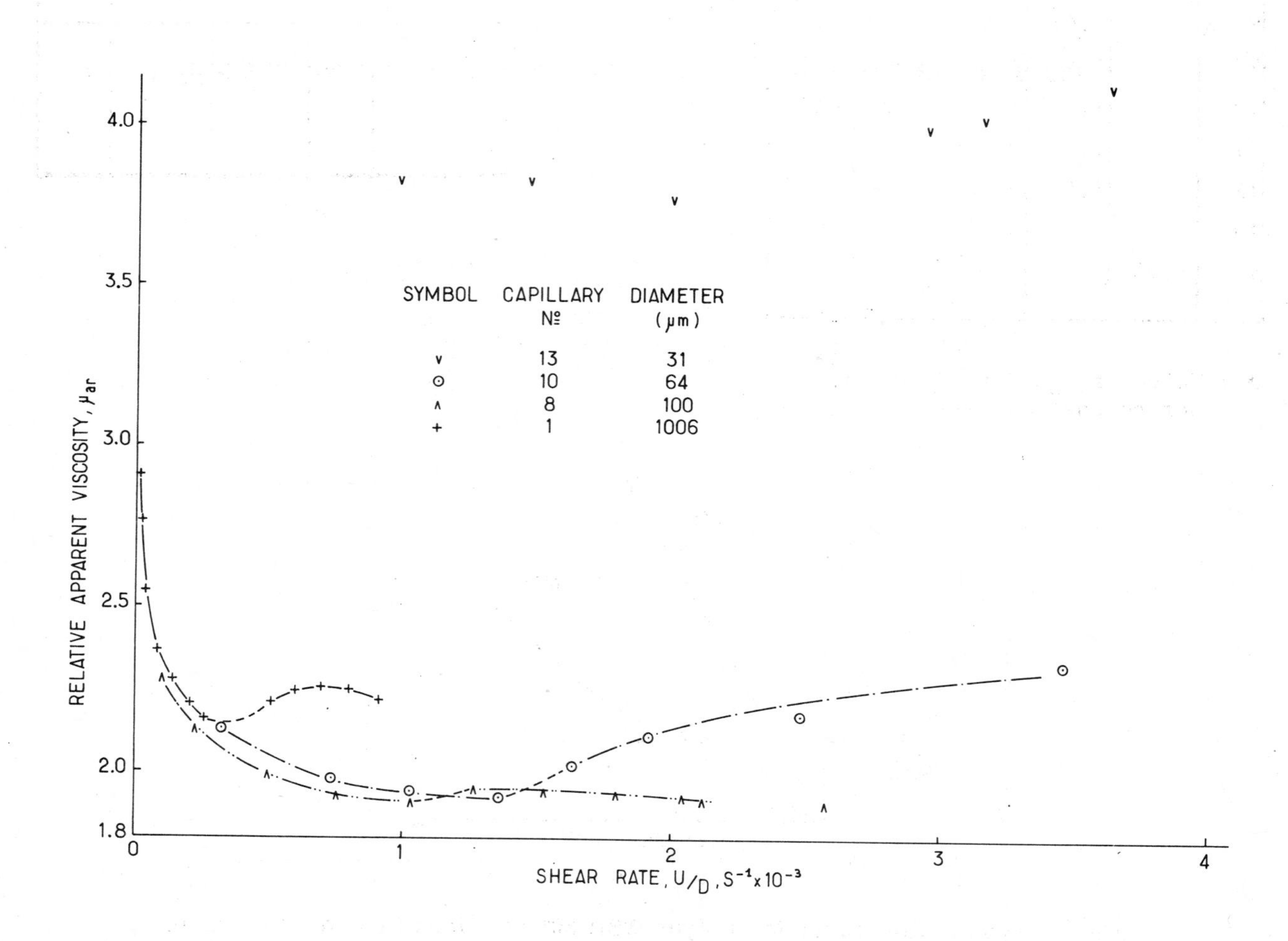

Fig.2 b) **Relative apparent viscosity versus shear rate for 600 ppm**

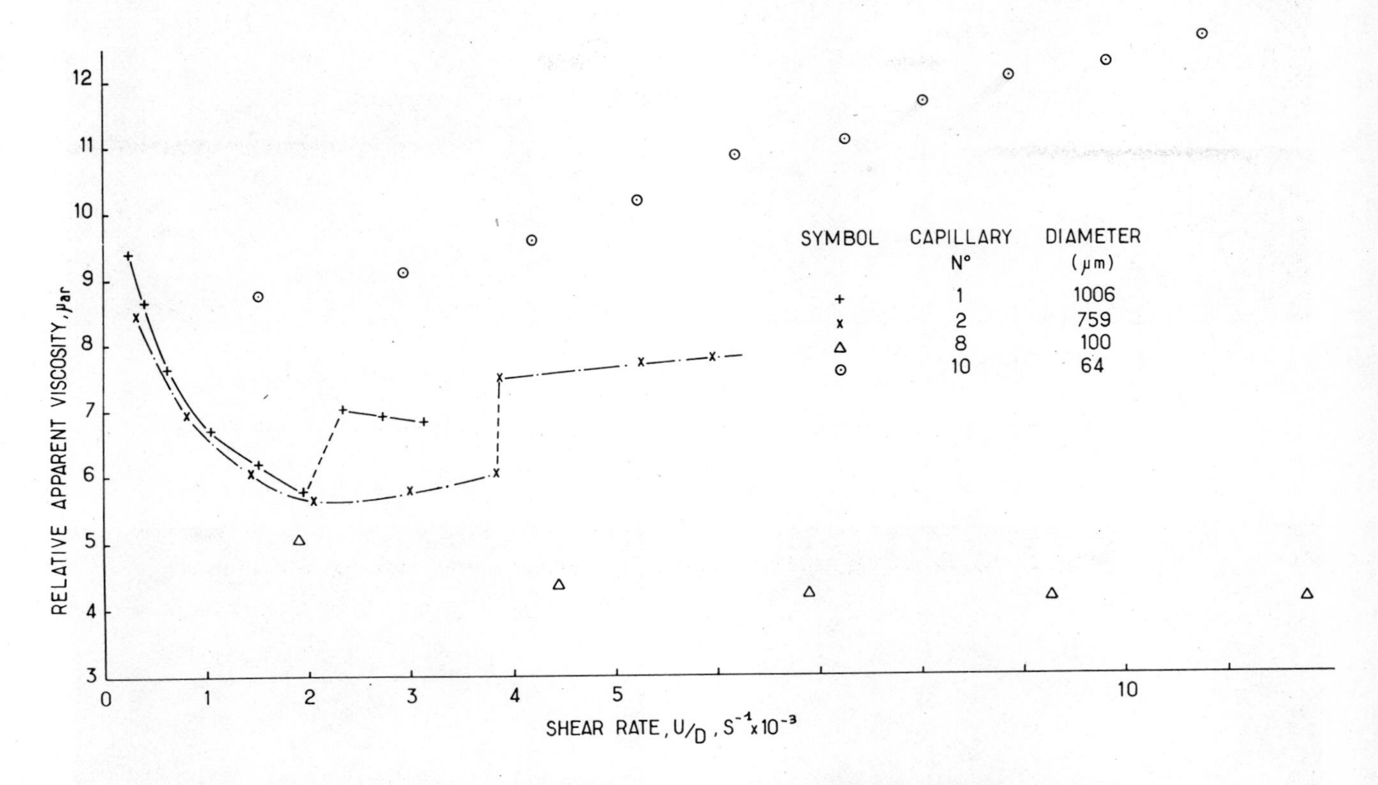

Fig.2 c) Relative apparent viscosity versus shear rate for 2000 ppm

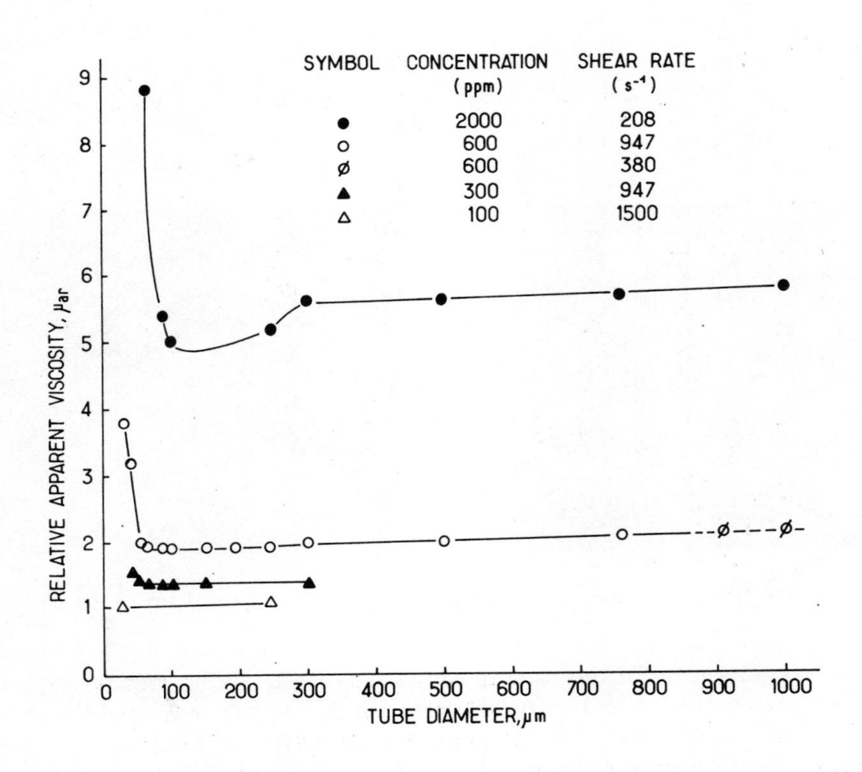

Fig.3 Relative apparent viscosity versus capillary tube diameter

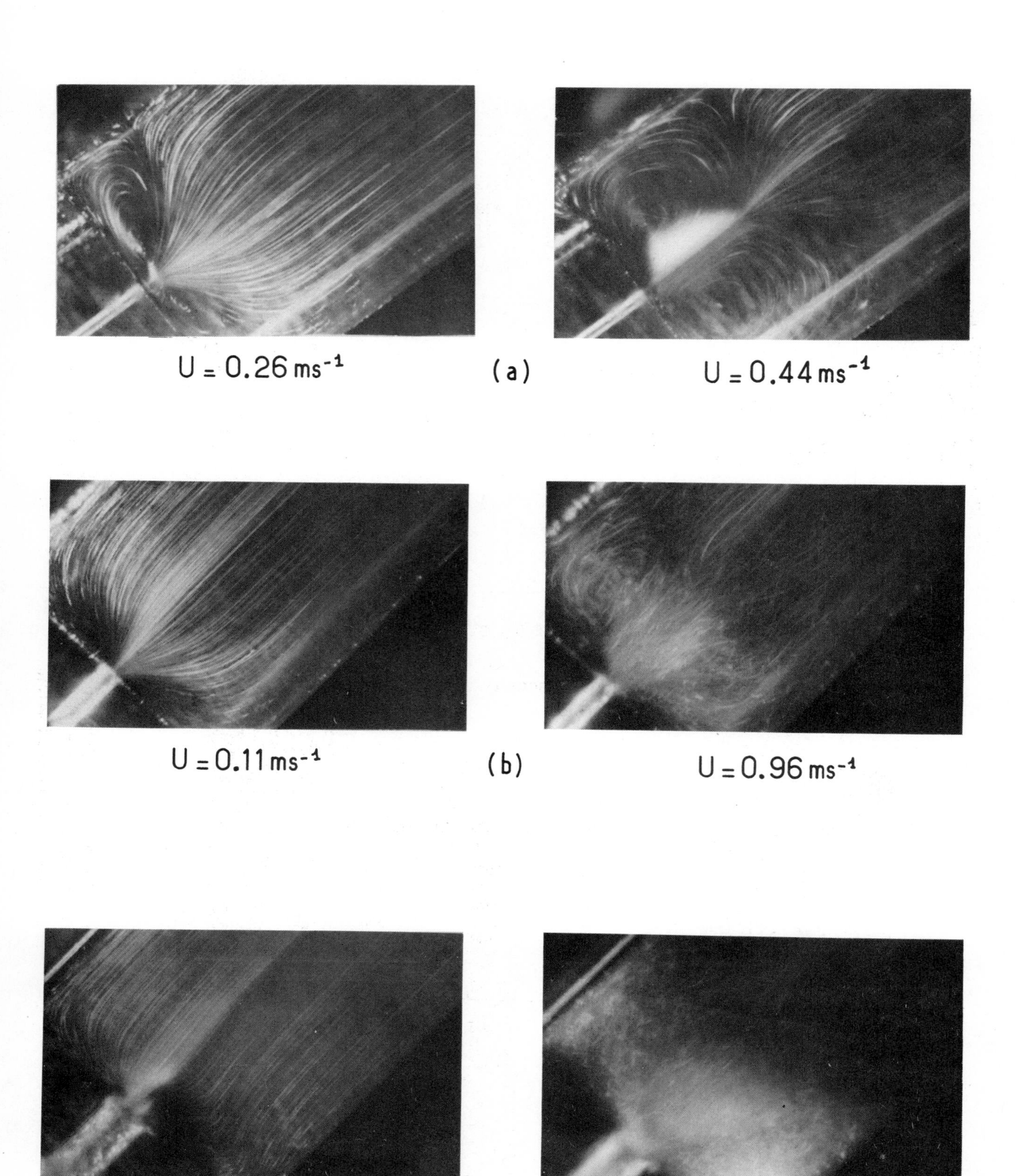

Fig.4 Visualisation of the flow in the steady (left) and unsteady (right) flow regimes: a) D = 0.246 mm; b) D = 0.497 mm and c) D = 1.006 mm and 600 ppm Polyox WSR 301 solution.

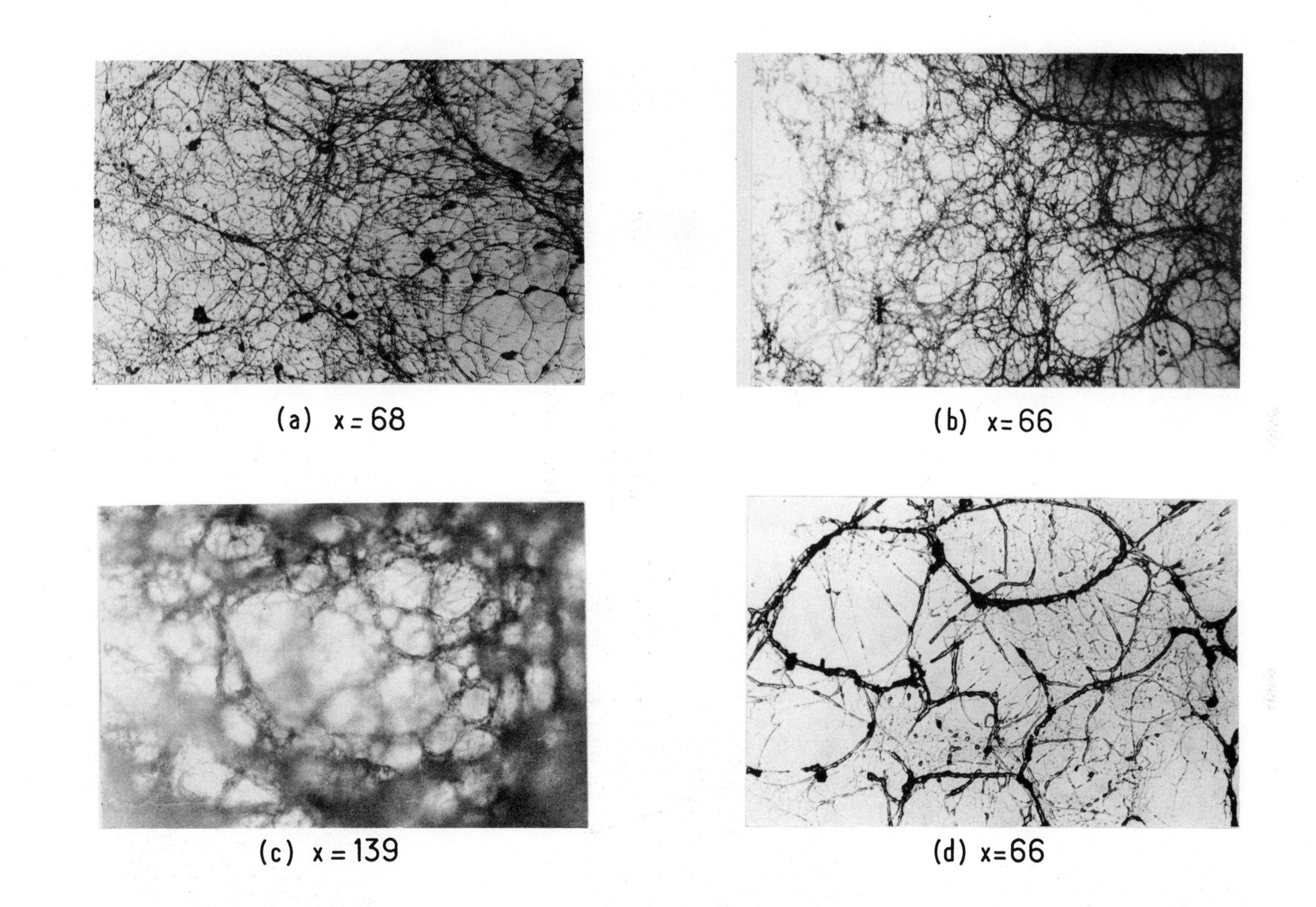

Fig.5 Visualisation of polymer fibrous structure: Polyox WSR 301, a) 10 ppm; b) 50 ppm; c) 600 ppm and d) PAM AP 30 : 50 ppm; X is the power magnification.

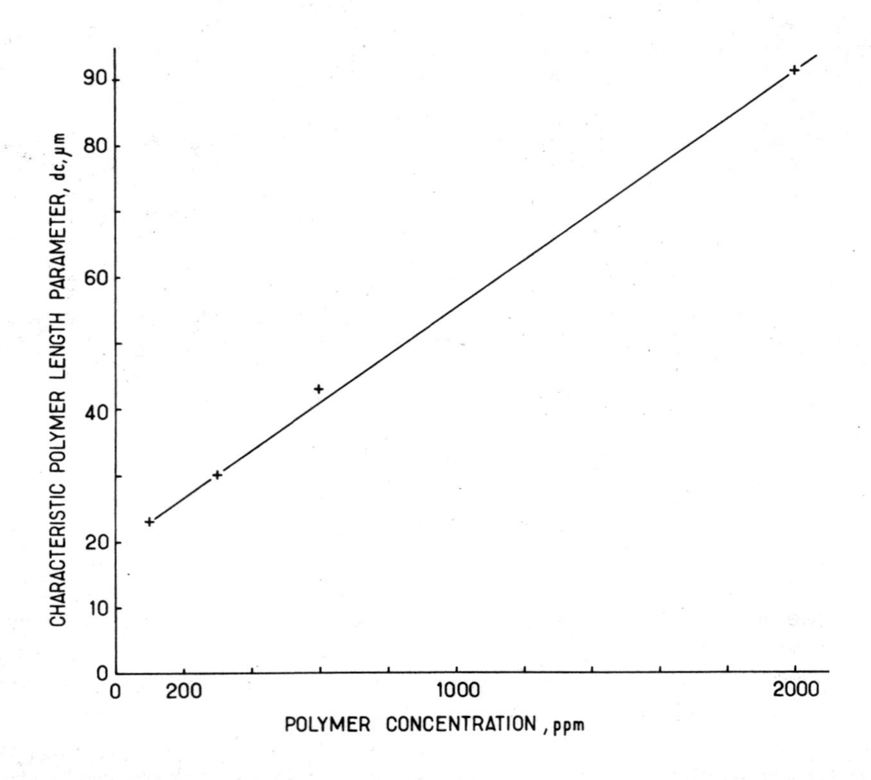

Fig. 6 Characteristic polymer length parameter as a function of concentration.

Second International Conference on

Drag Reduction

August 31st - September 2nd, 1977

PAPER G1

MODIFICATIONS TO LAMINAR AND TURBULENT BOUNDARY LAYERS DUE TO THE ADDITION OF DILUTE POLYMER SOLUTIONS

J.R. Bertshy and F.H. Abernathy,

Harvard University, U.S.A.

Summary

Two-dimensional boundary layer flows of dilute solutions of Polyox, WSR-301, have been investigated on a free surface water table. The natural disturbance level was low enough so undisturbed water flows were laminar. The flows had wall shear velocities up to 4.2 cm/sec and were at concentrations up to 32ppm as determined by testing samples in a vortex viscometer. Observations were made of the free surface, injected dye, streak spacing and the time-dependent velocity in the mean flow direction as a function of polymer concentration, wall strain rate and disturbance. Nearly continuous and very localized velocities were measured using a laser-Doppler anemometer and a novel method of selective seeding. Velocity spectra were also obtained.

Undisturbed polymer flows at sufficient concentration and strain rate exhibited velocity fluctuations distinct from turbulence. We have termed these as polymer induced fluctuations and believe that they are most likely responsible for the phenomenon of so called "early turbulence". Studies of turbulent polymer flows (disturbed artificially) indicated significant drag reduction and were all consistent with the view that hairpin eddies are the major mixing mechanism in a turbulent boundary layer, and that these eddies are suppressed by the addition of polymers. When strain rate and/or concentration were increased somewhat, polymer fluctuations were found to coexist with the turbulent motions. At still higher strain rates and/or concentrations, the polymer fluctuations dominated and the turbulent flow reached suppressed condition. No distinction between artificially disturbed and undisturbed flows at the same conditions could be observed.

Held at St. John's College, Cambridge, England.
Organised and sponsored by BHRA Fluid Engineering, Cranfield, Bedford, MK43 0AJ.

NOMENCLATURE

u = time dependent velocity in the mean flow direction

u'^{+} = rms fluctuating velocity divided by u_τ

U^{+} = mean velocity divided by u_τ

u_τ = wall shear velocity

y^{+} = $y\, u_\tau/\nu$, y = distance from the wall ν=solvent viscosity (water)

λ^{+} = $\lambda u_\tau/\nu$, λ = mean streak spacing

INTRODUCTION

The turbulent drag reducing capability of dilute polymer solutions has been known for a number of years. The two review articles by Lumley (ref. 1,2) and the extensive surveys of Hoyt (ref. 3) and Virk (ref. 4) review and place in perspective the known aspects of dilute polymer flows. Yet, mechanisms by which the turbulent mixing is reduced are still poorly understood. For example, Rudd (ref. 5), Logan (ref. 6) and Kumor and Sylvester (ref. 7) have each reported that the magnitude of the fluctuation component of the velocity in the mean flow direction, u'^{+}, in a turbulent boundary layer very near the wall increases with the addition of polymers. This has been interpreted as the polymers working primarily to thicken the viscous sublayer. On the other hand, Reischman and Tiederman (ref. 8) have reported a decrease in the maximum intensity of u'^{+} near the wall and (along with ref. 9) found a broad peak from y^{+}= 25 and beyond. They have interpreted this as an indication that polymers in drag reducing flows primarily alter the buffer zone (ref. 4). They reported finding no indication of a thickening of the sublayer. The present investigation was designed to explore the cause of these observed velocity fluctuations in dilute polymer boundary layers and to determine in what ways they differed from those observed with Newtonian fluids.

At the beginning of the present investigation, we held the view that the major turbulent mixing mechanism near a solid boundary was due to three-dimensional flows induced by hairpin eddies aligned in the flow direction. The addition of dissolved polymers to the flow should inhibit the formation and hence the mixing arising from such vortex pairs. At high enough polymer concentrations, the flow near the wall should be essentially laminar. The results of our experiments required only minor modification of this initial view of the turbulent boundary layer.

The experiments were designed to investigate boundary layer flows, at high enough Reynolds numbers to achieve turbulence if artificially disturbed, but at low enough free stream disturbance levels to enable a laminar boundary layer flow to exist in the absence of such disturbances. This enabled us to observe what, if any, velocity fluctuations might arise from the polymers themselves.

A free surface water table was selected as the optimal flow apparatus. In addition to fulfilling the prerequisites mentioned above, this apparatus is well suited for optical probing, since all surfaces are planar and the flow is easily accessible. The water table also has the tremendously useful benefit of providing information concerning where and when transition to turbulence or some other unsteady flow has occurred by visual observation of the free surface.

EXPERIMENTAL FLOW ARRANGEMENT AND EQUIPMENT

A diagram of the flow arrangement is shown in Fig. 1. Basically, filtered tap water is pumped from a sump reservoir through an overflow standpipe to a water table inclined to the horizontal. It then returns to the sump via a weighing tank. Polymer solutions were injected for short periods. It has been assumed that the polymers have been severely degraded before they return for a second time to the water table, especially since a large depth filter (5μ nominal pore size) was interposed in the flow circuit. In addition, the sump reservoir is so large (approximately 17,000ℓ) that effects of polymers returning for a second pass were neither expected nor observed. Concentrated (200-500 ppm) polymer solutions were injected by gravity feed from constant pressure tanks (two 208ℓ drums) into the piping over 38 diameters from the water table. Thus, the final dilution to the test concentration was accomplished by turbulent mixing in this pipe (10 cm dia.).

The water table has a deeper inlet section equipped with large cell honeycomb flow straighteners and surface wave calming devices. There is a smooth transition to the glass working section which is 1.22m x 1.83m. Typically, the flow was run 1.9 to 3.5 mm deep and the table's inclination was from 1.6^{o} to 4.2^{o}. All undisturbed water flows presented in this paper were laminar. To trip this flow to turbulence, a bead-chain of 2.63 mm beads was inserted in the flow near the leading edge of the glass working section.

Velocity measurements were made using a LDA manufactured by Thermo Systems, Inc. (TSI) oriented as shown in Fig. 2. The laser used had a 1 mW polarized output. Signal processing was accomplished by a TSI model 1090 tracker. It is normally difficult to resolve the local velocity in a thin shear flow using LDA techniques since the scattering volume of the intersecting laser beams is relatively large (here, 0.15mm dia. by 1.7mm length). We have overcome this difficulty by introducing light scattering contaminants to the flow only at a selected distance from the wall. This is accomplished by generating bubbles by electrolysis from a fine wire held parallel to the wall and normal to the flow. Most other scattering particles are filtered out of the flows. Hence, using a forward scatter, fringe mode LDA with the long axis of the scattering volume normal to the wall, we can obtain laminar and turbulent velocity profiles relatively rapidly by moving the bubble generating wire normal to the wall (cf. Fig. 2).

The bubble wire used was 80%Pt20%Ir, 0.0025 cm in diameter and 5 cm long. It was supported across two insulated, L-shaped, stainless steel tubes which entered the flow downstream of the scattering volume. The wire was always positioned at least 300 diameters upstream of the scattering volume so that its wake would not be detected (ref. 10). The flow was always such that the local Reynolds number of the wire was less than 25, so instability of the wire wake was never a problem. By adding a salt to the water (Na_2SO_4, anhydrous) we were able to maintain a sufficiently high bubble production rate so that the measured velocity was practically continuous (from 1000 to 6000 velocity samples/sec.) while the operating voltage was reasonably safe (<300V).*

To obtain mean velocities and rms turbulent intensities, a signal analysis box developed at the lab was used. It measured the mean voltage of the tracker's output by a continuously integrating op-amp with an accuracy of 0.1%. This voltage was electronically subtracted from the tracker output and then fed to a Burr-Brown 4128 rms module which determined true rms of this difference. Its accuracy is about 0.3% from 0.1 Hz to 20 kHz. Readings were taken using a 3½ digit Data Precision digital multimeter. Due to the one-pass nature of the experiments, short time constants were used and the measurements were considerably more uncertain than the potential accuracy of this instrumentation.

Velocity spectra were taken using a Tektronix 7L5 Spectrum Analyzer. It operates digitally and was set to filter with a 10 Hz bandwidth. Since it could not accept a dc signal from a high impedance source, a capacitor coupled op amp of gain 1 was interposed between the tracker and the spectrum analyzer.

Depth measurements were taken using a needle, point gauge accurate to 0.001 cm for flows with a smooth free surface. The water table's inclination was monitored by a 254 cm long liquid level located along the table side. A thermometer in the stilling section read temperature with an accuracy of $0.1^{o}C$.

---

* Cambridge City water contains typically 100-150 ppm of dissolved salts. The addition of 100-150 ppm of Na_2SO_4, to increase the electrical conductivity, did not effect the occurrence of polymer induced fluctuations as determined by viewing the free surface. While no quantitative measurements of the dependence of polymer effects on salt concentration were made, we have no evidence to suggest the results were influenced by the salts.

Polymer Preparation and the Rheological Test of Concentration

Batches of 0.1% solution (151ℓ) of PEO, WSR-301, were prepared by sprinkling the dry powder on the surface and hand mixing gently with a large paddle. This solution was allowed to sit for two days before using. The dry polymer was not dispersed in an alcohol slurry since the flow over the water table is shallow enough that it might be sensitive to surface tension changes. Surface tension measurements indicated that polymer solutions of much higher concentrations than those used in the water table exhibited negligible deviation from that of water.

The polymer concentration of a given flow was determined by withdrawing a sample during each run and observing its behavior in a large scale vortical flow similar to that of Balakrishnan and Gordon (ref. 11). They have demonstrated that measuring the minimum polymer concentration needed to achieve "vortex inhibition" is a more sensitive test for degradation than intrinsic viscosity measurements. So, we used a modified form of their bathtub vortex experiment as a rheological test for polymer concentration. Briefly the procedure was as follows: 20ℓ of the fluid to be tested was placed in a cylindrical container (29.2 cm inside dia.) with a central drain. The container was rotated until solid-body rotation was achieved (50 rpm) and then quickly stopped. Draining was commenced immediately and the vortex was allowed to form under the maximum draining rate. The drain was connected to a long flexible tube whose outlet elevation could be freely varied.

Since nearly all the flows we ran over the water table had concentrations below "vortex inhibition" (25 ppm for WSR-301) the air-liquid interface reached to the drain and there was two-phase flow in the exhaust tube. As soon as the vortex was securely established, the end of the exhaust tube was raised until two-phase flow no longer occurred in it. This elevation was then compared to a calibration curve made using dilute solutions made from a single batch of 0.1% stock solution. Thus, we were able to judge the concentration compared with this single stock solution to about 2 ppm independently of how poorly prepared or mishandled the sample had been. Typically, samples after passing over the water table behaved in the vortex as if their concentration had been 25% to 50% less than that injected.

The effective concentration of polymer determined by the vortex viscometer was in agreement with the water table flow observations. All data is presented with the concentration as determined in this fashion. In the near future, we expect to undertake a detailed study of this vortex flow.

Velocity Data

In so far as time allowed, complete profiles of u-velocity and turbulence intensity, were measured for each polymer run (45 minutes or less). Since the LDA would occasionally lose track of the Doppler signal, only short time averaging (about 8 sec.) was possible to ensure quick recovery. In addition, bubble production was reduced in polymer flows so constant attention was necessary to obtain data.

All velocity and depth measurements were taken at a station 146 cm downstream of the transition from the stilling section to the test section. Since the flows were typically 0.2 cm to 0.3 cm deep, this corresponds to a test location between 730 and 500 flow depths downstream of the transition.

Wall shear stress and hence u_τ was inferred from the depth measurement and the table inclination and assuming flow equilibrium. The flow accelerates as it progresses down the table and wall shear stress accordingly increases. A turbulent water flow comes to an equilibrium depth (and hence velocity) within 60 cm on the flat test section, while laminar water flow has not quite come to equilibrium even at the measuring station 146 cm downstream. Estimates of the acceleration of laminar water flows similar to those in which polymers were tested indicate that at most, u_τ might be decreased by 4% from the value used. Flows indicating any fluctuation and hence an increase in momentum exchange from laminar water flows were certainly much closer to equilibrium. Uncertainty in depth and angle measurements lead to an uncertainty in u_τ of $\pm$ 0.02 cm/sec for laminar flows and $\pm$ 0.06 cm/sec for turbulent flows.

The position of the bubble wire relative to the wall was also determined by the depth and angle measurements. Average velocity data from a laminar flow was plotted with relative uncertainty in position of only $\pm$ 0.001 cm. The wall shear stress measure (assuming solvent kinematic viscosity) was then used as an indication of the slope this velocity data had to have at the wall. The two were then compared and an absolute measure of distance from the wall to the bubble wire was determined with an uncertainty of approximately $\pm$ 0.005 cm.

All data is presented using the solvent kinematic viscosity as determined by the temperature. There is evidence that this may not be the best method to correlate the data; however, most previous investigators have presented their data using solvent viscosity. We do so to facilitate comparison, but point out that Reischman and Tiederman (ref. 8) did not follow this practice and allowances should be made accordingly.

One further flow visualization experiment was performed to obtain the streak spacing. Grains of Na_2SO_4 were sprinkled into the flow. Since the salt is denser than water, it tended to travel downstream along the bottom glass wall of the water table. The grains are small enough, however, to be lifted by turbulent motions and streaks were readily observed. Since the terminal velocity of these salt grains in still water is about 1 or 3 cm/sec, it is likely that the turbulent motions must occasionally exceed this velocity for streaks to be detected in this manner.

EXPERIMENTAL RESULTS

The philosophy used throughout this investigation has been to compare the dilute polymer flows with those of pure water at comparable strain rates and depths of flow. Following such a program required a careful investigation of the flow of water, measuring all of the variables with the same techniques to be used with dilute polymer flow.

The dimensionless mean velocity U^+ is plotted versus dimensionless distance away from the wall y^+ for both laminar and turbulent flow of water in Fig. 3. The measured results for the turbulent boundary layer flow of water show very close agreement with the "law of the wall" results for Newtonian fluids from both boundary layer and pipe flow. In fact, the water table turbulence data is nicely bracketed by the extremes commonly quoted in the literature for this universal profile (ref. 12). Departure from the universal velocity profile naturally does occur near the free surface since the slope $\frac{dU}{dy}$ must go to zero there. The laminar profile is a reasonable approximation to the calculated asymptotic limit. The nondimensional rms value of the fluctuating component of u velocity, u'^+ is plotted in Fig. 4 for turbulent water flow along with the laminar value, which results from the noise level of the LDA. (All u'^+ values presented have been corrected for this level.) The general shape and the location of the peak of u'^+ at $y^+ = 15$ conforms with the published measurements (ref. 12); however, the overall scale of u'^+ is lower for the water table flow. This should be expected because the water table flow extends only to y^+ on the order of 100, while pipe flow and boundary layer flow generally have a log layer flow several orders of magnitude larger. The limited log layer of water table flow does not effect the mean profile (Fig. 3) nor the shape of u'^+, it only decreases the latter's scale.

The mean and fluctuating velocity profiles for water (Figs. 3 and 4) were obtained using the selective seeding LDA technique. Nearly continuous velocity measurements can be frequency analyzed, and Fig. 5 shows the oscilloscope output from the spectrum analyzer at $y^+ = 14$, 48 and 58 for a water turbulent boundary layer. These three values of y^+ are in the neighborhood of those that will be presented for polymer flows. The fluctuations in spectral intensity are due to the manner in which the analyzer operates. Its 10 Hz pass band slowly and automatically sweeps across the frequency range of 0 to 500 Hz. When centered at a particular frequency, the signal then entering the instrument is digitally analyzed. Successive averaging of such spectra would naturally decrease the fluctuations in Fig. 5, resulting in a smooth curve over the 0 to 500 Hz range. The three spectral traces in Fig. 5 show no major variations from $y^+ = 14$ to $y^+ = 54$, suggesting that the same disturbances are active throughout the buffer layer. A large fraction of the fluctuation is at low frequency, below 75 Hz. Assuming these fluctuations are convected with the local velocity suggests that their wavelength is greater than the thickness of the layer, which is verified by visual observations of the free surface.

The object of the experiments was to determine how polymer concentrations and wall strain rate modify laminar flows and then to induce transition to turbulence and again measure the same parameter dependence. Since the wall shear stress as of turbulent and laminar flow at constant mass flow and table inclination differ in general, a patch of turbulent flow will be deeper than a laminar region and therefore easily seen by looking at the free surface (Fig. 6a). It was this property which led to the discovery by Emmons (ref. 13) of spots on a water table. Such spots can be reproducibly generated by injecting a pulse of fluid from a hypodermic tube upstream of the entrance contraction. The injected fluid creates a three-dimensional disturbance (mostly a vortex ring) which triggers a localized transition to turbulence. The spanwise extent of the turbulence is many times the depth of the flow. In water the width of the spot along the span increases as the spot is convected downstream. Recently (ref. 14) spots in an air boundary layer have been studied and the velocity profile in the spot away from the leading and trailing edges has been found to be fully turbulent.

With the addition of polymers to the flow, spot formation is altered dramatically (Fig. 6b). In contrast to a spot in pure water the lateral growth rate is greatly reduced and the depth of a spot is not much different from the laminar flow surrounding it. As the shear stress or polymer concentration increases, fluctuations on the free surface of the otherwise laminar flow appear and can be seen in the background flow of Fig. 6c. These fluctuations are distinct from those due to turbulence, since transition to turbulence can still be induced (Fig. 6c). An increase in wall shear (above the laminar value) in pipe flows at Reynolds numbers below the expected turbulent transition has been observed (ref. 15) and this was called early turbulence. It appears likely that this increase in drag can be attributed to the velocity fluctuations visible in Fig. 6c. We have called these velocity fluctuations polymer induced fluctuations because they differ from turbulence. The distinction between polymer induced velocity fluctuations in laminar flow and turbulence fluctuations is sharp at moderate concentrations and strain rates and becomes less distinct at higher concentrations and strain rates. Hence, it will be important to distinguish between laminar-like properties and turbulent-like properties. For laminar flow:

(1) The free surface is smooth and free of waves (background of Fig. 6a, away from the spots).

(2) The u-velocity trace is relatively smooth; its spectrum is broadband, and u'^+ is low and essentially independent of y^+ (Fig. 4).

(3) Dye introduced at the wall is convected along the wall with little lateral spread; salts introduced in the flow show no evidence of streaks.

(4) By introducing disturbances, either by a trip cord or by injecting fluid, an instability can be generated which causes transition to turbulence.

(5) The mean velocity profile can be distinguished from the turbulent profile (Fig. 3 for water).

For turbulent flow:

(1) The free surface is wavy with many wavelengths greater than depth (the spot in Fig. 6a).

(2) The u-velocity has a large fluctuating component (Fig. 4) and spectra (Fig. 5) have large low frequency components and are essentially independent of y^+.

(3) Dye introduced at the wall is rapidly dispersed; salts introduced at the wall give evidence of streaks with a spanwise spacing (λ^+) of 116 $\pm$ 15.

(4) Introducing disturbances has no noticeable effects on the flow downstream of the disturbance.

(5) The mean velocity profile follows the universal velocity distribution for turbulent flow (Fig. 3).

Laminar Flow with Polymers

Laminar flows with dissolved polymers exhibit fluctuations on the free surface only for certain values of concentration and wall strain rate (dU/dy), as shown in Fig. 7. No attempt was made to define the region precisely as it was enough for this initial investigation to determine the general boundaries of the region. The form that this boundary curve assumes is consistent with observations for the onset of drag reduction. Berman and George (ref. 16) suggest a time scale criterion which for WSR-301 predicts an onset strain rate of about 320 sec^{-1}*, certainly less than the strain at the knee 720 sec^{-1} of Fig. 7. While drag reduction has been observed near the limit of infinite dilution, the onset strain rate increases as concentration decreases (ref. 17). This too is in accord with the trends of Fig. 7, suggesting that polymer fluctuations are closely related to the mechanisms of polymer drag reduction.

Graphs of u'^{+} vs y^{+} for polymer fluctuations at several strain rates and concentrations are shown in Fig. 8. At first glance, they are similar to those for water turbulence in Fig. 4. For all cases except for the highest concentration and strain rate, u'^{+} peaks at $y^{+} \simeq 15$ and then falls off to a much lower value at higher y^{+}. The strong difference between the polymer fluctuation and water turbulence can be seen in the spectra shown in Fig. 9 which correspond to conditions 2 in Fig. 8. These spectra show the features common to all cases of polymer fluctuations, namely strong spectral variation with y^{+} while the low and high frequencies are suppressed compared to water turbulence. Compared to laminar flow of water, there is the addition of fluctuating components in the 50 Hz to 150 Hz range. With these fluctuations confined in a region around $y^{+} = 15$, it is natural to assume they are convected with the local mean velocity implying a wavelength on the order of the layer thickness or larger. An inspection of the free surface (cf. Fig. 6c) shows wavelengths on the order of the layer depth. It is a characteristic of laminar layers with polymer fluctuations to have the more intense fluctuations confined to a region near the wall. As either the concentration or strain rate increases, the vertical extent increases, out to $y^{+} \simeq 35$ at the highest strain rate and concentrations tested. In all cases the polymer fluctuations as detected in u'^{+} were in the frequency range of 50 to 150 Hz and could be detected only near the wall (cf. Fig. 10). In addition, it appears that u'^{+} at large y^{+} tends to increase with concentration.

The mean velocity profiles U^{+} versus y^{+} were quite sensitive to polymer concentration. The velocity curves 1 and 2 in Fig. 10 correspond to conditions 1 and 3 on Fig. 8. At lower concentration, for example 13 ppm, where polymer fluctuations can first be detected, the measured profiles fall between water laminar and curves 1 and 2. If the mean velocity curves had been corrected for the kinematic viscosity's dependence on polymer concentration and strain rate, then the profiles would have fallen closer to the water laminar curve. The measured mean velocity profiles for polymer fluctuations are similar to those of the laminar flow of water, with an apparent higher viscosity. This holds true up to effective polymer concentrations of 25 to 32 ppm where it is impossible to distinguish between polymer fluctuations and polymer turbulence if the strain rate is high enough for polymer effects to be observed (cf. Fig. 7).

For flow described as polymer fluctuations, attempts were made to find evidence of streaks in the flow direction but none were observed. For water turbulence, streaks were observed, and their mean spacing in dimensionless wall variables was 116 ± 15, in agreement with measurements in air and water of many others using many different techniques.

Turbulent Flows with Polymers

The same type of experiments and measurements already reported for the polymer fluctuations were repeated for turbulent flows. It was necessary to introduce a

---

*Berman and George (ref. 16) used WSR-N80 and determined onset for glycerine-water polymer solutions. We have adjusted their value to account for the higher molecular weight and intrinsic viscosity of WSR-301 to arrive at this figure for onset.

larger disturbance to cause transition to turbulence at higher polymer concentrations than it was for pure water transition. The larger disturbances were either longer time pulses of fluid to cause transition in spots (cf. Fig. 6b) or larger diameter rods that spanned the flow at the entrance station. Hansen et al (ref. 18) on the basis of a linear stability analysis of pipe flow, using a single time scale viscosity model for dilute polymers, calculated that the growth rate of disturbances would be altered by the polymers. From our observations of the water table flows, we can report that larger amplitude disturbances are definitely necessary for transition, and that the spanwise growth of turbulent spots is also greatly reduced. In fact, three-dimensional spanwise variations in disturbances appear to be required to insure reproducible turbulent flow. A bead-chain (in fact an aluminium electric light pull chain) was used which completely spanned the flow at the entrance station. The bead-chain seemed satisfactory for all flows. Observations were made of the free surface with and without the chain. Additional disturbances such as cylinder 2 or 3 cm in dia. placed in the flow downstream of the chain resulted in no observable effects for more than a few diameters. It was therefore concluded that turbulent conditions with polymers had been achieved, and it is this condition which was used for the reported flow measurements.

The visual appearance of free surface is different for polymer turbulence and polymer fluctuations at concentrations below 25 ppm as can be seen in Figs. 6b and 6c. The amplitude and wavelength of the free surface fluctuations are much larger for turbulence than for polymer fluctuations, as the spectrum of u'^+ reveals. In addition, the turbulent patch is deeper than the surrounding flow since the wall shear stress under the patch is greater.

The measurements of u'^+ vs y^+ for several polymer concentrations and strain rates are plotted in Fig. 11 along with that of pure water for reference. At the lowest concentration there is an increase in the magnitude of u'^+ over that of pure water, even though the spectral distribution (Fig. 12) is quite similar to that of pure water. At this concentration, no polymer fluctuations were detected in the laminar flow. Curve 2 in Fig. 11 shows a double peak in intensity, and its u-velocity spectra at four different values of y^+ are shown in Fig. 13. The first peak near $y^+ = 15$ appears to be associated with polymer fluctuations both on the basis of its y^+ location and the u-velocity spectrum. The increase in spectral intensity at lower frequency is definitely present. The second peak near y^+ of 30, on the basis of the spectra at $y^+ = 32$ and 41, is probably associated with the turbulence process.

From the wide range of the u'^+ versus y^+ data as a function of polymer concentration and strain rate, it is apparent that a simple characterization of the fluctuation dependence is not possible. Our results for a two-dimension mean flow indicate a strong dependence on wall strain rate over the entire range of polymer concentration from low to intermediate, and to high. Curve 1 of Fig. 11 indicates that u'^+ is augmented over the solvent values at low concentrations; curves 2, 3 and 4 are all for a moderate concentration of ~16 ppm and show the richness of the strain rate dependence. The lower strain rate (curve 3; $u_\tau = 3.25$ cm/sec) has u'^+ below the solvent value for all y^+. The u velocity spectra at the next highest strain rate ($u_\tau = 3.52$ cm/sec) in Fig. 13 shows evidence of strong polymer fluctuations near $y^+ = 15$, the first peak in the u'^+ intensity. Increasing u_τ to 3.85 cm/sec lowers the intensity of u'^+ everywhere, the spectrum near the peak shown in Fig. 14 being suggestive of polymer fluctuations. Curves 5 and 6 of Fig. 11 are representative of what we have called asymptotic conditions or suppressed turbulence. The effective concentrations and strain rates are so large that no distinction can be made between the flow with and without the bead-chain disturbance. Spectra for curve 5 of Fig. 11 are shown in Fig. 15. The flow is inactive except for the fluctuations near the wall which which are suggestive of polymer fluctuations.

The turbulent mean velocity profiles for concentrations below 21 ppm are plotted in Fig. 16. The profiles for the asymptotic cases are the solid circles and diamonds on Fig. 10 which are plotted with other polymer fluctuation flows. The asymptotic cases fit into an ordered sequence, as concentrations and strain rates are increased when viewed as laminar flows with polymer fluctuations. When viewed from the water turbulent limit, the mean velocity profiles increase their difference from the Newtonian turbulence as concentration increases, but then fall back as the asymptotic conditions are approached and polymer fluctuations play a dominant role.

The mean velocity profiles plotted in the standard wall coordinates are indicative of the effective viscosity in the different ranges in the flow. In the sublayer, $y^+ < 8$, the mean velocity profiles for laminar and turbulent water are the same and there is no evidence in our experiments that polymer flows with either polymer induced fluctuations or turbulence departs significantly from this solvent behavior. As far as the sublayer thickness is concerned, our results are in agreement with those of Reischman and Tiederman (ref. 8); however, the laminar water flow on the table departs from $U^+ = y^+$ in the neighborhood of $y^+ \simeq 8$ (cf. Fig. 3) because the shear stress is not constant throughout the flow. The laminar water flow mean velocity profile represents what is expected when the viscosity is constant throughout the flow. On the other hand, the turbulent water flow profile is the result of augmentation of molecular viscosity by the turbulent mixing process. Departures from this profile for turbulent polymer flows imply changes in the mixing process. Following this line of reasoning, Fig. 16 shows a departure from the universal profile for 9 ppm at $u_\tau = 3.72$ cm/sec and the maximum departure at 14 ppm and $u_\tau = 3.77$ cm/sec. At this latter concentration, polymer fluctuations were observed at $u_\tau = 3.71$ cm/sec,which is near the boundary curve for Fig. 7. Increasing polymer concentration and strain rate appears to inhibit normal turbulent mixing; however, beyond the moderate concentration of 14 ppm fluctuations in the u-velocity are induced in the laminar flow resulting in an effective increased viscosity, at least in the region where the fluctuations are observed. The resulting turbulent flows at concentrations of 21 ppm and beyond appear not only to consist of suppressed turbulent mixing but also have augmented viscosity due to polymer fluctuations. Those turbulent flows with the lowest peak values of u'^+ appear to be dominated by polymer induced fluctuations observable at y^+ near the peak (Fig. 15) rather than by normal turbulent mixing processes which show a similar profile of u'^+ (cf. Fig. 5 for pure water of Fig. 12 for 9 ppm of polymer). Increasing the polymer concentration from 25 to 32 ppm increases the maximum of u'^+ relative to its value at 15 ppm and shifts the resulting mean turbulent velocity profile back toward the water turbulent curve. The shift is due to an increase in polymer fluctuations rather than due to increased turbulent mixing. The asymptotic flow situations exhibit more laminar features than turbulent ones. Spectra for this flow are shown in Fig. 15. The fluctuations are confined to the wall region, outside of which the velocity gives the appearance of a laminar flow with possibly presence of some very low frequencies.

Measurements of the average spanwise streak spacing, λ^+, were made of the polymer turbulent flows using the salt crystal technique. A few examples illustrate the results. For pure water $\lambda^+ = 116 \pm 15$; at concentration of 9 ppm and $u_\tau = 3.72$ cm/sec, $\lambda^+ = 161 \pm 115$; at 9 ppm and $u_\tau = 4.21$ cm/sec, $\lambda^+ = 200 \pm 18$; at 16 ppm and $u_\tau = 3.52$ cm/sec $\lambda^+ = 190 \pm 20$. For the flows at and near asymptotic conditions there was no evidence of streaks using this technique.

DISCUSSION OF RESULTS AND CONCLUSIONS

The flows investigated were two-dimensional laminar and turbulent shear flows on a free surface water table. These basic flows were characterized in detail. Measurements of U^+, u'^+ and u-velocity spectra versus y^+ have been reported along with observations of the free surface and measurements of spanwise streak spacing for turbulent flows. The laminar flow is nearly an equilibrium flow and in good agreement with the calculated profile assuming the body force is equal to the local wall shear stress. The turbulent flow mean velocity profile is in excellent agreement with the universal velocity profile even though the flow extends only to y^+ on the order of 100. The peak of u'^+ occurs at y^+ of 15 in agreement with the results of other investigations made in other turbulent boundary layers and in pipes, while the absence of the outer flows appears to just reduce the overall level of u'^+. The spanwise streak spacing is in agreement with published values of other investigators. The absence of the outer flow allowed the near wall region to be studied without the extraneous disturbances imposed by the outer flow. As a result, it has been possible to establish features of the inner flow which have not been revealed in other investigations.

At modest polymer concentrations and strain rates large enough to be comparable to the time scale of the polymer, effects on the velocity very near the wall become noticeable in an otherwise laminar flow. The mean intensity of polymer induced fluctuations has a spatial distribution in y^+ similar to what one finds in Newtonian turbulence, though the intensity is lower and spectral distribution is quite different.

The distinction is further confirmed by velocity time traces and the free surface appearance. No evidence of streaks in the flow were found at the wall. Transition to a polymer-influenced turbulence would be triggered by a large enough three-dimensional disturbance in the flow.

The polymer induced velocity fluctuations are clearly strain rate dependent as shown in Fig. 7. Since the strain rate where they are first observed is of the same order as the inverse of the calculated response time for the polymer, it is natural to assume these fluctuations are associated with at least partial expansion of the polymer. Observations of an increase in laminar wall shear stress in capillary flows at similar maximum strain rates have been ascribed to "early turbulence" (ref. 15). We have also observed an increase in wall shear stress when polymer induced fluctuations are present; however, this flow state is distinct from turbulence, since at modest polymer concentrations a separate and distinct transition can be induced. In addition, the spatial variation of u-velocity spectra shows that polymer induced fluctuations are primarily confined to the region near the wall with the fluctuations primarily in the 50 to 150 Hz range.

Turbulent flows of polymer solutions at low and modest concentrations or strain rates did exhibit streaks in the sublayer. The spacing between streaks increased with concentration and strain rate. In this regime, turbulent fluctuations were similar throughout the layer, and the spectra were similar to those of water. By comparing the depth of the polymer flow to that of water at the same mass flow rate and table inclination, drag reduction was definitely established. In addition, the mean velocity profiles showed a departure from the universal velocity profile, suggesting a change in turbulent mixing or effective turbulent viscosity. Polymer effects on the mean turbulent velocity profile and on wall shear stress were observed at polymer concentrations and strain rates below those needed to cause polymer fluctuations. Though no attempt was made to define an onset curve for polymer effects on turbulence similar to the one for polymer fluctuations in laminar flows, the experimental data suggest that the onset line lies to the left and below the polymer fluctuation curve of Fig. 7. Such a result is consistent with our view of hairpin eddies being the major mixing mechanism in a Newtonian turbulent boundary. The maximum strain rate in an eddy aligned with the flow would be higher than that at the wall, since the vortex tube forming the eddy is stretched by the mean flow. In addition, the maximum strain rate occurs where rotational effects are small, outside the vortex core. Hence, expansion of the polymer coil could occur at a lower strain rate than it could in the laminar shear flow at the wall where strain and rotation are in balance (ref. 2). For both of these reasons, it should be expected that polymers would begin to inhibit turbulent mixing at lower strain rates than in laminar flows. At a fixed concentration, strain rate dependence should also be expected. Polymers such as WSR-301 contain molecules of different lengths, and the largest expand at lower strain rates than the shorter molecules. Independent of the exact structure of the three-dimensional mixing mechanism, some of the structures will involve higher strain rate motions than others. It is these more intense structures which will be inhibited at the lowest strain rates, leading to an increase in turbulent streak spacing as strain rate is increased. Such reasoning is consistent with the experimental result presented. It is further confirmed by comparison of the u'^{+} spectrum across the flow as a function of concentration and strain rate. Fig. 5 is the water spectra and Fig. 12 is for 9 ppm, a concentration below polymer induced fluctuations; the spectra are quite similar, though drag reduction is observed and the mean velocity profile has been influenced indicating a change in mixing. Increasing concentration and strain rate, Figs. 13 and 14 show the velocity fluctuations substantially reduced except for $y^{+} < 20$, where polymer induced fluctuations are found in laminar flows. With further increase in strain rate and concentration, an asymptotic regime is reached where most of the turbulent features are suppressed but the polymer induced velocity fluctuations remain (Fig. 15). Under these conditions, streaks were never observed. These suppressed turbulent flows have mostly laminar characteristics, and it was difficult to distinguish these flows from those of polymer induced fluctuations at similar strain rates and concentrations.

On the basis of the flows so far investigated, it is possible to speculate on what would happen if concentration and strain rate were both substantially increased. It is believed that the flow would remain suppressed but the level of u'^{+} would increase. If this should be the case, then perhaps the very substantial difference between the u'^{+}

maxima measured by Reischman and Tiederman (ref. 8) and those measured by Rudd (ref. 5) could be explained.

Of the other investigations which obtained extensive velocity measurements (using LDA) in the near wall region of a turbulent drag reducing flow, Reischman and Tiederman (ref. 8) operated at essentially the same range of strain rates as we did, Kumor and Sylvestor (ref. 7) were very slightly higher, and Rudd (ref. 5) was nearly an order of magnitude higher ($\sim 10^4$ sec^{-1}). (Logan [ref. 6] did not report strain or u_τ.) All of these investigators used 100 ppm or more of a similar or longer more effective drag reducing polymer than WSR-301. Thus, we expect that each of these investigations measured flows for which the near wall layers could be characterized as suppressed turbulence and that the u'^+ measurements reported were composed mainly of polymer induced fluctuations and fluctuations imposed by the outer flow which were not involved in mixing near the wall. Since the water table experiments reported here extended only to y^+ of 100, direct comparison is difficult and somewhat conjectural. Closed comparisons would have been possible if velocity spectral data had been measured by the other investigators.

The selective seeding of the flow with hydrogen bubbles, which allowed detailed spatial resolution and spectral measurement of the velocity to be made, is believed to be novel. Its use in conjunction with visual observations of the free surface disturbances enabled the laminar-like polymer induced velocity fluctuation flow to be identified.

Comparisons with other research have been made on the basis of wall strain rates. If one were to determine Reynolds numbers appropriate for water table flows, using thickness as the length parameter, they would certainly be much lower than the Reynolds numbers of previous investigations (Refs. 5-9), yet the very suitability of drawing conclusions from a Reynolds number comparison is questionable due to the inherent differences of the water table flow from those in pipes or over a flat plate. Since it is fairly well established that drag reduction due to polymer additives is a wall phenomenon, using wall variables appears to be the most reasonable approach for comparison.

ACKNOWLEDGEMENTS

It is with pleasure and gratitude that we acknowledge the financial support of the Division of Engineering and Applied Physics, Harvard University, Cambridge, Mass. and the U.S.A. Office of Naval Research and the Fluid Mechanics Program of the USA National Science Foundation Grant #NSF-ENG77-01478

REFERENCES

1. Lumley, J. L. "Drag reduction by additives" Ann. Rev. Fluid Mech., 1, pp.367 ff. (1969)

2. Lumley, J. L. "Drag reduction in turbulent flow by polymer additives" J. Polymer Sci. Macromolecular Reviews 7, pp. 263-290 (1973)

3. Hoyt, J. W. "The effect of additives on fluid friction" J. Basic Engg. D94, pp. 258-285 (1972)

4. Virk, P. S. "Turbulent kinetic energy profile during drag reduction" Phys. Fluid, 18, pp. 415 ff. (1975)

5. Rudd, M. J. "Velocity measurements made with a laser Dopplermeter on the turbulent pipe flow of a dilute polymer solution" J. Fluid Mech. 51, pp. 673-685 (1972)

6. Logan, S. E. "Laser velocimeter measurement of Reynolds stress and turbulence in dilute polymer solutions" Am. Inst. Aeronautics and Astronautics J. 10, pp. 962 ff. (1972)

7. Kumor, S. M. & Sylvestor, N. D. "Effects of a drag-reducing polymer on the turbulent boundary layer" AIChE Symposium Series, Drag Reduction, pp. 1 ff. (1973)

8. Reischman, M. M. & Tiederman, W. G. "Laser-Doppler anemometer measurements in drag-reducing channel flows" J. Fluid Mech. 70, pp. 369 ff. (1975)

9. Scrivener, O. "A contribution on modifications of velocity profiles and turbulence structure in a drag reducing solution" Proc. Int. Conf. on Drag Reduction, BHRA Fluid Engg. pp. C66 ff.(1974)

10. Schraub, F. A., Kline, S. J. Henry, J. Runstadler, P. W. Jr., Littell, A. "Use of hydrogen bubbles for quantitative determination of time-dependent velocity fields in low-speed water flows" J. Basic Engg. ASME Trans. Series D87, pp. 429 ff.(1965)

11. Gordon, R. J. & Balakrishnan, C. "Vortex inhibition: a new viscoelastic effect with importance in drag reduction and polymer characterization" J. Appl. Pol. Sci. 16, pp. 1629 ff. (1972)

12. Hinze, J. O. "Turbulence" 2nd edition, McGraw Hill, pp. 627-628 , 657 (1975)

13. Emmons, H. W. "The laminar-turbulent transition in a boundary layer" Part I. J. Aero Sci. 18, pp. 490-498 (1951)

14. Wygnanski, I. Sokolov, M. & Friedman, D. "On a turbulent 'spot' in a laminar boundary layer" J. Fluid Mech. 78, pp. 785 ff. (1976)

15. Forame, P. C., Hansen,R. J. & Little, R. C. "Observations of early turbulence in the pipe flow of drag reducing polymer solutions",AIChE J. 18, pp. 213 ff. (1972)

16. Berman, N. S. & George, W. K. "Onset of drag reduction in dilute polymer solutions" Phys. of Fluids, 17, pp. 250-251(1974)

17. Paterson, R. W. & Abernathy, F. H. "Turbulent flow drag reduction and degradation with dilute polymer solutions" J. Fluid Mech. 43, pp. 689-710 (1970)

18. Hansen, R. J., Little, R. C., Reischman, M. M. & Kelleher, M. D. "Stability and the laminar-to-turbulent transition in the pipe flows of drag-reducing polymer solutions", BHRA, Int. Conf. on Drag Reduction (1974)

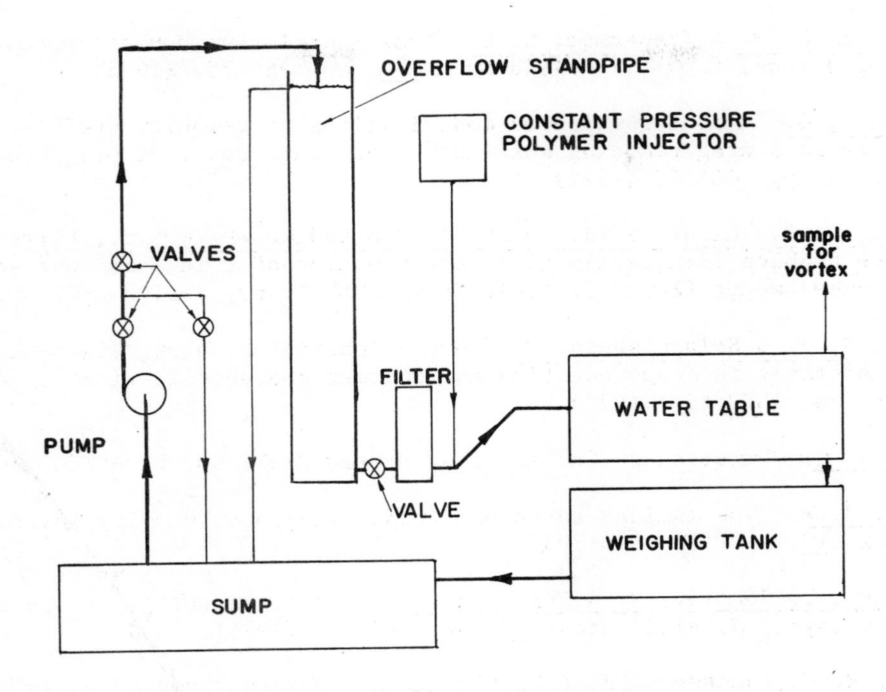

Fig. 1 Schematic flow arrangement

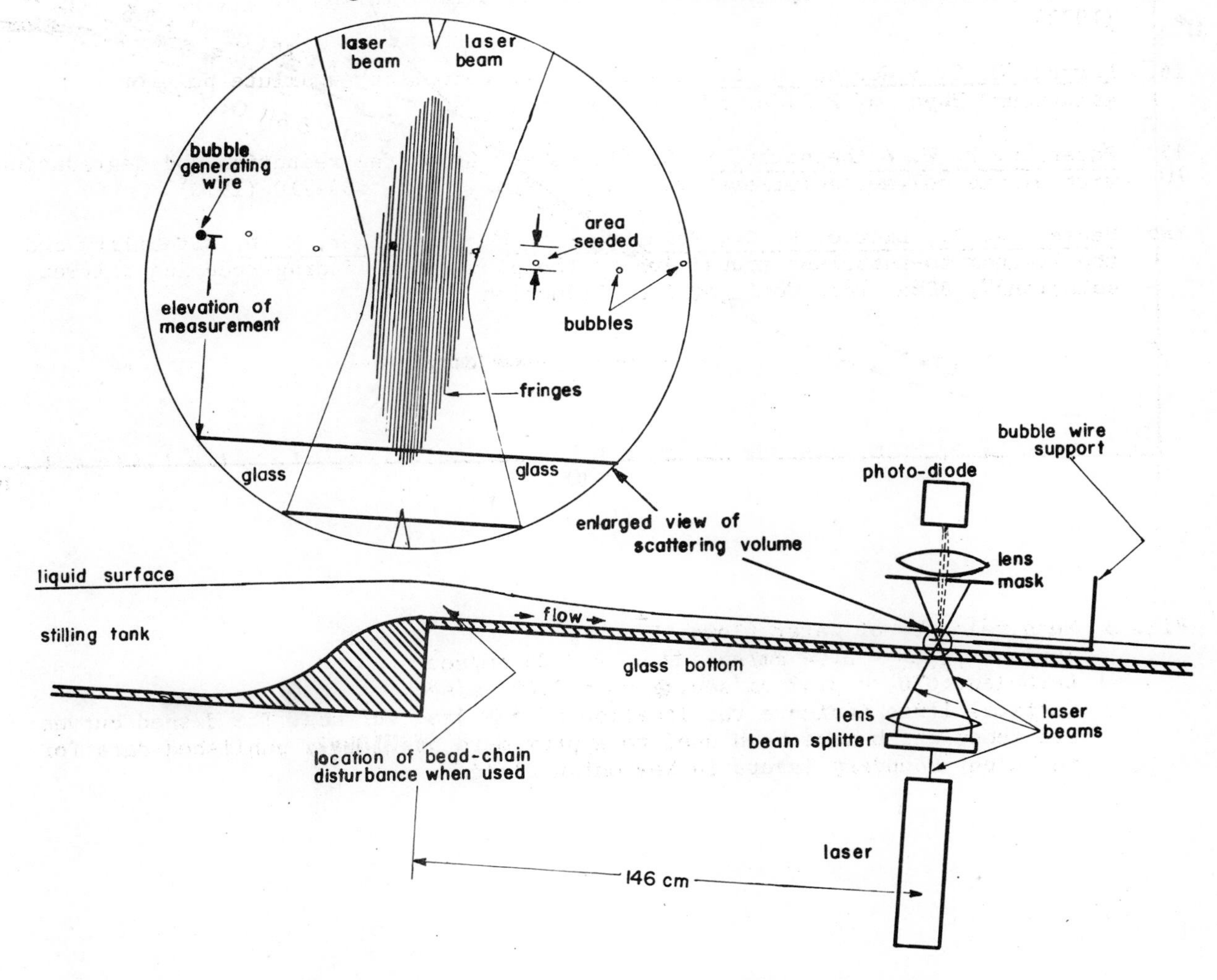

Fig. 2 Side view of water table showing the orientation of the laser-Doppler anemometer. The detailed view depicts the selective seeding method in which bubbles generated by electrolysis and introduced at a narrow range of y are used with the LDA.

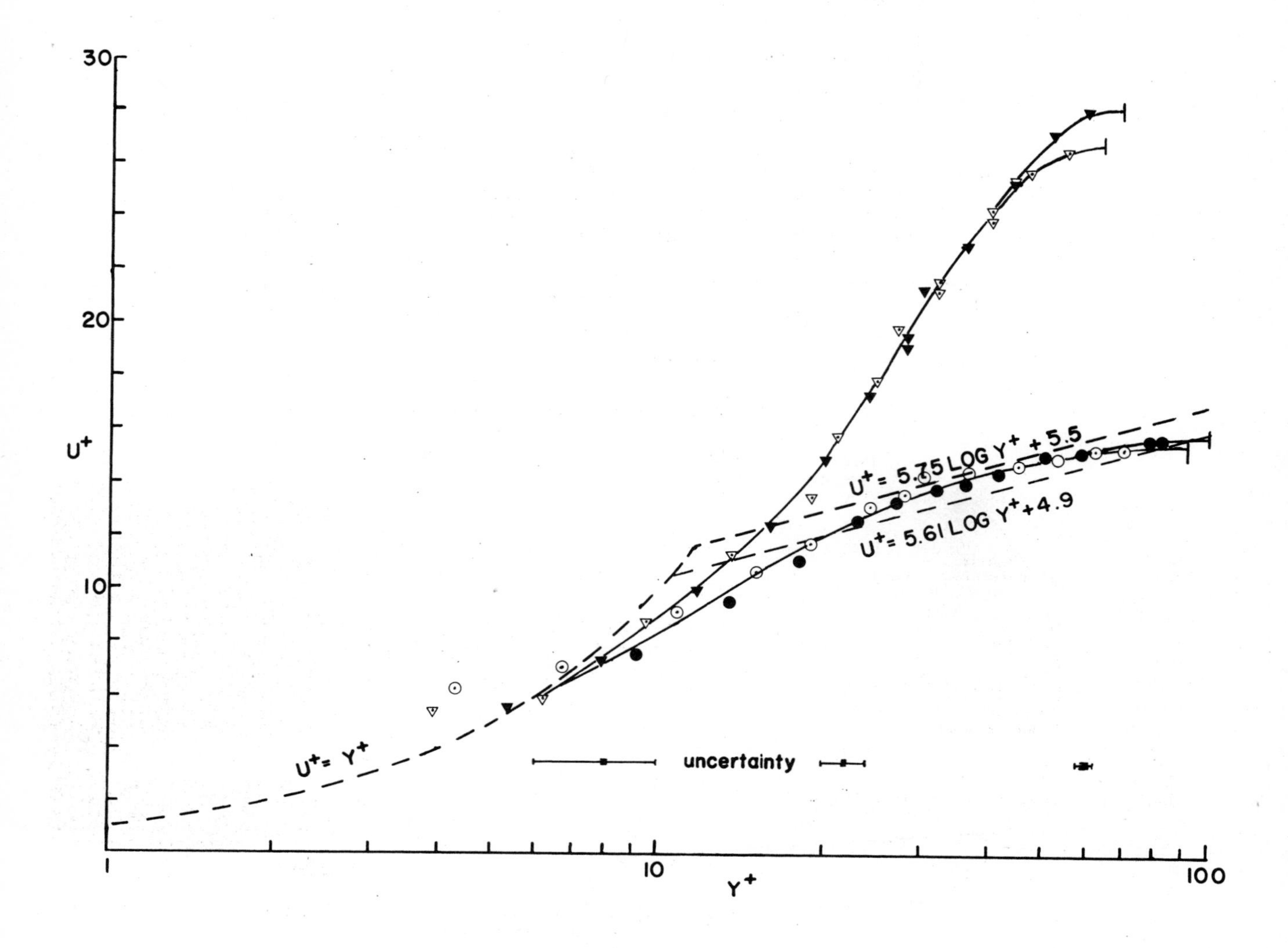

Fig. 3 Mean velocity of water flows, U^+ vs y^+
laminar: ▼ u_τ = 3.04 cm/sec, ▽ u_τ = 3.25 cm/sec;
turbulent: ⊙ u_τ = 3.42 cm/sec, ● u_τ = 3.68 cm/sec.
Vertical lines indicate the location of the free surface. The dashed curves are those which have been used to approximate previously published data for turbulent boundary layers in Newtonian fluids.

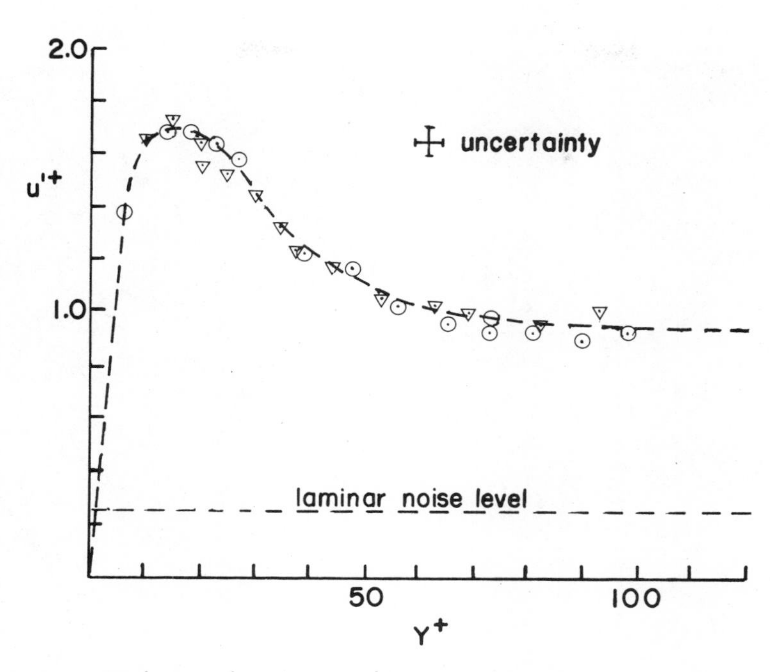

Fig. 4 Turbulent water u'^{+} vs y^{+}. All u'^{+} data has been corrected for the laminar LDA noise whose uncorrected level is shown.
⊙ u_{τ} = 3.42 cm/sec. ▽ u_{τ} = 3.83 cm/sec.

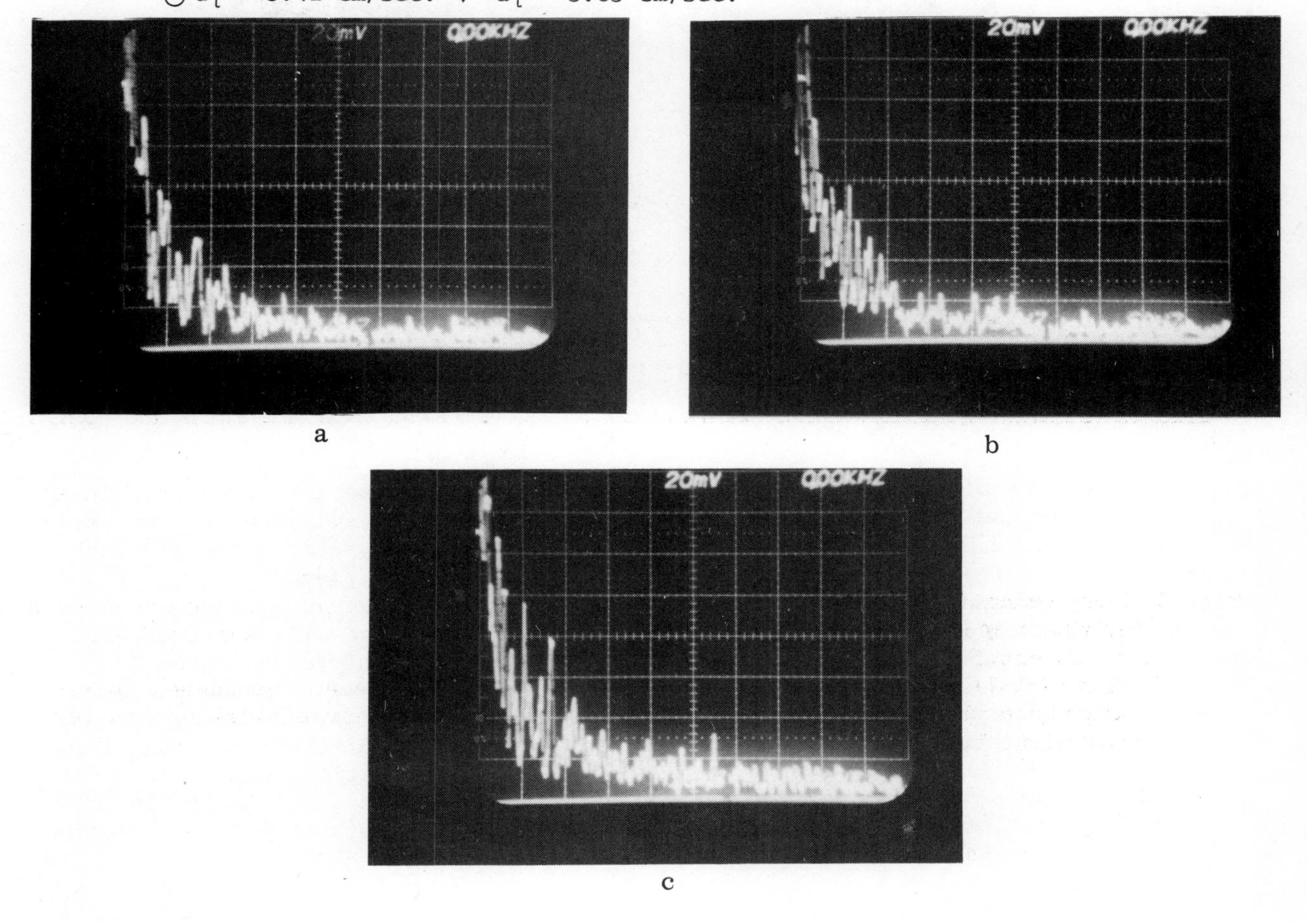

Fig. 5 Water turbulence spectra: as for all spectral data presented (Figs. 9, 12-15) full scale is 500 Hz and the vertical amplitude is 20 mV rms per division. u_{τ} = 3.71 cm/sec. (a) y^{+} = 58 (b) y^{+} = 48 (c) y^{+} = 14. The water turbulence spectra are essentially similar to these over the range of u_{τ} and y^{+} investigated.

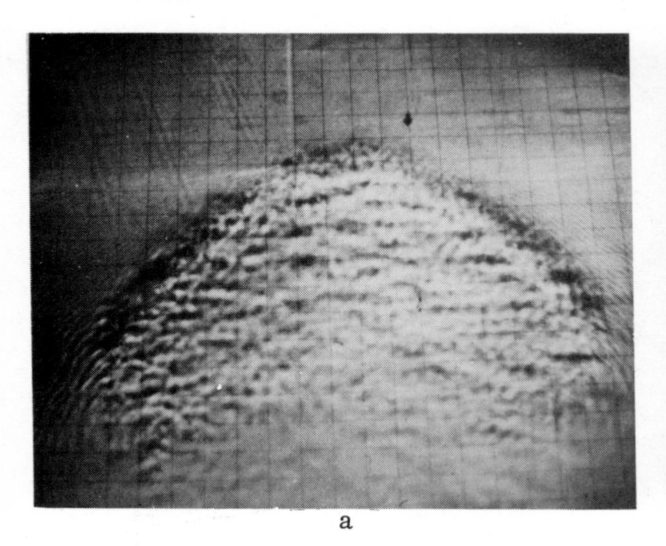

a

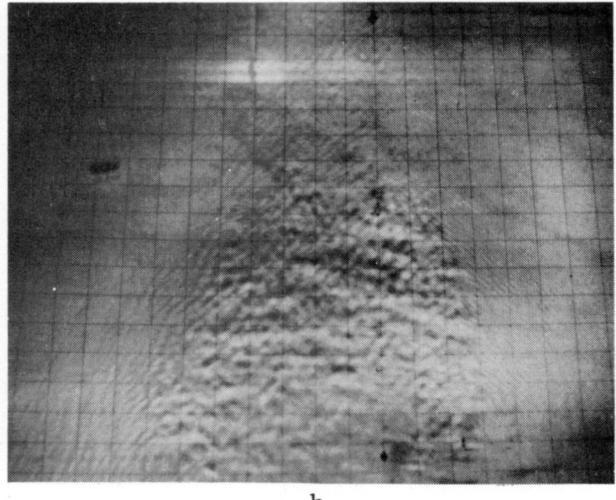

b

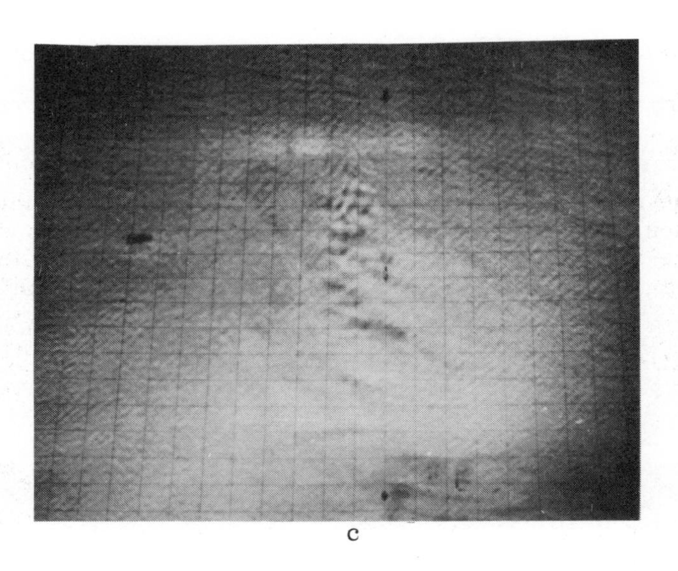

c

Fig. 6 Photographs of the free surface when turbulent spots have been triggered in an otherwise undisturbed flow over the water table. The heavy lines in the background are 2.54 cm apart and the flow depth is typically 2 or 3 mm. In each case the fully turbulent depth corresponds to $u_T \simeq 3.6$ cm/sec.
(a) Pure water. Over the range studied the general appearance, size and growth rate of the spot are essentially independent of depth and wall shear stress.
(b) 11-13 ppm WSR-301 solution. The wall strain rate of the background laminar flow ($\simeq 1130$ sec^{-1}) is smaller than that necessary to cause polymer fluctuations. The turbulent depth is 2.8 mm while the laminar one is 2.4 mm.
(c) 11-13 ppm WSR-301 solution. The wall strain rate of the background flow ($\simeq 1210$ sec^{-1}) is high enough to sustain polymer fluctuations which are visible as ripples on the free surface. While a spot is still identifiable, it is much less distinct than in (b). The turbulent depth is 2.0 mm while the laminar one is 1.9 mm.

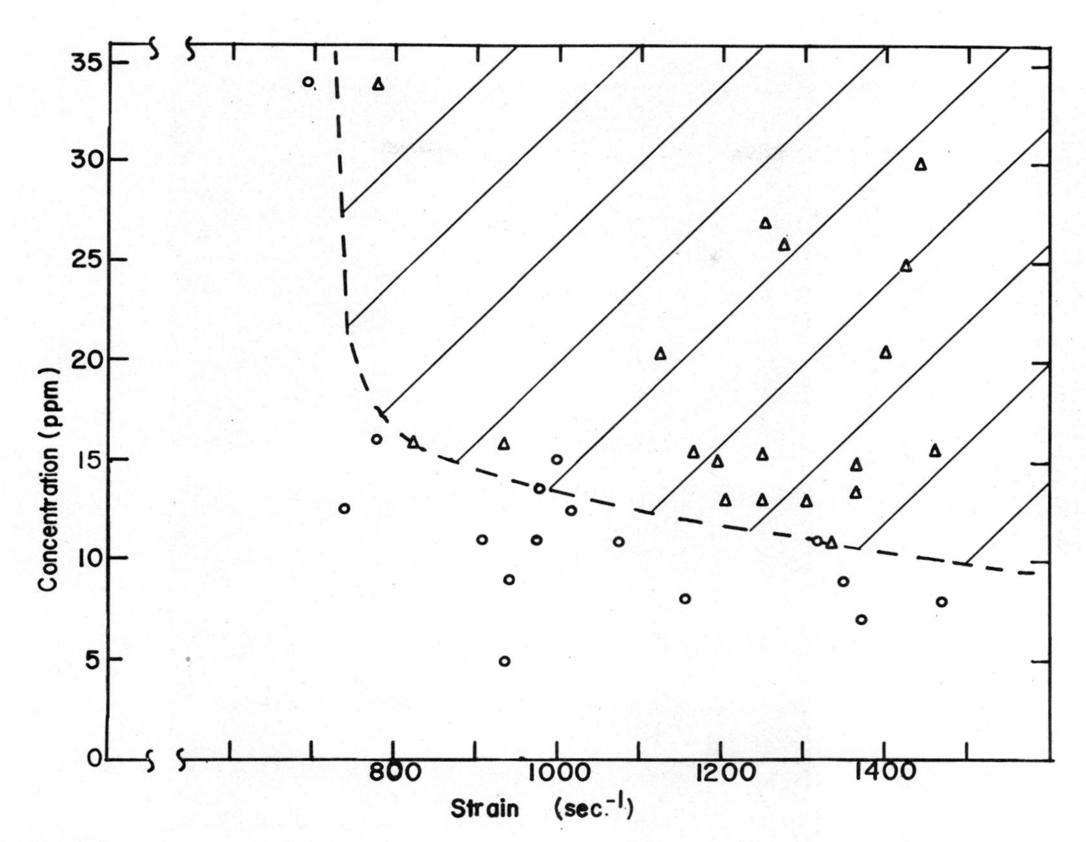

Fig. 7 Observations of polymer solution flows without disturbance as a function of concentration and strain rate (dU/dy). ○ smooth free surface: ▽ rippled free surface indicating the presence of polymer fluctuations (cf. the background of Fig. 6c). Fluctuations are expected in the shaded region.

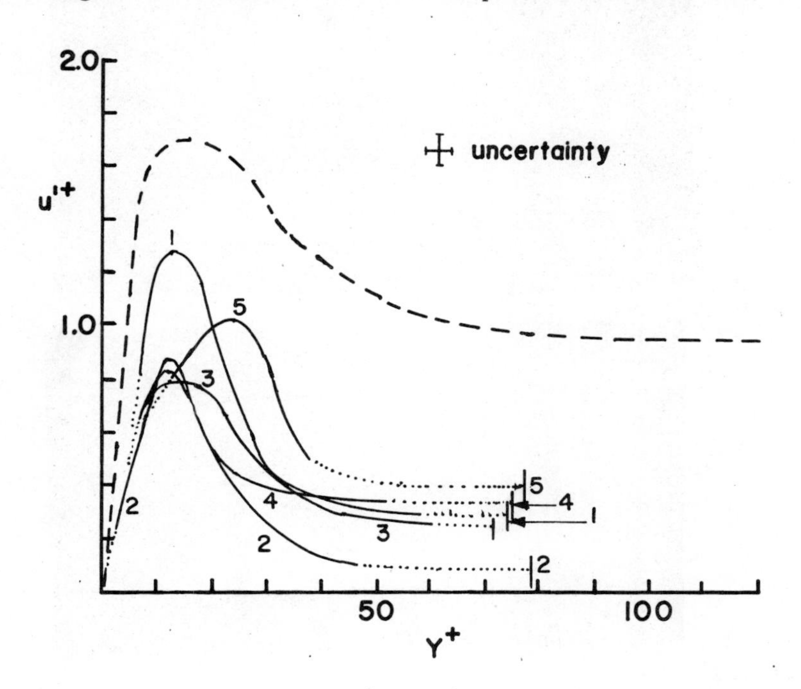

Fig. 8 Polymer induced fluctuations, u'^{+} vs y^{+} (in rough order of increasing activity). (1) 14 ppm, u_{τ} = 3.71 cm/sec. (2) 16 ppm, u_{τ} = 3.36 cm/sec. (3) 15 ppm, u_{τ} = 3.65 cm/sec. (4) 25 ppm, u_{τ} = 3.73 cm/sec. (5) 32 ppm, u_{τ} = 3.81 cm/sec. The dashed curve is u'^{+} vs y^{+} for turbulent water. A vertical line indicates the location of the free surface. Dotted lines indicate extrapolation beyond data points.

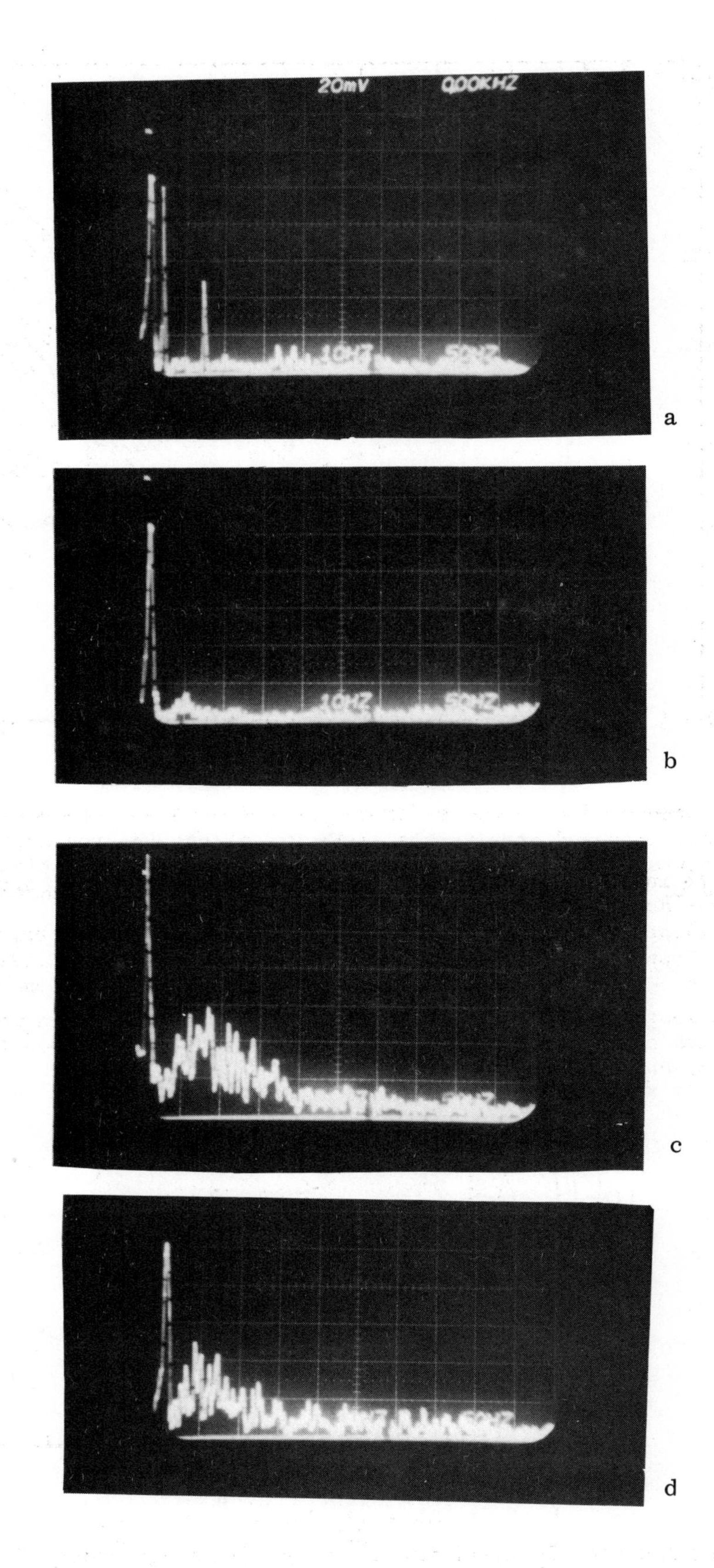

Fig. 9 Polymer fluctuation spectra: 16 ppm, $u_\tau = 3.36$ cm/sec.
(a) $y^+ = 57$ (b) $y^+ = 39$ (c) $y^+ = 12$ (d) $y^+ = 8$
The enriched spectral amplitude at 50-150 Hz is clearly visible near the wall.
This is the same flow as curve 2 in Fig. 8

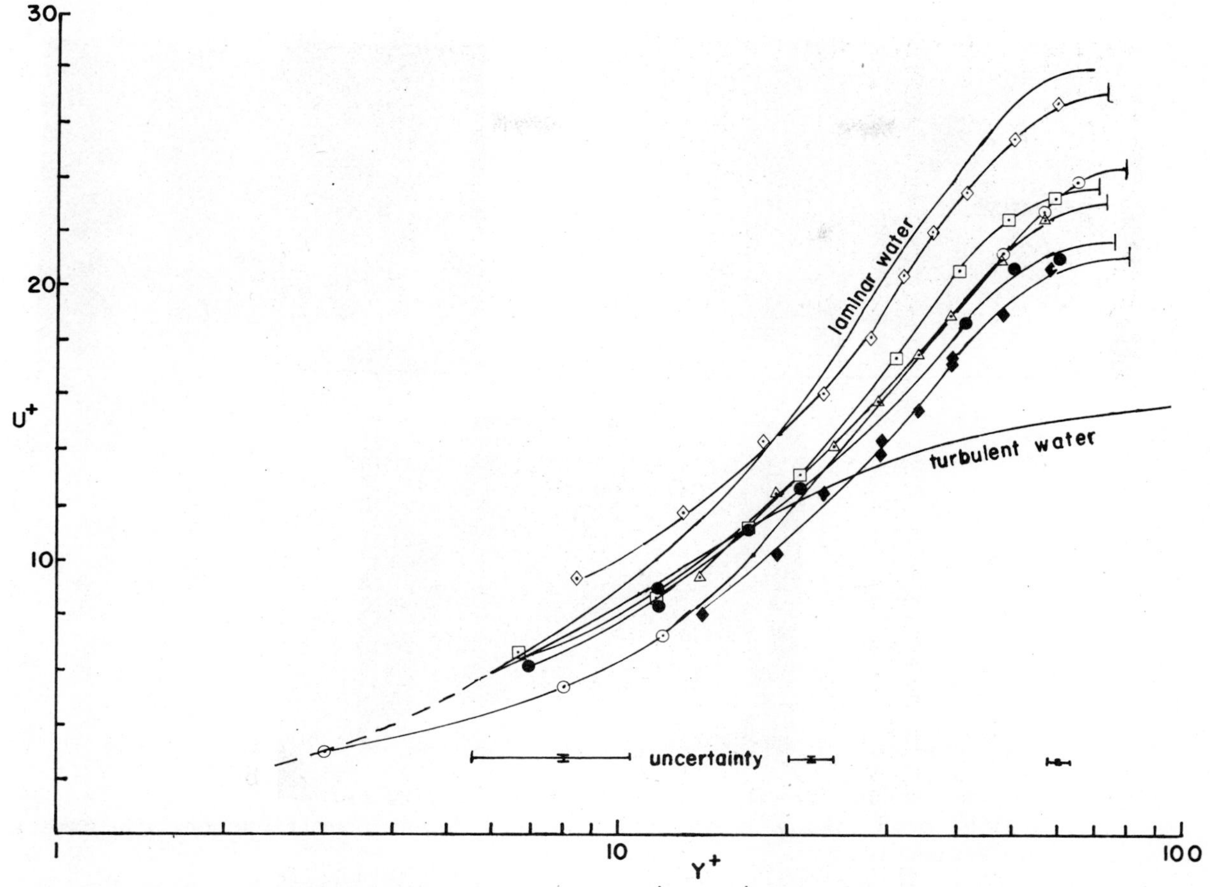

Fig. 10 Polymer induced fluctuation flows, U^+ vs y^+, roughly in order of increasing polymer activity.
◇ 14 ppm, u_τ = 3.71 cm/sec. △ 21 ppm, u_τ = 3.72 cm/sec.
□ 15 ppm, u_τ = 3.65 cm/sec. ● 25 ppm, u_τ = 3.73 cm/sec.
⊙ 16 ppm, u_τ = 3.36 cm/sec. ◆ 32 ppm, u_τ = 3.85 cm/sec.
These last two cases are at the asymptotic limit and the mean velocity profiles are practically identical to those of suppressed turbulence at the same u_τ and concentration.

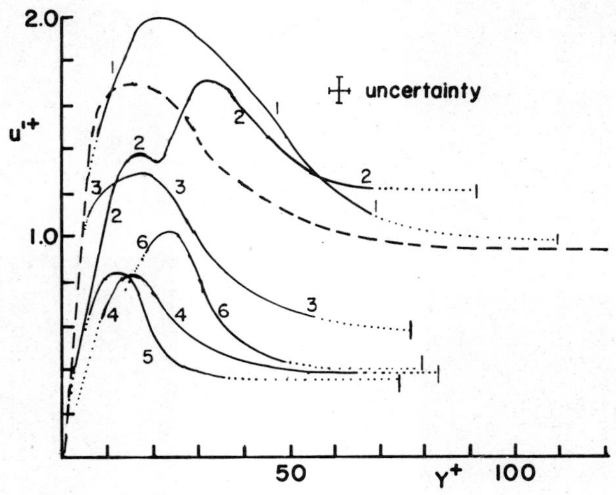

Fig. 11 Polymer turbulence u'^+ vs y^+
(1) 9 ppm, u_τ = 3.72 cm/sec. (2) 16 ppm, u_τ = 3.52 cm/sec.
(3) 15 ppm, u_τ = 3.25 cm/sec. (4) 16 ppm, u_τ = 3.85 cm/sec.
(5) 25 ppm, u_τ = 3.73 cm/sec. (6) 32 ppm, u_τ = 3.85 cm/sec.
A vertical line indicates the location of the free surface, and dotted lines are extrapolations beyond data points.

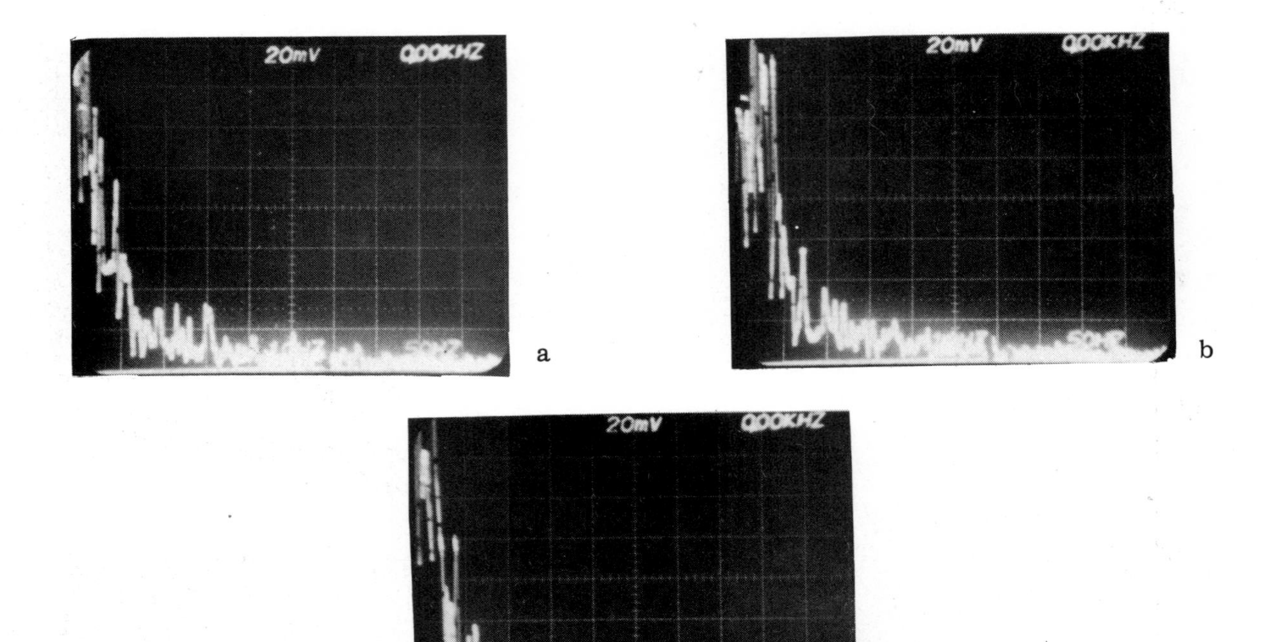

Fig. 12 Polymer turbulence spectra; 9 ppm, u_τ = 3.72 cm/sec.
(a) y^+ = 58 (b) y^+ = 48 (c) y^+ = 14. These conditions are the same as presented for water in Fig. 5.

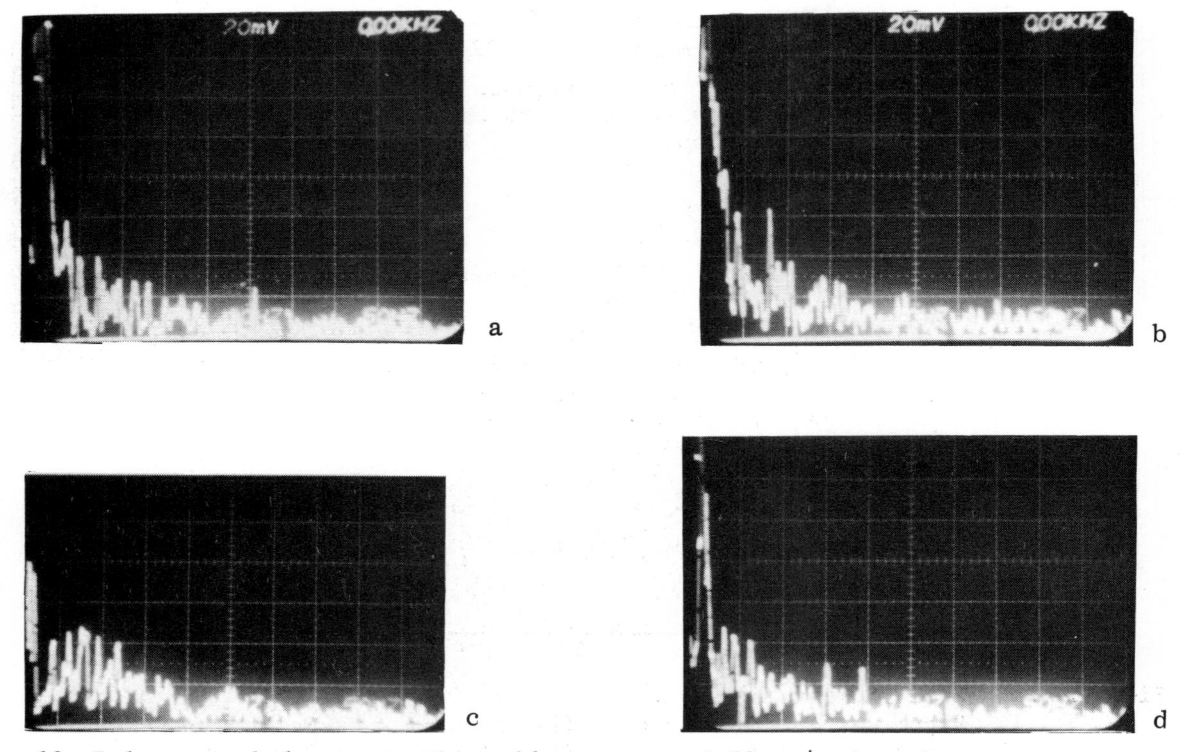

Fig. 13 Polymer turbulence spectra: 16 ppm, u_τ = 3.52 cm/sec.
(a) y^+ = 41 (b) y^+ = 32 (c) y^+ = 13 (d) y^+ = 8
This is the same flow as for curve 2 of Fig. 11. Note that at y^+ = 13, the spectrum is similar to those for polymer fluctuations. (Fig. 9).

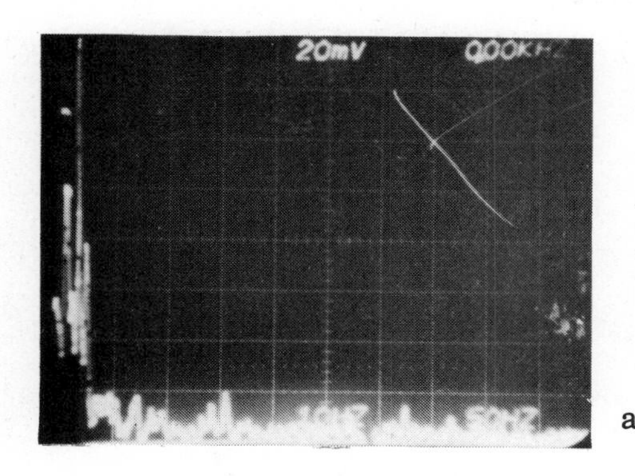

a

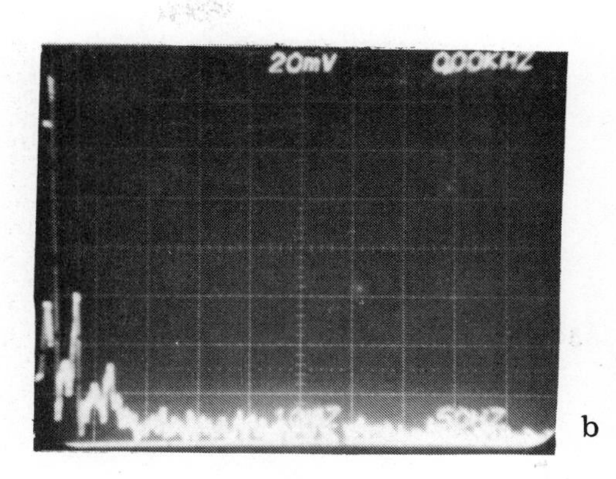

b

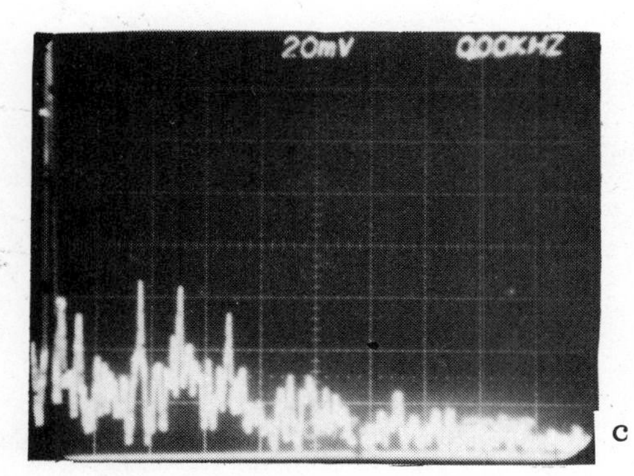

c

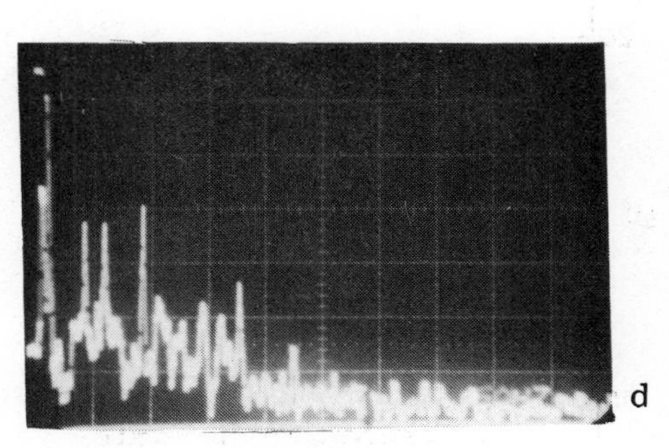

d

Fig. 14 Polymer turbulence spectra: 16 ppm, u_τ = 3.85 cm/sec. (a) y^+ = 54 (b) y^+ = 35 (c) y^+ = 19 (d) y^+ = 14. This is the same flow as for curve 4 of Fig. 11.

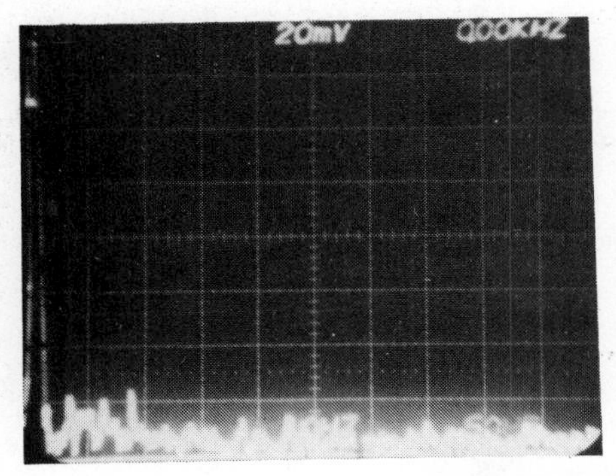

a

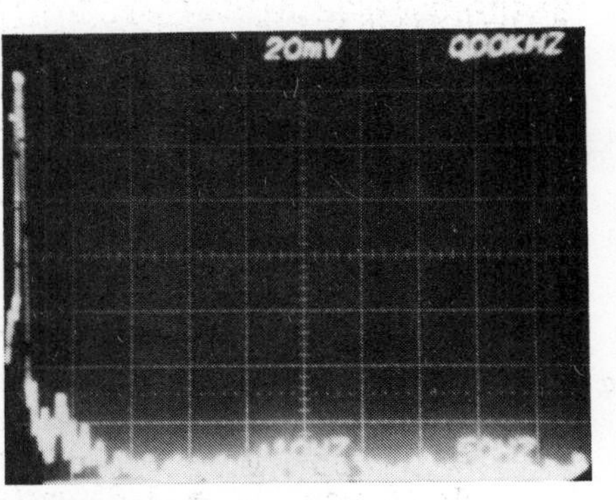

b

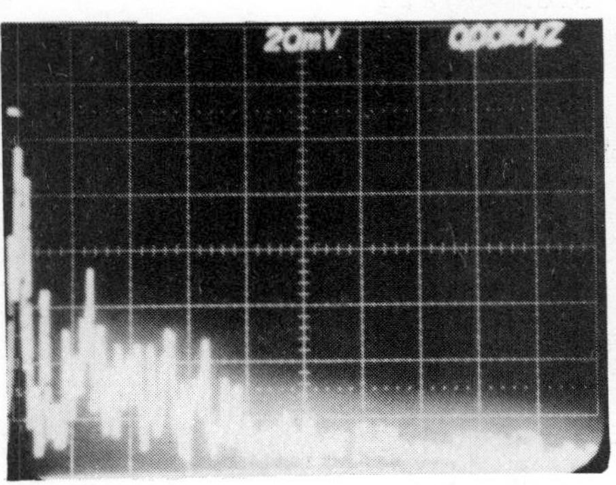

c

Fig. 15 Polymer turbulence spectra: 25 ppm, u_τ = 3.73 cm/sec. (suppressed turbulent regime) (a) $y^+ \cong 50$ (b) y^+ = 26 (c) y^+ = 12. This is the same flow as curve 5 of Fig. 11. The laminar counterpart of this flow had essentially identical spectra.

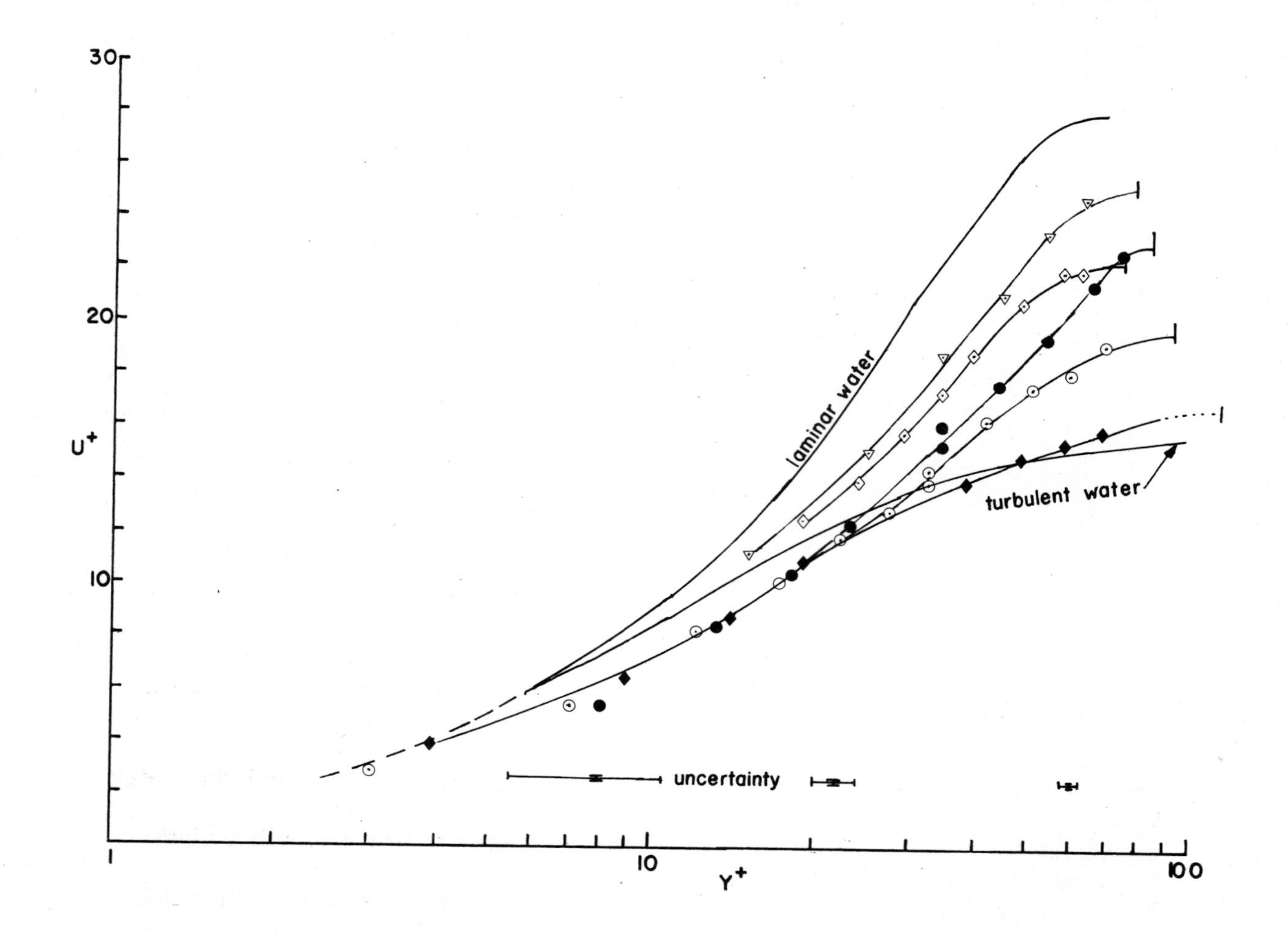

Fig. 16 Mean velocity of polymer turbulence, U^+ vs y^+

◆ 9 ppm, u_τ = 3.72 cm/sec.
⊙ 16 ppm, u_τ = 3.52 cm/sec.
● 16 ppm, u_τ = 3.85 cm/sec.
▽ 14 ppm, u_τ = 3.77 cm/sec.
◇ 21 ppm, u_τ = 3.73 cm/sec.

This last case is an example of the suppressed turbulence regime and is practically identical to the polymer fluctuation mean velocity profile under the same conditions.

Second International Conference on

Drag Reduction

August 31st - September 2nd, 1977

PAPER G2

EXPERIMENTS WITH SOLUBLE POLYMERIC DRAG-REDUCING COATINGS

M.T. Thew and Y.T. Lee,

University of Southampton, U.K.,

R.F. Long,

Admiralty Materials Laboratory, U.K.,

and

R. Bragg,

U.M.I.S.T., U.K.

Summary

To compare and develop soluble drag-reducing coatings three series of preliminary experiments were mounted:-

(i) coated plates fitted in one wall of a rectangular duct,

(ii) plates coated on both sides and mounted in the working section of a recirculating water channel, and

(iii) dynamometer mounted rotating discs carrying a coating on one or both sides, with flow through the surrounding housing.

Brief descriptions and interpretations are given for the first two series which exhibited zero drag reduction and about 5% reduction, respectively.

The discs showed very variabled results but did produce torque reductions up to about 30% with Polyox and polyacrylamide. More detailed explanations are then given for a further series of disc experiments which explored the significance of shaft speed, coating thickness, throughflow rate and particle size.

An outline is given of the method used for coating preparation.

Held at St. John's College, Cambridge, England.

Organised and sponsored by BHRA Fluid Engineering, Cranfield, Bedford, MK43 0AJ.

Nomenclature

Symbol	Unit	Definition
b	m	Disc thickness
C_M	-	Torque coefficient $= M/\frac{1}{2}\rho\omega^2 R^5$
C_{MO}	-	Torque coefficient for a disc of zero thickness when $Q = 0$
C_{MQ}	-	Increment above C_{MO} due to throughflow
C_{MT}	-	Increment above C_{MO} due to disc rim
ΔE	MJ/kg	Energy saving per unit mass $= \frac{\omega}{m}\int_o^t \Delta M\,dt$
FPTR	-	Flat plate test rig (hydraulic mean diameter 51 mm)
m	gram	Coating mass
M	Nm	Torque to rotate primed but uncoated disc
ΔM	Nm, %	Reduction in torque due to coating
PDRTM	-	Portable drag reduction test meter
$\overline{ppm}$	-	Average concentration (by mass) $= (m/\rho Q\Delta t) \times 10^6$ parts per million
Q	litre/s	Throughflow
R	m	Disc radius
Re	-	Rotational Reynolds Number $= \omega R^2/\nu$ (Re for FPTR $= \frac{\bar{u} \times \text{hydraulic mean dia.}}{\nu}$)
s_a	m	Disc axial clearance
s_t	m	Disc tip clearance
$\bar{t}$	s	Average residence time = net housing volume / Q
Δt	s	Run time until $\Delta M = 0$
Π	-	Throughflow Number $= (Q/2\pi\omega R^3) \cdot Re^{1/5}$
$\bar{u}$	m/s	Average velocity in working section of FPTR
ϵ	m	CLA roughness height on primed disc surface
ν	m/s^2	Kinematic viscosity
ρ	kg/m^3	Density
τ	N/m^2	Average shear stress $= \frac{3}{4\pi R^3}\left(M - \frac{\Delta E}{\omega\Delta t}\right)$, if ΔM is small $\simeq \frac{3M}{4\pi R^3}$
ω	rad/s	Disc speed

Effective concentration = concentration in ppm (mass) of freshly prepared pre-mixed solution giving the same effect in the PDRTM

1. INTRODUCTION

1.1. Why coatings?

In exterior flows with once-through systems the supply of drag-reducing additive to the boundary layer must constantly be renewed, either continuously or in a pulsed manner. With a mobile system there is a requirement for minimum total system weight and low volume,that precludes additive storage in the form of solution or suspension. If the additive is to be injected into the boundary layer as a solution then in-situ preparation is necessary, entailing some form of solids dispenser, a mixing chamber and ducting. This may be feasible in a large installation, as in the trials with HMS HIGHBURTON (Ref. 1), but on a small scale becomes unattractive. An apparently attractive alternative is the coating that releases additive at the right rate in the areas subject to high shear stress. Additives may vary the separation point on bluff bodies, thereby altering the form drag but this aspect is not considered here.

The coating could release additive by leaching from a porous matrix or by direct dissolution; only the latter arrangement is discussed here. The soluble coating is simple, does not require any alteration to the interior of a vehicle and providing it remains dry, has a much better storage life than some polymeric solutions. In dry film form some of the common polymeric additives e.g. Polyox, polyacrylanide, guar gum and cellulosic derivatives are mechanically quite robust. Against these attractions the renewal of a coating is more difficult than renewed injection from a solution system, but would not be a disadvantage for all potential applications.

The open literature contains little information on soluble drag-reducing coatings. In his 1972 review on drag reduction Hoyt (Ref. 2) gave only one reference (3), though this was encouraging as it reported drag reduction from a coating in the nose region of an axi-symmetric body. Since Hoyt's review little more has been published on coatings (Ref. 2a).

1.2. Coatings as compared with solutions

Though much remains to be understood about drag reductions, viz papers at this Conference, a considerable body of knowledge is becoming defined. A recent comprehensive review, restricted to pipeflow, has been given by Virk (Ref. 4). Further phenomena to be considered in the use of soluble coatings are:

(i) adherence of coating to the substrate,
(ii) surface roughness of the coating before and during dissolution,
(iii) the dissolution process.

Under (i) the coating must not flake off under impact while in the dry state, and must remain adequately bonded during dissolution.

Depending on the length of time for the hydration of the surface, the surface roughness of the dry coating may be important. During dissolution the surface topography must not become effectively rough. Hydration may cause a compliant surface and it is undesirable for this to be of a consistency to allow energy absorption by surface distortion.

The dissolution process must release material at the correct rate to give an adequate coating life. This material must be effective for drag reduction soon after being freed from the coating, though it may well be advantageous for it to be in the form of entangled molecular agglomerates. The concentration gradient both normal and axial, must fall within a range of required values. The concentration gradient will be interdependent on the structure of the boundary layer.

The dissolution process will depend on the way the coating is prepared. Both during preparation and dissolution degradation by reduction of molecular weight must be avoided. It is also undesirable for polymer to be released from the coating in partially hydrated agglomerates.

2. AIMS OF PROGRAMME

The programme has been essentially practical, and modest in scope. The overall objectives were:

(i) to compare different methods for the experimental evaluation of coatings,
(ii) to find the method best suited to valid, simple, cheap and quick tests,
(iii) to develop effective coatings.

After work had been in progress for some time a more modest interim objective emerged: to discover a coating and test method that exhibited at least some drag reduction on a specimen of appreciable surface area! (In earlier work, solution from a small coated duct a few millimetres across had reduced the pressure drop along a small bore pipe in series downstream from the duct.)

3. PRELIMINARY EXPERIMENTS

3.1. The flat plate tester

It is known that results from tests with additives in small bore pipes are relatively insensitive to molecular agglomerates or to the presence of large molecules at the upper end of the size distribution. Large rigs are expensive to build, run and maintain. As a compromise a rig to test coatings was constructed that uses a plate of 18 inches (.46 m) x 3½ inches (.09 m), with direct measurement of the shear force on this plate. For general ease of handling and measurement of concentration, the plate is set in the longer wall of a duct of rectangular cross section even though an exterior configuration might have been more representative of some applications. In its current guise the duct is 127 x 32 mm and a maximum mean flow velocity of about 16 m/s can be attained.

With the use of a pre-coat of primer on the aluminium plate, coatings of Polyox and polyacrylamide were developed that adhered to the plate. They dissolved steadily and apparently evenly at about the right rate overall, though at a faster rate near the front of the plate as the velocity profile was not fully developed and the flow there was disturbed by the gap of c.1 mm between the front edge of the plane and the surrounding wall. However, in tests over a wide range of flowrates no consistent drag reduction was ever achieved with coatings, though pre-mixed solutions behaved normally. Drag enhancement was traced to increased roughness on some coatings.

Coating break-up was definitely not the cause of zero drag reduction. Many runs were stopped when the coating was only partially dissolved and the residual film was dried. The coatings were found to dissolve evenly and at a fairly constant rate.

Provision was then made to put coated plates about 1 m upstream from the drag plate and to install LDA.

The LDA system was a single component, forward-scatter, reference beam type. By direct measurement traversing a wire of 5 μm diameter in air through the scattering volume it was found to be approximately 1.3 mm long (transverse to the plate) by 0.09 mm in diameter. To obtain velocity measurements to within 0.1 mm of the plate surface on its longitudinal centre line it was found necessary to bevel the surface by 1 mm either side of a central ridge 10 mm wide.

After the first test series with the flat plate tester (FPTR) five possible reasons were deduced for the absence of drag reduction:

(i) unsuitable flow conditions over the drag measuring plate,
(ii) increased surface roughness or distortion of the coating leading to a greater skin friction,

(iii) too short a residence time over the plate for the material removed from the coating to become effective, a problem possibly aggravated by partially solvated lumps being torn from the gelatinous surface,

(iv) excessive concentration of polymer in the near-wall region,

(v) polymer degradation during dissolution under high shear stress or earlier in coating manufacture.

Following the modifications, with the coated plate mounted upstream from the drag plate, no drag reduction was detected. Possible cause (ii) could be discounted under these conditions, while the LDA measurements produced no evidence of (i). Concentration measurements were made on water samples obtained immediately downstream from coated plates at heights of 0.5, 1 and 2 mm above the plate, in the Portable Drag Reduction Test Meter (PDRTM). This instrument times the discharge of a known volume of solution through a long, small bore tube under the action of a spring (Ref. 1). An effective concentration of about 1 ppm (by mass) was indicated for each station, a figure well below that indicated by calculations based on the overall solution rate. This result tended to suggest that possibility (iv) was untenable but could also lend credence to the removal of some polymer in partially solvated, large (on a molecular scale) lumps. An attempt to detect such lumps with a Coulter Counter was unsuccessful because of contaminant particles in the rigwater, but it would be worthwhile to try again.

Power spectra were also obtained from the LDA measurement in the near-wall region for water flow, with coatings and with pre-mixed solutions. Comparing coatings with water, there was a tendency for the power at the higher frequencies to be reduced(even though the drag was unchanged). This trend has been noted by other workers and suggests that at least some polymer was being effective. The picture was confused because pre-mixed solutions, that did reduce drag, had the opposite trend to the coatings and in addition coatings gave very variable spectral results. There was some evidence though, that coatings had even less effect on the power spectra when moved to the furthest upstream position. This supports the hypothesis that insufficient effective material in the critical region of the boundary layer is an important reason for the failure of the coatings, both for PEO and PAM.

Results from the disc tests, discussed later in sections 3.3 and 7, indicate that (v) degradation is an unlikely cause for the lack of success in the flat plate tester.

Further details of the work with the FPTR are in Ref.(5).

3.2 Double sided plates in a circulating water channel

The FPTR results, though not clearcut, indicated that insufficient effective material was present in the boundary layer on the drag plate. This might be remedied by allowing more time after dissolution and/or a greater solution rate. A longer coated plate could help and so tests were conducted on plates 2 m x 0.2 m deep and 6 mm thick which were coated on both sides with Polyox WSR-301. The plates were centrally mounted with their upper edge about 0.1 m below the surface in a working section 2 m wide with a water depth of 1 m.

The velocity profile in the working section is closely rectilinear. The velocity was measured by a large pitot tube on the central plane at a point just ahead and below the plate. Preparatory tests on an uncoated plate had shown results for overall drag coefficient versus Reynolds Number (Re_L = Length x $u_{centreline}/\nu$) that were an excellent fit to the I.T.T.C. curve. Results are given in table 1 below:-

TABLE 1 Drag tests on coated plates 2 m x .2 m

Water velocity	Av. shear stress	Re_L	Drag reduction
2 m/s	7.5 N/m^2	4.1 x 10^6	- 17 %
4	27	8.2	5
6	50	12.3	6

The drag increase in the first case was probably due to unfavourable surface topography* at the low overall shear stress. Accuracy was also poorer as the drag force was smaller. A test was continued until the drag reading had stabilised with time and this took about ten minutes in all cases. Unfortunately the water in the channel could not easily be drained between tests, but the final content is thought to have been below 1 ppm (by mass) and possibly less than 0.5 ppm; additionally some degradation would probably have occurred in the channel circulating pump.

At the scale involved, drag reduction by the coatings was thought to be real even though the water gradually became a very dilute solution. The value of Re_L cannot be compared directly with Re in the FPTR, since the latter was based on average velocity and hydraulic mean diameter. As an indicative value, Re_{MAX} was about 8×10^5. The shear stress averaged over the whole (2 m) plate at the greatest water channel velocity corresponded to about 4.6 m/s for the same average shear stress in the FPTR. At this velocity the maximum drag measurement error in the FPTR was about 5%, so that comparable results to those in the channel could have been missed.

In the water channel, as in the FPTR, the coatings were unavoidably wetted for about half a minute before the test began. This could have led to a higher initial solution rate.

A further increase in residence time for additive in the boundary layer (above the channel tests) was achieved by changing the geometry to rotating coated discs.

3.3 Coatings on rotating discs

A high throughflow was used in a close-fitting housing to obtain limited residence time and partially simulate external flow. The timescale was less than that of Cox et al (Ref. 5a) by a factor ~10^2 - 10^4; in their work a ϕ0.23 m disc was driven at 840 rev/min in a 580ℓ tank and torque/time curves obtained after powdered polymer was added to water.

A variable speed drive of 3 kW, dynamometer mounted on steel strip flexures, was used with a housing of one litre capacity and discs of 175 mm diameter run at speeds from 500 to 4500 rev/min. Flow was metered through a venturi fed from pressurised cylinders. The minimum mean residence time in the housing was about one second. (Ref. 6 had demonstrated a very rapid interchange of liquid between the two sides of the disc, so that the whole volume was considered in the calculation of $\bar{t}$).

Even with the throughflow used, experiments with Newtonian fluids (Ref.7) indicate recirculation is likely, with a separated boundary layer regime. Recirculated solution entering the boundary layer over the disc will have been diluted but its residence time will be greater than for the newly dissolved material flowing spirally outwards. The spirally outflowing boundary layer will enter regions of higher shear stress, in contrast to the decreasing stress level on the plates in the water channel or the roughly constant stress level in the FPTR.

After some variations in the method of coating manufacture, torque reductions up to 30% were attained with both PEO and PAM, though when a guar gum coating was tried it broke off in flakes when coated direct on the primer and again when with a PEO intermediate layer. Broadly similar results were obtained with both PEO and PAM for discs coated on one side or two sides, indicating that the drag reduction mechanism does not change markedly over a smooth metal surface as compared with a dissolving polymer coating.

An attempt to vary the technique of preparing the coatings could not be properly evaluated because of uncontrolled and pronounced variations in coating behaviour. Fig. 11 illustrates a type of torque reduction curve obtained from single sided PEO coating, but other coatings gave their maximum torque reduction at the earliest reading with a continuous fall thereafter. These unexpectedly

*See Appendix 2

large variations in performance with nominally similar coatings emphasised the variability noted with the power spectra. It was decided to conduct a further disc test series with all stages of coating preparation standardised as far as possible.

The outflow from the housing during the run illustrated on Fig. 11 was periodically checked with the PDRTM. Using a calibration curve derived with fresh pre-mixed PEO solutions the effective ppm data on Fig. 11 was then produced. The concentrations derived fit the $\overline{ppm}$ band, though this was broad due to uncertainty over the coating mass, and are reliable in indicating the relative variation during the run. There is moderate agreement between the ΔM and $\overline{ppm}$ curves except at the beginning. The test technique meant that the coating(s) were partially wetted for some 30 to 60 seconds before the disc reached the set speed. This could have produced the early peak in concentration as the gelatinous layer was stripped off. It was further decided to modify the rig so that the time of wetting would be known precisely, the disc would be spinning when wetted and the torque would be measured more quickly. The modifications are described in the next section.

Following the discovery of the pronounced variability in coating performance, some PEO samples were examined with X-rays and microscopy: both optical and electron beam. The crystallinity was found to be consistently over 90%; this compares with a figure of $\gg$95% estimated for the as-produced material by the manufacturer (Ref. 8). However, the size distribution of the spherulites was found to differ appreciably, as was the size distribution and spatial distribution of voids between them. The nature of voids would be expected to alter the solution rate, especially in the early life of a coating. Details of the preliminary test series conducted with the disc rig are given in Ref.(9).

4. THE ROTATING DISC RIG

Modifications to the rig feed system were simple and of low cost: see Fig. 1. The torque was still read manually from the (statically calibrated) deflection of the flexure mounted drive unit cradle. The calibration was linear over the working range.

The water feed system was calibrated and the venturi was then removed from the system. When the quick-opening shut-off valve was opened, the water flowrate built up to the set value within about two or three seconds. The disc was set to the correct speed - allowing for the small drop as the spinning disc became immersed in water - before the flow was started. A first torque reading could be obtained about 10 seconds after initial wetting, but to ensure reliability, increments of 15 seconds were used for torque recording. Testing was continued at constant shaft speed and constant throughflow until the water was exhausted or until the torque value stabilised. Coatings of PEO or PAM are virtually invisible when wetted, but one or two had enough left at the end of a test to confirm that the solution rate increased with radius. The water temperature was 15 - 20°C.

For simplicity all discs tested in the second series had single sided coatings. Geometric details of the discs and housing are given in appendix 1. The figure given there for the relative roughness of primer surface (exposed when a coating had dissolved) is based on a small sample and not on all ten. Some discs were probably slightly rougher in some areas. Because the torque reduction was based on the baseline figure for each disc, minor variations in roughness were not significant. Over the Reynolds Number range used, the relative roughness figure quoted is not hydrodynamically rough (Ref. 10). Details of the modified rig and of the test procedure are given in Ref.(11).

5. COATING PREPARATION

A general description is given below; further details are given in appendix 2 for the PEO coatings which have been used in the majority of tests. The process for PAM is somewhat similar.

All coatings for the PEO series were prepared from one batch of polymer. To give fast acting coatings the as-received powder was then milled down in a Kek pin-disc machine in one operation. For evaluating the effect of powder size, fine and a coarse milled materials were produced and subsequently sieve sized. To aid the milling process 1% of fume silica (Ref. 12) was added to the PEO, except for half the coarse milled powder. The silica coats the surface of the powder grains to prevent agglomeration during and after milling.

Milled and sieved powder was then manually sprinkled on a primed disc surface through a fine sieve until the pre-determined mass had been put on. The horizontal disc was then immersed in a solvent vapour bath for a few minutes to partially fuse the powder grains together. The same period was used for all coatings in this series. After removal from the bath the coating was dried in a ventilated oven for a set period. The torque reduction tests were generally performed on a group of coatings within two to three days of making a coating.

On Fig. 2 two photomicrographs show the top surface of a fine milled PEO film. Though not from the latest disc series, they are typical. At the lower magnification the surface roughness is accentuated by the illumination. With higher magnification the surface voids, which permit more rapid wetting, become visible. Under static water the solvation process would tend to close these voids, but it is not known what happens under high shear stress.

6. RIG VALIDATION : EXPERIMENTS WITH UNCOATED DISCS

After the rig modification the performance with uncoated but primed discs was compared with standard results. Referring to Fig. 3 the C_{MO} curve is taken from the results of Daily and Nece (Ref. 13) at an axial spacing ratio (s_a / R) of 0.217, which though slightly lower, is adequately applicable. The increment C_{MT} allowing for shear on the disc rim was found from Ref. (14), while a further increment C_{MQ} for the throughflow was estimated from Ref. (7). Allowing for the variation in the primed surfaces and the differences in exit geometry between Ref. (7) and this work the agreement is satisfactory. Fig. 4 compares C_{MQ} as measured and as predicted by Ref. (7). Though only one shaft speed was used the agreement was adequate and the investigation moved on to coated discs.

7. DISCUSSION OF RESULTS FROM COATED DISCS

7.1. Coating behaviour and comparative criteria

Two typical sets of results from fine and coarse PEO illustrating the variation in torque reduction during coating life were re-plotted on Fig. 5 from the torque : time readings. This figure also shows shaft speed influence, noting that the actual ΔM increased with speed. The coarse milled material's results are similar to the curve from the first test series shown on Fig. 11. Note that the later tests had no pre-wetting or slow acceleration of a wetted coating. Small amounts of air trapped in the housing on water entry formed a ring round the shaft, but the small radius rendered the torque reduction insignificant. Whereas the fine milled powder achieved its maximum reduction within a few seconds, perhaps even 1 or 2 seconds, the coarse milled (with and without silica) took much longer to reach its maximum. What was totally unexpected was that the $\Delta M(\%)$ cross-over points occurred within a few seconds, with the coarse material not only lasting longer but reducing torque much more.

The rising duration of the fine with increasing speed, in contrast to the coarse, is puzzling and may be linked to a flatter surface under higher shear see Appendix 2.

The long tail with the coarse material is a consequence of the falling overall solution rate, when the coating near the disc rim has all gone. With the less effective (fine milled) coatings, extrapolation of ΔM back to zero was difficult. Because of the tail to the coarse $\Delta M : t$ curves, the determination of the end of coating life for them and hence ΔT, was uncertain. There would be some argument for ending Δt at a point where ΔM falls to say

2 or 3%, but this was not done.

In addition to comparison on a ΔM basis the concept of total energy saved per unit mass, ΔE, seemed useful as this is independent of the $\Delta M : t$ curve shape and allows for variation in coating mass. This was defined as,

$$\Delta E = \frac{\omega}{m} \int_0^{\Delta t} \Delta M \, dt.$$

Shear stress on the disc is radius dependent and with an effective coating is also time dependent. To make comparisons with the FPTR results an average shear stress was thus defined: $\bar{\tau} = 3M/4\pi R^3$ (M is total measured torque). This is based on both sides of the disc because it is thought that polymer reaches the uncoated side of the disc, and comes from the concept:

$$\text{Total torque (neglecting rim)} = 2\left(2\pi \int_0^R \tau \cdot r^2 \, dr\right) = 4\pi\bar{\tau} \int_0^R r^2 \, dr.$$

Allowing for torque reduction by introducing ΔE, the definition becomes:

$$\bar{\tau} = \frac{3}{4\pi R^3} \left(M - \frac{\Delta E}{\omega \Delta t}\right)$$

The solution rate per unit area will vary with time and radius and, unless the outflow from the housing is examined, can only be obtained as an average value $m/\pi R^2 \Delta t$. Similarly, only the average concentration can be determined and is defined (on a mass basis) as:

$$\overline{ppm} = (m/\rho Q \Delta t) \times 10^6, \text{ parts per million.}$$

7.2. Effect of coating thickness

Fig. 8 makes comparisons on a ΔE basis. More tests were run at 3000 rev/min than at other speeds and these averaged results are thought to be more reliable. Low mass (thin) coatings are soon exhausted and thus have less well defined $\Delta M : t$ curves, reducing the accuracy of ΔE. This trend is accentuated at higher speeds. The results at 2000 rev/min reflect the marked drop in ΔE with decreasing speed, which is discussed in section 7.3.

The indication is that, ΔE is not a function of coating mass and that the coating is homogenous. The effect of m on the solution rate is discussed in section 7.4. The above suggestion is based on fine material, but it is tentatively thought to be true for coarse milled PEO.

7.3. The influence of shaft speed

Both Re and $\bar{\tau}$ vary with shaft speed. Comparing results from the FPTR and small bore tube equipment at the same wall shear stress, the size of the apparatus was found to have a pronounced effect. It is therefore somewhat arbitrary to plot ΔE against $\bar{\tau}$ on Fig. 7, rather than Re or shaft speed. The rapidly rising value of ΔE suggests that degradation had not set in and the maximum value of $\bar{\tau}$ was greater than in the FPTR at its highest speed: 400 N/m^2. Comparison of overall solution rate between the two test methods is made in section 7.4.

The coarse material was considerably superior at all shear stresses, probably most usefully so at the lower end of the range. Deletion of the silica has a definitely beneficial, but not important effect.

The steep rise in ΔE at higher values of $\bar{\tau}$ is because both ΔM and ω are rising while the solution rate, hence Δt, is falling proportionately less.

For perspective, the maximum value of ΔE is 26% of the heat of combustion of n-Octane. Allowing for the thermal efficiency of an internal combustion engine, the additive is in sight of equality on a mass basis and there is almost no installation mass beyond that of the coating. The corresponding shear stress of $\sim$450 N/m^2 is a high value in general terms.

If all the coating had been at the maximum value of $\bar{\tau}$ near the disc rim, the value of ΔE could have been still greater.

7.4. Variation in $\overline{ppm}$ and solution rate

Fig. 11 confirms that the shape of the $\Delta M : t$ curve is largely governed by the variation in solution rate during a run, at least for coarse milled material if the apparent similarity between Figs. 5 and 11 is real.

Only time-averaged values can be compared from Fig. 9, but the fine milled material was x 5 - x 10 above the coarse milled. No coherent pattern could be discovered for the fine material and the falling gradient of the approximate envelope is not significant. The rising values of $\overline{ppm}$ for the coarse milled coatings with speed reflect the increasing value of $\bar{\zeta}$; the silica apparently reduces the solution rate by impairing solvation, though a corresponding growth in Δt leads to the slight rise in ΔE.

Merrill et al (Ref.15) show a peak in drag reduction at low concentration for Couette flow so that further 'instantaneous' ppm values will be obtained with the PDRTM, though their results may be inapplicable because of different geometry, zero throughflow and suppression of degradation by extrapolation back to $t = 0$.

The greater ΔM obtained with coarse at a lower $\overline{ppm}$ is discussed in section 7.7.

The average specific solution rates have been calculated and present a situation like that for $\overline{ppm}$. Fine material fell in the range 0.3 - 1.8 $g/m^2/s$, with pronounced and apparently random scatter of results. Coarse material varied between 0.3 to 0.6 $g/m^2/s$, tending to rise with $\bar{\zeta}$. For comparison, in the FPTR the range was 0.2 to 0.5 $g/m^2/s$ for a span of average wall shear stress 20 - 65 N/m^2 for one batch of coatings. With another batch a rate of about 0.5 $g/m^2/s$ in the FPTR corresponded to 260 N/m^2. The solution rates are reasonably similar in the two rigs for the same average shear stress. More detailed comparisons are pointless as coating preparation was not so closely specified for the flat plates.

7.5. Variation in throughflow

Average shear stress on a disc has been shown to depend principally on shaft speed, for the range of throughflow used. Therefore if Q is raised the solution rate is unlikely to change much but $\bar{t}$ will fall inversely with Q and also one may postulate that $\overline{ppm}$ will fall as Δt will vary only a small amount. This was shown to be valid by the results for fine material on Fig. 10. With only three points, the straight line may be coincidental but it may mean there is a discontinuity near the origin because the first part of solvation occupies $\sim$0.1 s.

As $\bar{t}$ is reduced the situation probably becomes more analogous to that in the FPTR but the welcome flattening of the ΔE curve offers hope. The ΔE values for the fine are so inferior to the coarse material, that it is intended to try variation in Q with the latter. If a stabilised promising value of ΔE at small $\bar{t}$ is obtained, then it could be worthwhile to try coarse materials in the FPTR.

At very small values of $\bar{t}$, say below 0.5 s, the interchange flow between the two sides of the disc may not equalise polymer concentration so that water entry on both sides with double sided coatings may be necessary. By insertion of filler blocks in the housing to reduce s_a, smaller values of $\bar{t}$ may be attained, since Q cannot be substantially increased.

7.6. Effect of insoluble material in the coatings

The 1% of fume silica makes possible the manufacture of the fine material. It might be thought to raise the solution rate by impairing the fusion between neighbouring powder grains, but results for the coarse with and without silica show only a very slight effect. It does seem to slightly reduce ΔE by truncating the top from the $\Delta M : t$ curve, but the overall effect is not large.

During the preliminary disc experiments a few percent of an insoluble fluorescent pigment, "Dayglo", was incorporated to aid perception of the radial distribution of solution rate. It was thought to have a grain size about half or a quarter of the PEO powder. Judging by Δt figures there was little change in $\overline{ppm}$. There was a significant increase in torque of some 30%, or probably more in the first minute when no readings could be obtained. These results were not followed up, but they could provide information on the dissolution process. The pigment particles may produce a great increase in roughness or in some way hinder the expansion of molecules or small aggregates during dissolution.

7.7. Particle size effect

The big difference between fine and coarse was unexpected. Earlier thinking had concentrated on fine or fairly fine material to promote rapid dissolution and the coarse ground material was inserted into the test plan to provide completeness, and tested last. In rigs other than the disc rig, the fine material may still be better, but in the disc rig it was used up more quickly and on the whole was less effective in reducing drag. At 2000 rev/min the fine was better for the first 15 seconds, if extrapolations are valid. Compared with pre-mixed solutions giving maximum values of torque reduction about 70%, the best coating was much inferior.

The superior performance of the coarse could be ascribed to:-

(i) less degradation during the milling process,
(ii) less degradation in the vapour bath,
(iii) greater effectiveness after leaving the coating.

Prior experiment with the PDRTM make (i) appear improbable, though it is less sensitive to the larger molecules that are important in bigger apparatus.

As the time in the vapour bath was identical in both types, (ii) appears unlikely. Incidentally, it is clear from the difference between fine and coarse that the fusion between grains was far from complete.

The most likely cause appears to be (iii) therefore and comparison of the solution rate and ppm results is telling evidence. At 4000 rev/min four of the fine results were between 1.1 and 1.7 $g/m^2/s$, and only one fell with the four coarse results at 0.3 to 0.5 $g/m^2/s$. [About 45 coatings were tested in this series]. This difference is interpreted as rapid removal of the fine grains in a partially solvated state.

The fine material produced a coating that was slightly smoother when unwetted.

8. CONCLUSIONS AND RECOMMENDATIONS

(i) The rotating disc rig is most suitable for evaluating coatings since it is quick and easy to use and did show up differences. The FPTR provides very unfavourable conditions for drag reduction while it is impracticable to use the recirculating water channel for frequent tests as there is too much water to change and treat.

(ii) The disc rig should be modified to reduce mean residence time to make the flow more analogous to exterior flowfields.

(iii) In the disc rig, PEO coatings become increasingly more effective with rising speed with no obvious signs of degradation at high shear stress. Coating thickness (for PEO/WSR-301 at least) has little effect on performance other than on life.

The coarser of the two powders was much superior, except for the first few seconds after wetting, producing longer life and greater drag reduction. However, even at best, the coatings produced a torque reduction (30%) less than half that obtained with pre-mixed solutions (70 - 75%).

(iv) Any further work on coatings should concentrate on the dissolution process and the effect of powder grain size on it. Effluent from fine ground powder coatings should be examined for evidence of partially solvated agglomerates, either with a Coulter Counter or by looking for improving time trends with the PDRTM.

ACKNOWLEDGEMENT

This work was partially funded by the Procurement Executive M.o.D. (U.K.) Views expressed are those of the authors and do not necessarily represent those of the M.o.D.

REFERENCES

1. Canham, H.J.S., Catchpole, J.P. and Long, R.F. 'Boundary layer additives to reduce ship resistance.' The Naval Architect (J.R.N.I.A.) No. 2. 187-213 (July 1971).

2. Hoyt, J.W. 'The effect of additives on fluid friction.' Trans. ASME 94D 258-285 (June 1972).

2a. White, A. and Hemmings, J.A.G. 'Drag reduction by additives: Review and bibliography.' Pub. BHRA, Cranfield, Bedford (1976)

3. Thurston, S and Jones, R.D. 'Experimental model studies of non-Newtownian soluble coatings for drag reduction.' J.Aircraft (AIAA) 2 (2) 122-126 (March/April 1965).

4. Virk, P.S. 'Drag reduction fundamentals.' A.I.Ch.E.J. 21 (4) 625-656 (July 1975).

5. Bragg, R. and Thew, M.T. 'Drag reduction coating test methods: Final report.' Mech. Eng. Dept., Southampton University, U.K. Report ME/76/15 (August 1976).

5a. Cox, L.R., Dunlop, E.H. and North, A.M. 'Role of molecular aggregates in liquid drag reduction by polymers.' Nature 249 243-245 (1974).

6. Anand, J.S. 'The effect of aqueous media drag reducers on the performance of the hydrodynamic disc seal.' Ph.D. Thesis, University of Southampton, U.K. (1974).

7. Daily, J.W., Ernst, W.D. and Asbedian, J.W. 'Enclosed rotating discs with superposed throughflow.' M.I.T. Hydrodynamics Laboratory, Report 64 (1964).

8. Anon 'Polyox water-soluble resins'. Union Carbide Corporation, New York, 44 pp (1968).

9. Bragg, R., Dodd, E.N. and Thew, M.T. 'Testing of drag reduction coatings using a rotating disc dynamometer.' Mech. Eng. Dept., Southampton University, U.K. Report ME/75/27 (December 1975).

10. Nece, R.E. and Daily, J.W. 'Roughness effects on the frictional resistance of enclosed rotating discs.' Trans. A.S.M.E. 82D 553-556 (1960).

11. Lee, Y.T. 'Drag reducing characteristics of polymer coatings on rotating discs.' B.Sc. Thesis, Mech. Eng. Dept., Southampton University, U.K. (1977).

12. Bode, R., Ferch, H. and Fratzscher, H. 'Properties of and applications of AEROSIL.' DEGUSSA Co., Frankfurt, W. Germany.

13. Daily, J.W. and Nece, R.E. 'Chamber dimension effects on induced flow and frictional resistance of enclosed rotating discs.' Trans. A.S.M.E. 82D 217-232 (1960).

14. Thew, M.T. 'Further experiments on the hydrodynamic disc seal'. Proc. 4th Int. Conf. on Fluid Sealing, Paper 39, Philadelphia, 1969. Pub. BHRA Cranfield (1970).

15. Merrill, E.W., Smith, K.A. et al 'Study of turbulent flow of dilute polymer solutions in a Couette viscometer.' Trans. Soc. Rheol. 10 335-351 (1966).

Appendix 1 Rig Geometry

Disc radius	R	87.5 mm	
Disc thickness	b	7.5 mm	
Axial clearance	s_a	20 mm	s_a/R = .229
Tip clearance	s_t	3.5 mm	s_t/R = .040

Primed surface relative roughness ε/R 1/3000

Net housing volume 1.03 litre

Appendix 2 Coatings

Preparation

Fine milled powder : passing sieve mesh size 45 μm

Coarse milled powder : sieve mesh band 125 - 150 μm

Fume silica : type A.300. Mean dia. 7×10^{-9}m

Priming coat for PEO put on at least a day or so before immersing powder in the vapour bath.

Coating mass : 0.5, 1, 2 and 4 g, determined gravimetrically before immersion in the vapour bath to ±0.1 g.

Topography when wet

The surface layer of a dissolving Polyox coating is a weak gelatinous solid which is inhomogenous because of the differential solvation between the partially fused, milled grains and the boundary material. Direct observation is very difficult as the solvated material is translucent, but it appears likely that the form of the surface is a function of the magnitude of the wall shear stress. As the magnitude rises, the thickness of the soft partially solvated material is reduced. It thus seems plausible that the variations in the strength of the wet coating produce greater irregularities at low shear stresses.

Replication of the surface has not been attempted, as the replica should have a similarly low shear strength to be meaningful.

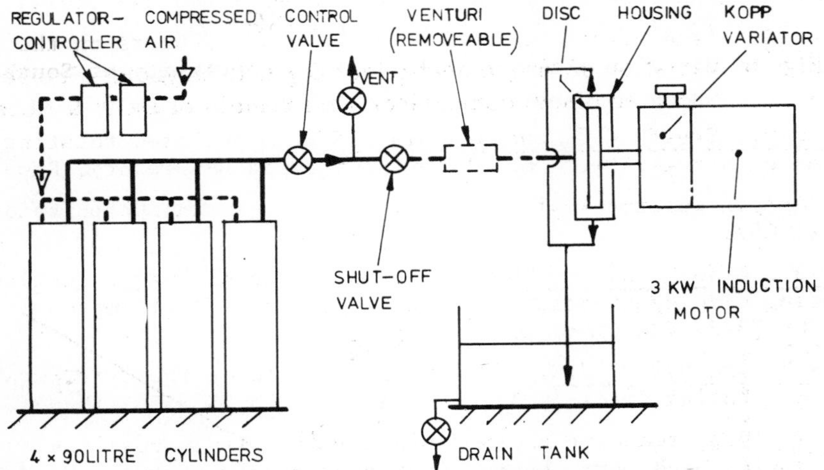

Fig. 1 Layout of rotating disc apparatus.

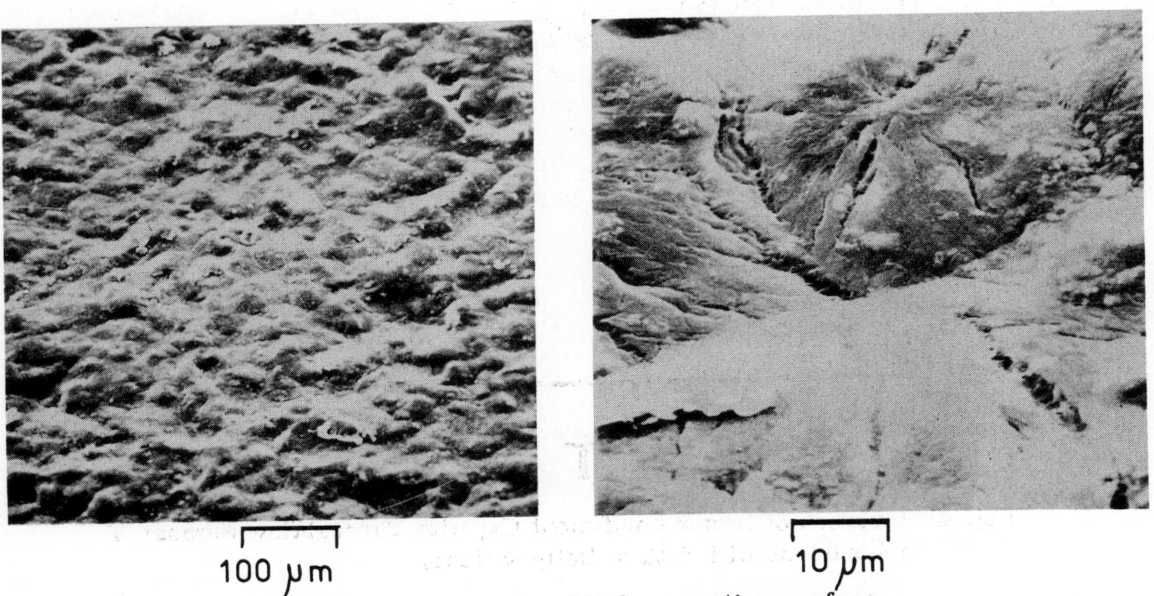

Fig. 2 Photographs of Polyox coating surface.

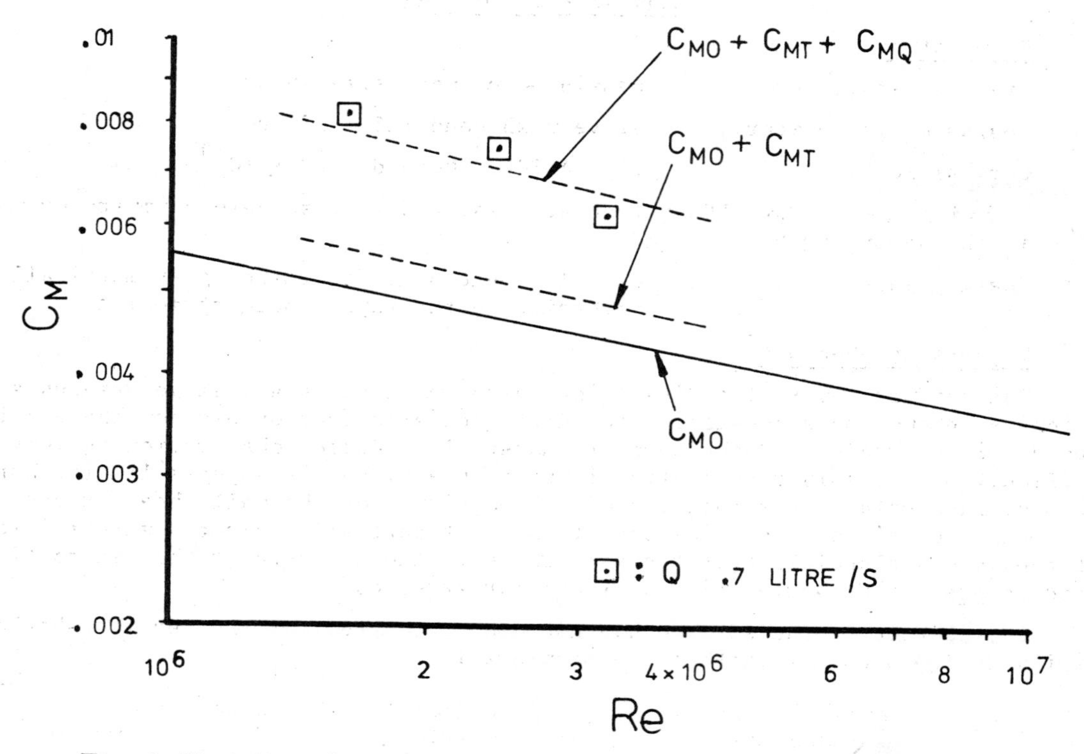

Fig. 3 Variation of torque coefficient C_M with Reynolds Number for uncoated discs: comparison with results of Daily & Nece.

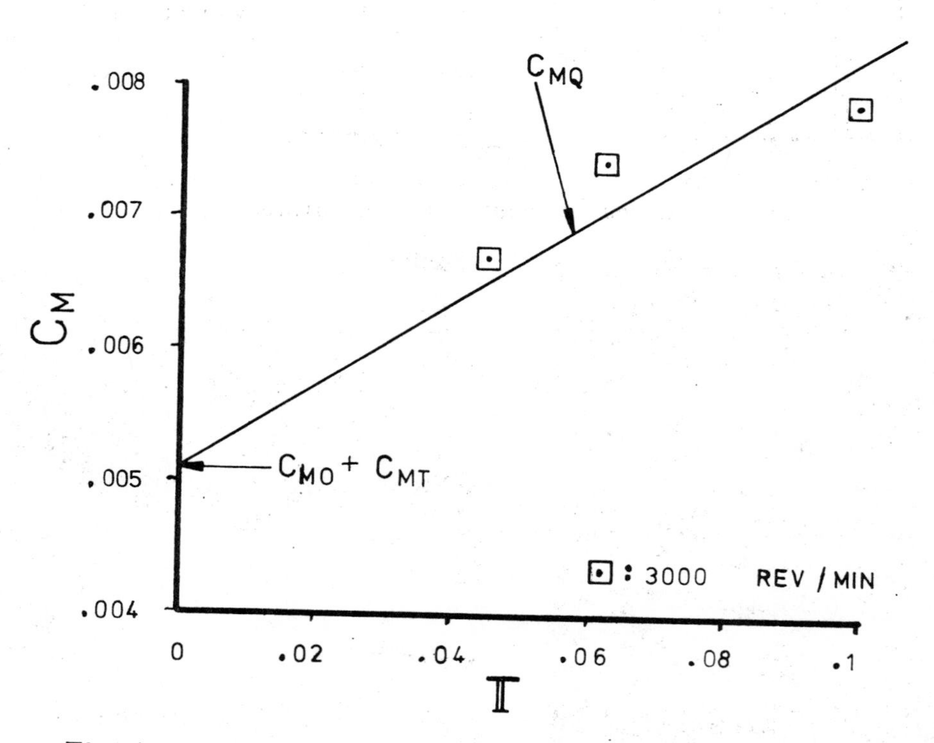

Fig. 4 Variation of torque coefficient C_M with Throughflow Number II comparison with data of Daily & Nece.

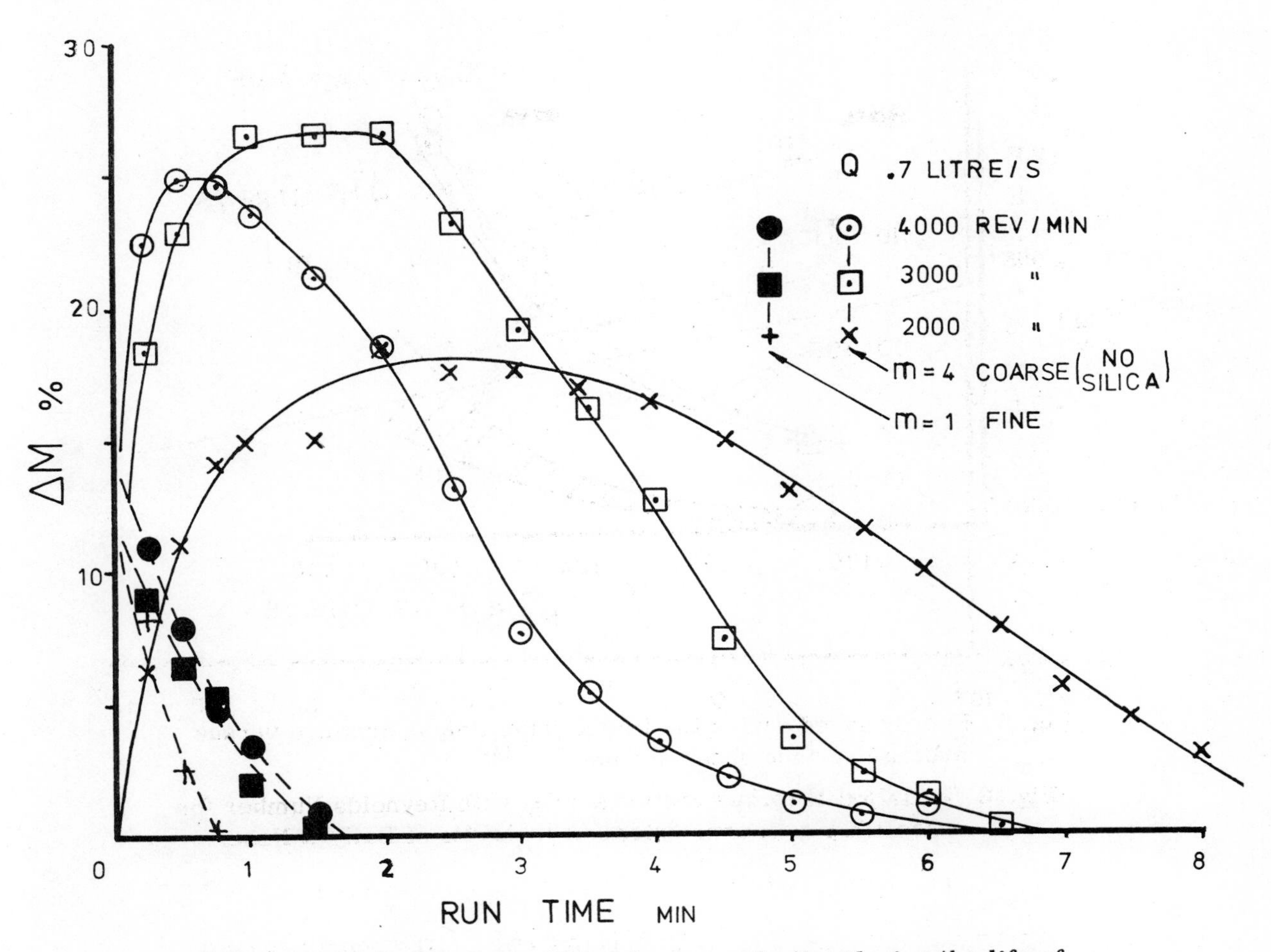

Fig. 5 Variation of percentage torque reduction during the life of some Polyox coatings

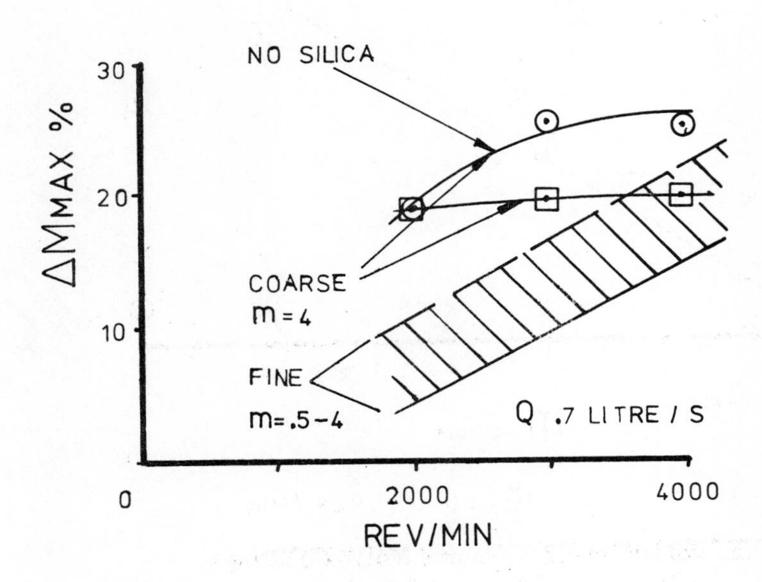

Fig. 6 Maximum percentage torque reduction versus shaft speed for some Polyox coatings.

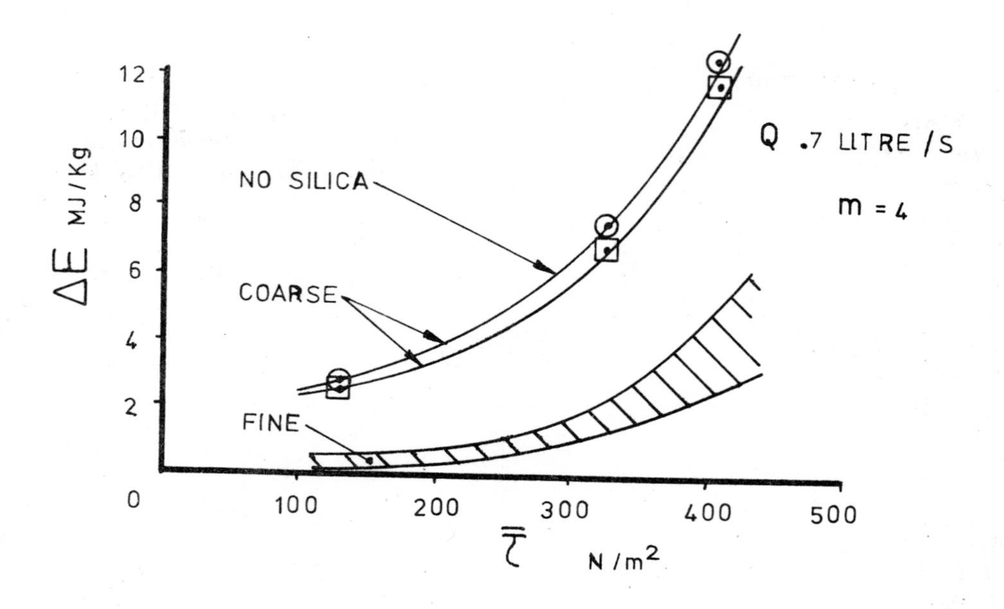

Fig. 7 Energy saved during the life of some Polyox coatings versus average surface shear stress.

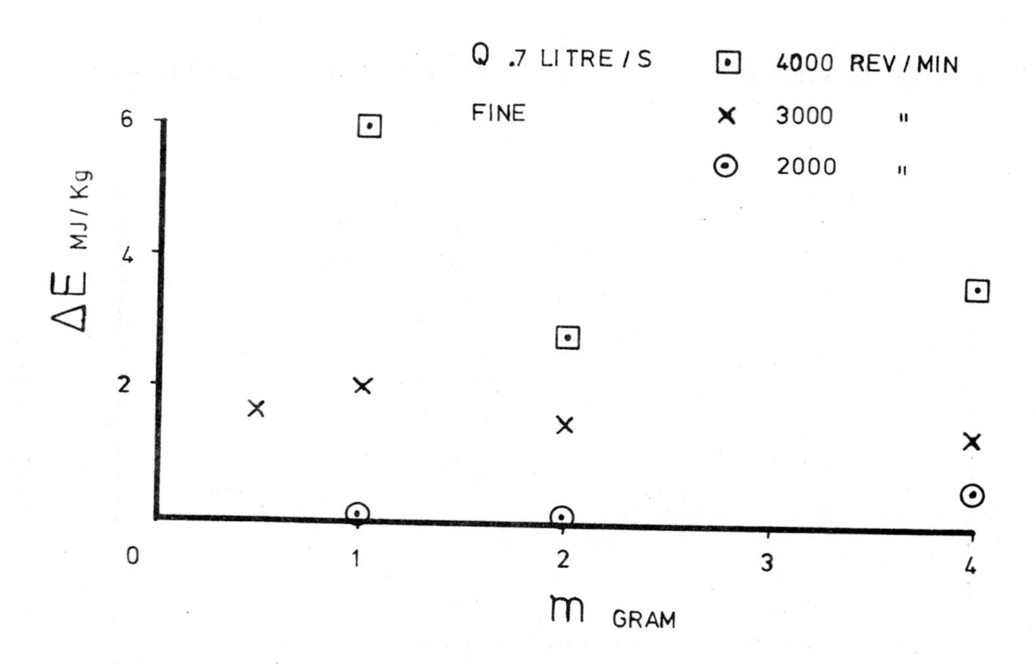

Fig. 8 Energy saving versus mass for some Polyox coatings.

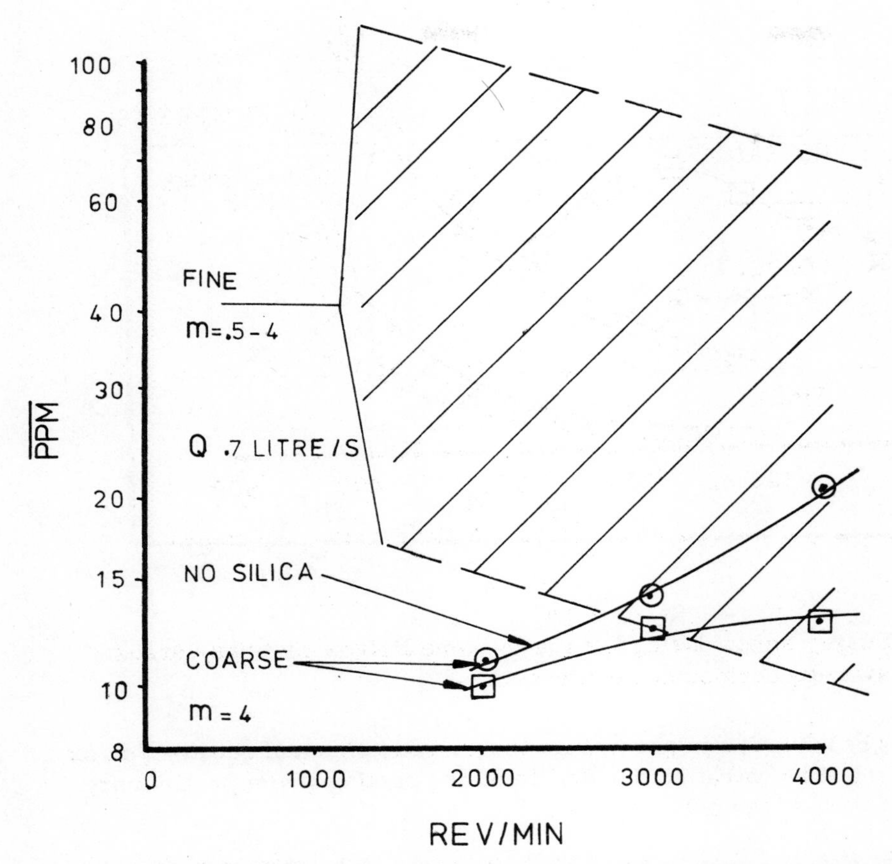

Fig. 9 Average concentration of Polyox in outflow from disc rig versus shaft speed.

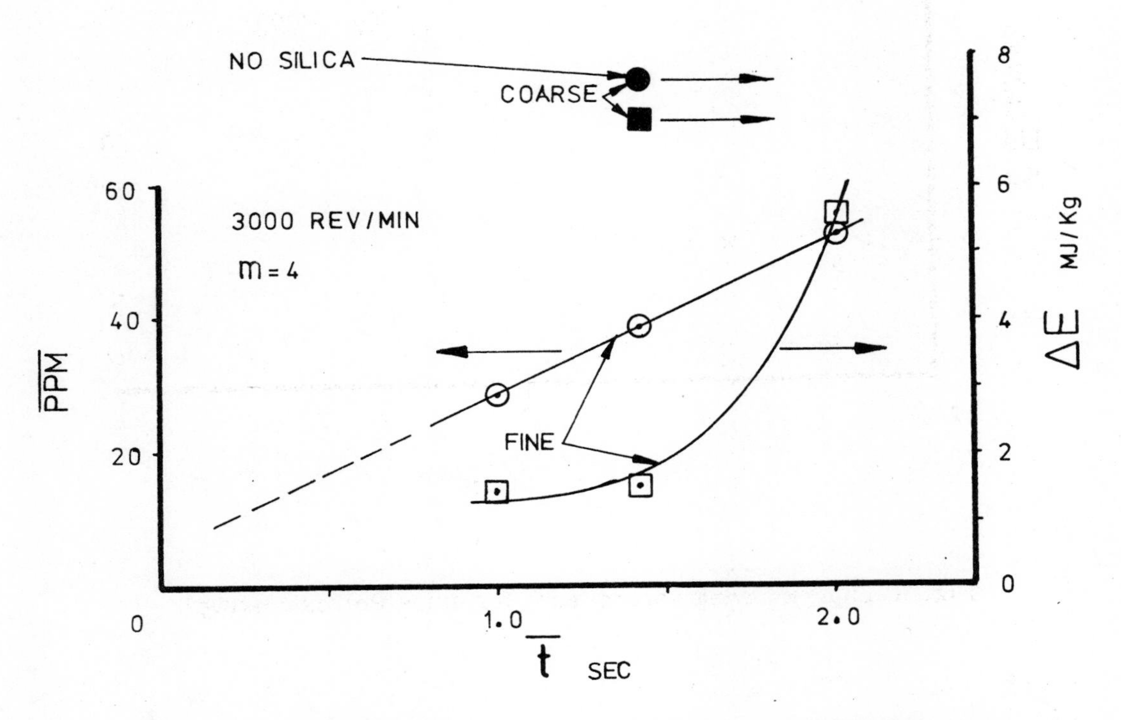

Fig. 10 Variation in average concentration and energy saving with average residence time.

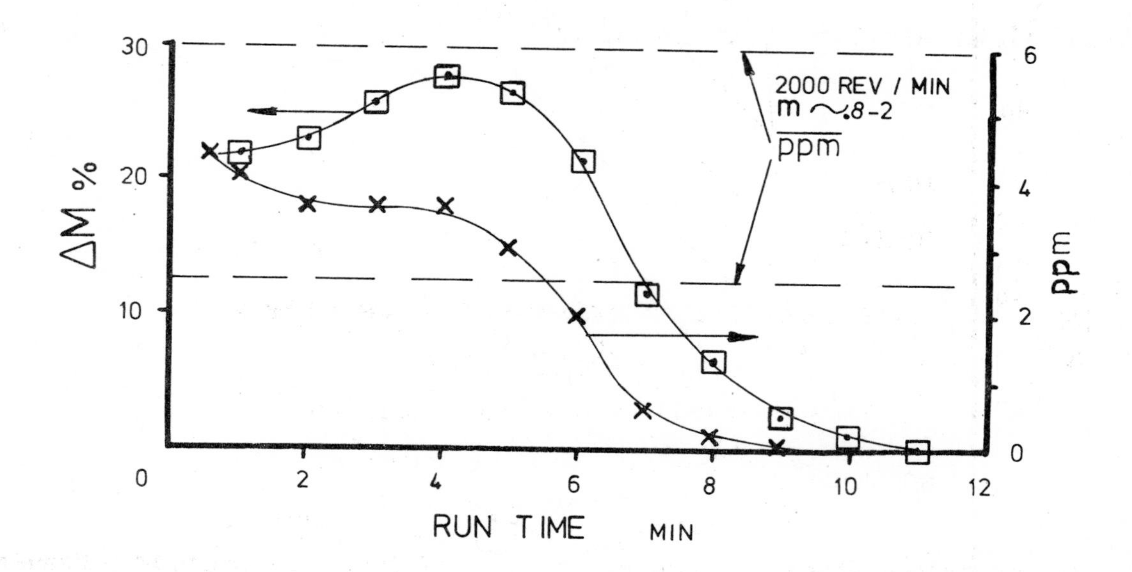

Fig. 11 Variation in percentage torque reduction and outflow Polyox concentration during life of one coating (from preliminary test series).

Second International Conference on

Drag Reduction

August 31st - September 2nd, 1977

PAPER G3

MASS TRANSFER AT MAXIMUM DRAG REDUCTION

P.S. Virk and T. Suraiya

Massachusetts Institute of Technology, U.S.A.

Summary

Trans-cinnamic acid, coated inside a tube of I.D. 0.295 cm and length 34.5 diameters, was dissolved into dilute polymer solutions flowing through under conditions of maximum drag reduction, the wall mass transfer rates being measured both by the classical weight loss method as well as by a new spectrophotometric technique. Six distilled water solutions of two polyethyleneoxide polymers, molecular weights 0.8×10^6 and 6.1×10^6, respectively, were studied. With a given polymer solution the fractional reduction in Colburn factor $StSc^{2/3}$ relative to Newtonian at the same Reynolds number Re increased with increasing Re. At constant Re, with solutions of the same polymer, the fractional $StSc^{2/3}$ reduction increased with increasing polymer concentration, up to a maximum asymptotic value. The concentration required to achieve maximum mass transfer reduction was significantly greater than that required to attain maximum drag reduction. The asymptotic maximum mass transfer reduction was described by:

$$StSc^{2/3} = 0.022\ Re^{-0.29};\quad 5 \times 10^3 < Re < 3.5 \times 10^4,\quad Sc \doteq 1000,$$

and under these conditions the diffusion sublayer thickness was about twice Newtonian. The present mass transfer measurements, combined with previous friction factor data, lead to an 'elastic sublayer' analogy for scalar transport during drag reduction; the predictions of this analogy at the two extremes of low and maximum drag reduction agree approximately with the available literature.

Held at St. John's College, Cambridge, England.

Organised and sponsored by BHRA Fluid Engineering, Cranfield, Bedford, MK43 0AJ.

1. Introduction

Drag reduction by dilute polymer solutions in turbulent flow appears bounded by two universal asymptotes. In pipes, the asymptotic friction factor relations are respectively the Prandtl-Karman law:

$$f^{-1/2} = 4.0 \log_{10} Ref^{1/2} - 0.4 \qquad (1)$$

and the maximum drag reduction asymptote:

$$f^{-1/2} = 19.0 \log_{10} Ref^{1/2} - 32.4 \qquad (2)$$

Relations (1) and (2) envelop the polymeric regime, so named because the friction factor relationships therein depend on both flow and polymeric parameters as described earlier (Virk 1971a). The present study of mass transfer at maximum drag reduction was undertaken to seek the corresponding limits of scalar transport reduction by dilute polymer solutions. Too, it was hoped that experiments at the high Schmidt numbers involved, $Sc \doteq 10^3$, might provide information about wall turbulence during maximum drag reduction.

No previous investigation of mass transfer during drag reduction appears to have been reported.

The related subject of heat transfer during drag reduction has been investigated by several workers (e.g. Gupta, Metzner & Hartnett, 1967, Marucci & Astarita 1967, McNally 1968, Smith, Keuroghlian, Virk & Merrill 1969, Debrule 1972). These studies, using a variety of aqueous polymer solutions with Prandtl numbers $Pr \doteq 10$, have established that in turbulent pipe flow the fractional reduction in the Stanton number is of roughly the same magnitude as the fractional reduction in friction factor when comparisons are made at constant Reynolds number. Analogies of the Karman (Wells 1968, Poreh & Paz 1968) and Colburn (Smith et al. 1969) types have been applied to these data and found to hold approximately, decisive comparisons being hampered by the unavailability of precisely matched friction and heat transfer data as well as by the variation of fluid properties with respect to both temperature and shear.

2. Experimental

Trans-cinnamic acid, coated inside a stainless steel tube to form a cylindrical surface of I.D. 0.295 cm and $L/d = 34.5$, was dissolved into distilled water and polymer solutions flowing through. Mass transfer rates were measured by two independent methods. In the well-known weight loss (abbreviated WL) method (Linton 1949, Linton & Sherwood 1950), a constant flow rate was maintained for a known time and the carefully dried test section weighed before and after to yield the corresponding amount of cinnamic acid dissolved. In the ultraviolet spectrophotometric method (abbreviated UV), developed during this work, the strong absorption, due to a $\pi \rightarrow \pi^*$ electronic transition, exhibited by the cinnamic acid at wavelength $\lambda_{max} = 272$ nm was employed to detect its concentration in the liquid leaving the test section relative to that (always zero) in the liquid entering. Flow rate was measured to within ±1 percent by bucket and stop-watch, and the corresponding bulk cinnamic acid concentration was obtained from the absorbance at λ_{max} of samples of bucket liquid. An absorbance versus concentration calibration was separately established on the spectrophotometer used for all measurements, a Bausch and Lomb Spectronic 505 with 1 cm path length quartz cell. This yielded $\lambda_{max} = 272 \pm 2$ nm, absorbance linear with concentration, and a molar extinction coefficient $\varepsilon_{max} = (1.95 \pm 0.15) \times 10^4$ liter mol^{-1} cm^{-1} in distilled water. These compare favorably with literature values of $(\lambda_{max}, \varepsilon_{max}) = (272, 1.59 \times 10^4)$ and $(272, 2.25 \times 10^4)$ in ethanol and methanol respectively (Dyer 1965, Sadtler Standard Spectra 1965). UV spectra were unaffected by addition of polymer up to the maximum amounts $c = 1000$ wppm, employed in this work. The cinnamic acid saturation ratio C/C_{sat} could be measured to within ±10 percent at worst, i.e, at the low end of the range $10^{-3} < C/C_{sat} < 2 \times 10^{-2}$ encountered. A flow developing section, made from seamless stainless steel tube 0.295 cm I.D. and L/d = 129 with ends machined flat and square, was placed immediately upstream of the mass transfer section. Some UV runs were also made using two mass transfer sections in tandem (L/d = 69); both in laminar and turbulent flow these gave results in agreement with those obtained using a single section, and final measurements were made only with the latter. The developing and mass trnasfer test sections were incorporated into a once-through blowdown flow system, details of which are available elsewhere (Suraiya 1969). The two commercial polyethyleneoxide (PEO) polymers used were characterized by low shear viscometry to obtain their intrinsic viscosities in distilled water which was the solvent for all runs. Polymer solutions employed in the present study are noted in table 1, the relative viscosity of all except two being $\eta_{rel} < 1.20$ and the highest $\eta_{rel} = 1.70$. Cinnamic acid solubility and diffusivity in water, taken from the literature (Linton 1949), were assumed to be unchanged in the polymer solutions; there is some evidence (Fortuna & Hanratty 1971) that diffusivities of molecular solutes in the polymer solutions of interest here are indeed the same as in solvent. All of the UV and most of the WL series of experiments were at temperatures of 25.5±1.5 C. Finally, it is worth noting that in the time taken to obtain a single data point, respectively 20±10 min and 30±20 sec in the WL and UV methods, the increase in test section ID on account of dissolution was typically less than 1 percent in the WL experiments and entirely negligible in the UV experiments.

3. Results and Analysis

Friction factor results are shown in figure 1 on doubly logarithmic coordinates of f/2 versus Re, Reynolds number being formed with polymer solution viscosity. For Re < 2000, solvent and all polymer solutions agree with Poiseuille's law:

$$f/2 = 8\ Re^{-1} \tag{3}$$

while for Re > 3000 the solvent data are close to the Prandtl-Karman law (1) whereas all but one of the polymer solutions lie on the maximum drag reduction (abbreviated mdr) asymptote (2). The turbulent flow results have been fitted by power law expressions which read, respectively, for solvent:

$$f/2 = 0.0317\ Re^{-0.234} \quad ; \ 4 \times 10^3 < Re < 2 \times 10^4 \tag{4}$$

and for mdr:

$$f/2 = 0.284\ Re^{-0.582} \quad ; \ 4 \times 10^3 < Re < 4 \times 10^4 \tag{5}$$

and these are shown by solid lines in figure 1. Under the present experimental conditions, solutions of PEO W301 evidently attain mdr at 2 < c wppm < 5.

Mass transfer results are presented in figure 2 on doubly logarithmic coordinates of the Colburn factor $StSc^{2/3}$ versus Re. Stanton numbers are formed with an overall mass transfer coefficient while Schmidt numbers are formed with solution kinematic viscosity and the diffusivity of cinnamic acid in solvent alone, the latter justified earlier. Data obtained by the weight loss and ultraviolet spectrophotometric methods are respectively indicated by solid and hollow symbols. Consider first the solvent results (circles). In laminar flow, 80 < Re < 2000, the test section operated at $L_+ = (2(L/d)/ReSc) < 10^{-3}$ so that Leveque's well-known solution

$$StSc^{2/3} = 1.62\ Re^{-2/3}\ (L/d)^{-1/3} \quad ; \ Re < 2000,\ L_+ << 1 \tag{6}$$

should apply. The solid line marked '1' in figure 2 represents equation (6) for our L/d, and it can be seen that the results of both UV and WL methods are within about ±10 percent of it. In turbulent flow the two methods are in good agreement for Re > 4000 although both scatter ±10 percent, the WL rather more than the UV method. These data are compared with, and are evidently quite reasonably described by, the recent scalar transport correlation of Notter & Sleicher (1971)

$$StSc^{2/3} = 0.0149\, Re^{-0.12} \quad ; \; Re > 5000, \; Sc > 100 \tag{7}$$

which is shown as a solid line marked '2' in figure 2. The foregoing solvent results permit us to place some confidence in the present experimental techniques. Turning to the polymer solutions, in the laminar regime Re < 1000 all data agree with equation (6) within ±15 percent, being identical with solvent. However, in the region 1000 < Re < 2000 the polymer solution data lie systematically above (6) which is curious because the corresponding friction data all very closely adhere to Poiseuille's law (3). This may be related to the observation (Ohara 1968, Little & Weigard 1970) that, during laminar to turbulent transition of polymer solutions in pipe flow at mdr conditions, signs of turbulence appear at Re = 1500, rather lower than the lowest Newtonian critical Re ≐ 2000. In the region Re > 4000, both WL and UV methods were used for each of four polymer solutions, and in all cases the two sets of measurements agreed within the expected ±20 percent precision and could therefore reasonably be averaged. As an example of the polymer solution results consider 1000 wppm N3000 (hexagons): $StSc^{2/3}$ was reduced to 0.45 times Newtonian at Re = 9000 and the fractional mass transfer reduction increased with increasing Re, being respectively (0.37, 0.55, 0.64) at Re (4500, 9000, 22000). The effect of polymer concentration can be seen by following the data for W301 solutions. At $Re = 2 \times 10^4$ for example, fractional $StSc^{2/3}$ reductions relative to Newtonian were respectively (0.51, 0.59, 0.70, 0.73, 0.73) for c (3, 10, 30, 100, 300) wppm. Evidently the mass transfer reduction increases with increasing polymer concentration but eventually attains an asymptotic, maximum, value. The present results are not enough to define well the maximum mass transfer reduction asymptote but perusal of figure 2 does reveal a lower bound comprising data points from several PEO W301 solutions, namely, 300 wppm for Re > 4000, 100 wppm for $Re > 10^4$ and 30 wppm for $Re > 2 \times 10^4$, all of which seem to lie on a line of slope −0.3. Guided by this, and by information from the analysis which follows, we propose an approximate relation

$$StSc^{2/3} = 0.022\, Re^{-0.29} \quad ; \; 5 \times 10^3 < Re < 3.5 \times 10^4, \quad Sc \doteq 10^3 \tag{8}$$

for maximum mass transfer reduction (mmr); the asymptote (8) is depicted by the solid line marked '3' in figure 2. A comparison with friction factor results is also revealing. At $Re = 2 \times 10^4$ the fractional $StSc^{2/3}$ reductions quoted above for various polymer solutions contrast with a constant fractional drag, i.e., friction factor, reduction of 0.71 for all. Whereas all of the polymer solutions considered in figure 2 obey identically the same mdr friction factor relation (5), as seen in figure 1, yet they follow appreciably different $StSc^{2/3}$ versus Re relations. Three solutions, 1000 wppm N3000, 3 and 10 wppm W301, yield significantly less than maximum mass transfer reduction over the whole Re range while two, 100 and 300 wppm W301, appear to have attained (8). Notice too that, for PEO W301 in the present pipe and Re range, c = 3 wppm were needed to achieve mdr whereas c = 100 wppm are required for mmr. Finally in regard to figure 2, a laminar boundary layer analysis of the mass transfer expected from our test section when the friction is given by (5) instead of (3) yields a solution of the Leveque form with coefficient and Re exponent respectively (0.52, −0.53) versus (1.62, −0.67) in (6). This shows that at maximum mass transfer reduction the total mass transport from the test section, although much reduced relative to Newtonian turbulent flow, still significantly exceeded that arising from laminar convection alone.

Analysis of the foregoing results will be performed in terms of the quantity $K = (f/2)^{1/2}/StSc^{2/3}$ which, save for physical properties, is entirely determined by experiment. Theoretically, K is related to the 'analogy integral' C_0^+ by the usual definitions:

$$K = C_0^+ \, (C_B/C_0) \, Sc^{-2/3} \tag{9}$$

$$C_0^+ = \int_0^{R^+} (\varepsilon/\nu + 1/Sc)^{-1} \, dy^+ \tag{10}$$

$$(C_B/C_0) = \int_0^1 (C/C_0) \, U \, d(1-\xi)^2 \Big/ \int_0^1 U \, d(1-\xi)^2 \tag{11}$$

In the preceding, symbols have their usual meanings, C being the concentration of the diffusing species (i.e., the cinnamic acid) with subscripts W, B, and O, respectively denoting wall, bulk, and centerline values; $\xi = (y/R)$ represents radius-normalized distance from the wall, U is the mean velocity at location y, and superscript + implies inner scale normalization. In the present experiments the ratio of bulk to centerline concentrations (both taken wrt $C_W = 0$) was always very close to unity, so that the experimental K essentially reflects the variation of C_0^+ with R^+, Sc, and the prevailing eddy viscosity profile.

Table 2 summarizes results obtained with (a) solvent and (b) polymer solutions at mdr. In addition to the present data, some earlier heat transfer results (Keuroghlian 1967) are included in table 2; these were obtained using the same flow system and virtually identical PEO polymers as in the present work and merit analysis for information at Pr = 5. Solvent results, noted in some detail because all polymer solution data will be normalized thereby, showed K_n independent of R^+ and in good agreement with recent literature (Notter & Sleicher 1971). Among polymer solutions tested in the mass transfer experiments, essentially random variations in K were encountered at all R^+ in the cases of 10 wppm W301 (WL), K = 14.6±1.0 encompassing all 10 points $99 < R^+ < 381$, and of 100 wppm W301 (UV), K = 21.2±2.1 encompassing 8 points $176 < R^+ < 375$. The 30 wppm W301 (UV) did show a systematic K increase from 16.0 to 20.8 as R^+ increased from 90 to 181, but K was then constant at 23±2 for $265 < R^+ < 432$; also the 30 wppm W301 (WL) gave K = 20.5±1.5 with no trends for $133 < R^+ < 345$. Within the present experimental precision therefore there was no variation of K with R^+ for $R^+ = 300\pm150$ and table 2(b) gives mean values of all the points obtained for each polymer solution. Keuroghlian's (1967) heat transfer data at mdr comprise only four points for each solution and K = 30.5±2.5, $150 < R^+ < 400$, covers them all although K increased monotonically, but with decreasing rate, from 28 at $R^+ = 150$ to 33 at $R^+ = 400$.

The quantity K is evidently proportional to diffusion sublayer thickness, defined as the value of y^+ at which $\varepsilon/\mathcal{D} = 0.1$, say. Thus, the ratio K_m/K_n, derived from the experimental data, physically represents a ratio of diffusion sublayer thickness in a polymer solution at mdr to that in Newtonian turbulent flow. The dependence of K_m/K_n upon polymer concentration and Schmidt number is indicated in figure 3. The effect of polymer concentration is shown in figure 3(a) with the mass transfer data for PEO W301 solutions at mdr; each point represents an average of WL and UV results with vertical bars between the individual means quoted in table 2(b). The ratio K_m/K_n increases from 1.0 at c = 3 wppm to a maximum asymptotic value 1.9±0.1 achieved at c ≐ 100 wppm. The effect of Schmidt number is shown in figure 3(b) with heat and mass transfer results for each of two solutions, 1000 wppm N3000 and 10 wppm W301. Both are at mdr, both exhibit significantly less than maximum mass transfer reduction with $K_m/K_n \doteq 1.3$ at Sc = 1200, but both appear to exhibit maximum heat transfer reduction with $K_m/K_n \doteq 1.95$ at Pr ≐ 6. These data suggest that at maximum mass (heat) transfer reduction the diffusion (conduction) sublayer is about twice as thick as Newtonian. Independent information from rough pipe experiments (Virk 1971b) indicates that at mdr the viscous sublayer thickness, y^+ at $\varepsilon/\nu = 0.1$ say, is also about twice Newtonian. Thus, for the 100 wppm W301 solution, which is at both mdr and mmr, both viscous and diffusion sublayers are twice as thick as Newtonian. However, at the other extreme, the 3 wppm W301 solution, which is at mdr but yields $K_m/K_n \doteq 1$, seems to be in the schizoid situation of possessing a viscous sublayer thickness twice Newtonian but a diffusion sublayer thickness the same as

Newtonian. Intermediate between these, the 1000 wppm N3000 and 10 wppm W301 solutions are at mdr, yield the asymptotic maximum K_m/K_n for heat transfer but less than maximum K_m/K_n for mass transfer, so that their viscous and conduction sublayers are both twice twice but their diffusion sublayer only 1.3 times as thick as Newtonian. The progressive increase in K_m/K_n witnessed with increasing polymer concentration in figure 3(a) and with decreasing Sc in figure 3(b) would therefore seem to reflect the wallward penetration of a region of reduced eddy diffusivity, the asymptotic maximum K_m/K_n being reached when all locations contributing to C_0^+ are encompassed.

Discussion of the preceding result requires reference to current impressions of the basic polymer-turbulence interaction responsible for drag reduction (see, e.g., recent reviews of the subject by Hoyt 1972, Lumley 1973, Landahl 1973, and Virk 1975). As summarized in the last-named review, experimental measurements of near-wall turbulence structure, of mean velocity profiles, and of the onset of drag reduction under hydraulically rough and smooth conditions, suggest the 'buffer zone' $5 < y^+ < 100$ as the most likely site of initial interaction; physically, this implies possible interference by the macromolecules in the turbulent bursting process. It is therefore reasonable to presume that the polymer-turbulence interaction starts in the vicinity of the plane of peak turbulent energy production, $y^+ \doteq 15$. We now hypothesize that the region of interaction, i.e., the elastic sublayer, penetrates both inwards and outwards with increasing drag reduction. The outward growth of the elastic sublayer is rather well known from friction and mean velocity profile measurements, maximum drag reduction being achieved when the entire cross section to R^+ is thus occupied. Inward growth of the elastic sublayer has not hitherto been mooted. The region involved does not decisively affect momentum transport, and for gross flow purposes the inner edge of the elastic sublayer remains effectively fixed at the outer edge of the usual viscous sublayer. However, scalar transport must clearly be controlled by regions ever closer to the wall with increasing Sc, and the present experiments suggest that the maximum scalar transport reduction is achieved only after the inward growth of the elastic sublayer entirely encompasses the relevant diffusion sublayer. It is further interesting that for the same PEO W301 polymer, concentrations of c (3, 10, 100) wppm respectively were required to achieve the asymptotic (mdr, mhr, mmr) with corresponding Sc (1, 5, 1000). This result yields an approximate concentration dependence for inward growth of the elastic sublayer, $-\ln (y_{e,i}^+/y_v^+) \propto c^{1/2}$, which is very similar to the concentration dependence for outward growth, $\ln (y_{e,o}^+/y_v^+) \propto c^{1/2}$, the latter having been obtained from extensive friction factor data (Virk 1975).

4. An 'Elastic Sublayer' Analogy

The present mass transfer results along with the elastic sublayer model for drag reduction, developed earlier (Virk 1971a) from mean velocity and gross flow information, allow formulation of a scalar transport 'analogy' which will be recounted briefly (see Virk 1977 for details). The physical basis of the analogy is outlined first, followed by development of the analogy expressions; predictions of the analogy are compared with the available literature in the next section.

Experimental information concerning eddy-viscosity profiles during drag reduction (Virk 1975) is yet incomplete and we are led to hypothesize the following. At zero drag reduction, the usual Newtonian profile must prevail. At low drag reduction, we expect an essentially Newtonian eddy viscosity profile which is dimpled downwards in the neighborhood of $y^+ \doteq 15$ (where $\varepsilon/\nu \doteq 1$); under these conditions, the ratio of scalar/momentum) transport reduction can be obtained by a perturbational technique. With increasing drag reduction, the region of reduced eddy viscosity grows both outward and inward from $y^+ \doteq 15$, the profile asymptotically approaching a lower bound which represents the minimum turbulent transport. Finally, at maximum drag reduction the eddy viscosity profile appears similar to Newtonian in shape but with mixing length constant about one-fifth and viscous sublayer thickness about twice Newtonian.

In the two asymptotic regimes, of zero and maximum drag reduction respectively, the eddy viscosity profiles were taken to be similar form:

$$\varepsilon/\nu = \gamma y^{+3}/(1 + (\gamma/\chi)y^{+2}) \qquad (12a)$$

$$\rightarrow \gamma y^{+3} \; ; \; y^+ \rightarrow 0 \qquad (12b)$$

$$\rightarrow \chi y^+ \; ; \; y^+ >> 1 \qquad (12c)$$

The constants (γ,χ) in (12) have physical significance and are directly available from experiments. Thus, the expression could be rather well fitted to the data of Laufer (1954) which yielded the Newtonian constants $(\gamma_n,\chi_n) = (0.870 \times 10^{-3}, 0.40)$. Adaptation to mdr then followed from the experimentally established ratios, χ_m/χ_n and γ_m/γ_n. The former ratio of mixing length constants is well known, being, of course, the ratio of the slopes of the respective Prandtl-Karman relations (1) and (2), $(\chi_m/\chi_n) = (4.0/19.0)$. The latter ratio is estimated both from the asymptotic mass transfer reduction at high Sc, $(\gamma_m/\gamma_n)^{-1/3} = (K_m/K_n)_{Sc \rightarrow \infty} = 1.9\pm0.1$, obtained in the present experiments and from earlier experiments on the onset of roughness at mdr which provided $(\gamma_m/\gamma_n)^{-1/3} = (k^+_m/k^+_n) = 2.5\pm0.2$; we will use $(\gamma_m/\gamma_n)^{-1/3} = 2.0$, so that $(\gamma_m,\chi_m) = (0.109 \times 10^{-3}, 0.085)$. It should be pointed out that, whereas the outer limit, (12c), of the present eddy viscosity expression is quite well established for mdr flows, the inner limit, (12b) is conjectural. However, in Newtonian flows, the cubic (or higher) dependence of ε/ν on y^+ as $y^+ \rightarrow 0$ arises from continuity and the assumption of a viscous-dominated turbulence, both of which should also apply at mdr once asymptotic conditions have been fully established.

Numerical results, obtained by straightforward substitution of (12) in (10) and (11), are presented in tables 3 and 4 for both Newtonian and mdr flows. Table 3 contains the basic analogy parameters C_0^+ and (C_B/C_0) as functions of Sc and R^+. Bulk concentrations were obtained from mean velocity profiles generated internally using (12) with Sc = 1; corresponding calculated average velocities U^+_{Av} are also quoted in table 3 and agree closely, ±2 percent or better, with the respective friction factor relations (1) and (2). The latter were then used to convert the analogy results to the experimentally determinate dependence of the Colburn factor $StSc^{2/3}$ on Reynolds number, these predictions being shown in table 4 for Schmidt numbers 5 and 1000 which are respectively relevant to heat and mass transfer.

In the intermediate 'polymeric' regime, the significant parameter defining the relatively low drag reduction is the nondimensional 'velocity slip' S^+, given by

$$S^+ = (U^+_{0,p} - U^+_{0,n})_{R^+} = 2^{1/2}(f_p^{-1/2} - f_n^{-1/2})_{R^+} \qquad (13)$$

We now define an analogous 'scalar slip' M^+ by

$$M^+ = (C^+_{0,p} - C^+_{0,n})_{R^+} \qquad (14)$$

and it follows from (10) that

$$M^+ = \int_0^{R^+} [(\varepsilon/\nu,p + 1/Sc)^{-1} - (\varepsilon/\nu,n + 1/Sc)^{-1}] \, dy^+ \qquad (15)$$

Clearly $M^+ = S^+$ at Sc = 1. At low drag reduction, as noted earlier, we expect an essentially Newtonian eddy viscosity profile which is dimpled downwards in the neighborhood of $y^+ \doteq 15$, $\varepsilon/\nu \doteq 1$. Only this region will contribute to the scalar slip, the integrand in (15) being essentially zero everywhere else, so with Δy^+, $\Delta(\varepsilon/\nu)$ as appropriate arbitrary scales:

$$M^+ = [d(\varepsilon/\nu + 1/Sc)^{-1}/d(\varepsilon/\nu)]_{\varepsilon/\nu = 1} \; \Delta(\varepsilon/\nu)\Delta y^+ \qquad (16)$$

Dividing both sides of (16) by the corresponding expression for S^+ and evaluating the bracket yields the dependence of the (scalar/velocity) slip ratio upon Schmidt number in the limit of zero drag reduction:

$$\mathrm{Lim}_{S^+ \to 0} \ (M^+/S^+) = 4\ (1 + 1/Sc)^{-2} \tag{17}$$

Values of (M^+/S^+) predicted by (17) are presented in table 3.3 in the form of an experimentally accessible quantity $(M^+/S^+C^+_{0,n}) = (K_p - K_n)/S^+K_n$. While (17) should provide the form of the (M^+/S^+) variation with Sc, the levels of S^+ to which this limiting result applies remain to be established experimentally and adjustments are also to be expected on account of velocity profile shape changes (velocity profiles during low drag reduction are blunter than Newtonian but then become sharper as mdr is approached). It is interesting that with increasing Sc, M^+/S^+ increases to an asymptotic value 4.0 while $C^+_{0,n}$ decreases monotonically, as $Sc^{2/3}$ at high Sc, so that $M^+/S^+C^+_{0,n}$ exhibits a shallow maximum at Sc = 3 and then decreases toward zero, being (0.061, 0.0031) at Sc (5, 1000), respectively.

5. Literature Comparison

The predictions of the present analogy will be compared with the literature in each of the Newtonian, polymeric, and maximum drag reduction flow regimes.

The Newtonian case needs only brief mention. Results of tables 3.1 and 4.1 accord well with the correlation of Notter & Sleicher (1971) - and with the host of recent literature surveyed by them. This is the expected result because the eddy viscosity profile arrived at by these authors from various considerations is very similar to that of our expression (12) in the region, $0 < y^+ < 50$, of major importance.

In the polymeric regime, the present predictions can be tested rather directly as follows. The experimentally observed K_p for scalar transport are plotted against the corresponding drag reductions expressed as S^+. The intercept at $S^+ = 0$ is evidently K_n and, for good data, should equal the solvent value. The ordinate is now normalized by K_n and the slope $(d(K_p/K_n)/dS^+)_{S^+ \to 0}$ is identically the quantity $(M^+/C^+_{0,n}S^+)$ which is separately predicted by theory. This procedure was applied to the heat transfer data of Keuroghlian (1967) for two distilled water solutions, 10 and 100 wppm of a polyethyleneoxide, N3000, Mw = 0.76 x 10^6, and of Gupta, Metzner & Hartnett (1967), for a 500 wppm water solution of partially hydrolyzed polyacrylamide, ET 597. Figure 4, Cartesian coordinates of (K_p/K_n) versus S^+, illustrates the results. Keuroghlian's (1967) data yielded a zero intercept $K_n = 16.0\pm0.5$ in good agreement with the solvent alone; with normalized ordinate (K_p/K_n) the points are evidently linear with S^+ over the entire range, $0 < S^+ < 10$, yielding a line, shown dashed in Figure 4, which passes through (0,1) with slope 0.064±0.006. For these experimental conditions, $(R^+, Pr) = (300, 5.0)$, the theoretical value given in table 3.3 is $(M^+/C^+_{0,n}S^+) = 0.061$, indicated by the solid line '1' in Figure 4. The data of Gupta et al. (1967) were analyzed assuming a polymer solution $\eta_{rel} \doteq 2.0$, hence $Pr \doteq 10$, and smoothing both experimental friction factors and Stanton numbers versus Re to yield K_p versus S^+. The zero intercept $K_n = 18.5\pm2$ (rather higher than expected for this Pr) was then used to provide the points (circles) shown in figure 4. These points are distinct from the other data, being represented by the indicated dashed line through (0,1) with slope 0.040±0.010. This slope is to be compared with the predicted $(M^+/C^+_{0,n}S^+) = 0.046$ at $(R^+, Pr) = 1000, 10)$. Evidently the experimental results shown in figure 4 accord well with our theoretical prediction of the (scalar/velocity) slip ratio for 5 < Pr < 10. The analogy cannot as yet be tested at high Sc because the literature contains no reference to mass transfer during drag reduction in the polymeric regime.

At mdr conditions, there are two sets (Keuroghlian 1967, Debrule 1972) of pertinent heat transfer data. Keuroghlian's (1967) mdr results have already been considered in table 2. Two polymer solutions, 1000 wppm N3000 and 10 wppm W301, definitely achieved maximum drag reduction, but we cannot be quite certain that the asymptotic maximum heat transfer reduction was achieved because results are available for only one concentration of each polymer, and it is conceivable that higher concentrations might have caused further, albeit relatively small, reductions in $StSc^{2/3}$. Debrule (1972) tested 10 and 50 wppm solutions of PEO W301, $Mw = 5 \times 10^6$ over a small but significant range of Pr = 4,6,10 and $2 \times 10^4 < Re < 2 \times 10^5$. Friction factors for both 10 and 50 wppm solutions were nearly identical, indicative of asymptotic conditions, and lay parallel to but somewhat below our mdr relation (2) on Prandtl-Karman coordinates. For heat transfer, e.g., at Pr = 6, fractional $StSc^{2/3}$ reductions relative to solvent for c = (10, 50) wppm were respectively (0.77, 0.83) at $Re = 3 \times 10^4$ and (0.81, 0.84) at $Re = 10^5$. Although lack of information at higher concentrations prevents us from being certain, the 50 wppm solution should closely approach maximum heat transfer reduction. The above data are compared with the present analogy in figure 5 on doubly logarithmic coordinates, $StSc^{2/3}$ versus Re. The solid lines marked '1' and '2' respectively represent our analogy for Newtonian and mdr flows at Sc = 5, as detailed in table 4; their agreement with the the respective Newtonian and maximum heat transfer reduction data is better than ± 20 percent over the range $5 \times 10^3 < Re < 2.5 \times 10^5$. It will be noted that this comparison is free of any curve-fitting forced agreement, because the analogy was derived entirely from friction and mass transfer measurements. The literature contains no previous information regarding mass transfer at maximum drag reduction, and the present analogy will, of course, play back the present data. Thus, the predictions of table 4.2 for Sc = 1000, which represent all Sc > 100, are essentially synonymous with the experimental maximum mass transfer reduction asymptote (8).

Acknowledgements.

The experiments were supported by financial assistance from the Sloan Basic Research Fund. Mr. G. A. Russell conceived the 'UV' method during an undergraduate laboratory project with one of us.

References.

DEBRULE, P. M. 1972 Ph.D. Thesis, Calif. Inst. Tech., Pasadena, Calif.

DYER, J. R. 1965 'Applications of Absorption Spectroscopy of Organic Compounds' Prentice-Hall, New Jersey.

FORTUNA, G., & HANRATTY, T. J. 1971 Chem. Engng. Prog. Symp. Ser. No. 111, 67, 90.

GUPTA, M. K., METZNER, A. B., & HARTNETT, J. P. 1967 Intnl. J. Heat Mass Trans., 10, 1211.

HOYT, J. W. 1972 ASME J. Basic Eng., 94, 258.

KEUROGHLIAN, G. H. 1967 S.M. Thesis, Mass. Inst. Tech., Cambridge, Mass.

LANDAHL, M. T. 1973 Proc. 13th Intnl. Congr. Theor. and Appl. Mech., Moscow, pp. 177-199.

LAUFER, J. 1954 NACA Rep. No. 1174.

LINTON, W. H. 1949 Sc.D. Thesis, Mass. Inst. Tech., Cambridge, Mass.

LINTON, W. H., & SHERWOOD, T. K. 1950 Chem. Engng. Prog., 46, 258.

LITTLE, R. C., & WIEGARD, M. 1970 J. Appl. Polym. Sci., 14, 409.

LUMLEY, J. L. 1973 Macromol. Rev. 7, 263.

MARUCCI, G., & ASTARITA, G. 1967 Ind. Eng. Chem. Fundam. 6, 471.

MCNALLY, W. A. 1968 Ph.D. Thesis, Univ. of Rhode Island, Kingston, R.I.

NOTTER, R. H., & SLEICHER, C. A. 1971 Chem. Engng. Sci., 26, 161.

OHARA, M. 1968 S.M. Thesis, Mass. Inst. Tech., Cambridge, Mass.

POREH. M., & PAZ, U. 1968 Intnl. J. Heat Mass Trans., 11, 805.

SADTLER, S. P. 1965 'Standard UV Spectra,' SP Sadtler and Son, Inc., Philadelphia, Pa.

SMITH, K. A., KEUROGHLIAN, G. H., VIRK, P. S., & MERRILL, E. W. 1969 A.I.Ch.E. J., 15, 294.

SURAIYA, T. 1969 S.B. Thesis, Mass. Inst. Tech., Cambridge, Mass.

VIRK. P. S. 1971a J. Fluid Mech., 45, 417.

VIRK, P. S. 1971b J. Fluid Mech., 45, 225.

VIRK, P. S. 1975 A.I.Ch.E. J., 21, 625.

VIRK. P. S. 1977 Submitted.

WELLS, C. S. 1968 A.I.Ch.E. J., 14, 406.

TABLE 1 Experimental grid

Entry	Transport	Test ID cm	PEO Polymer	$[\eta]$ $cm^3 g^{-1}$	Mw $\times 10^{-6}$	Solutions c in wppm
1	Mass (WL)	0.295	W301	2000	6.0	3, 10, 30, 100, 300
2	Mass (UV)	0.295	W301	2000	6.0	3, 10, 30, 100
3			N3000	430	0.82	1000
4	Momentum	0.292	W301	1880	5.6	1, 2, 5, 10, 20, 100
5			N3000	390	0.76	10, 100, 1000, others

Notes.

1. $[\eta]$, Mw, respectively, denote polymer intrinsic viscosity, weight-average molecular weight.

2. Polymer concentration in solution c is expressed in parts per million by weight.

TABLE 2 Summary of results

(a) Solvent - distilled water

Transport Data Set	Mass (WL)	Mass (UV)	Heat (Note 3)
Points	21	15	8
R^+ Range	160-680	160-530	180-480
Sc or Pr	950±40	925±25	5.0±0.2
K_n mean	12.8	12.1	16.0
std. dev.	1.3	0.6	0.4
Literature (Note 4)	12.5	12.5	15.5

(b) Polymer solutions at maximum drag reduction

Polymer	c, wppm	Mean Values of K_m		
N3000	1000	-	17.2	30.4
W301	3	15.8	12.2	-
	10	14.3	16.4	30.9
	30	20.3	21.1	-
	100	24.1	21.3	-
	300	24.5	-	-

Notes:

1. $K = (f/2)^{1/2}/StSc^{2/3}$

2. Subscripts n, p, and m, refer to Newtonian, polymeric, and maximum drag reduction regimes, respectively.

3. Heat transfer results from Keuroghlian (1967).

4. Newtonian values from Notter & Sleicher (1971).

TABLE 3. The 'elastic sublayer' analogy for scalar transport during drag reduction

No.	Regime	$R^+ \downarrow$	U^+_{Av}	Sc →	1	3	5	10	30	100	1000
1	Newtonian	200	14.4	C^+_0	18.9	33.6	44.8	67.2			
				C_B/C_0	0.823	0.894	0.918	0.944			
		300	15.7		19.9	34.6	45.8	68.2	132.6	284.	1280.
					0.834	0.900	0.923	0.947	0.972	0.987	0.997
		500	17.1		21.1	35.9	47.1	69.5			
					0.844	0.906	0.927	0.950			
		1000	19.0		22.9	37.6	48.8	71.2	135.6	287.	1283.
					0.855	0.911	0.931	0.952	0.975	0.988	0.997
		2000	20.8		24.6	39.4	50.5	72.9			
					0.864	0.914	0.933	0.954			
2	Maximum Drag Reduction	200	28.1	C^+_0	44.7	77.8	101.6	148.			
				C_B/C_0	0.727	0.821	0.857	0.900			
		300	32.5		49.3	82.5	106.4	153.	285	575.	2588.
					0.744	0.832	0.866	0.904	0.947	0.974	0.994
		500	38.2		55.2	88.5	112.4	159.			
					0.761	0.842	0.874	0.909			
		1000	46.1		63.2	96.6	120.5	167	299.	605.	2602.
					0.781	0.852	0.880	0.913	0.951	0.976	0.994
		2000	54.0		71.3	104.7	128.7	175.			
					0.799	0.860	0.886	0.916			
3	Polymeric	300		$\frac{M^+}{S^+C^+_{0,n}}$	0.0504	0.0650	0.0607	0.0485	0.0283	0.0138	0.0031
		1000			0.0438	0.0598	0.0569	0.0464	0.0276	0.0137	0.0031

TABLE 4. Asymptotic Colburn factors at selected Sc and R^+

Regime	1. Newtonian			2. Maximum Drag Reduction		
Sc →		5	1000		5	1000
$R^+ \downarrow$	Re x 10^{-3}	$StSc^{2/3}$ x 10^3		Re x 10^{-3}	$StSc^{2/3}$ x 10^3	
120	-	-	-	5.32	1.65	1.77
200	6.00	4.74	5.23	11.3	1.19	1.39
300	9.60	4.32	4.90	19.7	0.966	1.18
500	17.3	3.88	4.53	38.8	0.767	0.999
1000	37.9	3.40	4.12	93.9	0.587	0.824
2000	82.7	3.00	3.78	220.	0.466	0.701
3000	130.	2.80	3.60	-	-	-

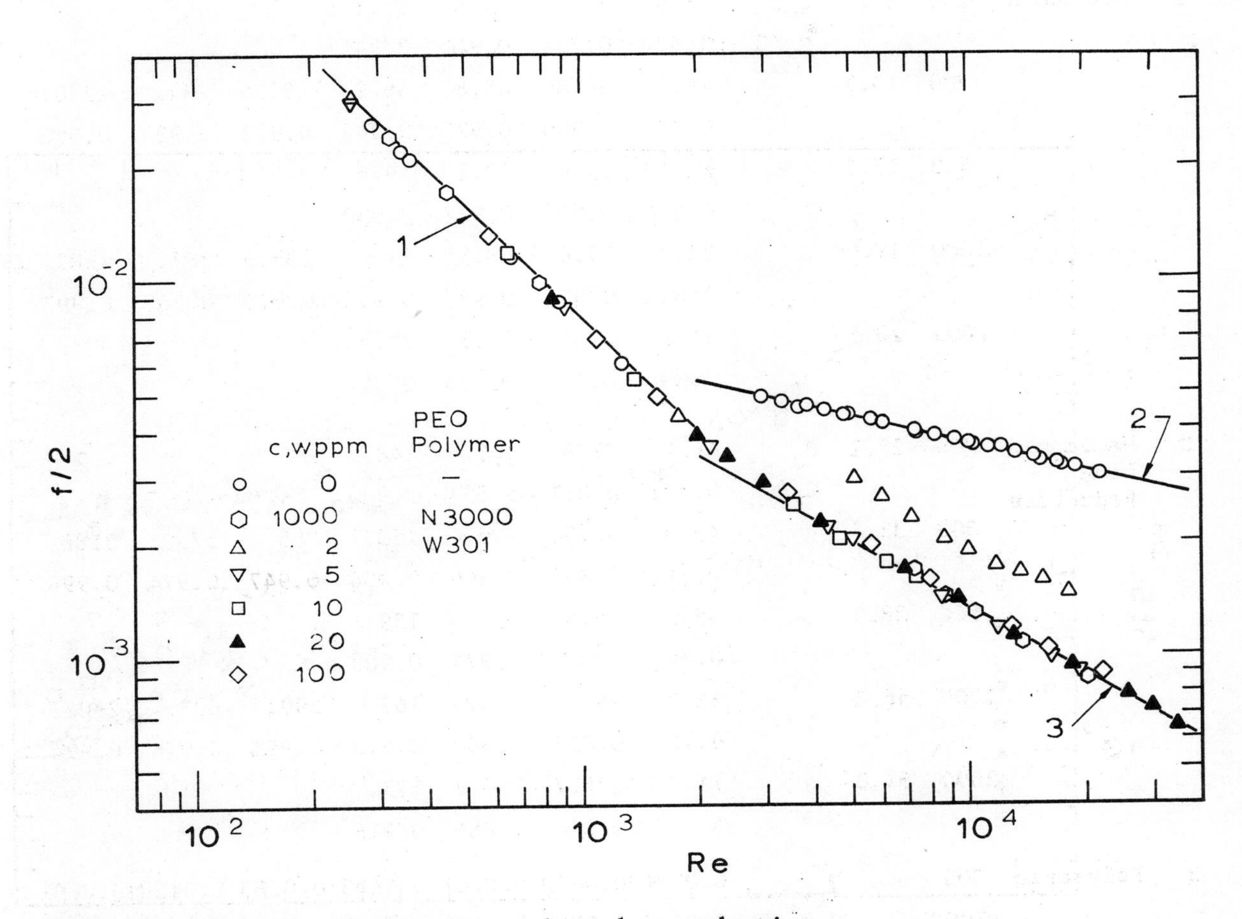

Figure 1. Friction factors at maximum drag reduction.

Solid lines marked 1, 2, and 3, respectively, represent equations (3), (4), and (5) in text.

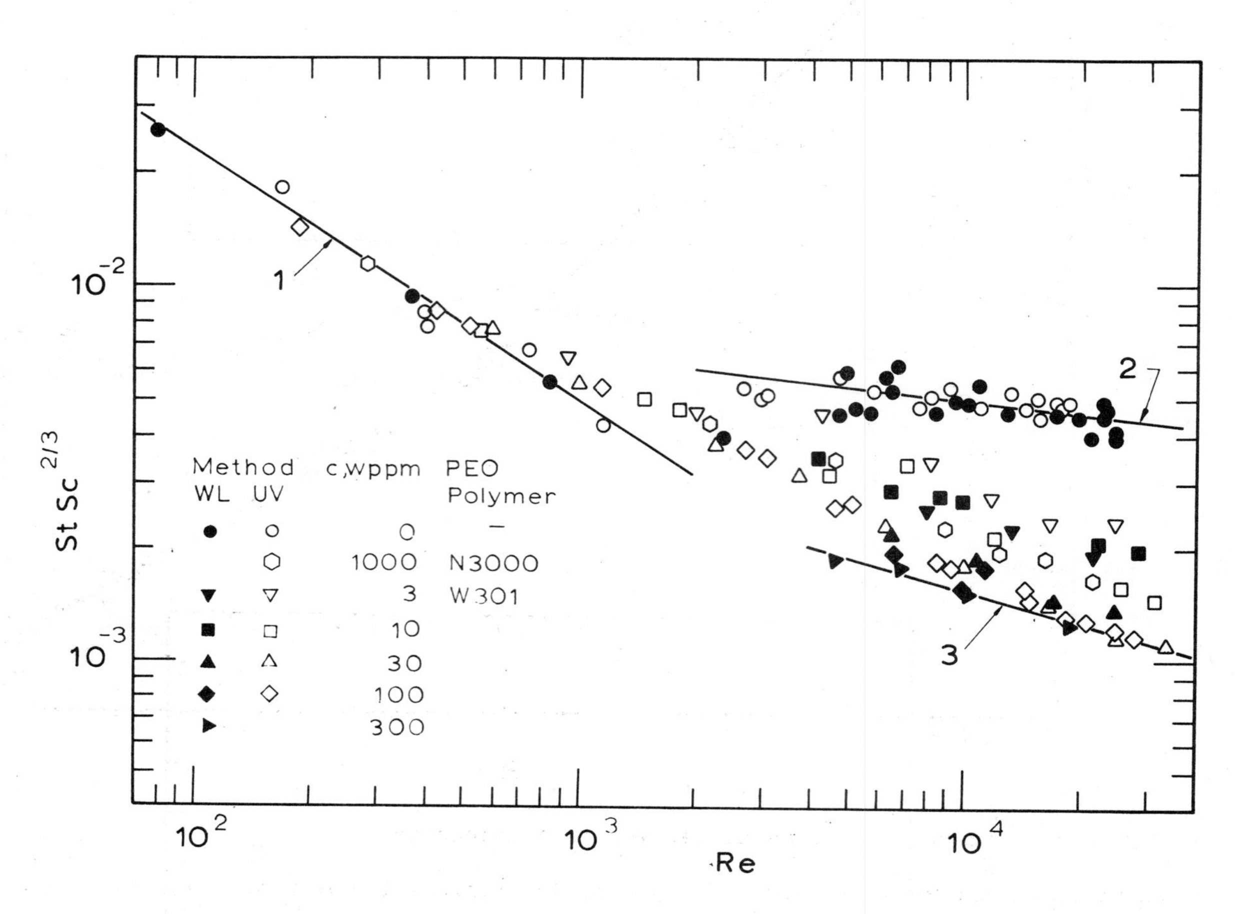

Figure 2. Mass transfer at maximum drag reduction.

Solid and hollow symbols respectively denote data from WL and UV methods (see text). Solid lines marked 1, 2, and 3, respectively, represent equations (6), (7), (8) in text.

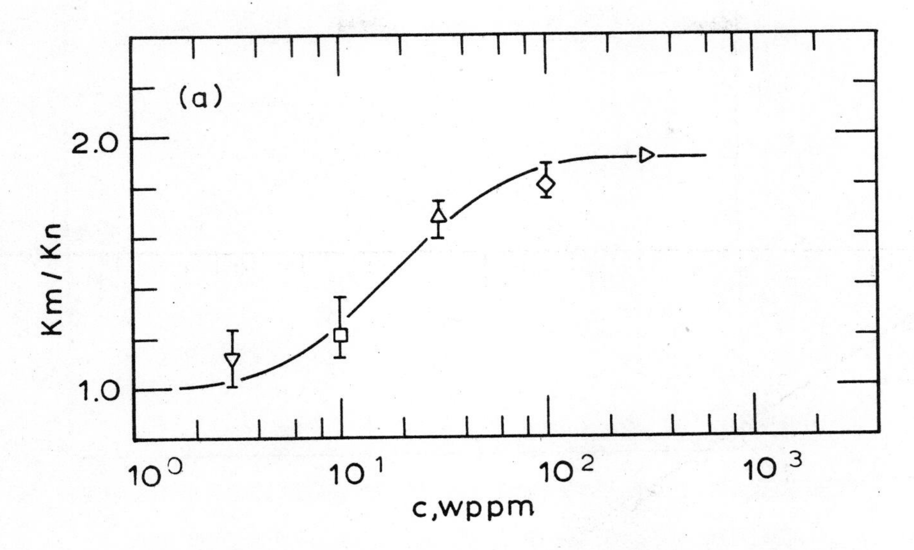

(a) Effect of polymer concentration,

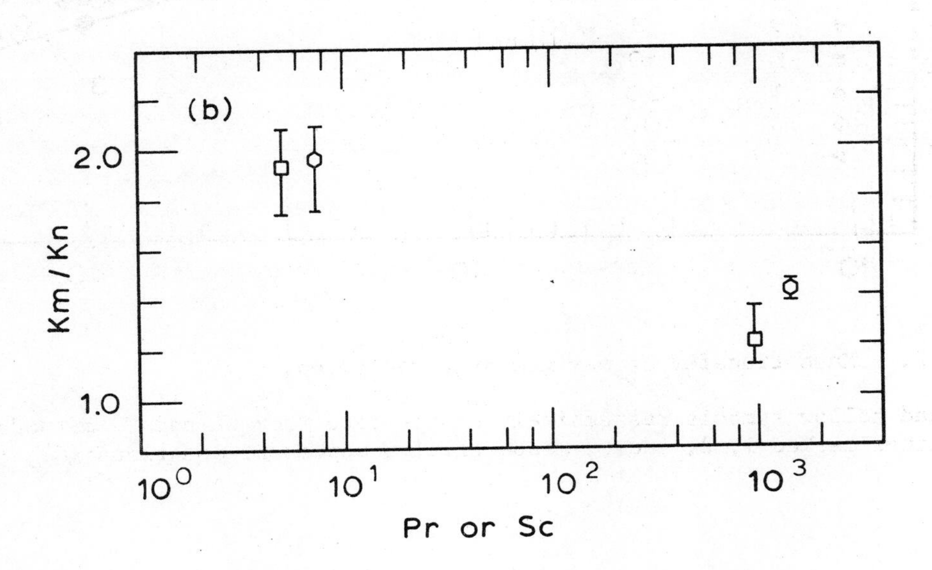

(b) Effect of Schmidt number.

Figure 3. Analysis of results by $K = (f/2)^{1/2}/StSc^{2/3}$:
All symbols as in Figure 2.

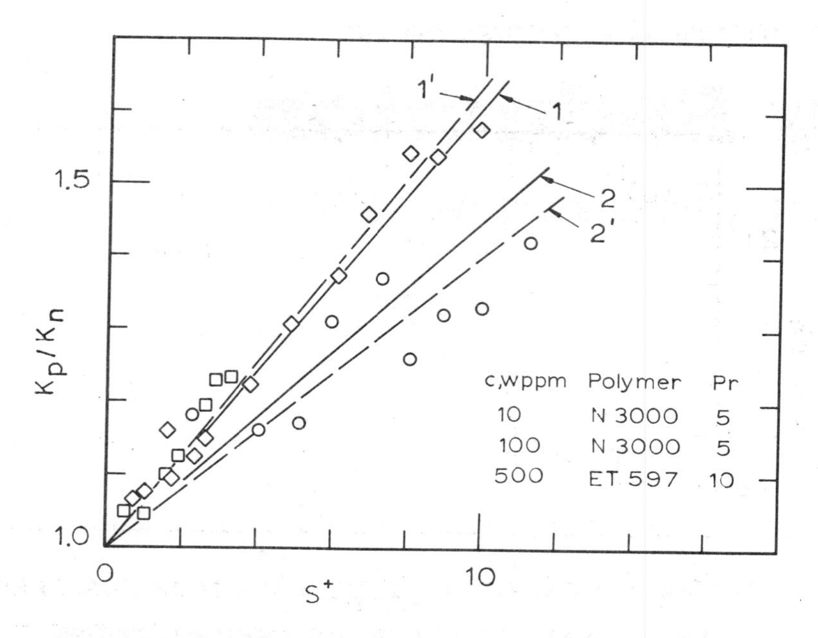

Figure 4. Comparison of present analogy with experiments in the polymeric regime.

Solid lines 1 and 2 are predicted by the analogy, table 3.3. Dashed lines 1' and 2' respectively are fitted to the data of Keuroghlian (1967) and Gupta, Metzner & Hartnett (1967).

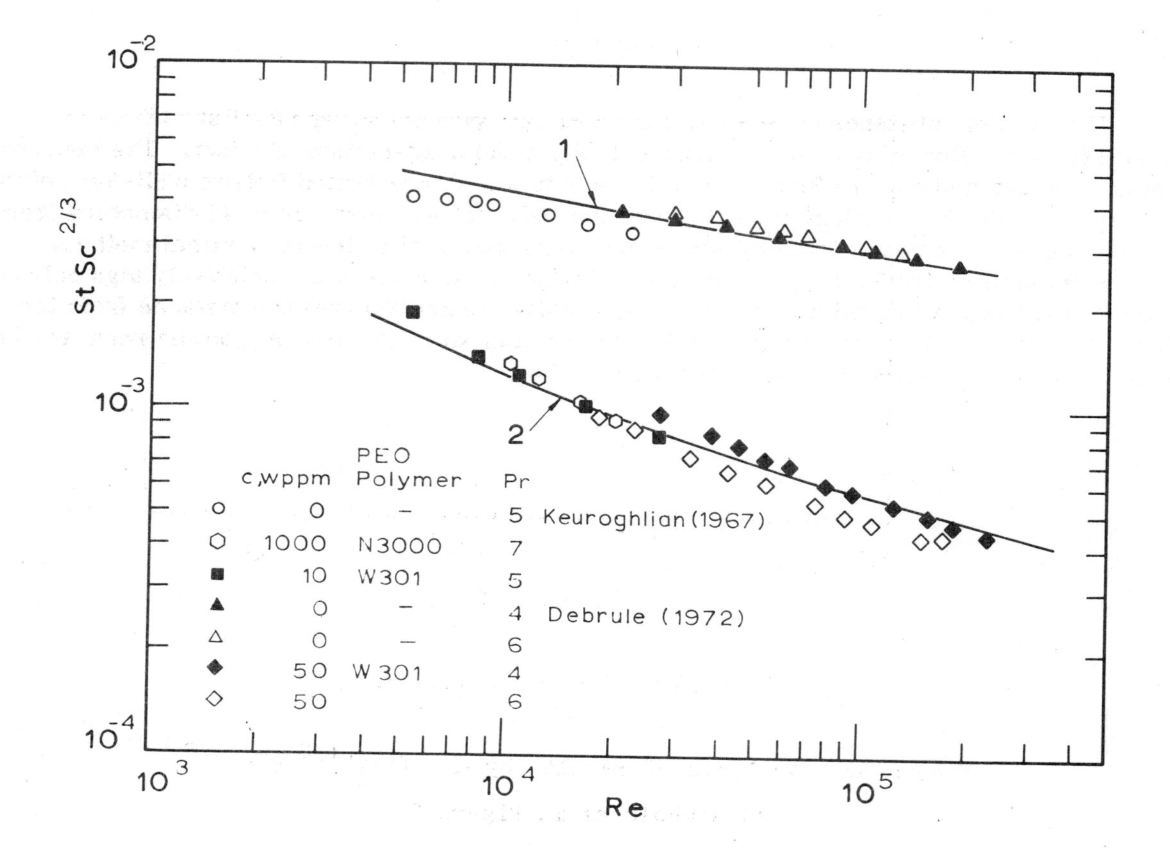

Figure 5. Comparison of present analogy with experiments at maximum drag reduction.

Solid lines marked 1 and 2, respectively, are predicted by the analogy for Newtonian and maximum drag reduction flows at Pr = 5, table 4.

Second International Conference on

Drag Reduction

August 31st - September 2nd, 1977

PAPER G4

SPECIFIC PROPERTIES OF TURBULENT MASS TRANSFER AT THE WALL IN DILUTE POLYMER SOLUTIONS

Z.P. Shulman, N.A.Pokryvailo, D.A.Prokopchuk and A.K.Nesterov

The Luikov Heat and Mass Transfer Institute, USSR

Summary

The laws of substance transfer in the turbulent boundary layer of a flat wall were considered for a flow of polymer solution exhibiting the drag-reduction effect. The measurements were carried out in a square 50 x 50 mm^2 tube. Concentrated Polyox WSR-301 solution was injected into the boundary layer through the wall slot at a distance of 42 diameters from the tube entrance. The concentration fields were measured by the electrochemical method.

For small distances downstream the injection place at the wall, relatively high polyox concentration was observed which lowered gradually the greater was the distance from the place of injection, For hydrodynamically passive admixture the anomaly downstream the flow disappeared more rapidly than for polymer solution.

Held at St. John's College, Cambridge, England.

Organised and sponsored by BHRA Fluid Engineering, Cranfield, Bedford, MK43 0AJ.

Introduction

The injection of drag-reducing polymers into the turbulent boundary layer of the solvent involves essential changes in the flow structure in a narrow wall region.Most effective proves to be the injection of polymer additives through narrow slots in the wall with solution around.

Optimum realization of the method requires the knowledge of the specific properties of convective polymer diffusion as well as of the polymer concentration fields in the turbulent boundary layer and especially at the wall,i.e. in the viscous sublayer region.

Nowadays there are scanty publications on source-to-wall turbulent diffusion of admixture in the boundary layer.They include Refs 1 and 2 on turbulent diffusion of admixture.Turbulent diffusion of hydrodynamically active admixture of the type of drag-reducing polymer additives has been studied in Refs 3-6.

As reported in Refs 4 and 5,the admixture concentration fields have been measured by the sampling method.This method,generally, yields acceptable accuracy of the measurements in the flow.As for the wall measurements,it does not provide reliable information on the concentration at distances less than the receiving orifice diameter because of the disturbances introduced by the intake device. The latter situation makes the concentration measurements too involved especially in very thin boundary layers.

Experiments

In the present work the electrochemical method of measuring the concentration of admixture injected into the boundary layer is grounded.The cathode-probes are used mounted aflush with the surface in a flow with the oxidization-reduction reaction $Fe^{-2} + e \rightleftarrows Fe^{-3}$ in the cathode-anode system (Refs 7 and 8).

In case of turbulent mixing the concentration field of the polymer admixture may be judged from the measured profiles of some low-molecular admixture injected into the flow together with the polymer.In our experiments ferro-ferricyanides $K_3Fe(CN_6)/K_4Fe(CN)_6$ are used.The procedure and equipment of the electrochemical approach are reported in detail in many publications (Refs 9-11). We shall consider this method from the point of view of extending the possibilities of the concentration field measurement in a turbulent flow.

As known,in case of the diffusion kinetics the intensity of substance transfer to the electrode surface is expressed as

$$i_o = jzF = zF\beta C_o$$

For homogeneous electrolyte solution with the known bulk concentration of the electrochemical reactant $C=C_o$ and with supply of the electrolyte of the same concentration C_o but with polymer additives in the region of the measuring electrode,the diffusion flux is determined by the expression

$$i_n^o = zF\beta_n C_o$$

If the bulk concentration of the reacting substance $C_o=0$,and polymer solution with the reactant concentration $C=C_o$ is injected into the region of the measuring electrode, the diffusion flux will be governed by the expression

$$i_n = zF\beta_n C_n$$

Using the relationship

$$\frac{i_n}{i_n^o} = \frac{C_n}{C_o}$$

one can identify the variation of the concentration of the polymer admixture along and across the flow beginning from the moment of its injection where $C=C_o$.

Mass transfer in polymer solutions is characterized by large $Sc \gg 1$. Therefore, on a short electrode the diffusion boundary layer proves to be much thinner than the laminar dynamic layer or viscous sublayer. And, hence, the electrochemical approach makes it possible to determine the concentration of the polymer admixture injected into the flow up to extremely small distances from the wall.

The tests were carried out in a square flow tunnel ($50 \times 50\ mm^2$) with the flow velocity 1.5m/sec and $Re = 1.5 \cdot 10^4$. The system used for polymer injection and measurements is shown in Fig.1. It provided tangential supply of polymer solution into the developed boundary layer through the slot 0.45mm wide. The diffusion flux was measured by electrochemical probes mounted aflush with the tube wall. The tests were made with a developed boundary layer at a distance 10 to 170 mm from the slot. The concentration profile was measured by an electrochemical probe with a wedge-like test surface. The minimum distance of possible approach to the wall was 1mm.

Experimental Results

The measured profiles of the passive admixture (ferricyanides) are presented in Fig.2 for the distances of 36.68 and 168mm from the slot. As seen, the concentration profile has a noticeable bend near the wall. The admixture content at the wall proves to be much greater than the concentration measured at a distance of 1mm from the wall. Such a phenomenon has not been observed by Poreeh and Cermak (Ref.1), Collins and Gorton (Ref.4) who have measured the concentration by way of fluid sampling with a tube-probe and through the wall orifice. However, disregard of the measured wall concentration in our experiments has yielded fine agreement between the concentration profile in the coordinates Y/λ and C/C_o and the data of Refs 1 and 4. The comparison of the results is given in Fig.3. Finally, in our experiments a fresh solution of Polyox WSR-301 with concentration 200-1000 ppm was used as a hydrodynamically active admixture, ferricyanide served as an indicator whose concentration was measured. Remember, that in Ref.4 rhodamin is reported to be such an indicator.

Strictly speaking, in similar cases the concentration measurement results should be related only to the indicator. We shall point out some conditions when the transport properties of these two different substances may be identified. The first condition holds for that portion of the boundary layer where the turbulent diffusion intensity is higher than the molecular one. Here the difference in the diffusion coefficients of the indicator and polymer admixture will not cause any essential discrepancies in the results. The validity of the second condition (certainly, at equal viscosity and temperature of substance) being the equality of the molecular diffusion coefficients provides identity of the results over the whole test region,from the wall to the flow core.

In Fig.4 the concentration profiles of the ferricyanide solution injected without polymer and with WSR-301 into the boundary layer are presented.The discharge through the slot and the initial concentration of ferricyanides in both cases was equal,while the polymer concentration amounted to 1000 ppm.Polymer additives raised sharp increase in the admixture concentration in direct vicinity of the wall and noticeable decrease of the region of admixture propagation in the transverse direction.

The admixture concentration measurements in the longitudinal direction at y=0, Fig.5, have shown that the polymer injection into the flow slows down the concentration reduction along the flow.Thus, at a distance of 170mm from the slot,almost 10-fold increase of the ferricyanide concentration at the wall was observed.

Conclusion

At present there are few data on the relationship between the coefficients of turbulent and molecular diffusion in the viscous sublayer.Neither there are reliable data on the coefficients of molecular diffusion of Polyox WSR-301.All this does not allow the recalculation of the measurement data on ferricyanides for polyetylene oxide.However,since the coefficients of molecular diffusion of water-soluble polymers of the type of WSR-301 are lower than the diffusion coefficients of ferricyanides ($d = 1 \cdot 10^{-9}$ m2/sec),we may suppose that near the wall the concentration of the polymer admixture will have large values.

References

1. Poreh, M. and Cermak, J.E: "Study of diffusion from a line source in a turbulent boundary layer". Int. J.Heat Mass Transfer, 7 ,10 pp.1083-1095.(October, 1964).

2. Morkovin, M.V: "On eddy diffusivity,quasi-similarity and diffusion experiments in turbulent boundary layers". Int. J. Heat Mass Transfer, 8 ,1 pp.129-145. (January, 1965).

3. Tullis, J.P. and Ramu, K.L.V: "Drag reduction in developing pipe flow with polymer injection". International Conference on Drag Reduction.(September, 1974).

4. Collins, D.J. and Gorton, C.W : "An experimental study of diffusion from a line source in a turbulent boundary layer ". AIChE Journal, 22 , 3 pp.610-612 (1976).

5. Vanin, Yu.P. and Migirenko, G.S: "Experimental study of the polymer additives distribution in the boundary layer behind the injection place". In "Investigations on the Boundary Layer Control". Novosibirsk (1976).

6. Wy, J : "Suppressed diffusion of drag-reducing polymer in a turbulent boundary layer". J. Hydronautics, 6 , 46 (1972).

7. Popov, V.P., Pokryvailo, N.A. and Gleb, A.K : "Electrochemical measurement of the local unsteady-state liquid flow velocity". In "Unsteady-State Heat and Mass Transfer Study". Minsk (1966).

8. Popov, V.P. and Pokryvailo, N.A : "Experimental electrochemical study of unsteady-state mass transfer for a cone with an external liquid flow around". In "Unsteady-State Heat and Mass Transfer Study". Minsk (1966).

9. Schütz, G : "Untersuchung des Stoffaustausch-Anlaufgebietes in einem Rohr bei vollausgebildeter.Hydrodynamischer Strömung mit einer elektrochemischen Methode". Int. J. Heat Mass Transfer, 7, 10 pp.1077-1082. (October, 1964).

10. Popov, V.P., Gleb, L.K. and Prokopchuk, D.A : "Mass transfer with a liquid flow in the cathode channel of the electrochemical diffusion converter". In "Thermodynamics". Nauka i Tekhnika. Minsk (1970).

11. Hanratty, T.J : "The use of electrochemical techniques to study fluid flows and mass transfer rates". International Seminar. Herceg-Novi (1969).

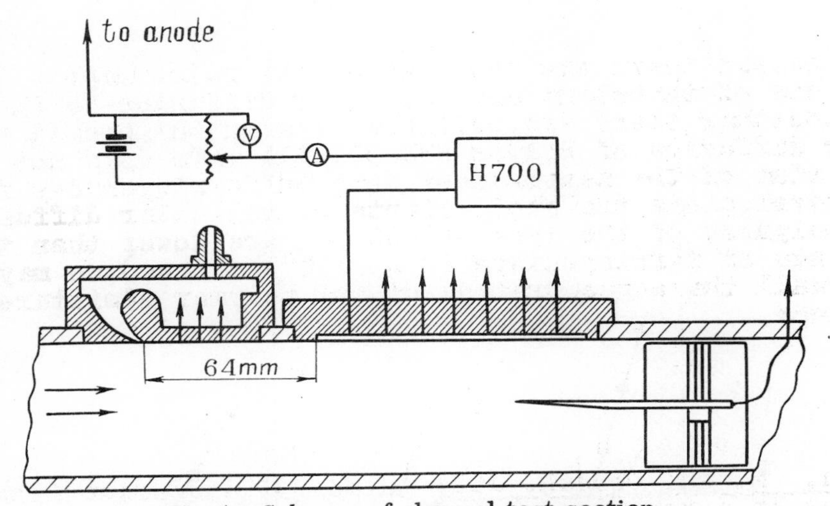

Fig.1 Scheme of channel test section

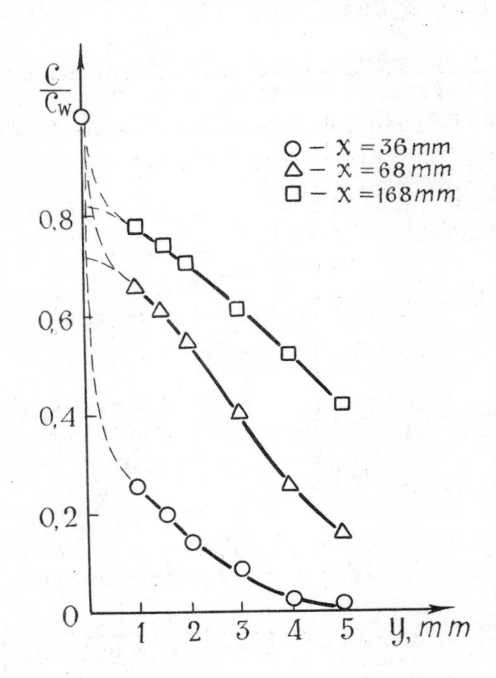

Fig.2 Concentration profiles of "passive" admixture

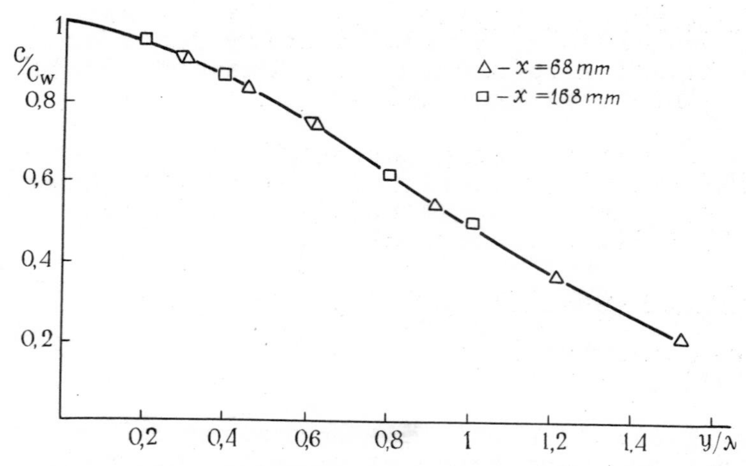

Fig. 3 Comparison of authors' data with results of Ref. 1 (solid curve)

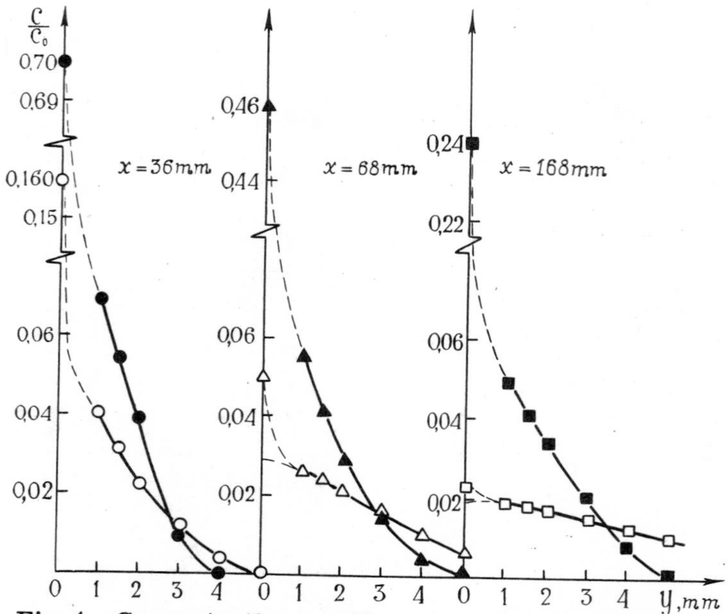

Fig. 4 Concentration profiles of "passive" admixture (light circles) and of polymer solution (dashed circles)

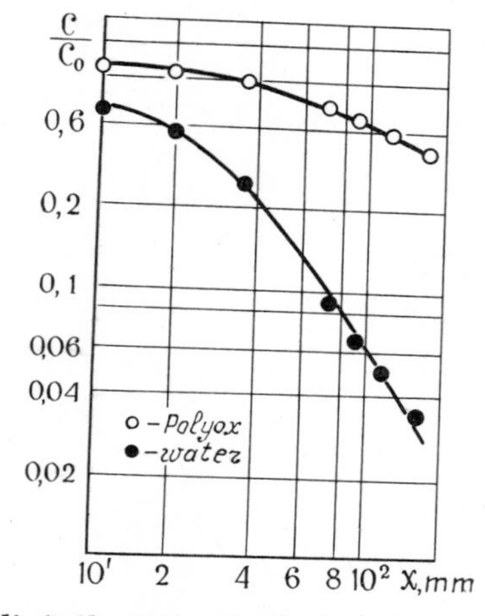

Fig. 5 Wall distribution of admixture concentration